Biophysical and Structural Aspects of Bioenergetics

RSC BIOMOLECULAR SCIENCES

EDITORIAL BOARD

Professor Stephen Neidle, *The School of Pharmacy, University of London, UK* (Chairman)
Dr Simon F Campbell FRS
Dr Marius Clore, *National Institutes of Health, USA*
Professor David M J Lilley FRS, *University of Dundee, UK*

This Series is devoted to coverage of the interface between the chemical and biological sciences, especially structural biology, chemical biology, bio- and chemo-informatics, drug discovery and development, chemical enzymology and biophysical chemistry.

Ideal as reference and state-of-the-art guides at the graduate and post-graduate level.

TITLES IN THE SERIES:

Biophysical and Structural Aspects of Bioenergetics
Edited by Mårten Wikström, *University of Helsinki*

Visit our website on www.rsc.org/biomolecularsciences

For further information please contact:
Sales and Customer Services
Royal Society of Chemistry
Thomas Graham House
Science Park, Milton Road
Cambridge CB4 0WF, UK
Telephone +44 (0)1223 432360, Fax +44 (0)1223 426017
Email sales@rsc.org

Biophysical and Structural Aspects of Bioenergetics

Edited by

Mårten Wikström
University of Helsinki, Helsinki, Finland

RSC Publishing

ISBN 0-85404-346-2

A catalogue record for this book is available from the British Library

Published by The Royal Society of Chemistry,
Thomas Graham House, Science Park, Milton Road,
Cambridge CB4 0WF, UK

Registered Charity Number 207890

For further information see our web site at www.rsc.org

Typeset by Macmillan India Ltd, Bangalore, India
Printed by Biddles Ltd, King's Lynn, Norfolk, UK

Preface

Molecular bioenergetics is a strongly interdisciplinary field of biochemistry, biophysics and molecular biology. It is concerned with how energy derived either from sunlight or from cellular respiration is primarily transduced into an electrochemical proton gradient across a coupling membrane, and how this gradient is subsequently utilised for energy-requiring reactions such as ATP synthesis or active transport. This field thus intimately involves biological membranes and membrane proteins. Bioenergetics has traditionally also included the cellular and tissue levels of organisation, as well as diseases. Pathophysiological, or 'systems' bioenergetics, is indeed today a very active research field, which includes studies of reactive oxygen species ('oxygen radicals'), apoptosis, mitochondrial diseases, and the process of aging. However, this book focuses on the physicochemical core of bioenergetics, which in the last 20 years has advanced greatly by the resolution of 3D structures of several of the protein complexes that catalyse key energy-transducing reactions in chloroplasts, mitochondria, and bacteria.

Molecular bioenergetics is a unique field in many ways. Due to its intrinsic multidisciplinarity, it is an excellent research arena for students who desire a broad education in biochemistry and biophysics, including molecular and structural biology. Such a comprehensive education is unfortunately becoming rare at this time, perhaps due to the current strong research focus on genomics and cell biology. Nevertheless, the necessity of ultimately understanding biological structures and reactions at the atomic level is obvious, which may give some consolation.

This book is by no means a comprehensive treatise, but rather a 'snapshot' of recent research and the state-of-the-art of this field. Yet, the reader can get a much broader insight into this and related fields from the extensive citations in each chapter. The book is comprised of 16 articles written by a group of active and authoritative researchers. The emphasis is on structure, and particularly on how the molecular structures may 'come alive' during their bioenergetic function. A functional description (mechanism) on the atomic level goes beyond static structures of single conformational states, and therefore requires a deeper level of understanding, which can be obtained by time-resolved biophysical methods, by quantum-mechanical and classical dynamics calculations and simulations, or by trapping series of transient intermediate states and solving their 3D structures, as for the case of bacteriorhodopsin (see cover picture). Such research is often strongly hypothesis-driven: the formulation of mechanistic models is a strong driving force in science the importance of which must not be underestimated.

I am very grateful to all the authors for their contributions to this book, to Dr. Janos Lanyi for kindly providing the cover picture, and especially to Mrs. Annie

Jacob and Miss Katrina Turner of the Royal Society of Chemistry for their patience and humour, which made the editor's work pleasant.

Mårten Wikström
Helsinki, May 2005

Contents

CHAPTER 1

Principles of Molecular Bioenergetics and The Proton Pump of Cytochrome Oxidase

ROBERT B. GENNIS

University of Illinois, Department of Biochemistry, 600 South Mathews Street, Urbana, IL 61801

1 Introduction: General Principles of Bioenergetic Systems

All of the bioenergetic enzymes described in this book can couple an exergonic or free-energy yielding reaction to the electrogenic movement of charged species across the membrane, generating a protonmotive force. In the case of bacteriorhodopsin, the driving reaction is the absorption of a photon, for the bc_1 complex, the oxidation of ubiquinol by cytochrome c is the driving force, and for the respiratory oxidases, the reduction of O_2 to H_2O provides the impetus. In this book, the principles utilized by a number of these systems are detailed with an emphasis on recent structural studies.

It is convenient to classify two classes of mechanisms used to generate a trans-membrane voltage:

(1) Mechanisms utilizing an oxidoreduction loop.
(2) True ion (proton) pumps.

1.1 Oxidoreduction Loops

The principle of coupling different chemical reactions is central to biology and is accomplished in a number of ways. Many of the systems that generate a protonmotive force can be understood in terms of Mitchell's chemiosmotic oxidoreduction loop.[1] This is illustrated by the example shown in Figure 1, which shows a redox loop formed from the anaerobic respiratory system comprised of formate dehydrogenase and nitrate reductase enzymes from *E. coli*. Recently, the structures of each of these two enzymes were determined.[2,3] The topology of the catalytic active sites assures that the

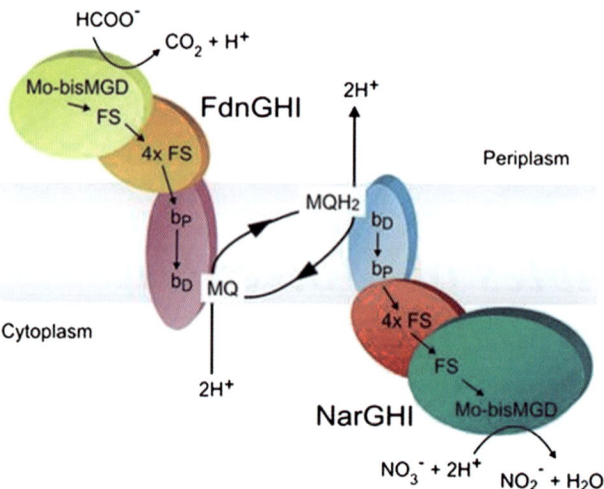

Figure 1 *Proposed mechanism for the protonmotive force generating redox loop by nitrate reductase (NarGHI) and formate dehydrogenase (FdnGHI) from* E. coli.[3] *MQ and MQH₂ are menoquinone and menaquinol, respectively; b_D and b_P indicate the distal and proximal hemes, respectively; FS indicates [Fe-S] clusters; Mo-bisMGD is the molybdenum cofactor. Note that only electrons cross the membrane resulting in the transmembrane voltage*

net reaction results in the generation of a protonmotive force.[3] Formate dehydrogenase oxidizes formate on the periplasmic side of the membrane (the positive or P-side) and electrons are delivered through a series of metal centers to a menaquinone reductase site located near the cytoplasmic surface (the negative or N-side of the membrane). The formate dehydrogenase, thus, separates the oxidative and reductive half-reactions on opposite sides of the membrane. Protons are released in the periplasm upon formate oxidation and protons are taken up from the cytoplasm upon the reduction of menaquinone. The actual charge crossing the chemiosmotic barrier is the electron.

Reduced menaquinol is a neutral, hydrophobic compound and can diffuse freely within and across the membrane bilayer. The nitrate reductase enzyme has a menaquinol oxidation site located near the periplasm, whereas the site where nitrate is reduced to nitrite is located on the opposite side of the membrane. Electrons are transferred across the membrane between these active sites to couple the two half-reactions catalyzed by the enzyme (see Figure 1). The full reaction of nitrate reductase, therefore, is coupled to the release of protons in the periplasm, the uptake of protons from the cytoplasm and the transfer of charges, in the form of electrons, across the membrane.

The net reaction of both of these enzymes together results in the transfer of four protons from the cytoplasm to the periplasm for each formate oxidized and nitrate reduced. Points to note are

(1) The actual charges crossing the membrane are electrons and not protons.
(2) The net transfer of protons is due to the vectorial placement of the enzyme active sites so that the oxidation and reduction half-reactions occur on opposite sides of the membrane.

(3) The protons are directly involved in the substrate chemistry.

(4) The two enzymes are coupled by a neutral, hydrophobic hydrogen carrier, in this case, menaquinol.

(5) The generation of the protonmotive force cannot be decoupled from the chemical reaction without changing the 'wiring'. The topology of the active sites, located on opposite sides of the membrane, and the uptake/release of protons from/to the N/P side of the membrane enforce this coupling so that the chemistry cannot proceed without generating a transmembrane voltage.

These are general features of Mitchell's initial proposal[1] for how the protonmotive force is generated, and the structural and functional studies on these and other systems have supported this proposal.

The photosynthetic reaction center[4–8] and the bc_1 (and b_6f) complex[9–13] can be understood as variations of this same general principle, illustrated schematically in Figure 2. In the case of the reaction center, the absorption of a photon results in electron transfer across the membrane leading to charge separation. The reductive and oxidative reactions that follow occur on opposite sides of the membrane. The direction of electron flow is from the P-side to the N-side of the membrane, as is also the case for nitrate reductase and formate dehydrogenase.[3] The reduction of ubiquinone in the bacterial reaction center utilizes protons from the N-side of the membrane (bacterial cytoplasm for the prokaryotic enzyme) and the protons are delivered to the buried Q_B active site through a proton-conducting channel.[14] In essence, the proton is transferred electrogenically towards the P-side of the membrane to 'meet' the electron at the quinone reduction site. Both proton and electron transfers contribute to the voltage generation by the reaction. The charge movement

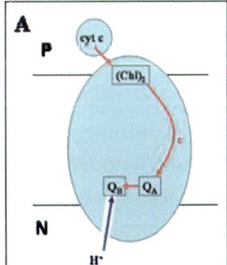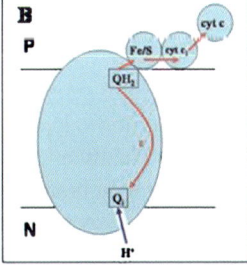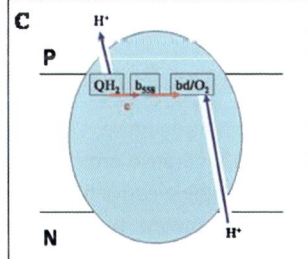

Figure 2 *Schematics showing the topology, the active sites, and electron and proton transfers for (A) the bacterial photosynthetic reaction center; (B) the bc_1 complex; and (C) cytochrome bd quinol oxidase. In the reaction center (A), light generates a charge separation at a chlorophyll special pair $(Chl)_2$ and electron transfer eventually to the Q_B site. Protons are transferred to the Q_B site, also contributing to the electrogenic processes. In the bc_1 complex, the oxidation of ubiquinol at the Q_o site by the Rieske Fe/S generates a highly unstable semiquinone species that reduces a quinone at the Q_i site via the 'wire' provided by the two hemes bs (not shown). In cytochrome bd, electron transfer from the quinol oxidation site to heme d where O_2 is bound, is not electrogenic in itself. Rather, proton transfer to the oxygenated species appears to be the voltage-generating step*

across the membrane is, thus, comprised partially by the electron transfer from the P- to the N-side of the membrane and partially by the proton being transferred in the opposite direction.

The topological constraint of the bc_1 center is that the quinol oxidation site and the site where cytochrome c is reduced are both located on the P-side of the membrane (Figure 2). Simple electron transfer between these sites would not be coupled to the generation of a protonmotive force. The enzyme has evolved a remarkable 'Q-cycle' to solve this problem.[11] The enzyme transfers one electron from reduced quinol to cytochrome c and uses the free energy of this reaction to drive the second electron across the membrane (P-side to N-side) to a quinol reductase site, located on the opposite side of the membrane. For every two quinols oxidized, one quinol is regenerated, two electrons cross the membrane, four protons are released to the periplasm and two protons are taken up from the cytoplasm. Although more complicated, the principles of how the protonmotive force is generated are those of the oxidoreduction 'loop'. The idea of the 'Q-loop' modification was also proposed long before the structures of the bc_1/b_6f complexes were known,[15,16] but the structural data fully support the functional model.

Finally, although the structure of the cytochrome bd quinol oxidase is not known, recent evidence[17,18] suggests that this represents yet another variation of the oxidoreduction loop (Figure 2). In this case, the two half-reactions, quinol oxidation and O_2 reduction to water, also occur on the same side of the membrane. However, although the active site where O_2 reduction occurs is near the P-side of the membrane, the protons are delivered to this active site from the opposite side of the membrane.[18] Hence, the proton delivery pathway enforces the coupling between the chemistry catalyzed by the enzyme and the generation of a transmembrane voltage. The working model is that the voltage is generated virtually entirely by protons crossing the bulk of the membrane to 'meet' the electron at the O_2 reduction site.

1.2 Proton Pumps

The transhydrogenase, ATP synthase, bacteriorhodopsin and cytochrome oxidase each also generate a protonmotive force principally or entirely by the transfer of protons across the electrochemical barrier of the membrane. Each of these systems must also have pathways that facilitate proton transfer across the membrane. However, in contrast to the systems described in the previous section, the transhydrogenase, ATP synthase, bacteriorhodopsin and cytochrome oxidase use a mechanism in which the protons transferred across the membrane are not directly involved in the chemical reactions driving the charge separation. One consequence of this is that it is possible in each of these enzymes to find mutants that decouple the proton pump from the chemistry.

In each of these systems, the driving reaction forces the protein into a transient, unstable state with a high chemical potential. This state then relaxes back to the ground state *via* a specific kinetic pathway which couples the dissipation of the transiently stored free energy to the generation of a protonmotive force. A high energy state of the protein, or a sequence of such states, serves as an intermediate to capture the free energy of one process (*e.g.* reduction of O_2 to water) and then utilize this free energy to drive a second process (*e.g.* electrogenic proton pumping). We will reserve the term 'proton pump' for the bioenergetic enzymes that operate in this manner and distinguish them from the systems that generate a protonmotive force through an oxidoreductive loop.

By far the best understood proton pump is bacteriorhodopsin.[19–21] It is useful to briefly review the mechanism of proton pumping by bacteriorhodopsin in order to put the cytochrome oxidase studies in perspective.

Bacteriorhodopsin has at least four protonation sites along a proton-conducting channel (D85, D96, Schiff's base, proton-release cluster). Each of these groups undergo large pK_a changes (5 to 10 pH units) as the protein goes through the sequence of intermediate states after absorption of a photon by retinal. At an early stage following the light-induced isomerization of retinal, an internal proton moves from the Shiff's base to D85, and a proton bound at the release cluster at the periplasmic (P-side) surface is ejected to the aqueous phase. Only after this has occurred is the internal proton on D96 transferred to the Schiff's base and then refilled through a channel from the cytoplasmic aqueous medium. Changes in the proton affinity at different sites within the protein during the photocycle is of obvious importance to understand why the protons shift between internal sites and are taken up/released from the bulk medium at specific times in the photocycle. In addition, the kinetics of proton transfer is equally important, since it is this feature that assures that the pump is unidirectional. The kinetics assure that the Schiff's base is reprotonated from the cytoplasmic side and not the periplasmic side of the membrane. The kinetics, in turn, are determined by structural features that facilitate or retard specific proton transfer reactions.

Recent structural and spectroscopic studies of bacteriorhodopsin have highlighted the role of internal water molecules.[19–23] As the protein changes conformation, water molecules shift positions, and water may also be taken up from or lost to the bulk medium. The changing hydogen-bonding patterns of internal water molecules not only play a major role in the proton affinities of ionizable residues within the protein, but also provide pathways for proton transfer within the protein. The elemental step of proton transfer is a shift of a proton from a hydrogen-bond donor to a hydrogen-bond acceptor.[24] Transfer over large distances requires movement along a series of hydrogen bonds. Because of this, changing the orientation or position of a single water molecule could have a very large influence on whether a proton pathway is functional. For example, it appears that the movement of water molecules essentially creates a proton-conducting channel from D96 to the Schiff's base in bacteriorhodopsin late in the photocycle, apparently gating proton transfer to re-protonate the Schiff's base.[19]

Although much less is known about the transhydrogenase[25,26] and ATP synthase[27–29] proton pumping enzymes, it is clear that each of these proteins also undergoes conformational changes during the catalytic cycle. In the case of the ATP synthase, this is in the form of the rotation of the C-subunit ring and its associated γ-subunit, which distorts the active site where ATP is synthesized. The details of the conformational changes at the active site are provided by the X-ray structure of the F_1 portion of the ATPase.[27]

1.3 Cytochrome Oxidase Uses Both Mechanisms of Energy Coupling

Cytochrome oxidase is a unique bioenergetic enzyme in that it generates a trans-membrane voltage using both the oxidoreductive loop principle and a true proton pump mechanism.[30–32] For each electron passing through the enzyme from reduced cytochrome c to O_2, two charges cross the membrane.

The mitochondrial enzyme is just one of a large superfamily referred to as the heme–copper oxidases.[33] All of the heme–copper oxidases catalyze the reduction of O_2 to H_2O at a bimetallic active site containing a heme and a Cu atom. The heme component has an open coordination site where O_2 can bind to the ferrous heme Fe and, subsequently, be reduced to H_2O.

$$O_2 + 4e^{-1} + 4\,H^+ \leftrightarrows 2\,H_2O \tag{1}$$

Studies with representatives of each of the major sub-classes of the heme–copper oxidases strongly suggest that all the heme–copper oxidases also are proton pumps. The invariant aspects for all these enzymes are as follows:

(1) The heme-copper active site is buried within the membrane, approximately halfway across the dielectric barrier. In the mitochondrial enzyme, the active site components are heme a_3 and Cu_B. The 'a' denotes the chemical structure of the heme and the subscript '3' has come to mean that this is the oxygen binding site.

(2) Electrons are directed to the O_2 reductase site from a second active site located at or near the P-side of the membrane where the reduced substrate is oxidised. Most of the superfamily members oxidize ferrous cytochrome c, but a number of heme–copper oxidases oxidize quinol. The most extensively studied quinol oxidase is cyochrome bo_3 from *E. coli*.[34,35]

(3) The electrons are transferred between the two active sites by a series of metal redox centers, the last of which is always a six-coordinate heme. Hence, all the heme–copper oxidases contain two hemes. These two hemes are located within the same subunit, with the Fe–Fe distance near to 12 Å and the heme edges as close as 5 Å. In the mitochondrial oxidase, the second heme is heme a, so this enzyme is called an aa_3-type cytochrome c oxidase.

(4) Proton input channels assure that all the protons that are used to generate water from O_2 come from the N-side of the membrane (mitochondrial matrix or bacterial cytoplasm). Proton channels or pathways also are critical to the transfer of protons all the way across the membrane, which is necessary for the proton pump. Hence, one proton must traverse about half of the distance across the membrane to 'meet' the electron at the oxygen at the heme–copper active site (an oxidoreduction loop), whereas the second proton is driven all the way across the membrane (a proton pump). This is shown schematically in Figure 3. We can take this topology and proton pumping into account in the actual reaction catalyzed by the enzyme.

$$O_2 + 4e^{-1} + 8\,H^+_{in} \leftrightarrows 2\,H_2O + 4\,H^+_{out} \tag{2}$$

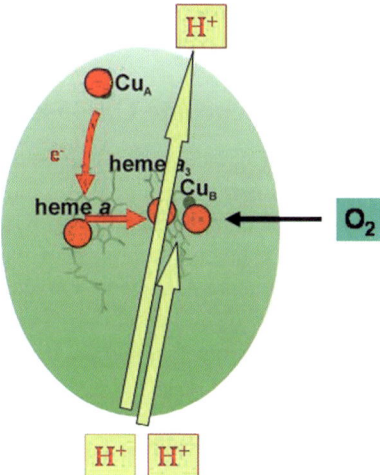

Figure 3 *A schematic illustrating that cytochrome oxidase utilizes both an oxidoreduction loop and a proton pump to generate a transmembrane voltage. Protons from the N side of the membrane 'meet' electrons from the P side of the membrane at the active site where O_2 is bound. In addition, a proton is pumped across the membrane. Hence, for each electron transfer to reduce O_2 to H_2O, two full charges cross the membrane*

This chemical reaction indicates that three processes are coupled together by the enzyme: (i) electron transfer to the active site; (ii) proton delivery to the active site; (iii) proton pumping. Furthermore, the reaction occurs in multiple steps. An understanding of the molecular mechanism of the proton pump can only follow an understanding of the chemistry that occurs at the enzyme active site. A brief review of the enzyme structure and the chemical mechanism by which O_2 is activated and reduced to water is a necessary prerequisite to discussing the proton pump.

2   The Structure of Cytochrome Oxidase

The structures of several heme–copper oxidases have been determined by X-ray crystallography. These include the cytochrome c oxidases from the bovine heart mitochondrion,[36–38] and from the prokaryotes *Paracoccus denitrificans*,[39,40] *Rhodobacter sphaeroides*,[41] *Thermus thermophilus*[42] and the quinol oxidase from *Escherichia coli*.[43] All except the oxidase from *T. thermophilus* are 'A-type' oxidases,[33,44] meaning they are from the majority sub-group of the oxidases defined by sequence similarities. Virtually all the mechanistic and mutagenesis studies have been performed with A-type oxidases, but it is useful to keep in mind that representatives of the B-type and C-type oxidases also have been shown to pump protons.[45,46] It is a reasonable assumption that the basic mechanism of the proton pump will be common for all the heme–copper oxidases, and the structural differences between the different oxidase sub-groups may, at some point, provide useful insights. For now, all the data refer to work done with the A-type oxidases.[33,44] We will refer to the *R. sphaeroides* oxidase residue numbers in the following discussion, but homologous residues are present in each of the other oxidases.

2.1 The Active Site of Cytochrome Oxidase

Figure 4 is a view of the heme a_3/Cu_B active site. Cu_B is ligated by three histidines (His284, His333 and His334), and heme a_3 is ligated on the distal side to a histidine (His419). The distance between the heme Fe and Cu atom is about 4.5 Å to 5 Å, and O_2 binds to the Fe of heme a_3. The heme component of the binuclear center in the A- and B-type oxidases is either a heme A or heme O (see Figure 5). In the C-type oxidases, the heme type appears always to be heme B. The significance of this variation in heme type is not known. (Note that the A, B, C classification[33,44] is not related to the heme type present, A, B or O)

A unique structural feature of the heme–copper oxidases is the presence of a crosslink between one of the Cu_B histidine ligands (His284) and a tyrosine (Y288)[37,40] (Figure 4). The tyrosine hydroxyl is a critical participant in the catalytic mechanism, providing a proton to promote breaking the O–O bond. It has also been proposed that the tyrosine can serve as a reductant and provide an electron during the O–O bond cleavage reaction, under some circumstances. The crosslink has been speculated to form spontaneously during an initial turnover of the oxidase by the generation of a putative tyrosyl radical by the strong oxidant at the active site.[40] It is not clear whether the crosslink alters the chemical properties of the tyrosine in an essential manner to facilitate its functional role or whether it is critical for structural rigidity and proper assembly of the enzyme.[47]

The six histidines that are metal ligands (three to Cu_B, one to heme a_3 and two to heme a) are totally conserved in all the sequences of the heme–copper oxidases, and

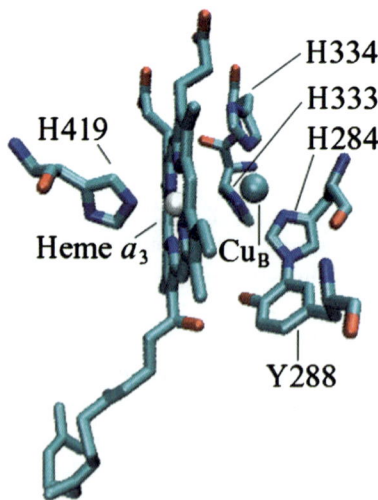

Figure 4 *The heme a_3-Cu_B binuclear center of cytochrome c oxidase. The heme is ligated on the distal side to H419. Cu_B is about 5 Å from the heme Fe and is ligated to H333, H334 and H284. The crosslink between H284 and Y288 is also shown*

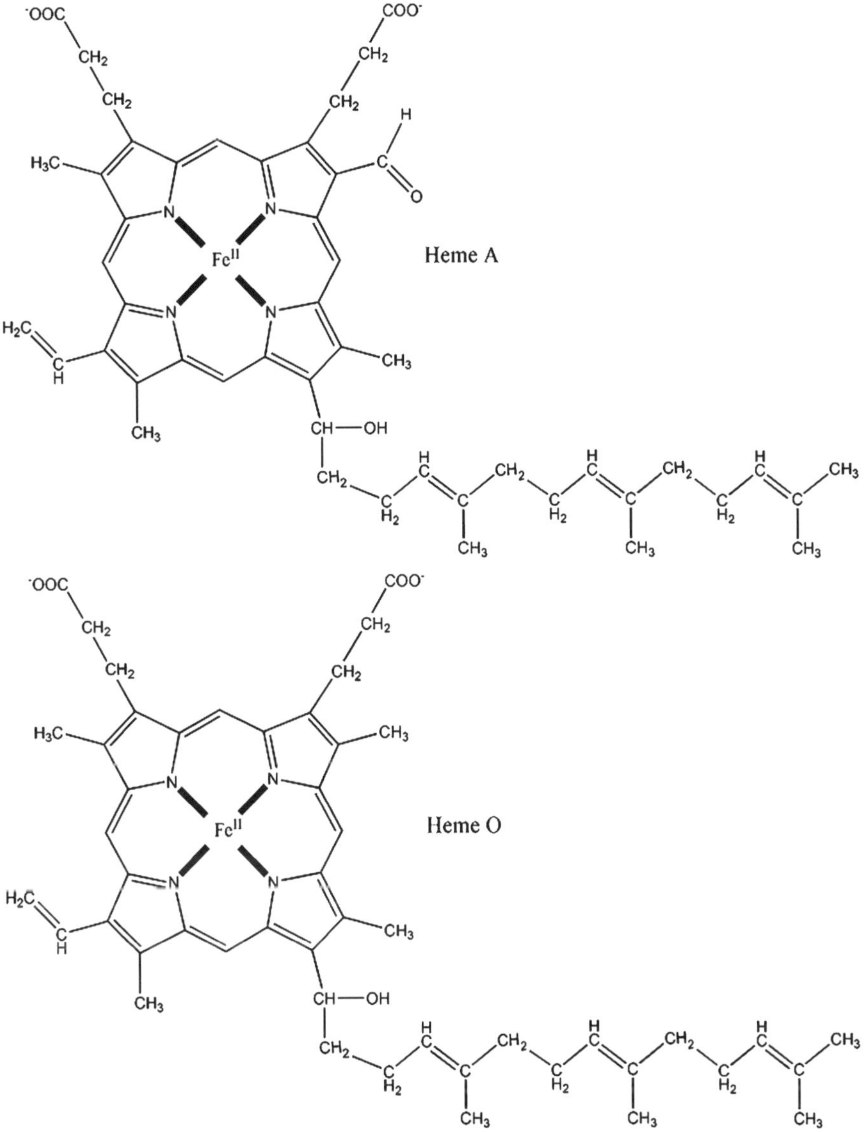

Figure 5 *Structures of heme O and heme A, which are found in variants of the heme–copper oxidases*

the tyrosine is totally conserved in both the A- and B-type oxidases. The equivalent of Y288 is, however, absent in all of the C-type oxidases (also known as the cbb$_3$-type oxidases). However, molecular modeling studies suggest that a tyrosine from a different helix can occupy the same physical location in the three-dimensional structure and may fulfill the same role in catalysis.[48]

2.2   Wiring for Electron Transfer

All of the heme–copper oxidases have a six-coordinate heme component, also located within subunit I, which provides electrons to the heme–copper center. Whether the role of this heme goes beyond simple electron transfer and it plays a role in the proton pumping mechanism is not clear. Several recent proposals hypothesize a role for heme a in proton pumping,[49–51] but other proposals completely ignore it aside from its role as a way-station for electrons.[52,53] This heme is ligated to two histidine residues (His121 and His421) which are totally conserved in all the heme–copper oxidases. In the A-type and B-type oxidases, there are examples of the heme in this location being heme A or heme B, whereas in the C-type oxidases there is only heme B at this location.

The electrons are provided to heme a from the Cu_A redox center in all of the A-type or B-type cytochrome c oxidases.[44] The Cu_A center actually contains two Cu atoms but is a one-electron donor to heme a. Cu_A is located in the hydrophilic domain of a different subunit, subunit II, which is found in all of the A- and B-type cytochrome c oxidases. Some of the A- and B-type oxidases contain an additional domain on subunit II on the P-side of the membrane which contains a c-type cytochrome. Presumably, the substrate cytochrome c binds to this domain and reduces the cytochrome c component of the oxidase which, in turn, reduces Cu_A and then heme a. Such oxidases are denoted as 'caa$_3$-type' oxidases.

Some of the oxidases in the A- and B- classes are quinol oxidases. These quinol oxidases lack Cu_A, and they must have a quinol oxidase site located near the P-side of the membrane. The quinol oxidation site has been identified in the *E. coli* cytochrome bo$_3$ quinol oxidase.[54,55]

The C-type oxidases do not contain either Cu_A or subunit II, but most contain two unique subunits that contain cytochrome c redox components. These subunits are the CcoO (or FixO) protein, which contains one heme C, and the CcoP (or FixP) subunit, which has two hemes c.

Regardless of the specifics, all of the heme–copper oxidases utilize electrons from the P-side of the membrane that are directed to the heme–copper binuclear center.

2.3   Proton-conducting Channels

Two proton-conducting pathways, the D-channel and the K-channel, have been identified in the prokaryotic oxidases. These channels ensure that all the protons used for the chemistry at the active site come from the N-side of the membrane and, thus, contribute to the proton motive force. Figure 6 shows some of the key residues within these channels in the *R. sphaeroides* oxidase.

2.3.1   The K-channel

This channel leads from the surface of the protein at the N-side, to the tyrosine (Y288) at the heme–copper center.[56–60] The K-channel is present in the structures of both the A-type and B-type oxidases.[44] Besides the active-site tyrosine, the K-channel in the A-type oxidases contains a lysine, from which the name is derived

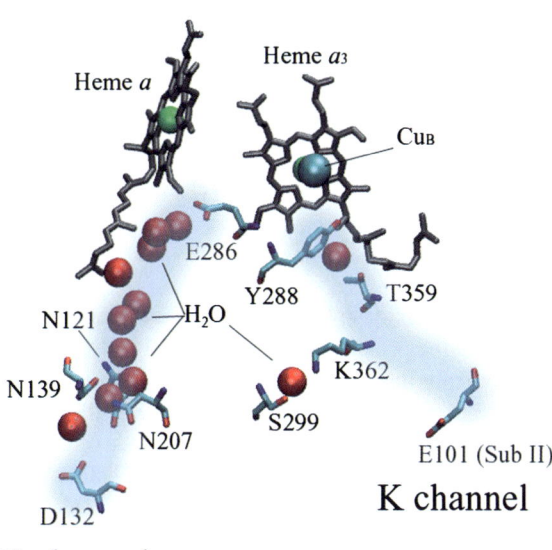

Figure 6 *Key residues implicated in the two proton-input pathways of the R. sphaeroides cytochrome c oxidase. The D-channel appears to be used for all of the pumped protons and for two of the four protons directed to the active site (substrate protons). The K-channel is necessary for both the protons taken up along with the electrons during the $H\rightarrow E_H$ and $E_H\rightarrow R_2$ transitions*

(Lys362), a threonine (Thr359) and a glutamate at the protein surface (Glu101 in subunit II). The proton-delivery function in the K-channel in the A-type oxidases has been confirmed by site-directed mutagenesis studies. Mutagenesis of Lys362 to methionine completely eliminates oxidase activity. The equivalent of the K-channel is also present in the B-type oxidases, although the lysine residue is not conserved. There is no structure of a C-type oxidase yet, but sequence homologies and molecular modeling suggests an equivalent channel with hydrophilic residues that are conserved within the C-type oxidase sub-group.[44,48]

2.3.2 The D-channel

This is one of the most striking features in the X-ray structures of the oxidases,[36,37,39,41,42] and is marked by a string of water molecules connecting the N-surface to a region near the heme–copper active site (Figure 6). Most of the A-type oxidases have acidic residues at either end of the channel: an aspartate (Asp132) at the surface of the protein and glutamate (Glu286) located about 10 Å from the heme–copper center. Mutagenesis of either of these residues essentially eliminates turnover of the oxidase. There is a sub-group of the A-type oxidases (denoted A2-type) in which the Glu286 is replaced by an alanine, but in all these cases, there is a conserved tyrosine and serine located nearby that apparently fulfills the same function.[44] This has been demonstrated by mimicking this arrangement by mutagenesis in the *P. denitrificans* oxidase and showing that the mutant oxidase is functional and pumps protons.[61] An equivalent to the D-channel is present in the B-type oxidase from *T. thermophilus*, which also contains a chain of internal water molecules.[42] However, the equivalents of Asp132 and Glu286 are not present and there is no pattern of conserved hydrophilic residues within the B-type oxidases. Similarly, there is no pattern of conserved hydrophilic residues in the C-type oxidases that would suggest where the equivalent of the D-channel is located.

2.3.3 The H-channel

The structure of the bovine oxidase has revealed an additional putative proton channel that is not present in the prokaryotic oxidases. This 'H-channel' provides a possible pathway leading from the N-side to the P-side of the membrane and has been proposed to be central to proton pumping, specifically in the mammalian oxidases. The channel leads to Asp51 (bovine), located on the P-side of the membrane.[49,62] The conformation of Asp51 (bovine) differs depending on whether the enzyme is in the reduced or oxidized state.[63] It has been proposed that the accessibility of Asp51 in the bovine oxidase to either the H-channel or to the bulk aqueous medium depends on the redox state of heme a, and that the changing redox state of heme a can drive or gate the proton pump in this manner.[49,62] A mutation of Asp51 has been made in the mammalian oxidase and the results are consistent with the elimination of proton pumping.[49] This work cannot be replicated in the prokaryotic oxidases since the bacterial oxidases all lack the key residues, including Asp51, and the H-channel does not appear in the structures of the prokaryotic oxidases.[64] Although it seems unlikely that the mammalian oxidases utilize a unique proton-pumping mechanism, this cannot be ruled out. It is not possible that the mechanism of proton pumping that involves the H-channel would apply to the prokaryotic enzymes.

2.3.4 The Exit Channel

Little is known about the pathway of pumped protons beyond the level of the hemes to the P-side of the membrane. Experimental evidence suggests that the pumped protons all use the D-channel to get to Glu286, and are then transferred to one or perhaps a group of proton acceptors in the exit pathway.[31,65–69] A role for Arg481, which forms a salt bridge with one of the propionates of heme a_3, has been proposed.[70] Changing this arginine to an asparagine in cytochrome bo_3 decouples proton pumping from the oxidase activity. Hence, the heme a_3 propionate has been proposed as one of the proton binding sites in the pathway of protons being pumped to the P-side of the membrane.[53] It has also been proposed that His334 is utilized as a proton-binding site in the exit pathway.[71] The pathway for water exiting from the binuclear center appears to lead to the vicinity of the Mn^{2+}/Mg^{2+} site at the interface between subunits I and II.[72] Whether this pathway is involved in proton exit is not known.

3 Hysteretic Properties of Cytochrome Oxidase

Early biochemical studies of isolated bovine cytochrome oxidase were plagued by multiple forms of the oxidized forms of the enzyme. The initial rate of the oxidase reaction and the rates of binding of ligands to the heme–copper center depend very much on the history of the enzyme, *e.g.* which preparative protocol was used or which storage buffer.[73,74] There are various '*resting*' forms of the oxidized state of the enzyme, some depending on the binding of exogenous ligands such as halides at the heme–copper center, and others apparently due to a spontaneous rearrangement of endogenous groups at the active site which alters the optical spectra of the hemes.[73,75,76] These resting forms can all be converted to a form with consistent properties by fully reducing the enzyme and then reoxidizing it just prior to use in experiments. This form is called the '*pulsed*' enzyme.[74]

More recently, this phenomenon or one very similar to it has been shown to be an important consideration in studies of the proton-pumping behavior of the enzyme.[66,77] The state of the enzyme immediately after it is fully oxidized by the reaction with O_2 is called the *H* state of the enzyme. The *H* state is presumably the oxidized form of the enzyme during steady-state turnover conditions and, therefore, the physiologically relevant form during turnover. It is proposed, but not yet demonstrated, that the *H* state of the enzyme is in a conformation in which a significant fraction of the free energy released by the reaction of the fully reduced enzyme with O_2 is stored, and can be utilized to drive proton pumping upon further reduction. Hence, the *H*-state is a 'high energy' state of the enzyme, similar to the form of bacteriorhodopsin after absorption of a photon. Unlike bacteriorhodopsin, however, proton pumping does not spontaneously occur, but requires electron transfer to the heme–copper center. If the *H* state of the enzyme is not re-reduced, the stored free energy is presumably dissipated unproductively, forming the *O* (oxidized) state of the oxidase, which is the form normally encountered experimentally. The lifetime and characteristics of the *H* state are not yet known and, indeed, its existence has only been deduced. This is one area of active investigation (and, therefore, subject to revision). There is substantial evidence that, under certain circumstances, the simple reduction of the binuclear center metals of cytochrome oxidase (prior to the reaction with O_2) is coupled to proton pumping.[66,69,77,78] The mechanism and energetics of this phenomenon remain to be clarified.

4 The Mechanism of Oxygen Reduction to Water

There has been considerable progress in our understanding of the catalytic mechanism and intermediates that are formed as O_2 is reduced to water at the heme–copper center.[79] All the electrons are provided by the sequential oxidation of four molecules of ferrous cytochrome c, which bind at a site located on the hydrophilic domain of subunit II on the P-side of the membrane. The sequence of electron transfer is

$$\text{cytochrome c} \rightarrow Cu_A \rightarrow \text{heme a} \rightarrow [\text{heme a}_3/Cu_B] \tag{3}$$

The following outlines the sequence of events, summarized in Figure 7. We will assume the existence of a 'high energy' or activated form of the oxidized enzyme (*H*) either

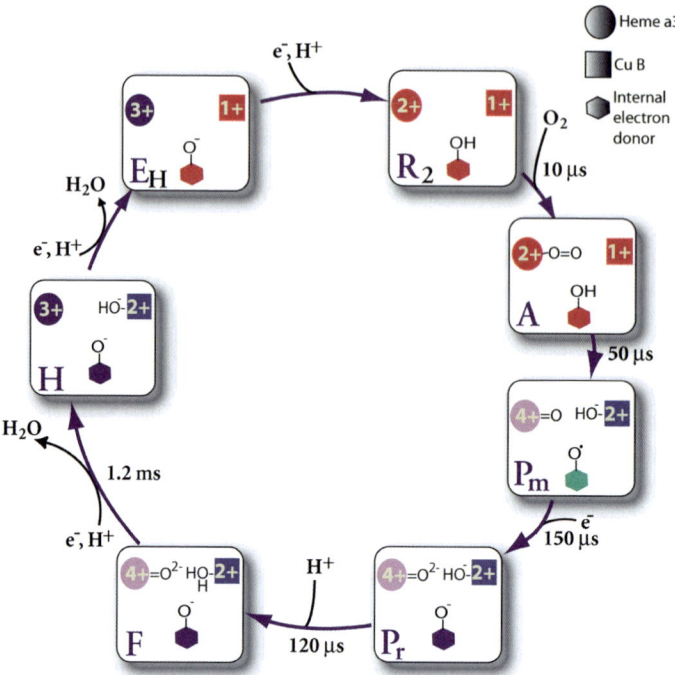

Figure 7 *The catalytic cycle of cytochrome c oxidase. In order to emphasize the energetics, the starting point at the top is indicated as the R_2 state of the enzyme, which is two-electron reduced. This state is not competent to pump protons. The reaction with O_2 forms the P_m state of the enzyme, in which requires the reaction with O_2 which forms the P_m state of the enzyme in which is stored sufficient electrostatic or chemical free energy to drive pumping for four consecutive electron transfer reactions. If an electron is present on heme a at the time that O_2 reacts, the product formed is the P_R state in which the electron is transferred to the active site from heme a. This is followed by a proton transfer ($P_R{\rightarrow}F$) which is coupled to proton pumping. If an electron is not on heme a at the time of the reaction with O_2, then the addition of the next electron from cytochrome c occurs concomitantly with proton transfer, forming the F state ($P_m{\rightarrow}F$), bypassing the P_R state. Finally, electron transfer to the F state converts the enzyme to the putative high energy form of the oxidized enzyme, the H state, speculated to have a high potential Cu_B. The next two electron transfer reactions are coupled to proton pumping, bringing the enzyme back to the R_2 state*

similar to the 'pulsed' form or identical to it. There may also be a high energy form of the one-electron reduced enzyme (E_H) which would be competent to pump a proton upon transfer of a second electron to the binuclear center, even in the absence of O_2: [31]

(1) $R_2 \rightarrow P_m$: We will start with the two-electron reduced form of the oxidase (R_2), since this is the form which will react rapidly with O_2. The location of the two electrons in the enzyme will depend upon how it is formed, but there is a rapid reaction with any two-electron reduced species with ferrous heme a_3. If we assume that there is a high energy form of the one-electron reduced enzyme (E_H), then the product, R_2, will have one electron on each heme a_3 and on Cu_B.

The initial spectroscopically detectable intermediate is called the A state of the enzyme, in which O_2 is liganded to ferrous heme a_3, much like the oxymyoglobin complex.[80] Presumably, the binding of O_2 stabilizes the form of the enzyme with the two electrons on Cu_B and heme a_3. The A state of the enzyme forms within 10 µsec (with 1 mM O_2) but is unstable and rapidly (50 µsec) converts to state P_m. The P_m state is one of the well-characterized intermediates in the reaction cycle.[79,81] This intermediate state has a visible absorption that has a peak at about 607 nm in the P_m-minus-O difference spectrum, and it is often referred to as the 607 nm species.[80]

Both chemical and spectroscopic studies have clearly demonstrated that the O–O bond is broken in the P_m state of the enzyme.[79] One oxygen atom is bound to heme a_3 as an oxoferryl group, with the iron in a high valence state ($F^{4+}=O^{2-}$), and the second oxygen atom is found as a hydroxide liganded to Cu_B. The P_m state is also associated with an amino acid free radical.[82-85] The $R_2 \rightarrow P_m$ reaction is essentially a concerted four-electron reduction of O_2 in which both oxygen atoms are reduced to the valence state of water.[79] This eliminates the threat of forming reactive oxygen species such as superoxide and peroxide at the active site of the enzyme. Three of the four electrons necessary are derived from the metals ($Cu_B^{+1} \rightarrow Cu_B^{+2}$ and heme a_3 $Fe^{+2} \rightarrow Fe^{+4}$) and the fourth electron comes from an amino acid at or near the active site. The most reasonable postulate is that the active site tyrosine (Tyr288) provides both a proton and an electron to catalyze the splitting of the O–O bond:

$$[\text{Tyr-OH}] \ Fe^{+2}\text{–}O_2 \ Cu_B^{+1} \rightarrow [\text{Tyr-O}^\bullet] \ Fe^{+4}{=}O^{2-} \ HO^- \ Cu_B^{+2}$$

Time-resolved EPR spectroscopy has shown that about 50 µsec after the reaction of O_2 with the fully reduced enzyme, a free radical is formed that is located on one or more tryptophan residues, and not on the active site tyrosine.[84] The initial radical could, conceivably, be located on the active site tyrosine (Tyr288), before its appearance on the tryptophan(s). Alternatively, there could be an equilibrium between the two radicals (tryptophan and tyrosine) with only the tryptophan appearing in the EPR spectrum due to spin coupling of the putative Tyr288 radical and the metal centers. The prime candidate for the tryptophan radical is Trp280, a totally conserved residue in all sequences of the heme–copper oxidases, which is stacked with the imidazole ring of His334, one of the Cu_B ligands. The radical formed is a cationic tryptophan, indicating that the tryptophan does not deprotonate upon oxidation. If there is an electron on heme a, as is the case if one starts with the fully reduced enzyme, then the radical signal vanishes as the electron from heme a is transferred to the active site. If an electron is not available to rapidly reduce the initially formed radical, the radical migrates to another location, definitively identified as Tyr175,[83] which is hydrogen bonded to Trp280.

Importantly, the $R_2 \rightarrow P_m$ step does not generate a transmembrane voltage and is not associated with proton uptake or release. This conclusion is based on studies of the reaction of O_2 with either the fully reduced enzyme or with the CO mixed valence form of the enzyme, in which heme a_3 and Cu_B are initially reduced. The $R_2 \rightarrow P_m$ reaction is essentially a rearrangement of the electrons and protons already at the active site of the enzyme. The resulting form of the enzyme is in a high-energy state and must have sufficient stored free energy to drive eight charges across the membrane against a voltage gradient during the subsequent four electron transfers that bring the system back to the R_2 state.

If we do not assume the existence of a high energy form of the one-electron reduced enzyme, then the two electrons in R_2 are very unlikely to be located on heme a_3 and Cu_B. A more populated state is likely to be one electron at the binuclear center and one electron on heme a. When this species reacts with O_2, an electron will be transferred from heme a to the binuclear center accompanied by a proton, most likely *via* the K-channel. The reaction with O_2 traps the two-electron reduced form of the binuclear center. In this sequence, the $R_2 \rightarrow P_m$ reaction will incorporate electron tranfer to the binuclear center, and would be expected to be coupled to proton pumping.

(2) $P_m \rightarrow P_R$: If an electron is present on heme a at the time P_m is formed, rapid electron transfer (100 to 150 μsec) ensues which quenches the free radical signal and, therefore, can be deduced to reduce the amino acid radical(s). It is not evident that the P_m state is a necessary intermediate form under these circumstances, since rapid electron transfer from heme a to the binuclear center can effectively compete with electron transfer from Tyr288 at the active site. The existence of P_m as a kinetic intermediate in the reaction of the fully reduced enzyme with O_2 relies on the time-resolved EPR spectra.[84] Species P_R and P_m have identical optical spectra. Electron transfer from heme a to the active site of the enzyme (Trp280/Tyr288) is not associated with any charge movement across the membrane or with any proton uptake or release. Hence, the P_R state is also a maximally energized state of the enzyme. It is very unstable and, at room temperature, can only be observed transiently. The P_R state can also be generated by the same reaction at low temperatures, and can be shown to have a Cu_B EPR signal, demonstrating that at least a portion of the population has Cu_B^{2+} that is electronically decoupled from the heme a_3 oxoferryl group.[86]

The P_R state is anomalous in the reaction cycle since it is the only state where electron transfer to the active site is clearly separated in time from the coupled proton transfer. Since the P_R state is observed only if one starts with the fully reduced enzyme, it is possible that the protonation of internal sites in the fully reduced enzyme kinetically favors this electron transfer, which otherwise would be concomitant with proton transfer.

It is also interesting that if one forms the P_R state by the reaction of the fully reduced oxidase with O_2, the electron that is on Cu_A is not transferred to heme a during the $P_m \rightarrow P_R$ transition.[87] The re-reduction of heme a ($Cu_A \rightarrow$ heme a electron transfer) only occurs in the subsequent step.

(3) $P_R \rightarrow F$: The active site in the P_R state must have a very high proton affinity, effectively coupling the transfer of the electron to form P_R to a subsequent proton transfer to the active site. The fact that the proton transfer occurs substantially more slowly than the electron transfer (120 μsec) indicates an intrinsic rate limitation for proton transfer from Glu286 to the active site. The pH-dependence of this proton transfer reaction reflects the pK_a of Glu286.[88] At pH < 9 the rate of this step is virtually pH-independent. Only above pH 9 is

the reaction slowed down, reflecting the deprotonation of Glu286. Presumably, the proton is transferred to convert the Cu_B^{2+} ^-OH to $Cu_B^{2+}H_2O$. This alters the hydrogen bonding in the vicinity of the heme a_3 $Fe^{4+}=O^{2-}$ moiety, which alters the optical spectrum. The product is the F species, which has a diagnostic spectrum with a peak near 580 nm in the F-minus-O difference spectrum.

The proton transfer from Glu286 to the active site is coupled to a series of subsequent proton movements. Importantly, the $P_R \rightarrow F$ transition is coupled to proton pumping. This has now been demonstrated both by electrometric measurements[89] and by the appearance of a proton at the P-side of the membrane coincident with this step[90]. The re-protonation of Glu286 occurs *via* the D-channel, and this proton transfer is strongly coupled to the electron transfer from $Cu_A \rightarrow$ heme a. Mutants that prevent the reprotonation of Glu286 also prevent the $Cu_A \rightarrow$ heme a electron transfer reaction if Glu286 is deprotonated.[87]

(4) $F \rightarrow H$: The delivery of the next electron from heme a to the binuclear center reduces the oxoferryl heme a_3 to the ferric heme ($Fe^{4+}=O^{2-} \rightarrow Fe^{3+} - ^-OH$). The proton is again delivered from Glu286. This step is also coupled to proton pumping. This is the slowest step in the reaction cycle, and takes about 1 msec to complete. The $F \rightarrow H$ transition has a very high H/D solvent kinetic isotope effect (5 to 7) which indicates that a proton transfer is rate limiting.[91,92] The product of this reaction is the high energy state of the fully oxidized enzyme, the H state described previously. This state is competent to pump protons upon further electron transfer to the heme–copper center.[66]

(5) $H \rightarrow E_H$: It has been proposed that the high energy oxidized H state has a high potential Cu_B^{2+}. The addition of one electron to the high energy form of the fully oxidized enzyme (H) is speculated to create a one-electron reduced state of the enzyme, which we will assume is also in a 'high energy' configuration (E_H state). The one-electron reduction of the heme–copper center in the fully oxidized enzyme, in either the O or H state, is accompanied by the delivery of a proton through the K-channel, probably converting hydroxide (bound to Cu_B) to water. The electron in the E_H state has been suggested to reside on Cu_B^{3+}, but no published data have verified this. Experimentally, however, the $H \rightarrow E_H$ transition is associated with pumping one proton across the membrane.[66] The one-electron reduced enzyme does not react with O_2, primarily because the electron does not reside on heme a_3, and O_2 does not bind to ferric heme a_3.

The one-electron reduction of the dissipated or low-energy form of the fully oxidized enzyme (O) results in the electron residing primarily on heme a, without substantial electron transfer to the heme–copper center. The one-electron reduction of the O state is not coupled to proton pumping.[66]

(6) $E_H \rightarrow R_2$: The addition of a second electron to E_H creates a two-electron reduced form of the enzyme (R_2). Little is known about this transition, but it has been deduced that this electron transfer into the heme–copper center also triggers the pumping of one proton across the membrane.[78]

If the two-electron reduced enzyme is formed from the dissipated form of the oxidized enzyme (O), then the reduction by the second electron (R_2) is not coupled to proton pumping. Simply reducing the resting, oxidized form of the enzyme reconstituted in vesicles does not pump protons.

5 Some Considerations Concerning the Mechanism of the Proton Pump

5.1 Four Pumping Steps per Cycle but One Common Mechanism

The realization several years ago that there is pumping during the reductive part of the catalytic cycle ($H \rightarrow E_H \rightarrow R_2$) makes it very likely that protons are pumped accompanying each electron transfer step into the enzyme. This has been verified recently. The four pumping steps are:

(1) $H \rightarrow E_H$.
(2) $E_H \rightarrow R_2 \rightarrow P_m$.
(3) $P_m \rightarrow P_R \rightarrow F$.
(4) $F \rightarrow H$.

Each of these steps requires delivery of both an electron and a proton to the enzyme active site, though the electron and proton acceptors are different in each case. Step 3 is unique insofar as the electron and proton delivery are separated in time, with the electron transfer occurring signficantly before the proton transfer to the active site. This is informative since the proton pumping does not initiate until the proton transfer reaction ($P_R \rightarrow F$). Each of the four steps listed above is expected to be accompanied by the uptake of two protons from the N-side of the membrane and the release of one proton on the P-side of the membrane. This is shown schematically in Figure 3. Evidence is solid that all four protons taken up by the enzyme during steps 3 ($P_m \rightarrow P_R \rightarrow F$) and 4 ($F \rightarrow H$) enter through the D-channel. In contrast, the protons delivered to the heme–copper center in step 1 ($H \rightarrow E_H$) and probably also step 2 ($E_H \rightarrow R_{2H} \rightarrow P_m$), apparently come through the K-channel. If the mechanism of proton pumping is the same in all four steps, then it is likely that the pumped protons associated with steps 1 and 2 also utilize the D-channel.

Our working model is that all the pumped protons use the D-channel and are transferred from Glu286 to a site(s) within the exit pathway; the protons that are transferred to the heme–copper center and eventually consumed to form water come from both the K-channel (steps 1 and 2) and the D-channel (steps 3 and 4). The reason why there are two input channels is not known but may be related to the fact that the K-channel is operative in steps prior to the reaction with O_2 to form the oxoferryl form of heme a_3.

5.2 What Do the Decoupled Mutants of Cytochrome Oxidase Tell Us?

The substitution of acidic residues into positions near the entrance of the D-channel (N139D or N207D) completely eliminates proton pump activity by the oxidase.[52,67,93,94] The existence of mutations that decouple the proton pump without compromising the oxidase activity is very signficant. The enzyme still generates a protonmotive force because the oxidoreduction loop part of the mechanism is still operable. Since the protons involved in the oxidoreduction loop are substrate protons, it is not possible to separate the chemical reaction from the generation of the protonmotive force. The protons that are transferred across the membrane by the separate pumping mechanism are not substrate protons and the proton pump machinery can clearly be compromised without altering the catalytic activity of the oxidase.

Although the mechanism of decoupling is not known, the apparent pK_a of Glu286 is raised by about 1.5 pH units during the $P_R \rightarrow F$ transition by the N139D mutation, located 25 Å away[52]. At the very least this indicates that a relatively slight shift in the rate of some process that is essential for the pump to operate will disable the pump.

5.3 The Operation of the Proton Pump Depends on Kinetic Considerations

Cytochrome oxidase is forced into a high energy state by the driving reaction with O_2, followed by relaxation of the system by four subsequent electron transfers, coupled to the proton pump. Thermodynamics determines how much work can be derived from the operation of the pump. The free energy available from the chemical reaction represents the maximum possible work. Formally, this maximum work is an ideal representing what could be achieved for a perfectly reversible reaction following a pathway operating near equilibrium. In fact, the oxidase does function at a point near equilibrium, and the efficiency is quite high. As pointed out by Wikström,[31] the chemical reaction provides about 500 mV per electron of driving force and this is coupled to moving two charges against a protonmotive force that can be as high as +220 mV, *i.e.* 440 mV work output. This system operates with near 90% efficiency and the molecular mechanism must be efficiently protected against proton back leaks.

The operation of the pump, however, is strictly a matter of kinetics, which determines the pathway by which the energy is dissipated. The enzyme is postulated to exist in a series of conformations. The natural starting place is the P_m state of the protein, in which the maximum free energy from the O–O bond splitting is stored within the protein as conformational or electrostatic strain. It now appears that there are four consecutive steps in which a proton plus an electron are transferred to the active site, and that each of these steps is coupled, in turn, to a sequence in which a site in the exit channel is protonated from the N-side and deprotonated to the P-side of the membrane. Addition of an electron to the P_m state forms P_R. The proton transfer from Glu286 to the active site ($P_R \rightarrow F$) accompanies or possibly triggers the proton-pumping.[90]

Let's now postulate a series of states that form between P_R and F. The initial state that forms is M_{Ia} (where the subscript Ia indicates the heme–copper center is one-electron reduced compared to state P_m, and the protein is in conformation 'a'). One plausible model that can be used to illustrate the formalism follows. The proton transfer reaction $P_R \rightarrow M_{Ia}$ increases the pK_a of a group, X, in the exit pathway. Futhermore, the conformation of M_{Ia} facilitates rapid proton transfer to X from the N-side of the membrane, through the D-channel, forming species M_{Ib}. The kinetic barriers for the proton transfer reaction $M_{Ia} \rightarrow M_{Ib}$ favor the proton transfer from the N-side of the membrane. Species M_{Ib} is unstable and converts to another

conformation, M_{Ic} in which the pK_a of group X is lowered and in which the kinetically favored pathway for deprotonation is to the P-side of the membrane. Possibly
the $M_{Ib} \rightarrow M_{Ic}$ transition is coincident with the re-protonation of Glu286 through the
D-channel. The species M_{Ic} is ready to receive the second electron, and an analagous
sequence of events follow. We can write the species $M_{Ic} \rightarrow [M_{IIa} \rightarrow M_{IIb} \rightarrow M_{IIc}]$.
Species M_{IIc} is equivalent to the high–energy form of the fully oxidized enzyme, H.
The addition of two additional electrons, however, take the protein through
analagous conformational states: $M_{IIc} \rightarrow [M_{IIIa} \rightarrow M_{IIIb} \rightarrow M_{IIIc}] \rightarrow [M_{IVa} \rightarrow M_{IVb} \rightarrow M_{IVc}]$, where M_{IVc} is the fully relaxed, two-electron reduced protein. Reaction with
O_2 forms species P_m and the sequence repeats.

   This formalism has limited value and can be adapted to any particular molecular
model of how the pump operates. It is useful, however, to emphasize the sequence
of electron and proton transfer reactions that occur and the necessary properties of
this minimum number of conformational states.

5.4   The Energetics of Proton Pumping

Even if all four electron transfer steps are coupled to proton pumping, this does not
mean that the free energy available for work is equal for these four steps. The overall reaction yields a work potential of about -2000 mV.[30,31] Divided equally, this
would be -500 mV per step, which is sufficient to drive two charges per step against
a transmembrane potential of $+220$ mV. The availble free energy available from
each of the four proton pumping steps is not known with certainty. In principle, most
of the free energy of the overall reaction could be associated with the two electron
transfer steps following the formation of the P_m species at the active site. In this case,
the proton pumping associated with electron transfer to the enzyme active site prior
to the reaction with O_2 would be thermodynamically unfavorable, but pulled along
by the highly favorable subsequent exergonic steps, i.e., $P_m \rightarrow P_R \rightarrow F$ and $F \rightarrow H$.
The data demonstrating that the $H \rightarrow E_H$ and $E_H \rightarrow R_2 \rightarrow P_m$ steps are associated
with proton pumping are from measurements of voltage generation, but the driving
force for each of these steps is not known. Based on the equilibrium midpoint potentials of the metal centers, electron transfer from cytochrome c to the Cu_B would be
favorable by about -100 mV in the absence of a membrane potential. However, a
proton motive force of $+220$ mV would still make this unfavorable by $+340$ mV,
since two charges cross the membrane. It would appear implausible that the catalytic
cycle could turn over at a rate of 10^3 s^{-1} with such an unfavorable intermediate. This
is the reason for postulating a transient 'high energy' form of H in which Cu_B has a
very high potential. The subsequent electron transfer to the binuclear site, $E_H \rightarrow R_2$,
would be similarly disfavored by as much as $+340$ mV but this reaction is coupled
to the reaction of O_2: $R_2 \rightarrow P_m$. The binding of O_2 by itself is relatively weak and,
under ambient concentrations of O_2, will not drive the reaction forward. The energetics of converting bound O_2 to the oxoferryl P_m species is not known. In principle,
the energetics from this reaction could stabilize the P_m state sufficiently to drive the
$E_H \rightarrow R_2 \rightarrow P_m$ reaction even in the presence of an opposing protonmotive force.
The P_m state has been shown qualitatively to be relatively stable in solution,[81] but this
has not been quantified.

6   Where Do We Go From Here?

The concept of a high energy oxidized state of the enzyme (*H*) remains a postulate. This state needs to be shown to exist and then characterized. The model assumed in this chapter that the four electron transfers are equally engaged in proton pumping is also not fully established or universally accepted, and this is in need of further experimental confirmation. The last decade has provided information about the chemical events defining the nature of the P_m, P_R and *F* states of the enzyme. However, this information does not include important details such as where internal protons are bound. What is the protonatable group in the exit channel, or is there more than one such group? At what points in the reaction cycle is this group protonated? We don't know. Many experiments utilize forms of the enzyme (*e.g.* fully reduced, CO-mixed valence, peroxide-generated *F* species) that have unknown internal proton distributions that could, and likely do, have a major impact on the reactions that are observed. Caution is required before extrapolating results to steady-state turnover conditions.

The states of cytochrome oxidase, such as the P_R and *F* states are the beginning and end points of the reaction sequences during which proton pumping occurs. Nothing is known about the what happens to the enzyme in between these states. The key question is what is the nature of the 'a', 'b' and 'c' conformational states as outlined in the previous section. How is the free energy of the chemical reaction stored in the protein and used to drive the sequence of events four times in succession? The answers are not known. There are virtually no experimental data, though speculations are not lacking. Wikström has suggested that water orientation is one critical feature in favoring rapid proton transfer from the N-side to the a group in the exit channel, possibly the heme a_3 propionate.[53] Several authors have speculated that the redox state of heme a plays a critical role in the proton pump, by being coupled to proton uptake, changing the pK_a of groups involved in proton pumping.[49,50,95] Stuchebrukhov has emphasized the total charge at the heme–copper center being coupled to the pK_a of one of the Cu_B histidine ligands.[71] Brzezinski has suggested a protein conformational change due to deprotonation of Glu286.[52,96]

It is necessary to develop experimental methods to probe these proposed intermediate states of the enzyme. Electrometric studies of the one-electron reduction of state *F* and state P_m have revealed that these reactions have clear, time-resolved steps associated with charge movements normal to the plane of the membrane.[65,94,97] This makes it highly likely that at least one intermediate accumulates transiently during these one-electron reactions.

Considering the historical progression in understanding bacteriorhodopsin, it seems likely that one important development for further understanding the oxidase will be determining the structure of the enzyme trapped in intermediate states by X-ray crystallography. Although structures alone will not provide a unique explanation, such structures are essential to interpret the functional data. In bacteriorhodopsin, all the events in the protein are the result of the isomerization of a double bond in retinal, which pushes groups within the protein by about 1 Å. The strain inherent in this change is dissipated through a series of conformational states that (i) alter the pK_a of internal groups by as much as 10 pH units; (ii) change the location of internal water molecules; and iii) assure that protons are taken up from one side of the membrane and released on the opposite side. In cytochrome oxidase, we have energetic strain introduced by the splitting of the O–O bond, and this is dissipated over four separate electron transfer reactions. We can expect similar conformational substates as observed in bacteriorhodopsin, and much more going on than we are presently aware of. Identifying and trapping these states is the experimental challenge of future studies.

References

1. P. Mitchell, *Nature*, 1961, **191**, 141.
2. M. Jormakka, S. Tornroth, B. Byrne and S. Iwata, *Science*, 2002, **295**, 1863.
3. M.G. Bertero, R.A. Rothery, M. Palak, C. Hou, D. Lim, F. Blasco, J.H. Weiner and N. Strynadka, *Nature Struct. Biol.*, 2003, **10**, 681.
4. J. Deisenhofer and H. Michel, *Science*, 1989, **245**, 1463.
5. A. Kuglstatter, U. Ermler, H. Michel, L. Baciou and G. Fritzsch, *Biochemistry*, 2001, **40**, 4253.
6. A. Ben-Shem, F. Frolow and N. Nelson, *Nature*, 2003, **426**, 630.
7. S. Iwata and J. Barber, *Curr. Opin. Struct. Biol.*, 2004, **14**, 447.
8. P. Fromme, A. Melkozernov, P. Jordan and N. Krauss, *FEBS Letters*, 2003, **555**, 40.
9. D. Xia, C.-A. Yu, H. Kim, J.-Z. Xia, A.M. Kachurin, L. Zhang, L. Yu and J. Deisenhofer, *Science*, 1997, 277.
10. A.R. Crofts, B. Barquera, R.B. Gennis, R. Kuras, M. Guergove-Kuras and E.A. Berry, *Biochemistry*, 1999, **38**, 15807.
11. A.R. Crofts, *Annu. Rev. Physiol.*, 2004, **66**, 689.
12. G. Kurisu, H. Zhang, J.L. Smith and W.A. Cramer, *Science*, 2003, **302**, 1009.
13. D. Stroebel, Y. Choquet, J.-L. Popot and D. Picot, *Nature*, 2003, **426**, 413.
14. P. Ädelroth, M.L. Paddock, A. Tehrani, J.T. Beatty, G. Feher and M.Y. Okamura, *Biochemistry*, 2001, **40**, 14538.
15. K.M. Andrews, A.R. Crofts and R.B. Gennis, *Biochemistry*, 1990, **29**, 2645.
16. P. Mitchell, *FEBS Letters*, 1975, **59**, 137.
17. A. Jasaitis, V.B. Borisov, N.P. Belevich, J.E. Morgan, A.A. Konstantinov and M.I. Verkhovsky, *Biochemistry*, 2000, **39**, 13800.
18. I. Belevich, V.B. Borisov, J. Zhang, K. Yang, A.A. Konstantinov, R.B. Gennis and M.I. Verkhovsky, *Proc. Natl. Acad. Sci. USA*, 2005, **102**, 3657.
19. J.K. Lanyi, *Mol. Membr. Biol.*, 2004, **21**, 143.
20. J.K. Lanyi, *Annu. Rev. Physiol.*, 2004, **66**, 665.
21. K. Takeda, Y. Matsui, N. Kamiya, S. Adachi, H. Okumura and T. Kouyama, *J. Mol. Biol.*, 2004, **341**, 1023.
22. A. Maeda, M.A. Verhoeven, J. Lugtenburg, R.B. Gennis, S.P. Balashov and T.G. Ebrey, *J. Phys. Chem.*, 2003, **B108**, 1096.
23. F. Garczarek, L.S. Brown, J.K. Lanyi and K. Gerwert, *Proc. Natl. Acad. Sci. USA*, 2005, **102**, 3633.
24. J.F. Nagle and S. Tristam-Nagle, *J. Membrane Biol.*, 1983, **74**, 1.
25. J.B. Jackson, *FEBS Letters*, 2003, **555**, 176.
26. A. Singh, J.D. Venning, P.G. Quirk, G.I. van Boxel, D.J. Rodrigues, S.A. White and J.B. Jackson, *J. Biol. Chem.*, 2003, **278**, 33208.

27. R.I. Menz, J.E. Walker and A.G. Leslie, *Cell*, 2001, **106**, 331.
28. M. Muller, K. Gumbiowski, D.A. Cherepanov, S. Winkler, W. Junge, S. Engelbrecht and O. Pank, *Eur. J. Biochem*, 2004, **271**, 3914.
29. A. Aksimentiev, I.A. Balabin, R.H. Fillingame and K. Schulten, *Biophys. J.*, 2004, **86**, 1332.
30. P. Brzezinski, *TRENDS in Biochem. Sciences*, 2004, **29**, 380.
31. M. Wikström, *Biochim. Biophys. Acta*, 2004, **1655**, 241.
32. R.B. Gennis, *Frontiers in Bioscience*, 2004, **9**, 581.
33. M.M. Pereira, M. Santana and M. Teixeira, *Biochim. Biophys. Acta*, 2001, **1505**, 185.
34. M.R. Cheesman, V.S. Oganesyan, N.J. Watmough, C.S. Butler and A.J. Thomson, *J. Am. Chem. Soc.*, 2004, **126**, 4157.
35. P. Hellwig, B. Barquera and R.B. Gennis, *Biochemistry*, 2001, **40**, 1077.
36. T. Tsukihara, H. Aoyama, E. Yamashita, T. Takashi, H. Yamaguichi, K. Shinzawa-Itoh, R. Nakashima, R. Yaono and S. Yoshikawa, *Science*, 1996, **272**, 1136.
37. S. Yoshikawa, K. Shinzawa-Itoh and T. Tsukihara, *J. Bioenerg. Biomemb.*, 1998, **30**, 7.
38. S. Yoshikawa, K. Shinzawa-Itoh and T. Tsukihara, *J. Inorg. Biochem.*, 2000, **82**, 1.
39. S. Iwata, C. Ostermeier, B. Ludwig and H. Michel, *Nature*, 1995, **376**, 660.
40. C. Ostermeier, A. Harrenga, U. Ermler and H. Michel, *Proc. Natl. Acad. Sci. USA*, 1997, **94**, 10547.
41. M. Svensson-Ek, J. Abramson, G. Larsson, S. Tornroth, P. Brzezinski and S. Iwata, *J. Mol. Biol.*, 2002, **321**, 329.
42. T. Soulimane, G. Buse, G.P. Bourenkov, H.D. Bartunik, R. Huber and M.E. Than, *EMBO J*, 2000, **19**, 1766.
43. J. Abramson, S. Riistama, G. Larsson, A. Jasaitis, M. Svensson-Ek, L. Laakkonen, A. Puustinen, S. Iwata and M. Wikström, *Nature Struct. Biol.*, 2000, **7**, 910.
44. M.M. Pereira and M. Teixeira, *Biochim. Biophys. Acta*, 2004, **1655**, 340.
45. A. Kannt, T. Soulimane, G. Buse, A. Becker, E. Bamberg and H. Michel, *FEBS*, 1998, **434**, 17.
46. E. Arslan, A. Kannt, L. Thöny-Meyer and H. Hennecke, *FEBS Letters*, 2000, **470**, 7.
47. T.K. Das, C. Pecoraro, F.L. Tomson, R.B. Gennis and D.L. Rousseau, *Biochemistry*, 1998, **37**, 14471.
48. J. Hemp, C. Christian, B. Barquera, R.B. Gennis and T.J. Martinez, *Biochemistry*, 2005, **44**, 10766.
49. T. Tsukihara, K. Shimokata, Y. Katayama, H. Shimada, K. Muramoto, H. Aoyama, M. Mochizuki, K. Shinzawa-Itoh, E. Yamashita, M. Yao, Y. Ishimura and S. Yoshikawa, *PNAS*, 2003, 15304.
50. H. Michel, *Biochemistry*, 1999, **38**, 15129.
51. S. Papa, N. Capitanio and G. Capitanio, *Biochim. Biophys. Acta*, 2004, **1655**, 353.
52. A. Namslauer, A. Pawate, R.B. Gennis and P. Brzezinski, *PNAS*, 2003, **100**, 15543.
53. M. Wikström, M.I. Verkhovsky and G. Hummer, *Biochim. Biophys. Acta*, 2003, **1604**, 61.
54. J. Abramson, S. Riistama, G. Larsson, A. Jasaitis, M. Svensson-Ek, L. Laakkonen, A. Puustinen, S. Iwata and M. Wikström, *Nature Struct. Biol.*, 2000, **7**, 910.

55. P. Hellwig, T. Yano, T. Ohnishi and R.B. Gennis, *Biochemistry*, 2002, **41**, 10675.
56. M. Brändén, H. Sigurdson, A. Namslauer, R.B. Gennis, P. Ädelroth and P. Brzezinski, *PNAS*, 2001, **98**, 5013.
57. C. Pecoraro, R.B. Gennis, T.V. Vygodina and A.A. Konstantinov, *Biochemistry*, 2001, **40**, 9695.
58. M. Brändén, F.L. Tomson, R.B. Gennis and P. Brzezinski, *Bichemistry*, 2002, **41**, 10794.
59. R.B. Gennis, *Biochim. Biophys. Acta*, 1998, **1365**, 241.
60. O.-M.H. Richter, K.L. Dürr, A. Kannt, B. Ludwig, F.M. Scandurra, A. Giuffre, P. Sarti and P. Hellwig, *FEBS Journal*, 2004, **272**, 404.
61. C. Backgren, G. Hummer, M. Wikström and A. Puustinen, *Biochemistry*, 2000, **39**, 7863.
62. S. Yoshikawa, *FEBS Letters*, 2003, **555**, 8.
63. S. Yoshikawa, K. Shinzawa-Itoh, R. Nakashima, R. Yaono, E. Yamashita, N. Inoue, M. Yao, M.J. Fei, C.P. Libeu, T. Mizushima, H. Yamaguchi, T. Tomizaki and T. Tsukihara, *Science*, 1998, **280**, 1723.
64. H.-M. Lee, T.K. Das, D.L. Rousseau, D.A. Mills, S. Ferguson-Miller and R.B. Gennis, *Biochemistry*, 2000, **39**, 2989.
65. A.A. Konstantinov, S. Siletsky, D. Mitchell, A. Kaulen and R.B. Gennis, *Proc. Natl. Acad. Sci. USA*, 1997, **94**, 9085.
66. D. Bloch, I. Belevich, A. Jasaitis, C. Ribacka, A. Puustinen, M.I. Verkhovsky and M. Wikström, *PNAS*, 2004, **101**, 529.
67. A.S. Pawate, M. J.A. Namslauer, D.A. Mills, P. Brzezinski, S. Ferguson-Miller and R.B. Gennis, 2002.
68. M. Rultenberg, A. Kannt, E. Bamberg, K. Fendler and H. Michel, *Nature*, 2002, **417**, 99.
69. M. Ruitenberg, A. Kannt, E. Bamberg, B. Ludwig, H. Michel and K. Fendler, *PNAS*, 2000, **97**, 4632.
70. A. Puustinen and M. Wikström, *Proc. Natl. Acad. Sci. USA*, 1999, **96**, 35.
71. D.M. Popovic and A.A. Stuchebrukhov, *FEBS Letters*, 2004, **566**, 126.
72. B. Schmidt, J. McCracken and S. Ferguson-Miller, *PNAS*, 2003, **100**, 15539.
73. A.J. Moody, *Biochim. Biophys. Acta*, 1996, **1276**, 6.
74. E. Antonini, M. Brunori, A. Colosimo, C. Greenwood and M.T. Wilson, *Proc. Natl. Acad. Sci. USA*, 1977, **74**, 3128.
75. A.J. Moody, C.E. Cooper, R.B. Gennis, J.N. Rumbley and P.N. Rich, *Biochemistry*, 1995, **34**, 6838.
76. A.J. Moody, C.E. Cooper and P.R. Rich, *Biochim. Biophys. Acta*, 1991, **1059**, 189.
77. M.I. Verkhovsky, A. Tuukkanen, C. Backgren, A. Puustinen and M. Wikström, *Biochemistry*, 2001, **40**, 7077.
78. M. Ruitenberg, A. Kannt, E. Bamberg, K. Fendler and H. Michel, *Nature*, 2002, **417**, 99.
79. M. Fabian, W.W. Wong, R.B. Gennis and G. Palmer, *Proc. Natl. Acad. Sci. USA*, 1999, **96**, 13114.
80. S. Ferguson-Miller and G.T. Babcock, *Chem. Rev.*, 1996, **7**, 2889.
81. K. Oda, T. Ogura, E.H. Appelman and S. Yoshikawa, *FEBS Letters*, 2004, **570**, 161.

82. D.A. Proshlyakov, M.A. Pressler, C. DeMaso, J.F. Leykam, D.L. DeWitt and G.T. Babcock, *Science*, 2000, **290**, 1588.

83. K. Budiman, A. Kannt, S. Lyubenova, O.-M. H. Richter, B. Ludwig, H. Michel and F. MacMillan, *Biochemistry*, 2004, **43**, 11709.

84. F.G.M. Wiertz, O.-M.H. Richter, A.V. Cherepanov, F. MacMillan, B. Ludwig and S. de Vries, *FEBS Letters*, 2004, **575**, 127.

85. P.R. Rich, S.E.J. Rigby and P. Heathcote, *Biochim. Biophys. Acta*, 2002, **1554**, 137.

86. J.E. Morgan, M.I. Verkhovsky, G. Palmer and M. Wikström, *Biochemistry*, 2001, **40**, 6882.

87. M. Karpefors, P. Ädelroth, Y. Zhen, S. Ferguson-Miller and P. Brzezinski, *Proc. Natl. Acad. Sci. USA*, 1998, **95**, 13606.

88. P. Ädelroth, M. Karpefors, G. Gilderson, F.L. Tomson, R.B. Gennis and P. Brzezinski, *Biochim. Biophys. Acta*, 2000, **1459**, 533.

89. A. Jasaitis, M.I. Verkhovsky, J.E. Morgan, M.L. Verkhovskaya and M. Wikström, *Biochemistry*, 1999, **38**, 2697.

90. K. Faxén, G. Gilderson, P. Ådelroth and P. Brzenzinski, *Nature*, 2005, submitted.

91. M. Karpefors, P. Ädelroth, A. Aagaard, I.A. Smirnova and P. Brzezinski, *Israel Journal of Chemistry*, 1999, in press.

92. A. Namslauer, A. Aagaard, A. Katsonouri and P. Brzezinski, *Bichemistry*, 2003, **42**, 1488.

93. U. Pfitzner, K. Hoffmeier, A. Harrenga, A. Kannt, H. Michel, E. Bamberg, O.M.H. Richter and B. Ludwig, *Biochemistry*, 2000, **39**, 6756.

94. S.A. Siletsky, A.S. Pawate, K. Weiss, R.B. Gennis and A.A. Konstantinov, *J. Biol. Chem.*, 2004, **279**, 52558.

95. S. Papa, N. Capitanio, G. Capitanio and L.L. Palese, *Biochim. Biophys. Acta*, 2004, **1658**, 95.

96. P. Brzezinski and G. Larsson, *Biochim. Biophys. Acta*, 2003, **1605**, 1.

97. S. Siletsky, A.D. Kaulen and A.A. Konstantinov, *Biochemistry*, 1999, **38**, 4853.

Proton Entry, Exit and Pathways in Cytochrome Oxidase: Insight from Conserved Water

MARTYN A. SHARPE, LING QIN AND SHELAGH FERGUSON-MILLER

Department of Biochemistry & Molecular Biology,
Michigan State University,
East Lansing,
Michigan, 48824-1319,
USA

1 Overview

In this study we have examined the three-dimensional crystal structures of cytochrome oxidase from five organisms: the aa_3 cytochrome c oxidase of *Rhodobacter sphaeroides* (RS) for which we have three different crystal structures (PDB entry 1M56[1]; two unpublished structures from our laboratory), the most recent aa_3 cytochrome c oxidase of *Bos taurus* (BT) resolved by Yoshikawa and co-workers (PDB entry 1V54[2]); the aa_3 cytochrome c oxidase of *Paracoccus denitrificans* (PD) (PDB entry 1QLE[3]); the bo_3 ubiquinol oxidase of *Escherichia coli* (EC) (PDB entry 1FFT[4]); and finally the ba_3 cytochrome c oxidase of *Thermus thermophilus* (TT), for which we have two structures (PDB entry 1EHK[5] and one supplied to us before publication by Professor James Fee; PDB entry 1XME). We have examined the common features of water molecules within the structure of this enzyme superfamily. We hypothesize that, as all the genes that code for these oxidases evolved from a common ancestor, the basic machinery of oxygen reduction and proton pumping should be common to all. Hence, comparison of the 13 subunit mammalian aa_3 oxidase and 3 subunit ba_3 oxidase from a thermophilic bacterium will show the same basic features that are required for enzyme function and, ultimately, for the survival of the organisms which express these proteins. This analysis focuses on crystallographically resolved water and reveals that there are regions of the protein where water positions are 'conserved', even though all the amino acids in the vicinity are not. These observations lead to novel conclusions regarding the path and gating of proton movement in the vicinity of the heme/Cu binuclear center.[*]

2 Heme–copper Oxidases, their Evolution and General Function

Fossil and geological evidence suggests that atmospheric oxygen began to be generated by water-splitting, photosynthetic cyanobacteria-like organisms some 2.5–2.2 billion years ago. Over the next 400 million years, organisms had to respond to an increase in oxygen, which had been less than 1% of present day levels for hundreds of millions of years, to levels some 15% higher than today.[6] Organisms had to evolve biochemical methods for dealing with the presence of this oxidant. One of these methods was to use oxidative respiration. The reduction of oxygen to water also had the advantage that the redox potential drop from the oxidation of organic and inorganic substrates to the reduction of oxygen could be coupled to the generation of a transmembrane chemiosmotic potential, by using substrate protons from the inside of the cell. It appears likely that heme–copper oxidases evolved from one class of nitric oxide reductases (NOR), present in the then ubiquitous archaebacterial flora.[7–10]

Many prokaryotic, archaebacterial and mitochondrial terminal oxidases involved in aerobic respiration belong to the large heme–copper oxidase superfamily. All members of this evolutionarily diverse family of enzymes contain two core subunits that are closely related in terms of amino acid sequence. Enzymes of this superfamily share a common structural feature, a binuclear center found in subunit I, formed by a five-coordinated heme–iron and adjacent copper called Cu_B (Section 4.4). At the binuclear center, oxygen is reduced to water and this redox reaction is coupled to the translocation of protons across the bacterial cell membrane or the mitochondrial inner membrane:

$$4 \, e^- \text{ outside} + 4 \, H^+ \, (s) \text{ inside} + 4 \, H^+ \, (p) \text{ inside} + O_2 \rightarrow$$

$$2 \, H_2O + 4 \, H^+ \, (p) \text{ outside} \quad (1)$$

(H^+ (s) are substrate protons used to make water and H^+ (p) are 'pumped' protons)

Four electrons are transferred from outside the membrane dielectric into the binuclear center, moving electrogenically (that is, generating an externally-positive membrane potential), where they reduce molecular oxygen to water. The protons required for this reaction normally come from the inside of the mitochondria/bacteria. It has been generally found that in addition to this consumption of substrate protons, four 'pumped' protons are moved from the inside of the mitochondria/bacteria to the outside, also

---

[*] Unless otherwise indicated, we use the amino acid numbering system of RS throughout. Similarly, the default three-dimensional crystal structure used is that of RS, obtained in our laboratory. Semi-conserved refers to waters present in three or more current structures, which overlap, within the van der Waals radius. Conserved waters are present in the structures of RS, BV, PD and TT.

electrogenically, contributing to the membrane potential. Thus, during a single turnover of the enzyme eight charges are moved across the membrane dielectric to generate a transmembrane electrical potential (positive outside), and eight protons are removed from the inner bulk phase, increasing the inside pH and thus generating a transmembrane proton gradient, ΔpH. This chemiosmotic gradient is used to drive thermodynamically unfavorable reactions, such as the phosphorylation of ADP to ATP. The superfamily of heme–copper oxidases can be further divided into two sub-families: the quinol oxidases, which oxidize quinols; and cytochrome c oxidases, which oxidize cytochrome c. Quinol oxidases and cytochrome c oxidases are most closely related in terms of the metal cofactors and the ligands to these metals, found in subunit I. In all cases they have two hemes and a copper center (Cu_B). In the case of the heme–iron/copper binuclear center, the Cu_B is ligated by three histidines and the heme by a single histidine. The histidines that ligate Cu_B are the histidine pair H333/H334 and H284. This latter ligand is post-translationally modified by a covalent bond from a nitrogen atom to a carbon atom of the ring of a tyrosine (H284, $N_{\varepsilon 2}$ and Y288, $C_{\varepsilon 2}$).[11]

The hemes present in the oxidases in this study are heme b, heme o and heme a. Heme o and heme a are synthesized from heme b. The synthesis of heme o begins with farnesylation of the vinyl group at carbon C-2 of heme b. In some archaebacterial oxidases, a hydroxyl-ethylgeranylgeranyl side chain is used instead of the hydroxyl-ethylfarnesyl side chain (see Lubben & Morand[12] for an overview of the use of various hemes in the cytochrome oxidases of many different species).

Heme a is synthesized in turn from heme o by the oxidation of a methyl to a formyl group on C-8. Although each of these different hemes may carry out the same redox chemistry, the nature of the different hemes changes the bonding of the heme to the protein. Each of these hemes has two propionate groups which are capable of forming both ionic and hydrogen bonds. The hydroxyl group of the farnesyl chains of heme o and heme a are also capable of hydrogen bonding, as is the formyl group of heme a. It is likely that this hydroxyl group is essential for enzyme function as it is found in the oxygen reduction center of all the members of the oxidase superfamily. Structural studies have shown that the hydroxyl groups of heme o and a_3 are hydrogen bonded to a water molecule. This water molecule is also hydrogen bonded to an amino acid residue at the top of a putative proton supply pathway, the K-channel.[13] In most species including RS this is a threonine residue (or serine residue in TT). Site directed mutagenesis of this residue to an alanine, T359A, causes a 75% reduction in turnover.[14]

3   Water and Protein Function

3.1   Water Interactions Within Protein Structures

Water has many interactions with the amino acid residues of proteins mainly mediated by hydrogen bonds. A hydrogen bond is a partially covalent interaction between a hydrogen atom that is covalently bonded to an electronegative donor atom, and another electronegative acceptor atom. Within proteins these electronegative atoms are normally nitrogen and oxygen atoms. Hydrogen bonds play an important role in defining the structure and function of biological macromolecular systems. They are a major factor in protein folding and stabilization of the secondary structure.[15]

Detailed studies of high definition X-ray structures have revealed the ability of water to form a wide variety of hydrogen bonds to other water molecules, peptide bonds and to the side chains of amino acids in a variety of geometries.[16]

In this study we will examine the ability of water molecules to facilitate the movement of protons from both the inner bulk aqueous phase to the binuclear oxygen reduction center and from the binuclear center to the outer bulk aqueous phase. It has been shown that protons move along one-dimensional hydrogen-bonded water chains some 40 times faster than in bulk water. These chains of hydrogen-bonded water molecules provide excellent conductors for protonic currents through pores across membranes and within proteins. Protons hop from one water molecule to the next *via* a Grotthuss mechanism, and no water molecules are displaced. The mobility of an excess proton along chains of hydrogen-bonded water molecules is particularly high in environments with restricted dimensions, such as the pore of gramicidin[17] and inside nano-tubes.[18] Defects, which stabilize a hydronium cation (H_3O^+) along the water chain, slow the rate at which protons are transferred and reduce the proton flux. Defects include the presence of small anions or cations or the presence of charged amino acids such as glutamate or arginine within the pore.

Although water, the universal solvent, is able to hydrogen bond to a wide variety of species, the maximum number of hydrogen bonds that an individual water molecule is able to make is four. In tetra-coordinated water, both hydrogen atoms of the water molecule are hydrogen bonded to electronegative atoms and the two-electron lone pairs of the oxygen atom are each bonded to a hydrogen atom. A protonated water molecule is a called a hydronium ion. In a hydronium ion the positive charge is delocalized over the whole molecule and so this species is capable of forming up to four hydrogen bonds with electronegative atoms. When tetra-coordinated by four hydrogen bonds both water and hydronium ions tend to have a pyramidal bonding geometry (*i.e.* a pure or distorted tetrahedral). Examples of this geometry for the hydrogen bonds of water and hydronium ions are to be found in azeotropic mixtures of water with alcohols and strong acids. An azeotropic mixture of ethanol and water (80 mole percent ethanol and 20 mole percent water) consists of cyclic pentamers composed of four ethanol molecules arranged around a single pyramidal, tetra-coordinated, water molecule.[19] This is the same geometry found for a hydronium ion in an azeotropic mixture of nitric acid and water, hydronium nitrate (50 mole percent nitric acid and 50 mole percent water), where the oxygen cation is bonded to three hydrogen atoms and a nitrate anion. These geometric restraints may play an important role in the vectorial translocation of protons in the oxidase mechanism.[20]

3.2 Water Interactions in Cytochrome Oxidase: Overall Features

The aa_3 family of oxidases has a number of representatives, the most complex being the mammalian enzyme with 13 subunits, 9 more than the simplest member of the bacterial oxidase family, and 10 more than the TT ba_3 enzyme. However, it appears that only subunits I and II are required for catalytic activity and proton pumping. We have examined the ligation and positions of water within subunits I and II in RS, BV, PD and TT. (Subunits II and III of TT are essentially identical to subunit II of the other oxidases and for the sake of the narrative will be dealt with as if they formed a single protein subunit). We have overlaid the different crystal structures of these evolutionarily diverse, but related, oxidases to examine the role of water in the catalytic activity of the oxidases of this superfamily. We have also attempted to elucidate the mechanism whereby these enzymes are able to move both substrate and 'pumped' protons against an electrochemical potential.

Subunit I is typically made up of 12 transmembrane helices, although some oxidases have more, including EC *bo*₃ ubiquinol oxidase where there are 14. Subunit II has two transmembrane helices and a large head group or cap which sits atop subunit I. There are striking similarities in the amino acid sequences and in the three-dimensional structures of subunits I and II of these enzymes, and also in some of the water molecules bonded within the structures of subunits I and II. Figure 1 shows the arrangement of conserved and semi-conserved waters within the structure of cytochrome oxidase. The subunit structures are outlined and positioned so that the bottom of the figure is 'the inside' of the bacteria or mitochondria and the top is the outside. Conserved water molecules (those present in two or more structures) that are within hydrogen-bonding distance of an amino acid or another water molecule are shown (at three-quarters of the van der Waals radius). The water molecules are color coded so as to provide a guide as to their putative role in the enzyme catalytic cycle. At the base of the oxidase are a large number of non-specifically bonded water molecules (in yellow). These may act as a 'proton antenna,' facilitating the movement of protons from the buffering system of the inner aqueous phase into the two proton channels that direct 'pumped' and substrate protons to the binuclear center.

Following the publication of the first crystal structures of cytochrome oxidase it became apparent that there were at least two water chains leading from the cytosolic/matrix aqueous bulk phase (in bacterial and mitochondrial oxidases, respectively) to the binuclear reaction center. The two pathways were named the D-channel and K-channel on the basis of D132 and K362 being key components of the pores leading from the aqueous phase to the binuclear center.[21] The waters that form the hydrogen-bonded proton chains in the D-channel are shown in green and the even more positionally conserved waters of the K-channel are shown in red.

In Figure 1 it is possible to see water chains that reach up to the interface of subunits I and II, seemingly in communication with the bulk aqueous phase. Water is generated during the oxygen reduction reaction and has to exit the enzyme. We have colored waters along a probable water egress pathway magenta, the exit being near the small pore formed by P270, D271 and R234[II].

It appears likely that 'pumped' protons leave the binuclear center *via* the Mg^{2+} ion (yellow), along the interface between the Cu_A binding domain of subunit II (shown in green) and the highly conserved loop between helices 11 and 12 (shown in red), near the Ca^{2+} ion (shown in green). The waters in this proton egress pathway are shown in light blue.

3.3   The Position of Metals and Water in Cytochrome Oxidases

Figure 2 shows more detailed views of the metal centers in relationship to each other, their ligands and water bonded to the metals. The ligands to the physiologically

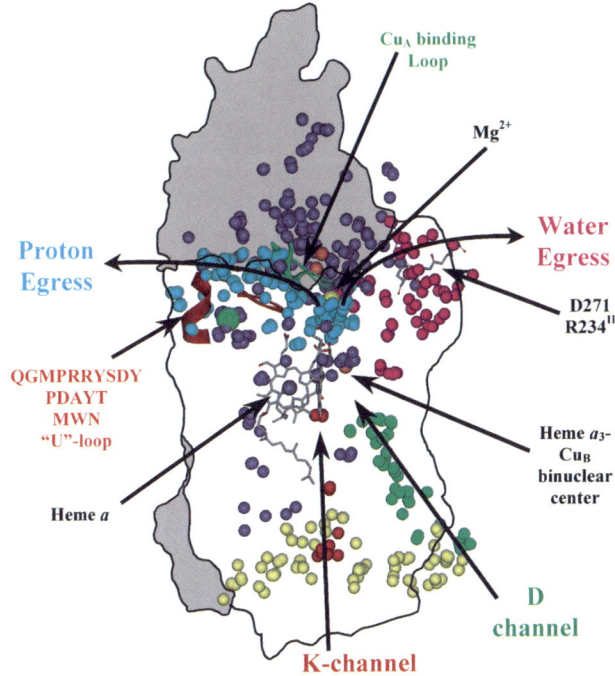

Figure 1 *The three dimensional structure of the two subunit oxidase. The overlaid RS oxidase structure is outlined (subunit I transparent, subunit II gray) along with conserved and semi-conserved water molecules. All five oxidases are approximately the same size and essentially the same shape. Only the "cap" of subunit II is shown; at the interface can be seen the centrally located Mg^{2+} ion (○) above the heme propionates and the Ca^{2+} ion on the left of the structure (●). Water molecules are identified by their function: "Antenna" water molecules bonded to amino acids at the base of the oxidase, ○; proton wire water molecules in the D-channel, ●; water molecules in the K-channel, ●; waters leaving the structure via the egress pathway, ●; waters of the proton egress pathway, ●; and unassigned and/or structural waters, ●. Amino acid residues thought to be important in the egress of water molecules between subunits I/II, D272 (stick) and R234^{II} (line), are indicated. The loop between helices 11 and 12, which facilitates the formation of hydrogen-bonded water chains, is in red, whilst the Cu_A binding domain of subunit II is shown in green*

relevant metals, for subunit I and II of each oxidase, are shown in Table 1. Figure 2 also shows the ligands to the non-redox metals, magnesium and calcium, and also the amino acids ionically or hydrogen-bonded to the hemes a and a_3. It can be seen that there is an extensive network of hydrogen bonded waters, some in chains, interacting with the four carboxylic acid groups of the hemes (the heme propionates) and the Mg^{2+} ion, and to a much lesser extent with the Na^+/Ca^{2+} ion (Figures 2A and 2B). Figure 2C shows the ligation of amino acids and conserved waters to the binuclear center. The chosen viewpoint shows the H284-H288 covalent bond and also the hydrogen bond between Y288 and the heme hydroxyl group. The propionates of heme a_3 are bonded to R481 and W172. This latter amino acid has been converted to a tyrosine (W172Y) in TT and, instead of the indole nitrogen, the phenol group of

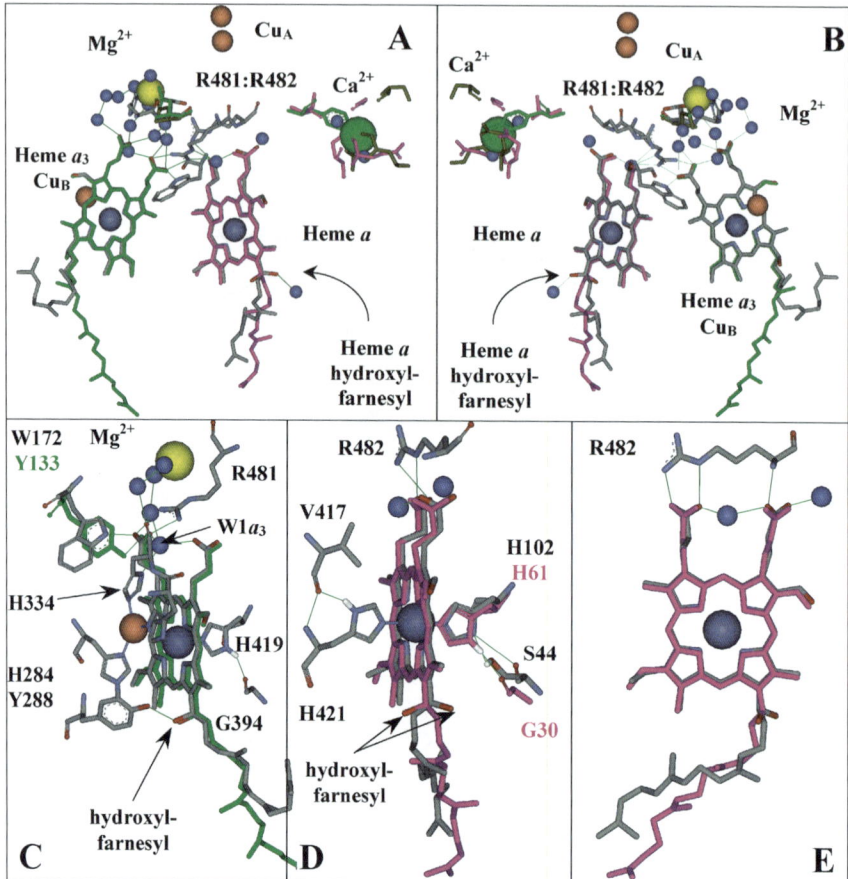

Figure 2 *The position of the metal centers and conserved waters. Figures 2A and 2B show conserved water molecules bonded to the metal centers of cytochrome oxidase. Waters directly bind Ca^{2+} (●) and Mg^{2+} (○). Water is extensively bonded to the propionate groups of both heme a and heme a_3. Heme a_3 and the Mg^{2+} ion are interconnected by water chains. W172 (Y133 in TT and many thermophiles) and R481 are hydrogen bonded to heme a_3 and R482 is bonded to heme a. Figure 2C shows the ligation of heme a_3 and Cu_B and the conserved water molecule, $W1a_3$. The bonds of $W1a_3$ are arranged in the form of a tetrahedral pyramid. $W1a_3$ will block the electrophoretic migration of hydroxide, generated in the binuclear center, through the protein. Figure 2D shows the ligands to heme a and conserved waters. The structures are color coded so that RS is in the elemental color scheme, BV is in pink, TT is in green and EC is in dark green. Note that the hydroxyl-farnesyl position in BV compared to RS. H102 is ligated in its imidazolate form by a serine in RS and PD. In the majority of other oxidases, including BV, the ligand is glycine, fixing the H61 as an imidazole. Figure 5E shows the three hydrogen bonds that connect R482 with the heme propionates of heme a. R481/R482, linked to the heme propionates, are central residues to the proton exit pathway and may regulate proton backleak*

Table 1 *The ligands to the metals found in cytochrome oxidase according to species*

RS	BV	PD	TT	EC	Notes
Subunit I					
H419	H376	H411	H384	H419	Heme a_3
G394	G351	G386	G359	G394	carbonyl=O H-Bond to imidazole-H, not EC
H102	H61	H94	H72	H106	Heme a
S44	G30	S46	G35	G44*	carbonyl=O/-OH H-Bond to imidazole-H
H421	H378	H413	H386	H421	Heme a
V417	V374	V409	P382	I417	carbonyl=O H-Bond to imidazole-H
Y288	Y244	Y280	Y237	Y288	Covalent link to H284
H284	H240	H276	H233	H284	Cu_B
H333	H290	H325	H282	H333	
H334	H291	H326	H283	H334	
H411	H368	H403	H376	H411	Mg^{2+}
					No Mg in TT or EC structure
D412	D369	D404	N377	N412	
E54	E40	E56		Q82	Ca^{2+}
A57	Q43	H59		L85	No Ca^{2+} in EC structure
G59	G45	G61		S87	No Ca^{2+} binding site in TT structure
I484	S441	I476	Y452	S484	Na^{2+} in Bovine
Subunit II					
H217[II]	H161[II]	H181[II]	H114[II]	N172[II]	Cu_A
C252[II]	C196[II]	C216[II]	C149[II]	S207[II]	Cu_A
E254[II]	E198[II]	E218[II]	Q151[II]	S209[II]	Cu_A/Mg^{2+}
C256[II]	C200[II]	C220[II]	C153[II]	S211[II]	Cu_A
H260[II]	H204[II]	H224[II]	H157[II]	F215[II]	Cu_A
M263[II]	M207[II]	M227[II]	M160[II]	M218[II]	Cu_A

* The first 51 amino acids in the EC structure are not resolved; however, an examination of the sequence data from the oxidase family leads us to assign the carbonyl oxygen of residue G44 as the hydrogen-bonding species to H106 of EC. This residue is highly conserved in cytochrome o ubiquinol oxidases.

Y133 is bonded to the heme a_3 propionate. The water chains shown in Figure 2C are found in RS, BV, PD and TT, and it appears that they are also present in EC. Figures 2C and D show the ligation of heme a. It can be seen that both histidine ligands are hydrogen bonded to other amino acids. In RS and PD, H102 is ligated by a serine or a threonine, while the much studied bovine enzyme has a glycine ligand. R482 is connected to heme a by three hydrogen bonds. There is very little conserved water around the hemes except in the vicinity of the propionates. The functional significance of these features is considered in Section 4.

3.4 Subunit II and Water

A cursory examination of the amino acid sequences of subunit II, in the absence of the three dimensional structure, suggests that the genes for this protein are very diverse. However, the peptide can be divided into three basic building blocks, which are closely related:

(a) Two transmembrane helices which bind the subunit to the membrane and to the transmembrane region of subunit I;

(b) Three main series of β-pleated sheets, the first two of which are highly hydrogen bonded within themselves and to each other;

(c) The Cu_A binding site, also containing β-pleated sheets, whose ligands are also bonded to the Mg^{2+} ion and to waters at the subunit I/II interface.

Figures 3A and 3B show the overlaid structures of subunit II, viewed from opposite sides, from RS (solid tube), BV (line ribbon), TT (solid ribbon), and EC (flat ribbon). The protein in Figure 3A and 3B is color coded so that loops are shown in yellow, transmembrane helices are shown in dark blue, α-helices are shown in light blue, linking peptides are shown in gray, the Cu_A binding region is shown in green and finally, the areas which consist of highly hydrogen-bonded β-pleated sheets are shown in red.

Viewed from the side, subunit II resembles a golf club, in which the transmembrane α-helices make up the shaft and a large strongly hydrogen-bonded extramembrane domain makes up the head. The upper part of the protein covers the top of subunit I and acts as a cap, providing a barrier to water from the outer aqueous phase and to water from the inside above the heme propionates. It can be seen that heme a_3 is linked to the Cu_A binding loop by water chains. At the top of these water chains which connect heme a_3 to Cu_A, is the Mg^{2+} ion (shown in yellow), which shares a ligand (E254 ^{II}) with the Cu_A copper pair.

Figure 3C shows the β pleated sheet 'cap' structure of subunit II. This basic structure is present in the other oxidases, although the actual size of the four blocks varies between species somewhat. Figure 3C includes the copper-binding region, conserved waters and the hydrogen bonds, both within the β-sheet regions and between regions. Eight water molecules are bonded within the structures of RS and BV, while six are bound in the PD structure. These appear to act structurally, hydrogen bonding and bracing together different β sheet-pleated regions of the protein. There are also 13 semi-conserved and conserved waters bonded to this cap and projecting into the subunit I/II interface. There is no water or proton exit pathway through the subunit II cap; there are far too many hydrogen bonds between amino acids for water to be able to diffuse rapidly from subunit I, *via* subunit II, into the bulk phase. In addition to stopping water migration, the cap will stop the migration of ions and other species dissolved in the water of the outer bulk phase into the water space at the subunit I/II interface.

3.5 Subunit I and Water

3.5.1 The D-channel

The D-channel consists of a region of the oxidase containing an ordered water chain extending from the inner side of the oxidase towards the binuclear center. This channel, appears to terminate at E286, the carboxylic acid group of which is some 11 Å from the binuclear center. An as yet undefined proton pathway (or pathways) must

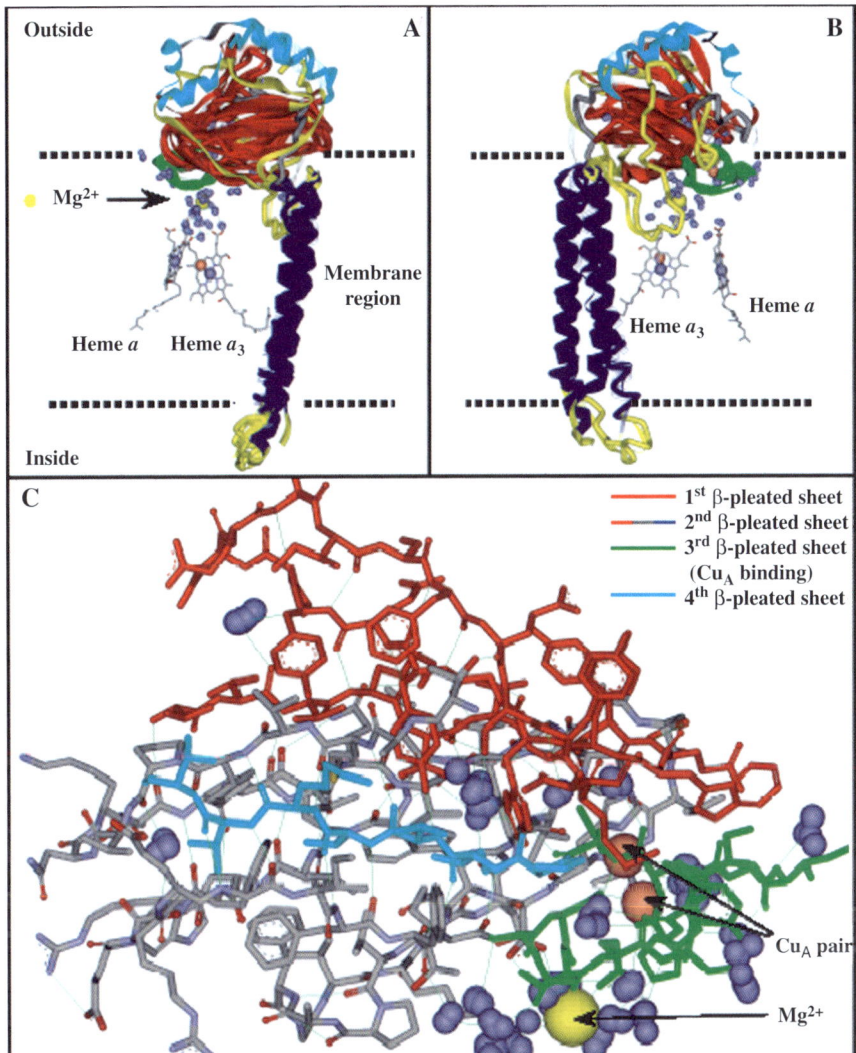

Figure 3 *The structure of subunit II. Figures 3A and 3B show the three-dimensional structure of subunit II and hydrogen-bonded waters. The protein is color coded so that loops are yellow, transmembrane helices are dark blue, α-helices are light blue, linkers are gray, the Cu_A binding region is green and β-pleated sheets are red (RS (solid tube), BV (line ribbon), TT (solid ribbon), and EC (flat ribbon)). Figure 3C shows the β-pleated sheet cap of subunit II, including the Cu_A binding site, waters at the I/II interface and conserved structural waters. The first β-pleated sheet is shown in red, the major β-pleated sheet is shown in the elemental color scheme, the Cu_A binding β-pleated sheet is shown in green and the final β-pleated sheet is in turquoise*

facilitate the movement of protons from the D-channel to the binuclear center. The main channel runs along the sides of helices I, II and IV. In this analysis we have identified a conserved water molecule, $W1_{V330}$, beyond the end of the D-channel. This water molecule, rather than the carboxylic acid residue (E286) normally considered to be a key gating species, is proposed to be responsible for the transfer of protons to the binuclear center during oxygen reduction catalysis.

The particular identities of the amino acids that bond the water molecules within this pore vary somewhat with the genotypes (see Table 2), but the essential structure of the water chain is common to all the species (Figures 4A and 4C). We have chosen the extremely divergent organisms, RS and TT, to demonstrate that water molecules themselves and not the water-bonding amino acids that line the D-channel are the proton transferring species. The D-channels of RS and TT only have one residue in common (S200 in RS, S155 in TT), but share the same water chains (Figure 4). The base of the pathway starts with the antenna waters in contact with the inner bulk aqueous phase. The amino acids at the base of helix I, most clearly T24, are hydrogen bonded to water and are at the start of water chains within the D-channel. Midway through the membrane the channel propagates sideways towards the binuclear center. Kinetic studies using a large number of site-directed mutants of this channel demonstrate that it transports 'pumped' and likely some substrate protons. The D132N and D132A mutations result in $< 5\%$ turnover and eliminate proton pumping, due to an almost complete block in proton uptake into the D-channel.[22] The residual activity appears to be the result of proton uptake from the periplasmic side of the enzyme, likely utilizing the exit channel normally reserved for pumped protons, but in the reverse direction.[14] These mutants can be chemically rescued by the addition of a lipophilic carboxylic acid, such as free fatty acids.[23]

Further into the channel, there is normally an asparagine, N139; in TT this is replaced by a tyrosine. It has been shown that the mutant N139D abolishes proton pumping, whilst at the same time increasing the steady-state enzyme turnover. This suggests that proton pumping has been 'decoupled' from oxygen reduction.[24] It also suggests that this substitution 'unlocks' the gating mechanism that normally allows one-way transfer of protons.

3.5.2 The Gate: E286 and Variants

At the top of the D-channel, in the majority of oxidases, is a highly conserved carboxylic acid residue, E286 (E278 in PD). Mutations that remove the carboxylic acid at this location essentially stop enzyme turnover and block proton pumping.[25,26] The crystal structures show that the carboxylic acid sidechain of this residue binds a conserved water molecule, $W1_{E286}$, Figure 4. However, there is no strict need for an acid group to be present in this area, either *in vivo* or *in vitro*. Oxidases present in a whole range of organisms are missing this residue and yet are kinetically competent and are also able to pump protons.[9] In many thermophiles, including TT, the equivalent of $W1_{E286}$ is not ligated by a glutamic acid residue, but by the indole nitrogen of a tryptophan residue two amino acids upstream of Y288 (W239 in TT, ligating $W1_{W239TT}$).

The majority of cyanobacteria and some archaea, including *Rhodothermus marinus* (RM), use a different ligand, a tyrosine two amino acids downstream from H284. The top of the D-channel of RM consists of the sequence WFYSHPAVY (280–288, where H284 is the Cu_B ligand and is covalently cross-linked to Y288), and this construction was reverse engineered into the PD oxidase (originally WFFGHPEVY). The double mutant G283S/E286A was essentially inactive with an activity

Table 2 *Water-binding amino acids involved with oxygen reduction catalysis and proton pumping*

RS	BV	PD	TT	EC	Notes (mutant activity %)
D-channel main pore structure					
T24	**T10**	T26			
H26	H12	H28	**E17**		Only TT
Y33	**Y19**	Y35	F24	Y61	PD; Y35F 55%
F116	M74	F108	**Q86**	I119	Only TT
N121	N80	N113	Y91	N124	PD; N113D 37%, N113V 16%. EC; N144D 56%
P131	**P90**	P123	M99	R134	
D132	**D91**	D124	**R100**	D135	RS and PD; D to N <5%. EC; D135N 45% (no pump)
M133	**M92**	M125	**P101**	V136	
A134	**A93**	A126	**N102**	A137	H-bonded to T210 *via* -NH- of peptide bond
N139	**N98**	N131	M106	N142	RS; N139D 150–300%. PD; N131D 99% (no pump)
N140	**N99**	N132	W107	N143	
S200	S156	S192	S155	G200	Only common residue in RS and TT.
S201	**S157**	S193	T156	T201	RS; S201A 35%. PD; S193A 73%
G204	G160	G196	S159	T204	PD; G196F,W 70%
N207	**N163**	N199	**I162**	N207	RS; N207A 75%, PD; N199D 55%, N199V 174%
T210	T166	T202	**D165**	V210	H-bonded to A134 *via* =O of peptide bond
Top of the D-channel					
M106	M65	**M98**	N76	M110	RS; M106Q <5%
M107	I66	M99	A77	I111	Carbonyl position conserved. RS; ligates $W1_{E286}$
F108	**F67**	F100	**I78**	F112	I78 is common variant in thermophiles.
V111	V70	**V103**	**T81**	A115	
I112	M71	I104	**Q82**	M116	PD; I104F 76%, I104W 45%
S197	**A153**	S189	F152	S197	PD; S189A 85%, S189F 67%, S189W 47%
Redox/Oxygen and proton pumping machinery					
W172	W126	W164	Y133	W170	ligates heme a_3 propionate. PD; W164F 25%
Y175	**Y129**	**Y167**	**Y136**	Y173	ligates heme a_3 propionate via $W1_{Y175}$ and $W1_{a3}$
Q276	**Q232**	**Q268**	**R225**	I276	No ligand in *o* type. R225 replaces $W1_{Q276}$
F281	**F237**	**F273**	**W230**	A281	Carbonyl position conserved. RS; ligates $W1_{F281}$
H284	H240	H276	H233	H284	RS; H284A inactive
P285	P241	P277	P234	P285	PD; P277G 2.5%
E286	E242	E278	I235	E286	RS; E286D 50%. E286Q <0.1%. PD E278D 60%
V287	V243	V279	V236	V287	PD; V279I 50%
Y288	Y244	Y280	Y237	Y288	RS; Y288F <0.1%. PD; Y280H <1%
I289	**I245**	I281	F238	I289	
I290	L246	I282	**W239**	L290	TT; ligates $W1_{W239}$
G327	**G284**	**G319**	**S276**	S327	Carbonyl position conserved. RS; ligates $W1_{V330}$
V330	**V287**	V322	V279	V330	Carbonyl position conserved. RS; ligates $W1_{V330}$
H333	**H290**	H325	**H282**	H333	RS; H333N, G, D, Y inactive. PD; H325N 3%
H334	**H291**	H326	**H283**	H334	RS; H334N, Y inactive.

(continued)

Table 2　　*(continued)*

RS	BV	PD	TT	EC	Notes (mutant activity %)
T337	**T294**	**T329**	**A286**	T337	
T352	T309	T344	T302	T352	H-Bond H333. RS; T352A <1%. PD T344D, V 1%
K-channel					
E101[II]	**E62**[II]	E78[II]	**E15**[II]	E89[II]	RS; E101[II] to Q,A,D 8-29%; EC; E89[II] to Q,A,D 10-60%
T359	**T316**	T351	**S309**	T359	RS; T359A 2%.
F295	**F251**	F287	Y244	F295	
S299	**S255**	**S291**	**Y248**	S299	PD; S291A 105%
H300	**H256**	H292	T249	E300	PD; H292L 115%
K307	**K264**	K299	**G257**	R307	
K308	**K265**	K300	K258	K308	
	E266				
P309	**P267**	P301	L259	R309	
L314	M271	L306	S261	T314	
K362	**K319**	**K354**	T312	K362	RS; K362A, M, R inactive. PD; K354M inactive

Residues shown in bold indicate the amino acid binds water in one of the structures

of < 0.1% WT. The triple mutant that included F282Y had 10% of the wild-type steady-state oxygen reduction activity and could also pump protons.[26] It appears that it is the Y282 that is the water ligating species because of two pieces of evidence:

(a) A serine or threonine is missing in the Y<u>X</u>HPXVY motif in a large number of proteobacteria, such as *Sulfolobus solfataricus P2*,[27] but there is no 'rescue' by a tryptophan residue two residues upstream of Y288, as in the case of TT;

(b) We have modeled the PD triple mutant that was used to demonstrate that the D-channel found in RM was able to pump.[26] In this mutant, the phenol group of Y282 is pointing down into the top of the D-channel and is in a position to ligate a water molecule occupying a very similar location to $W1_{E286}$ and $W1_{W239TT}$.

In Figures 4B and 4D we show how a conserved water molecule is ligated in the same position by three different amino acid residues found in different sub-families of oxidase.

Thus, it appears that critically placed water molecules, not amino acid residues *per se*, are responsible for the movement of protons along the D-channel. The amino acids that line the channel merely hold the water molecules in a position whereby they can form water chains, which in turn facilitate the movement of protons. In addition, changing the charge within the D-channel, and hence local water protonation state, affects the ability of protons to move along the water chains. These data match theoretical studies of proton movement in water chains in hydrophobic pores, such as carbon nano-tubes.[18]

At the top of the proton channel there must be a gate or one-way valve which favors the movement of protons against a membrane potential and proton gradient, in one direction only. The profound effect of E286 mutations has led to the concept that it may be an integral part of this gating mechanism. However, E286 itself is not the

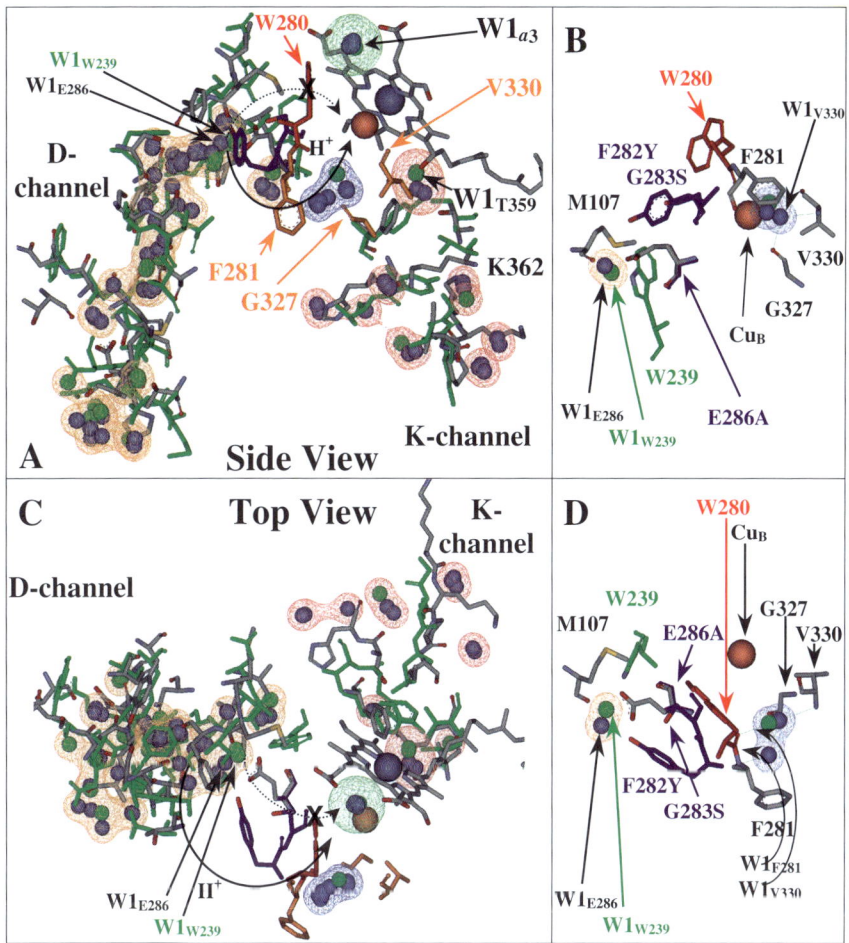

Figure 4 *Conservation of K and D water channels despite the divergence of amino acids which make up the lining of the pores. Figures 4A and 4C show side and top views of the D and K-channels in the distantly related RS (elemental color scheme) and TT (green). The D-channels of RS and TT only have one residue in common, but share the same water chains. Water molecules of RS, BV and PD are shown as (●) and waters in the TT structure are shown as (●). Residues F281, G327 and V330 are colored orange and ligate the conserved water molecules identified in this study, $W1_{F281}$ and $W1_{V330}$. W280, red, that blocks the direct movement of protons from the D-channel to the binuclear center as indicated by the broken arrow. The route from $W1_{E286}$ to the binuclear center along unresolved water molecules is shown by a solid arrow. The mutant PD, emulating the RM sequence that binds a water molecule corresponding to $W1_{E286}$ and $W1_{W239}$, is shown in blue. Figures 4B and 4D show close up views of the water molecules bonded by E286 in RS, by W239 in TT and by F282Y in the PD mutant*

proton gate, as indicated by the viability of oxidases which naturally lack this residue (TT, RM) or have been mutated to do so (the PD triple mutant). Indeed, Figures 4A and 4C show that protons have a long journey to make from the water bonded to E286 to the iron or the copper atom of the binuclear center (14 Å and 15 Å); protons must make this journey via water molecules. The mechanistic consequences of this structure are discussed in Section 4.

3.5.3 The K-Channel

The K-channel has been suggested to either begin at the base of helix VI at an opening centered near the S299:H300 pair where there is a well-conserved water molecule ($W1_{K362}$) hydrogen bonded to K362, or at $E101^{II}$ which is very near to this pair.[28] In all of the crystal structures there is a glutamate residue in the $E101^{II}$ locality (Table 2). A moderately conserved group of antenna amino acids (KKPI) is found on the loop between helix VI and helix VII that lies between $E101^{II}$ and the S299:H300 pair. The waters bonded by these amino acids may facilitate the entry of protons to $W1_{K362}$. TT is again aberrant utilizing a TY pair instead of the S299:H300 pair. The K-channel appears to provide at least two substrate protons during the reaction cycle. The movement of protons along the channel may involve mobile water molecules. Their mobility means that they are not resolved in the X-ray crystal structure; however, recently an extensive computer-aided analysis has been performed on this channel.[13] At the top of the K-channel (shown in red in Figure 4) is a semi-conserved water molecule bonded by the farnesyl hydroxide, $W1_{T359}$. It appears that a network of four water molecules can connect $W1_{K362}$ to $W1_{T359}$. The T359A mutant is able to function at 25% of normal turnover, so this residue is not an absolute requirement for enzymatic function, unlike Y288.[29] T359 is also hydrogen bonded to the hydroxyl of the farnesyl chain of heme a_3. This farnesyl-hydroxyl is also hydrogen bonded to Y288, of the covalently bonded H284–Y288 pair. $W1_{T359}$, at the top of the K-channel, appears unable to supply protons to the binuclear center as the passage of protons is blocked by the hydrogen bonding of Y288 and farnesyl hydroxide. The structures indicate that Y288 is hydrogen bonded to the hydroxide moiety of the hydroxyl-farnesyl in a manner similar to that observed in phenol/alcohol or phenol/water mixtures. In such systems the average bond length is between 2.86 and 2.93 Å.[30,31] The bond length between Y288 and the hydroxyl-farnesyl is shorter, 2.7 ± 0.22 Å, excluding EC (which lacks the resolution required to position the ligands with accuracy). Thus, it appears that this hydrogen bond between Y288 and the hydroxyl of the farnesyl chain would block the entry of protons from the K-channel to the binuclear center. Moving T359 may open this gate, allowing $W1_{T359}$ (and perhaps another water) to not only hydrogen bond between the farnesyl hydroxide, Y288 and T359, but also to change its protonation state (between a water and hydronium ion); $W1_{T359}$ could then donate protons to the binuclear center from the inner aqueous phase.[13] This may well be the case, but another possibility is that movement of Y288, not T359, breaks the Y288-farnesyl hydroxide hydrogen bond, allowing the three alcohol groups to bind $W1_{T359}$, creating a proton water wire between the inner aqueous phase and the binuclear center. Thus, Y288 can act as a proton gate to these (substrate) protons.

3.5.4 The Sodium/Calcium and Magnesium Binding Sites

The least conserved metal binding site is that which binds Na^+ in bovine oxidase and Ca^{2+} in RS and PD. In TT this ligation site has been completely lost. There is no evidence as yet that the binding of Na^+ or Ca^{2+} is a physiological modulator of oxidase function.

The Mg^{2+} center is well conserved, and although the crystal structures do not show a bonded Mg^{2+} in TT, the position of the ligands, amino acids and conserved water molecules suggest that one is present. It is not clear whether there is a functional role for the Mg^{2+} binding site in the cytochrome oxidase super family, as enzymes which normally have a Mg^{2+} present can be made to lose this ion, yet can still function to a large degree, although with loss of stability. Site directed mutagenesis studies on RS oxidase suggest that oxidase is able to function with an absent or distorted Mg^{2+} center. The $H411^I A$ and $D412^I N$ mutants resulted in a halving of maximum enzyme turnover, and no Mg^{2+} bound to the enzyme.[29, 32]

3.6   The Subunit I/II Interface

3.6.1   The Water Exit Pathway?

The structures show well-populated chains of hydrogen-bonded water that link the propionates of heme a_3 to the Cu_A ligands $E254^{II}$ and $H260^{II}$ *via* Mg^{2+}. Mn^{2+} can be substituted for Mg^{2+} in oxidase and because Mn^{2+} has an unpaired electron it can be monitored using electron paramagnetic resonance spectroscopy (EPR). The water chains between heme a_3, the Mg^{2+} ion and the bulk aqueous phase are consistent with recent studies that show that bulk water is in rapid equilibrium (ms timescale) with water bonded to the RS Mn/Mg center.[33] Mn^{2+} substituted oxidase was also used to examine the kinetics of water movement from the binuclear center to the Mg/Mn center, using $^{17}O_2$ as a spin probe. This study indicated that water produced by the binuclear center from the reduction of molecular oxygen, rapidly and preferentially migrates to the Mn/Mg center.[34] Thus, water migration from the binuclear center to the bulk phase *via* Mg^{2+} on a ms timescale can be assumed. A pathway at the subunit I/II interface is a likely route whereby waters generated by oxygen reduction exit the enzyme.

It is important to note that any pathway that allows water to exit the enzyme can also act as a route into the enzyme, potentially for anions and cations as well. Yet is seems likely that rapid access of protons and ions to the active site would be detrimental to the coupling function of the enzyme. The best understood group of proteins with the capacity to transport water but discriminate against ions is the aquaporins.[35] The transmembrane pore through aquaporin contains an aperture that precludes the movement of any species that is larger than a water molecule. The aperture consists of tandem repeats with the signature motif asparagine–proline–alanine (NPA). The hydrophobic/hydrogen-bonding nature of this NPA 'noose' prevents the movement of hydronium ions. There is no NPA motif present in any of the amino acid structures herein. However, there is a candidate for a water exit pathway that restricts ion import into the oxidase at the interface of subunit I/II on the opposite side of the oxidase to the suggested proton egress pathway (see below). The aperture involving the loop between helix 5 and 6 of subunit I consisting of a well conserved consensus sequence GGGDP. At the aperture, $D271^I$ and $R234^{II}$ hold the two subunits together, leaving only a small gap connecting the interior water reservoir to the bulk outer aqueous phase.[34] Another hydrophobic constriction in the vicinity of V231[II] also appears to provide a proton filter, as well as the Mg^{2+} ion itself.[34]

3.6.2   The Proton Exit Pathway?

Theoretical calculations show that many more waters are present in oxidase than are shown in the crystal structure and the majority of these are expected to be mobile.[13,36] A systematic examination of the different oxidases with respect to the presence of 'conserved' waters indicates that there may be a preferred pathway of proton exit. This pathway appears to involve amino acids along the subunit I/II interface, consisting of the top loop above helix 11 and the first part of helix 12 (Q477 to N494) in subunit I and the amino acids of the loop below the Cu_A binding site in subunit II (Figure 1).

The U-shaped loop connecting helix 11 and the first four residues of helix 12 is well preserved across many species and has the consensus sequence:

477 Q<u>G</u>MP<u>RRYS</u>DY--PD<u>A</u>Y---- ‖ <u>TMWN</u> 494

In the sequence shown above, amino acids in bold are bonded to waters in three of the structures analyzed, and those underlined in four. In the TT oxidase, G478 is replaced by an asparagine, which still ligates a water molecule using its sidechain, and is unique to TT. The RRYS loop region is hydrogen bonded through waters to the heme propionates, Cu_A ligands and the Mg^{2+} center, while the ends of the loop are bonded to water chains that connect to the bulk aqueous phase. The conservative mutations, R481K and R482K, in the equivalent residues in PD result in an enzyme that appears to differ little from wild-type, with no observed changes in proton pumping ratios or in maximal turnover.[37] In RS, both mutants pump well and have maximal activities close to the wild-type enzyme; however, R481K is very sensitive to inhibition by an electrochemical potential when reconstituted into lipid vesicles.[38] Less conservative mutations in this region cause varying degrees of dysfunction, but also instability of the enzyme due to disruption of the critical I/II interface, thus interpretation of the data is somewhat complicated.[38] The loop of subunit II that forms another part of the putative pathway includes six amino acid residues on each side of G257[II], among them H260[II], a ligand to Cu_A. The amino acids of this loop are semi-conserved and bind water in a number of the structures. Throughout the whole area we observe branched water chains indicating that waters can form and break hydrogen bonds and connect to water in the bulk aqueous phase.

4   Water and the Proton Pumping Function

4.1   Role of Heme *a*

Heme *a* (or heme *b* in EC and TT) is a low-spin heme maintaining two axial histidine ligands, H102 in helix 2 and H421 in helix 10. A perfunctory examination of the environments of the six coordinated hemes indicates that they are all very similar. Figure 2 shows the structure of heme *a*, its histidine ligands and the waters and

R482 coordinated to the propionate groups. The redox potential of heme *a* is pH sensitive[39] and reduction of heme *a* is accompanied by the uptake of a proton.[40] By specific ^{13}C labeling of the heme propionates, it has been possible to assign reduced-minus-oxidized FTIR spectral features to the hemes of PD oxidase[41] and to demonstrate that, upon reduction, one of these propionates changes its protonation state.[37] As there is one electron oxidation/reduction cycle of heme *a* for every proton that is pumped, it is possible that heme *a* has a role in gating proton exit in the final step of proton pumping, but it is unlikely to be a major player in the proton-pumping mechanism due to the lack of complexity of this redox center and its distance from water paths.

However a role for heme *a* in proton pumping has been suggested by a number of investigators.[42] Most recently it has been proposed that heme *a* could be involved in proton pumping *via* the putative H-channel.[43,44] The H-channel is proposed to be a water pore involved in proton pumping, that utilizes the formyl group of heme *a*. Site directed mutagenesis studies in the *R. sphaeroides* oxidase have not provided any confirmation of the existence of this path in the bacterial enzyme.[45] No conserved or semi-conserved water molecules were found in this region in our overlaid oxidase crystal structures. A more detailed examination of structures has given us an insight into the nature of the protonation of the histidine ligands to the low spin heme in the oxidase superfamily. Both histidine ligands are hydrogen bonded to other amino acids. In RS and PD one of the heme *a* ligands, H102 is itself hydrogen bonded to a serine, while in the much studied bovine enzyme the heme *a* ligand H61 is hydrogen bonded to the backbone carbonyl of glycine. This difference in interactions of the proximal histidine ligands may explain some of the differences in the EPR signals of this low spin center.[46] The EPR signals of the oxidized heme *a* of both the RS and PD oxidase are shifted from the values commonly found in BV oxidase. When the proximal histidine is hydrogen bonded to an alcohol group, the low spin heme *a* gives rise to signals at g = 2.83, 2.31, and 1.62, whereas the carbonyl oxygen hydrogen bonded BV enzyme has values of g = 3.03, 2.21, and 1.45. These shifts and other spectroscopic data suggest that heme *a* of RS is ligated by one histidine-imidazole and one histidine with more imidazolate character, similar to the histidine-imidazolate derivatives of the hemes of soybean met-leghemoglobin *a*[47] and met-myoglobin[48] which have *g* values nearly identical to those of heme *a* in RS and PD.

The pK_a for the deprotonation of histidine depends on its ligation status. When bonded to a metal cation, the pK_a is much less than its normal value in free aqueous solution (free solution ≈ 14.5, met-myoglobin ≈ 10.5 and in met-leghemoglobin ≈ 7.0).[47] The presence of an alcohol group next to the proximal ligand must stabilize the imidazolate. The ability of a metal cation to have an imidazolate ligand is arguably relevant to the behavior of Cu_B (see Section 5).

Figure 2 also shows that the hydroxyl group of the hydroxyl-farnesyl chain of heme *a* in BV is projecting in the opposite direction to that in the bacterial structures. In RS, a water molecule is ligated to this hydroxyl group, but no water is found in the BV structure. The comparative structures suggest that the farnesyl-hydroxyl of heme *a* is not functionally important, unlike the highly conserved positioning of the heme a_3 farnesyl-hydroxyl group (see below).

4.2 The Role of the Heme a_3/Cu$_B$ Center and Associated Water

Table II shows the amino acids that constitute the site of oxygen reduction and the likely site of proton pumping. Figure 5A shows the structure of the binuclear center, illustrating conserved amino acids in the pocket (using our primary RS structure as default, indicated by elemental colors; heme a_3 is dark purple). The water pathways have a color-coded space-filling overlay that shows the proton entry and exit routes into and out of the binuclear center. Protons are proposed to exit the binuclear center *via* the top egress pathway, which we have shown in green.

There is a water molecule, $W1a_3$, bonded by both heme a_3 propionates, by a nitrogen atom from the histidine ring of the Cu$_B$ ligand, H334, and by another water molecule, $W1_{Y175}$. $W1_{Y175}$ is ligated to the Mg^{2+} ion *via* another conserved water molecule, $W1_{Mg}$. We suggest that $W1a_3$ has two important roles in oxidase function: firstly, to stop the electrophoretic movement of hydroxide from the binuclear center; and secondly to act as a gate for the proton pump. During the catalytic cycle it is possible that the enzyme will generate hydroxide in the pocket which is only weakly bound to a metal center (for instance a_3^{2+}-OH⁻). In the presence of a large membrane potential, hydroxide must be trapped in the binuclear center and not allowed to electophoretically migrate through the enzyme into the outer bulk phase. $W1a_3$ is tetrahedrally coordinated and therefore cannot bond to hydroxide, blocking its movement.

The second and major role of $W1a_3$, is as a valve or gate in proton pumping. $W1a_3$ cycles between water/hydronium geometries so that protonation will cause the rearrangement of hydrogen bonds in the water chain leading, *via* $W1_{Y175}$ and $W1_{Mg}$, to the Mg^{2+} ion and the route out of the binuclear center for 'pumped' protons. Protonation of this water molecule is responsible for pumped proton efflux.

We have also examined the possible routes whereby protons can travel from $W1_{E286}$ to the binuclear center. There are essentially four possible routes along which water chains can join $W1_{E286}$ to the ligands of Cu$_B$ and heme a_3:

(a) The direct route is along a straight line connecting $W1_{E286}$ to Cu$_B$. This pathway is blocked both by the sidechain of E286 itself and by a hydrogen bond between the fully conserved residues H284 and V287;

(b) A second route is along the heme a_3 plane, skirting the heme a_3 π-orbitals. This appears to be blocked by the aliphatic sidechain of the <u>invariant</u> residueV287 (and also by W239 in TT);

(c) Protons may go above E286, around the carbonyl oxygen of G283 and towards Cu$_B$ and H334. This route has been proposed by Wikström and co-workers as the route by which hydroxide in the binuclear center is neutralized by protons.[49] This route is however partially blocked by the highly conserved W280;

(d) Finally, we have identified the <u>V330 route,</u> a path along the conserved water molecules $W1_{F281}$ and $W1_{V330.}$ (Figures 4 and 5 and see below.)

The protons that arrive at the binuclear center from the D-channel, shown in light orange, may populate the water pair, $W1_{F281}$ and $W1_{V330}$, hydrogen bonded by F281

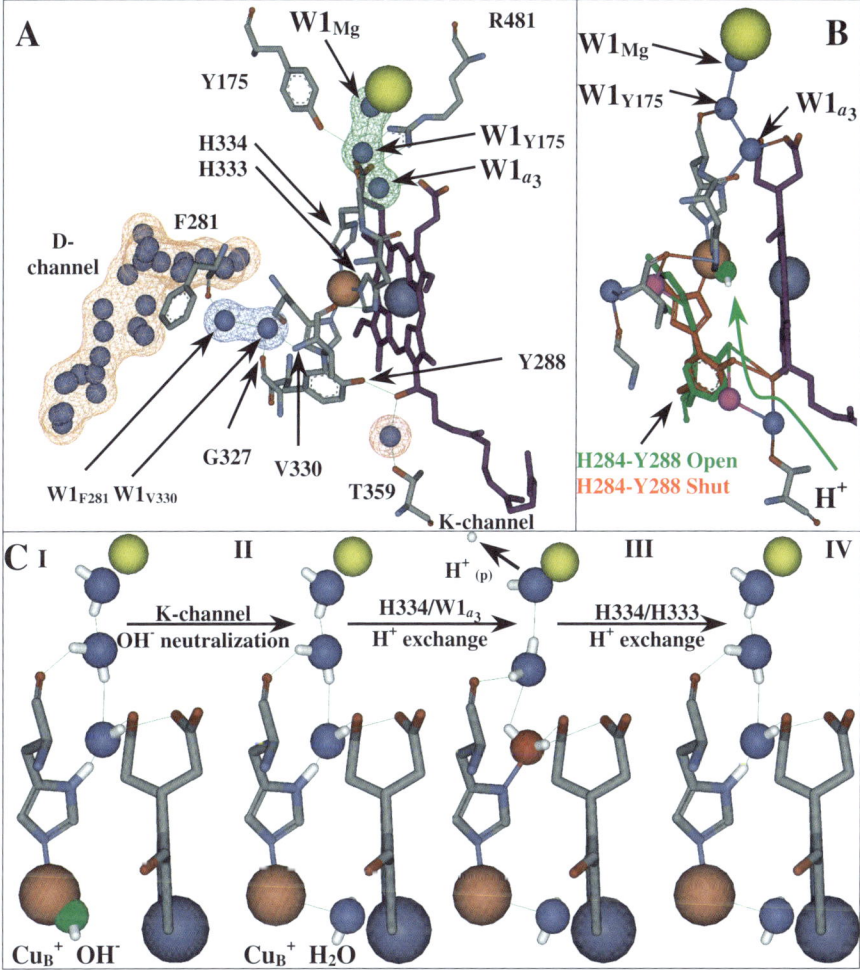

Figure 5 *The oxidase core and proton pumping. Figure 5A shows the amino acids involved in oxygen reduction chemistry and proton pumping. Figure 5B shows how the Y288 gate may function. Normally, Y288 is hydrogen bonded to the hydroxyl group of the hydroxyl-farnesyl of heme a_3, blocking protons from the K-channel. The binding of hydroxide to Cu_B^+ releases H284, which rotates and ligates a water molecule allowing Y288 to ligate $W1_{T359}$ and another water, opening a proton wire to the K-channel. Key; hydroxide* ●, *water in structure* ●, *water added by authors* ●. *Figure 5C shows the $W1a_3$ gate pumping a proton following the first electron entering the binuclear center. (I) Reduction of Cu_B results in a hydroxide being bonded to Cu_B^+. This ligand displaces the H284-Y288 pair, opening the K-channel. (II) With the K-channel open the hydroxide is neutralized by a proton from this channel. (III) Cu_B and ligands have an overall charge of plus 1. H334 deprotonates so as to restore electron neutrality by protonating $W1a_3$. The protonation of $W1a_3$ causes an ejection of a proton along the water chain, leading eventually to the outer bulk phase via a proton wire. (IV) H333 exchanges a proton with H334*

and the highly conserved pair of amino acids V330:G327. The latter of these waters, $W1_{V330}$ (shown in blue), may be the group that passes protons to the binuclear center from the D-pathway. Protons must make the final journey from $W1_{V330}$ to H333 on a partially mobile water, unseen in the X-ray crystal structures. Recent studies have examined the theoretical distribution of mobile water in the hydrophobic cavity around the catalytic center of the enzyme.[50,51] All the crystal structures contain such a cavity that connects $W1_{E286}$ to $W1_{V330}$, and could accommodate up to six water molecules capable of catalyzing the transfer of protons from $W1_{E286}$ to the binuclear center (Figures 4B and 4D).[51]

4.3 What Evolution Tells Us About the Minimal Structural Requirements for Oxidase to Pump Protons

The evidence suggests that cytochrome oxidases evolved from nitric oxide reductases (NORb). This ancestor contained two b type hemes, a non-heme iron in a peptide similar to subunit I, and a subunit II. The evolutionary transition from a cytochrome c NORb to cytochrome oxidase probably occurred starting with a NOR with minimal oxidase activity, similar to that of the PD cytochrome cbb NORb.[52] The next transition would be a proto-oxidase which contained an aqueous pore from the inner aqueous phase to the binuclear center, allowing substrate protons to be drawn from the opposite side of the enzyme to the redox electrons (presumably *via* a proto-K-channel). Such an oxidase would generate a chemiosmotic potential. The final stage would be an enzyme that proton pumps, increasing the bioenergetic efficiency of the oxygen reduction reaction by a factor of two. This latter stage requires:

(a) Formation of another proton pore into the binuclear center;
(b) The non-heme iron atom of the binuclear center to become a copper atom;
(c) The introduction of a hydroxyl group at the corner of the heme at the binuclear center (heme o or a);
(d) Redesign of the core of the binuclear center to allow the cross linkage of the H-Y pair.

The water at the top of the D-channel which is capable of transporting protons to the binuclear center, $W1_{E286}$, can also be bonded by another amino acid, such as tryptophan at position 290 or a tyrosine at 282 (underlined). A number of 'primitive' organisms appear not to have made the transition from NORb to proton-pumping oxidase: it appears that the core structure of TT falls between these two groups.

The major difference in these four classes of oxidase lies in the ligation of water at the top of the D-channel (underlined) and residues that can support proton pumping. The major types are shown below:

Classical (Eukaryotes)	276	QHLFWFFGHP<u>E</u>V<u>Y</u>IL 290	$H^+/e^- = 1$
Thermo-classical (thermophiles)		RTLWWFFGHPIVYF<u>W</u>	$H^+/e^- \approx 1$
Cyanobacterial		QHLFWF<u>Y</u>SHPAVYSH	$H^+/e^- = 1$
Non-classical (microaerophiles)		ALIQWWWGHNAVAFV	$H^+/e^- = 1$

There are essentially four core structures that evolved from NORb:

(a) The classical, found in eukaryotes and in some bacteria (RS, BV, PD and EC);
(b) The thermo-classical, mainly found in both the cytochrome *c* and quinol oxidases of thermophiles such as TT, *Aquifex, Pyrobaculum*; but also in alkalinophiles (*Natronobacterium pharaonis* and *Bacillus halodurans*) and halophile (*Haloarcula marismortui*). In these organisms, the ligation of water molecules by hydrogen bonds within oxidase will be weakened by either thermal or salt stress.
(c) The cyanobacterial core of cytochrome oxidase has recently been extensively reviewed;[9]
(d) The non-classical. This core is common in miroaerophiles such as *Pseudomonas pseudomallei* and *Helicobacter pylori*. Microaerophilic bacteria are optimized for growth in oxygen-limited environments, environments that would have been plentiful some billion years ago.[6,9] It is unlikely that this core can support proton pumping, but it does illustrate a way station along the evolutionary journey made from NORb to the classical oxidase.

The machinery required for proton pumping must come from commonalities that different proton pumping oxidases share, although there might be some variations in different sub-families of oxidase. These commonalities include a heme with a hydroxyl group at the corner, the histidine ligands to the heme a_3 and Cu_B, the amino acids of the core, and associated water molecules. Other residues within and around the core must certainly help guide protons or act as proton gates or structural supports for gates, but appear not to be vital for function.

4.4 The Importance of Metal Center Charge in Oxidase Function

The ligation and charge of the metals of the oxidase are of interest with respect to the mechanism of proton pumping. It has been suggested that changes in the redox state of the iron and copper in the binuclear center may be coupled to the movement of protons during the reduction of molecular oxygen.

Heme consists of an iron ion ligated inside a porphyrin ring. Prior to the insertion of iron into the ring, the center of the porphyrin has an overall charge of -2. Thus, the conjugated ring of ferrous (Fe^{2+}) heme has no overall charge. Upon oxidation of the iron to the ferric (Fe^{3+}) state the overall charge of the heme ring becomes $+1$. Thus, ferric heme a_3 is capable of binding small anionic ligands such as CN^-, N_3^-, NO_2^- and OH^-. In contrast, ferrous heme a_3 binds small neutral ligands such as O_2, NO, CO and H_2O.

Resonance Raman studies of cytochrome oxidase have followed the ligands to heme a_3, during the oxygen reduction reaction, starting with the CO reduced enzyme. Such studies indicate that heme a_3^{2+} ligates oxygen or water and that a_3^{3+} ligates a hydroxide. In the latter case the heme a_3^{3+}-OH^- is high spin, unlike most globins binding hydroxide, which are low spin.[53,54] This difference in the spin state in oxidase, as compared with proteins such as myoglobin, led to the proposal that there is a strong hydrogen bond between the proton on the hydroxide moiety and some other group (such as Cu_B or Tyr-288).[54]

EXAFS studies of the EC bo_3 oxidase indicate that in fully reduced cytochrome bo_3 the Cu_B contains two strongly bonded and one weakly bonded histidine ligands.[54] This study also tentatively suggested that a Cl^- or OH^- ligand could replace the weakly bonded histidine to Cu_B in the reduced enzyme.

It was originally suggested by Yoshikawa and co-workers that the fully oxidized enzyme contains a peroxy oxygen bridge between heme a_3 iron atom and Cu_B, based on the elongated electron density observed between these two metals, in their early X-ray crystal structure of the bovine enzyme.[44] The electron density in the heme a_3/Cu_B pocket of the early structure was too large for a single oxygen atom. A larger than expected ligand, the so called 'Fat oxygen' has also been found in many EXAFS studies of the oxidized binuclear center.[56]

Careful titrations of the oxidized enzyme demonstrated that only four electrons are required to fully reduce the bovine enzyme, thus eliminating a peroxy oxygen ligand in the oxidized enzyme.[57]

It has long been known that photolysis of the heme a_3-CO results in CO transiently binding to reduced Cu_B.[58,59] This suggests that Cu_B^+ may be a neutral species, like ferrous heme a_3. This would suggest that one of its nitrogen ligands may be an imidazolate, at least at physiological pH. H284 cannot be an imidazolate ligand because of its covalent crosslink to Y288, so the imidazolate ligand must be either H333 or H334.

An examination of the bond lengths between the Cu_B and its histidine pair of ligands (in RS, BV, PD and TT) indicates that they are some 0.17 Å shorter than the similar histidine pair found in amine oxidases (at 2.23 Å, PDB entry 1LVN).[60] This suggests that the H333/H334 pair of histidines have some imidazolate character, with H333 behaving most like an imidazolate. Yoshikawa and co-workers have a number of crystal structures of the BV enzyme which, as the electron density suggests, may contain a fractional number of oxygen atoms in the heme a_3/Cu_B pocket of the oxidized enzyme (PDB entry 1V54 (Remark three in PDB file: 'Shows a residual density between heme a_3 iron and Cu_B'); PDB entry 1OCC; PDB entry 2OCC).[2,43,44] We explain this electron density as due to the presence of hydroxide, more easily observed by X-ray studies, or water, more difficult to observe in X-ray studies, in the pocket of the binuclear center. Cu_B^{2+} can bind either water or hydroxide, whilst at the same time the heme a_3/Cu_B pocket remains neutral, because H333 and H334 can switch between imidazole/imidazolate.

In the 'CuB-hydroxyl' state, the enzyme has two hydroxides bonded to the metals of the binuclear center. In order to preserve the overall electroneutrality of the Cu_B center $[^-OH/Cu_B^{2+}]^{1+}$, it would have the equivalent of one imidazolate ligand.

In the 'Cu$_B$-aqua' state, water binds the Cu_B, $[H_2O/Cu_B^{2+}]^{2+}$, so both H333 and H334 are required to become imidazolates for this center to remain a neutral species. Coordinate bonds between imidazole and copper are longer than ionic bonds between imidazolate and copper. This transition in the average bond length, from imidazole coordinate bonds to imidazolate ionic bonds is seen starting with the EC amine oxidase, 2.23 Å (two imidazoles bonds); the oxidase 1OCC structure, 2.16 Å (one imidazole and one imidazolate bond); and the oxidase 1V54 structure, 1.98 Å (two imidazolates bonds).

In the Cu_B-hydroxyl state the oxygen atom of the hydroxide is strongly bonded to Cu_B, making the oxygen atom's position relatively easy to discern. In contrast, the oxygen atom of Cu_B-aqua is only weakly bonded to Cu_B and therefore its position is more difficult to resolve using X-ray techniques. This difference in the mobility of the oxygen atom in the Cu_B-aqua and Cu_B-hydroxyl forms may explain the 'Fat oxygen' seen in some X-ray crystal structure (PDB entry 2OCC) and EXAFS examinations of the binuclear center. The number of oxygens observed in the pocket will therefore range between >1 and <2 for the Cu_B-aqua and Cu_B-hydroxyl states, respectively.

The principle of electron neutrality in cytochrome oxidase has been used to explain how the movement of substrate electrons or protons into the binuclear center causes this center to take up or release pumped protons.[61–63]

5   Gating of the Pump

All pumps have a requirement for one or more one-way valves or gates. Our study strongly indicates the presence of two gates that are involved in the pumping of protons during oxidase turnover. The first gate, the H284-Y288 gate, appears to control the uptake of two substrate protons during the reductive phase of the enzyme cycle. The second gate, the $W1a_3$-H334 gate, appears to direct protons from the binuclear center to the proton wire above $W1a_3$, for each of the pumped protons.

5.1   The H284-Y288 Gate

EXAFS data suggest that OH^- may be able to replace H284 as a ligand to the reduced Cu_B.[55,56] The Cu_B under these circumstances will be ligated by H333, H334 and hydroxide. This change of Cu ligands is also found in mammalian Cu/Zn superoxide dismutases, where Cu^{2+} is ligated by three imidazole histidines, a water molecule and a histidine imidazolate ligand also ligated to Zn^{2+}. Upon reduction the imidazolate ligand is lost and the Cu^+ is ligated by three histidines.[64] In the oxidase, the displaced H284 will move away from its normal position, moving its covalent ligand, Y288, with it (Figure 5B). This will break the hydrogen bond with the hydroxyl-farnesyl, allowing protons to enter the binuclear center from the K-channel (see Section 3.4.2). Modeling studies using RS indicate that in some states the phenol alcohol of Y288, the hydroxyl of the heme a_3 farnesyl and $W1_{T359}$ allow protons to enter the binuclear center, following a change in the position of T359.[13] We propose that a similar access change occurs following the movement of Y288. Moving Y288, like moving T359, allows another water molecule to be hydrogen bonded in parallel to $W1_{T359}$, opening the K-channel. The T359A mutation has a spectroscopically normal binuclear center, pumps protons, but has an activity of only 20–35% of normal.[14] The equivalent mutant of the EC oxidase has only 15% of the wild type activity.[65,66]

When H284 is ligated to Cu_B, the Y288 is pointed at the hydroxyl group of the hydroxyl-farnesyl chain and these two alcohol (hydroxyl) groups are 2.6 Å apart. In this configuration, the hydrogen on the farnesyl-hydroxyl is constrained and no protons from the K-channel can travel to the binuclear center. Protons <u>can</u> travel to the H333 and H334 from the D-channel, *via* $W1_{V330}$.

When not hydrogen bonded to Cu_B, H284 rotates on its axis by $90°$ and forms a hydrogen bond with the conserved water $W1_{V330}$ or some other hydrogen bonding species. This closes the D-channel and allows a proton from the conserved water bonded to the farnesyl-hydroxyl and WI_{T359}, to go to the binuclear center. This gating mechanism explains why all oxidases of the family have a hydroxide on the heme in this position.

Figure 5B shows the open (green) and closed (red) gate positions.

5.2   The $W1a_3$-H334 Gate

Figure 5C shows a model of how $W1a_3$ and H334/H333 act as a proton gate, as discussed in Section 4.2. We show how the oxidase can pump during the reductive cycle (a full version of the proton pumping model is in preparation). This pump differs from previous histidine cycling pumps[61,62] by showing the involvement of water molecules in the gating of the proton flow and showing the H284-Y288 gate. Neutralizing a hydroxide bound to Cu_B results in this center having an overall charge of $+1$. The protonation of $W1a_3$ by H334 is a response to this change in the charge. We propose that the interaction of $W1a_3$ and H334/H333, the two Cu_B ligands, is central to proton pumping. We describe the rules that could govern the proton-pumping behavior of the oxidases:

(a) Cu_B is ligated so that, with ligands, it has an overall charge of zero; hence Cu_B^+ binds CO. Cu_B^{2+} is a tetrahedral pyramid with ligands H284, H333, H334 and either H_2O or OH^-. Either H333 and/or H334 can be an imidazolate, depending on whether H_2O or OH^- is the fourth ligand;

(b) Cu_B^+ tends toward a trigonal planar structure in alkali conditions (OH^-, H333 and H334) or towards a tetrahedral pyramidal structure under acid conditions (H_2O, H284, and H333, H334.) Either H333 or H334 can be an imidazolate, but H333 is the preferred imidazolate;

(c) H333 can get protons from the D-channel. H334 can get protons from H333 (possibly *via* a water molecule);

(d) When the H284-Y288 is ligated to Cu_B the D-channel is open. When not ligated to Cu_B, then the pair blocks the D-channel.

(e) The K-channel can only supply protons when the H284-Y288 pair is blocking the D-channel. Protons from the K-channel are driven by the local pH gradient to neutralize OH- in the binuclear center;

(f) Protons leave the H334 for $W1a_3$ when Cu_B and its ligands have an overall charge of $+1$. This occurs following the formation of Cu_B-aqua after the neutralization of Cu_B-hydroxyl.

6   Summary

The major role of water in the cytochrome oxidase superfamily is to modulate the flow of protons during the enzyme's turnover cycle. Conserved waters, rather than conserved residues, at the top of the D-channel are proposed to be critical to the pumping mechanism. Another conserved water molecule bonded by the heme a_3

propionates, is implicated in gating and connecting the binuclear center to the outer bulk aqueous phase. The alternative access of substrate and pumped protons from the D and K channels is suggested to be controlled by the ligation state of the cross-linked H284-Y288 Cu_B ligand. Analysis of the evolutionary changes in the amino acid sequences and in the structures of the heme-Cu superfamily supports the conclusion of a key role of water and Cu_B ligation in the proton pumping mechanism.

Acknowledgements

This work was supported by NIH GM 26916, Human Frontiers in Science Program and the MSU REF Center for Structural Analysis of Membrane Proteins (to SFM). The authors wish to thank Prof. James Fee for providing us with a pre-published TT oxidase crystal structure.

MAS would like to thank Denise Mills and Carrie Hiser for their help, Peter Nicholls for his continued advice and Mårten Wikström for his patience and for his input at the reviewing stage.

References

1. M. Svensson-Ek, J. Abramson, G. Larsson, S. Tornroth, P. Brzezinski and S. Iwata, *J. Mol. Biol.*, 2002, **321**, 329.
2. T. Tsukihara, K. Shimokata, Y. Katayama, H. Shimada, K. Muramoto, H. Aoyama, M. Mochizuki, K. Shinzawa-Itoh, E. Yamashita, M. Yao, Y. Ishimura and S. Yoshikawa, *Proc. Natl. Acad .Sci. U.S.A.*, 2003, **100**, 15304.
3. A. Harrenga and H. Michel, *J. Biol. Chem.*, 1999, **274**, 33296.
4. J. Abramson, S. Riistama, G. Larsson, A. Iasaitis, M. Svensson-Ek, L. Laakkonen, A. Puustinen, S. Iwata and M. Wikstrom, *Nat. Struct. Biol.*, 2000, **7**, 910.
5. T. Soulimane, G. Buse, G.P. Bourenkov, H.D. Bartunik, R. Huber and M.E. Than, *EMBO. J.*, 2000, **19**, 1766.
6. H.D. Holland, Early Proterozoic Atmospheric Change. in S. Bengtson, (ed.), Early Life on Earth. Columbia University Press, New York, 1994, 237.
7. M. Saraste and J. Castresana, *FEBS Lett.*, 1994, **341**, 1.
8. J. van der Oost, A.P. de Boer, J.W. de Gier, W.G. Zumft, A.H. Stouthamer and R.J. van Spanning, *FEMS Microbiol Lett.*, 1994, **121**, 1.
9. M. Paumann, G. Regelsberger, C. Obinger and G.A. Peschek, *Biochim. Biophys. Acta.*, 2004, **1707**, 231.
10. S.M. Musser and S.I. Chan, *J. Mole. Evol.*, 1998, **46**, 508.
11. G. Buse, T. Soulimane, M. Dewor, H.E. Meyer and M. Bluggel, *Protein Sci.*, 1999, **8**, 985.
12. M. Lubben, and K.J. Morand, *J. Biol Chem.*, 1994, **269**, 21473.
13. R.I. Cukier, *Biochim. Biophys. Acta*, 2005, **1706**, 134.
14. J.R. Fetter, J. Qian, J. Shapleigh, J.W. Thomas, A. Garcia-Horsman, E. Schmidt, J. Hosler, G.T. Babcock, R.B. Gennis and S. Ferguson-Miller, *Proc. Natl. Acad. Sci. U.S.A.*, 1995, **92**, 1604.

15. D. Bordo and P. Argos, *J. Mol. Biol.*, 1994, **243**, 504.
16. J.A. Ippolito, R.S. Alexander and D.W. Christianson, *J. Mol. Biol.*, 1990, **215**, 457.
17. R. Pomes and B. Roux, *Biophys. J.*, 1996, **71**, 19.
18. C. Dellago, M.M. Naor and G. Hummer, *Physic. Rev. Lett.*, 2003, **90**, 105902.
19. T.D. Ferris, M.D. Zeidler and T.C. Farrar, *Molec. Phys.*, 2000, **98**, 737.
20. M. Wikström, M.I. Verkhovsky and G. Hummer, *Biochim. Biophys. Acta*, 2003, **1604**, 61.
21. A.A. Konstantinov, S. Siletsky, D. Mitchell, A. Kaulen and R.B. Gennis, *Proc. Natl. Acad. Sci. U.S.A.*, 1997, **94**, 9085.
22. D.A. Mills, L. Florens, C. Hiser, J. Qian and S. Ferguson-Miller, *Biochim. Biophys. Acta*, 2000, **1458**, 180.
23. J. Fetter, M. Sharpe, J. Qian, D. Mills, S. Ferguson-Miller and P. Nicholls, *FEBS Lett.*, 1996, **393**, 155.
24. A.S. Pawate, J. Morgan, A. Namslauer, D. Mills, P. Brzezinski, S. Ferguson-Miller and R.B. Gennis, *Biochemistry*, 2002, **41**, 13417-13423.
25. A. Aagaard, G. Gilderson, D.A. Mills, S. Ferguson-Miller and P. Brzezinski, *Biochemistry*, 2000, **39**, 15847.
26. C. Backgren, G. Hummer, M. Wikström and A. Puustinen, *Biochemistry*, 2000, **39**, 7863.
27. Q. She, R.K.F. Singh, F. Confalonieri and 29 other authors, *Proc. Natl. Acad. Sci. U.S.A.*, 2001, **98**, 7835.
28. F.L. Tomson, J.E. Morgan, G.P. Gu, B. Barquera, T.V. Vygodina and R.B. Gennis, *Biochemistry*, 2002, **42**, 1711.
29. J.P. Hosler, S. Ferguson-Miller, M.W. Calhoun, J.W. Thomas, J. Hill, L. Lemieux, J.X. Ma, C. Georgiou, J. Fetter, J. Shapleigh, M.M.J. Tecklenburg, G.T. Babcock and R.B. Gennis, *J. Bioenerg. Biomembr.*, 1993, **25**, 121.
30. G. Berden, W.L. Meerts, M. Schmitt and K. Kleinermanns, *J. Chem. Phys.*, 1996, **104**, 972.
31. M. Gerhards, M. Schmitt, K. Kleinermanns and W. Stahl, *J. Chem. Phys.*, 1996, **104**, 967.
32. J.P. Hosler, M.P. Espe, Y.J. Zhen, G.T. Babcock and S. Ferguson-Miller, *Biochemistry*, 1995, **34**, 7586.
33. L. Florens, B. Schmidt, J. McCracken and S. Ferguson-Miller, *Biochemistry*, 2001, **40**, 7491.
34. B. Schmidt, J. McCracken and S. Ferguson-Miller, *Proc. Natl. Acad. Sci. U.S.A.*, 2004, **94**, 15539.
35. P. Agre, L.S. King, M. Yasui, W.B. Guggino, O.P. Ottersen, Y. Fujiyoshi, A. Engel and S. Nielsen, *J. Physiol.*, 2002, **542**, 3.
36. E. Olkhova, M.C. Hutter, M.A. Lill, V. Helms and H. Michel, *Biophys. J.*, 2004, **86**, 1873.
37. J. Behr, H. Michel, W. Mantele and P. Hellwig, *Biochemistry*, 2000, **39**, 1356.
38. J. Qian, D.A. Mills, L. Geren, K.F. Wang, C.W. Hoganson, B. Schmidt, C. Hiser, G.T. Babcock, B. Durham, F. Millett and S. Ferguson-Miller, *Biochemistry*, 2004, **43**, 5748.

39. D.F. Wilson, M. Erecinska and E.S. Brocklehurst, *Arch. Biochem. Biophys.*, 1972, **151**, 180.

40. N. Capitanio, T.V. Vygodina, G. Capitanio, A.A. Konstantinov, P. Nicholls and S. Papa, *Biochim. Biophys. Acta*, 1997, **1318**, 255.

41. J. Behr, P. Hellwig, W. Mantele and H. Michel, *Biochemistry*, 1998, **37**, 7400.

42. G.T. Babcock and P.M. Callahan, *Biochemistry*, 1983, **22**, 2314.

43. T. Tsukihara, H. Aoyama, E. Yamashita, T. Takashi, H. Yamaguichi, K. Shinzawa-Itoh, R. Nakashima, R. Yaono and S. Yoshikawa, *Science* 1996, **272**, 1136.

44. S. Yoshikawa, K. Shinzawa-Itoh, R. Nakashima, R. Yaono, E. Yamashita, N. Inoue, M. Yao, M.J. Fei, C.P. Libeu, T. Mizushima, H. Yamaguchi, T. Tomizaki and T. Tsukihara, *Science*, 1998, **280**, 1723.

45. H. Lee, T. Das, D. Rousseau, D. Mills, S. Ferguson-Miller and R.Gennis, *Biochemistry*, 2000, **39**, 2989.

46. J.P. Hosler, J. Fetter, M.M. J. Tecklenburg, M. Espe, C. Lerma and S. Ferguson-Miller, *J. Biol. Chem.*, 1992, **267**, 24264.

47. G. Sievers, P.M.A. Gadsby, J. Peterson and A.J. Thomson, *Biochim. Biophys. Acta,* 1983, **742**, 637.

48. P.M.A. Gadsby and A.J. Thomson, *FEBS Letts.*, 1982, **150**, 59.

49. M. Wikström, M. Verkhovsky and G. Hummer, *Biochim. Biophys. Acta.*, 2003, **1604**, 61.

50. R.I. Cukier, *Biochim. Biophys. Acta*, 2004, **1656**, 189.

51. M. Tashiro and A.A. Stuchebrukhov, *J. Phys. Chem.* 2005, **109**, 1015.

52. J. Hendriks, A. Warne, U. Gohlke, T. Haltia, C. Ludovici, M. Lubben and M. Saraste, *Biochemistry*, 1998, **37**, 13102.

53. C. Varotsis, G. Babcock, M. Lauraeus and M. Wikström, *Biochim. Biophys. Acta.,*1995 **1231**, 111.

54. S. Han, S. Takahashi and D. Rousseau, *J. Biol. Chem*, 2000, **275**, 1910.

55. M. Ralle, M.L. Verkhovskaya, J.E. Morgan, M.I. Verkhovsky, M. Wikström and N.J. Blackburn, *Biochemistry*, 1999, **38**, 7185.

56. Y.C. Fann, I. Ahmed, N.J. Blackburn, J.S. Boswell, M.L Verkhovskaya, B.M. Hoffman and M. Wikström, *Biochemistry*, 1995, **34**, 10245.

57. G.C.M. Steffens, T. Soulimane, G. Wolff and G. Buse, *Euro. J. Biochem.*, 1993, **213**, 1149.

58. J.O. Alben, P.P. Moh, F.G. Fiamingo and R.A. Altschuld, *Proc. Natl. Acad. Sci. U.S.A.*, 1981, **78**, 234.

59. B.R. Dyer, K.A. Peterson, P.O. Stoutland and W.H. Woodruff, *J. Am. Chem. Soc.*, 1991, **113**, 6276.

60. S. Kishishita, T. Okajima, M. Kim, H. Yamaguchi, S. Hirota, S. Suzuki, S. Kuroda, K. Tanizawa and M. Mure, *J. Am. Chem. Soc.*, 2003, **125**, 1041.

61. M. Wikström, A. Bogachev, M. Finel, J.E. Morgan, A. Puustinen and M. Raitio, *Bochim. Biophys. Acta.*, 1994, **1187**, 106.

62. J.E. Morgan, M.I. Verkhovsky and M. Wikström, *Bioenerg. Biomembr.*, 1994, **26**, 599.

63. P. Rich, *Aust. J. Plant Physiol.*, 1995, **22**, 479.

64. L.M. Murphy, R.W. Strange and S.S. Hasnain, *Structure*, 1997, **5**, 371.
65. J.W. Thomas, L.J. Lemieux, J.O. Alben and R.B. Gennis, *Biochemistry*, 1993, **32**, 11173.
66. M. Svensson, S. Hallen, J.W. Thomas, L.J. Lemieux, R.B. Gennis and T. Nilsson, *Biochemistry*, 1995, **34**, 5252.

CHAPTER 3

Structural Chemical Studies on the Reaction Mechanism of Cytochrome c Oxidase

SHINYA YOSHIKAWA

Department of Life Science,
University of Hyogo,
Kamigohri Akoh Hyogo 678-1297, Japan

1 Introduction

All physiological processes are comprised of chemical reactions, each driven by proteins. Thus, elucidation of the mechanism of any physiological process would be most accurately served by describing it in chemical terms. Most of these chemical reactions are driven by large and highly organized complex proteins such as cytochrome c oxidase which has highly specific reaction sites defined by various amino acid residues fixed in three dimensions. The unique anisotropies of such active sites, which have never been accurately reproduced in chemical model systems, define catalytic activities that are impossible to describe using only chemical terminology established thus far. Therefore, significant improvement in chemical terminology is required for elucidation of the mechanism of the catalytic activities. In other words, studies on these catalytic activities would stimulate not only biological sciences but also chemical sciences. For such studies, X-ray structural analysis of a given reaction site at high resolution is indispensable and obtaining an X-ray structure of a given protein is a time and labor-intensive effort which often acts as the rate-limiting step for investigations into its catalytic mechanism. A complicating factor in such investigations is that the chemical reactivity of a functional group is often greatly influenced by structural changes much smaller (0.1 Å) than those detectable in data obtained by the present protein X-ray crystallography. Vibrational spectroscopic methods are the best for analysis for such small structural changes in a functional group, although they do not provide three dimensional data.[1] Both X-ray crystallography and vibrational spectroscopy have been applied for studies of the reaction mechanism of cytochrome c

oxidase.[2] In this chapter, the reaction mechanism of cytochrome c oxidase will be discussed based on the recently determined X-ray structures of bovine heart cytochrome c oxidase at 1.8 Å resolution in the oxidized state and 1.9 Å in the reduced state.[3]

2 The Mechanism of O_2 Reduction

2.1 Chemistry of O_2 Reduction

Although the process of one electron reduction of molecular oxygen (O_2) in the ground (triplet) state is energetically unfavorable, the process of two electron reduction is favorable.[4] In fact, superoxide (O_2^-) is known to chemically reduce cytochrome c.[5] It has been proposed that this propensity of O_2 to disfavor the process of single electron reduction is the major factor contributing to stabilization of the O_2-bound form (the oxygenated form) of hemoglobins and myoglobins, since ferrous heme isolated in the protein has only one equivalent of electron available to interact with ferrous iron-bound O_2.[4,6] This property of O_2 suggests also that Cu_B, which is located close to the O_2-binding site (heme a_3), activates the bound O_2 by donation of the second electron for the two electron reduction process wherein O_2 is reduced to H_2O_2.[4] Long before the X-ray structure of the enzyme was solved, the observation of antiferromagnetic coupling between Cu_B and Fe_{a3} in the fully oxidized state suggested that Cu_B would be located close enough for coordination to O_2 when O_2 is bound to Fe_{a3}.[7] The postulated structure of the complex strongly suggests that the process of two-electron reduction of the bound O_2 occurs at a significantly faster rate than the rate of formation of the oxygenated species since the latter rate is limited by the process of O_2 transfer in the interior of the protein from the molecular surface to the O_2 reduction site and the former rate is limited by the process of electron-transfer from Cu_B to the bound O_2 which is located at a distance of about 2 Å from Cu_B.[4,8] On the other hand, the two-electron reduced O_2 (peroxide) species is relatively stable, since one-electron reduction of the peroxide (O_2^{2-}) to generate the highly reactive (unstable) oxide radical ($O^{\bullet-}$) is expected to be an unfavorable process. Therefore, the initially detectable intermediate is most likely to be a peroxide-bridge between Cu_B and Fe_{a3} (Cu_B^{2+}–O–O–Fe_{a3}^{3+}) and not to be the oxygenated form.[9,10]

2.2 The Intermediate Species Detectable during the Course of the O_2 Reduction

The best method available at present for identification of the intermediate species of the O_2 reduction process in the enzyme is resonance Raman spectroscopy.[1] As observed by time-resolved resonance Raman spectroscopy, the initial intermediate formed during the course of the O_2 reduction process has been assigned as the oxygenated form (Fe_{a3}^{2+}–O_2).[11,12,13] The band position (521 cm^{-1}) and the isotopic shift effects that occur upon replacement of $^{16}O_2$ with $^{18}O_2$ and $^{16}O^{18}O$ clearly indicate that O_2 binds in a bent end-on fashion which is similar to that observed for oxygenated complexes of hemoglobins and myoglobins.[14] The results indicate that there is essentially no interaction between Cu_B and the bound O_2. This conclusion is consistent with the remarkable stability of the oxygenated form, but clearly in contrast to the proposal that the oxygenated intermediate is too unstable to detect, based on the proximity of Cu_B to Fe_{a3}, as described above.[9,10]

Another unexpected finding obtained by the extensive time-resolved resonance Raman investigation is the identification of the second intermediate, which exhibits an Fe–O stretch band at 804 cm^{-1}, as an iron oxide species ($Fe=O^{2-}$) instead of a bridging peroxide species (Fe^{3+}–O–O–Cu^{2+}).[14] The second intermediate is known as the P-form and has a structure indicating that the O=O bond has been broken. Four electron equivalents must be donated to break the O=O bond. It has been shown that the P-form is the result of transformation from the two-electron reduced enzyme in which, amongst the four metal sites, only heme a_3 and Cu_B are in the reduced state (Fe^{2+} and Cu_B^{1+}).[15] It thus follows that the four electron equivalents required for the O=O bond break must originate solely from the O_2 reduction site.

The third detectable intermediate in the O_2 reduction process is one-electron reduced relative to the P-form and is known as the F-form. Resonance Raman spectra of the F-form exhibit a Fe–O stretch vibration at 780 cm^{-1} and isotope shifts indicate that this mode arises from an iron-oxide (Fe=O) vibrational mode similar to the 804 cm^{-1} band.[14] The large difference in the α-band absorption peaks between the P-and F-forms (604 nm *vs.* 580 nm) suggests the occurrence of an oxidation state change in Fe_{a3} upon transition from P to F ($Fe_{a3}^{5+}=O^{2-}$ to $Fe_{a3}^{4+}=O^{2-}$).[16] The highly stable P-form ($t_{1/2} \sim 70$ min) is thought to be more likely to have a $Fe_{a3}^{5+}=O^{2-}$ structure than to include a Tyr240 radical,[1] since the X-ray structure indicates that the protein environment near Tyr240 would not be supportive for a Tyr240 radical.[15] The possibility that this intermediate consists of a pentavalent heme (Fe^{5+}) is being investigated in organometallic model studies. The fourth intermediate following the F-form exhibits a Fe–O stretch at 450 cm^{-1} in resonance Raman spectra, which is indicative of an $Fe_{a3}^{3+}-OH^-$ structure.[14,16]

2.3   X-ray Structure of the O_2 Reduction Site

The X-ray structure of the O_2 reduction site of the fully reduced bovine heart cytochrome c oxidase at 1.9 Å resolution (Figure 1) shows that Cu_B is coordinated by three histidine imidazole groups and located essentially at the geometric center of the triangle plane of the three coordinating imidazole nitrogen atoms.[3,17] The trigonal planar coordination of cuprous copper is strong evidence that the Cu^{1+} center is a poor electron donor as well as a poor ligand acceptor.[18] The X-ray structure of the O_2 reduction site (Figure 1) shows that the Tyr244–OH is located near the O_2 reduction site and covalently bound to one of the three imidazoles coordinating to Cu_B.[3,17] The covalent bond is expected to serve as an efficient electron transfer pathway. However, examination of the possible configurations of the atomic model of O_2 fit to Fe_{a3} in the fully reduced X-ray structure at 1.9 Å resolution indicates that hydrogen-bond formation between Tyr–OH and the O_2 bound at Fe_{a3}^{2+} would be sterically obstructed by one of the three imidazole groups. The structure suggests that a small but significant conformational change is required for formation of the hydrogen bond between the bound O_2 and Tyr244–OH. Once the hydrogen bond is formed, the OH group would donate a hydrogen atom (or a proton and an electron) to the bound O_2 to yield a hydroperoxo intermediate (Fe_{a3}^{3+}–O–O–H) and a Tyr244 radical. The radical could readily extract a valence electron from Cu_B^{1+} *via* the covalent bond

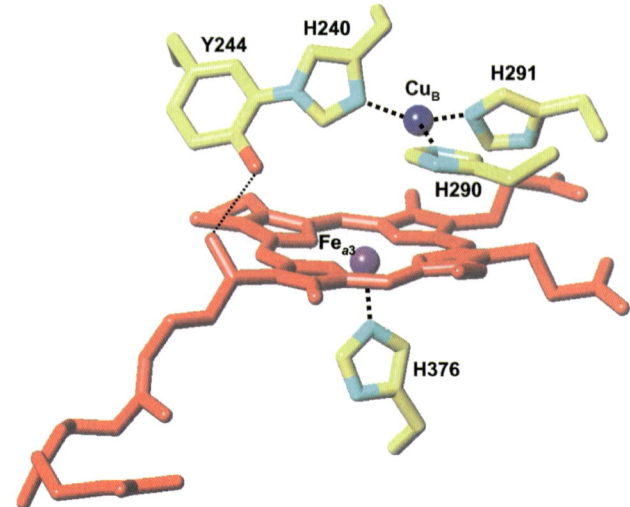

Figure 1 *X-ray structure of the O_2 reduction site of the fully reduced bovine heart cytochrome c oxidase at 1.9 Å resolution. Red structure denotes heme a_3. The broken lines are coordination bonds between the two metal ions and histidine imidazole groups. The dotted line shows the hydrogen bond between the two OH groups of Tyr244 and the hydroxylfarnesylthyl group of heme a_3*

between Tyr244 and His240. The Tyr244–O⁻, which is hydrogenbonded to the hydroperoxo intermediate, would be expected to trigger the second two-electron reduction step, causing cleavage of the O–O bond to yield the Tyr244 radical, $Fe_{a3}^{4+}=O^{2-}$ and OH⁻ which would most likely be bound to Cu_B^{2+}. The second two-electron reduction process which provides the iron oxide form ($Fe_{a3}^{4+}=O^{2-}$) as the second intermediate (detectable in the time-resolved resonance Raman measurement), occurs at a much faster rate than the first two electron reduction process. The oxidation equivalents on Tyr244 in the P-form could be interchangeable with the iron oxide form as $Fe_{a3}^{5+}=O^{2-}$, as described above.

These X-ray and resonance Raman results suggest that the stability of Cu_B^{1+} and the steric hindrance which prevents formation of a hydrogen bond between Tyr244 and the O_2 bound at Fe_{a3} contribute to the stability of the oxygenated form. The covalent bond between Tyr244 and His240 enables efficient electron transfer from Cu_B to Tyr244 radical to promote the four-electron reduction process.

3   The Mechanism of Proton Pumping

3.1   Structural Requirements for Proton Pumping by Cytochrome *c* Oxidase

In general terms, proton pumping is known to be driven by redox-coupled conformational changes that result in changes in the pK_a of the proton pump site and in the accessibility of the site to the aqueous phase (intermembrane or matrix sides in the case of mitochondrial cytochrome *c* oxidase). Another structural requirement for the proton-pumping process mediated by cytochrome oxidase is that the enzyme must have a system for sorting the protons employed in the proton-pumping process from the protons used for water formation. Without such a system, most of the protons taken up from the matrix phase would be used for water formation, since protons have strong affinity to the O_2 reduction intermediates formed at the O_2 reduction site deeply buried inside the transmembrane region.[19] The system should be stable since the overall rate of cytochrome *c* oxidase is relatively slow, possibly because the slow rate contributes to effective energy transduction.

3.2 Redox-coupled Conformational Changes in the X-ray Structure of Bovine Heart Cytochrome *c* Oxidase

The improved X-ray structures indicate that a significant conformational change occurs for Asp51, an acidic residue located near the molecular surface facing the intermembrane phase (Figure 2).[3] In the oxidized state, the carboxyl group of Asp51 is hydrogen bonded to two Ser–OH groups and two peptide N–H groups, while three water molecules and one Ser–OH group are detectable in the hydrogen-bonding distance of the carboxyl group of Asp51 upon reduction of the enzyme (Figure 2B).[3] The X-ray structures indicate that the dielectric environment of the Asp51 carboxyl group in the oxidized state is equivalent to that of a methanol solution and in the reduced state is equivalent to that of an aqueous solution. It is well known that the pK_a of a given organic acid is greatly influenced by the solvent species.[20] The alteration of the Asp51 environment which occurs upon oxidation of the enzyme is expected to induce a carboxyl group pK_a increase of about 5 pH units from its pK_a of 4.8 in aqueous solution. The electron density of a hydrogen atom cannot be identified in a 1.8/1.9 Å resolution X-ray structure. However, the hydrogen-bonding structures determined at these resolutions clearly reveal the protonation state of the carboxyl group.

Asp51 is connected to the matrix phase by a proton transfer pathway as schematically shown in Figure 3. The upper half of the pathway (the intermembrane side half) is a hydrogen-bond network that extends from Asp51 to Arg38 and the formyl group of heme *a*. The formyl–Arg38 system is connected to the matrix phase by a water-channel through which the bulk water molecules in the matrix space are accessible. In Figure 3, the ovals denote the spaces where the presence of at least one water molecule is expected. No detectable electron density in the spaces suggests the existence of mobile water molecules. The channel section is not wide enough to accommodate water molecules without conformational changes, but the flexibility of the protein is expected to allow water transfer through the channel. Thus, the overall water capacity of the channel is essentially determined by the total volume of the cavities. The hydrogen-bond network includes a peptide bond, two water molecules, Tyr371, Arg38 and the propionate group of heme *a*, which is hydrogenbonded to one of the water molecules located between Tyr371 and the peptide carbonyl group. The hydrogen-bond network is disconnected between Tyr371 and the water molecule which is hydrogenbonded to Arg38. The disconnection could provide a unidirectional proton

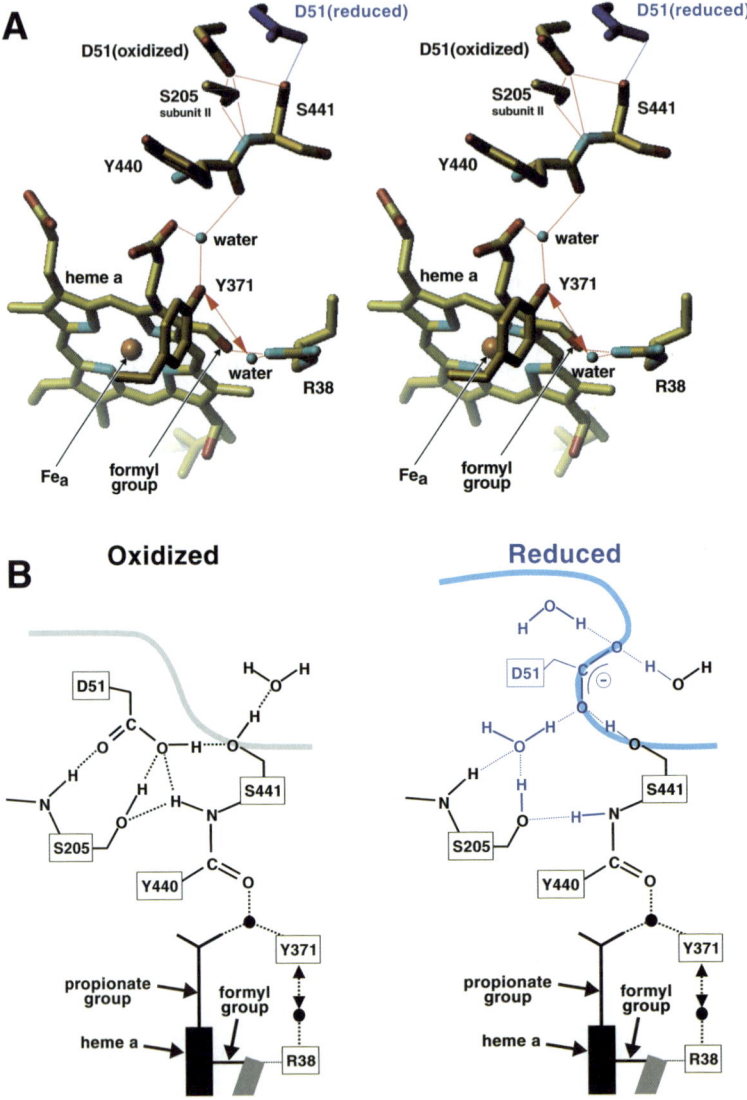

Figure 2 *Redox-coupled conformational changes in Asp51. (A) Stereoscopic drawing of the hydrogen-bond network in H-pathway in the fully oxidized and reduced states. The conformation of D51 in the reduced form is shown in blue. The X-ray Structure of other moiety of the reduced form is identical to that of the oxidized form within the experimental error. (B) Schematic representation of the redox-coupled conformational changes in the hydrogen-bonding structure of Asp51. The thick curves indicate the water-accessible surface facing the intermembrane space. The conformational changes upon reduction are given by blue structures in the right panel. The blue and black balls in panels A and B respectively, denote the fixed water molecules. The dotted lines represent hydrogen bonds. The double-headed dotted arrows indicate spaces where the water molecule hydrogenbonded to Arg38 could move toward Tyr371 to form a new hydrogenbond after breaking the hydrogen-bond to Arg38*

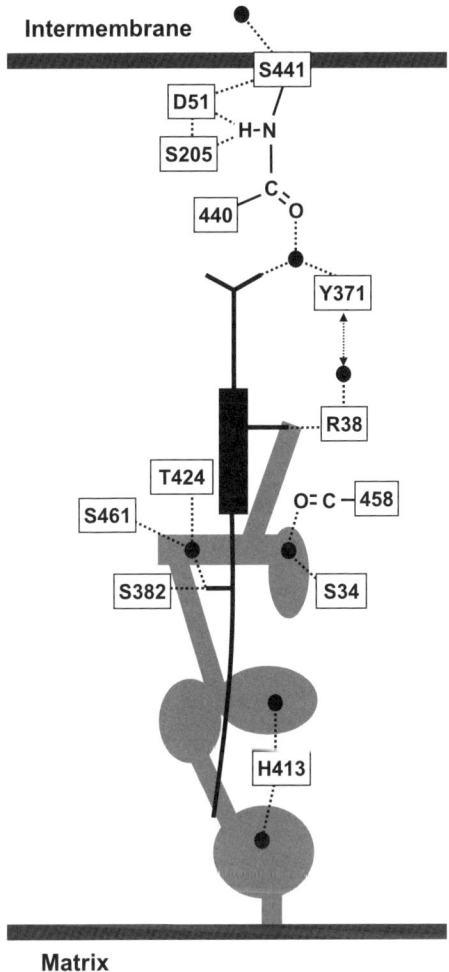

Figure 3 *A schematic representation of H-pathway in the fully oxidized enzyme. The dotted lines and black balls are hydrogen bonds and the fixed water molecules. The gray zone represents the water-channel where the bulk water molecules in the matrix space are accessible. The ovals are the spaces in which at least one water molecule is trapped. The black rectangle denotes a side view of the porphyrin plane of heme a. The long slightly curved line extending toward the matrix surface represents the hydroxylfarnesylthyl group and the short branch on the long line is its OH group. The short stick and the Y-shaped stick on the porphyrin are schematic representations of the formyl and the propionate groups, respectively*

transfer, if the water molecule in the protonated state has much stronger affinity to Tyr371 than to Arg38.

As shown in Figure 2B the hydrogen-bond network known as the H-pathway, connects the intermembrane space with the matrix space in both oxidation states, which seems to indicate that Asp51 is always in a position accessible to both spaces.

However, the peptide bond located between Asp51 and the second water molecule would be expected to block the reverse proton transfer process effectively. Proton transfer through a peptide bond has been reported to occur forming an intermediate, called the imidic acid ($-C(OH)=N^+H-$),[21] as given in Scheme 1.

$$-CO-NH- + H^+ \rightarrow -C(OH) = N^+H- \rightarrow -C(OH)=N- + H^+ \qquad \text{(Scheme 1)}$$

The enol form of the peptide bond ($-C(OH)=N-$) will tautomerize readily to the keto form ($-CO-NH-$). The extremely weak basicity of the peptide amide group ($-NH-$) provides unidirectional character for proton transfer through the peptide bond. A reverse reaction, as illustrated in Scheme 2, is unlikely to occur on a physiologically relevant timescale, since the rate of the initial step of the reaction sequence (the protonation of the $-NH-$ group) is expected to be very slow.

$$-CO-NH- + H^+ \rightarrow -CO-N^+H_2- \rightarrow -C(OH)=N^+H- \rightarrow -C(O^-)=N^+H- + H^+$$

$$-C(O^-)=N^+H- \rightarrow -CO-NH- \qquad \text{(Scheme 2)}$$

Proton transfer through the peptide bond is driven by the two downhill reactions: (i) active proton transport to the peptide carbonyl group through the hydrogen bond network (see below); and (ii) proton extraction from the imidic acid by Asp51 which has a high pK_a value while the enzyme is in the oxidized state. It is clear that leakage of proton transfer is blocked by the peptide bond.

Heme *a* interacts with the hydrogen-bond network by forming two hydrogen bonds between the formyl group and Arg38 and between the propionate group and the water molecule located between Tyr371 and the peptide carbonyl group (Figure 3). The heme is in the six-coordinated low-spin state in both oxidation states. Thus, in the reduced state, it has no net charge because the two positive charges on the iron are counterbalanced by the two negative charges of the porphyrin. Upon oxidation it obtains one equivalent of net positive charge since no counter ion is available inside the protein. The positive charge is readily delocalized in the porphyrin π electron system which includes the formyl group. The formyl group is revealed to be coplanar with the porphyrin π electron system in the improved X-ray structure. Resonance Raman spectra have indicated, *via* a large shift of the formyl C–O stretch vibrational band from 1610 cm^{-1} to 1650 cm^{-1} upon oxidation, that the positive charge is delocalized.[22,23] The 40 cm^{-1} shift is the biggest redox-coupled resonance Raman shift thus far observed within cytochrome *c* oxidase.[1] The positive charge increase (or the electron density decrease) must decrease the effective pK_a of Arg38 which is hydrogenbonded to the formyl group. Driven by the pK_a decrease, the protons taken up from the water molecules or hydronium ions transferred from the matrix space through the water channel are transferred through the hydrogen-bond network up to the peptide carbonyl group. The propionate group also could be acidified upon oxidation of heme *a* to promote the active transport of protons through the hydrogen-bond network, although the extent of the acidification is greatly influenced by the conformation of the propionate group. These structures provide a convincing argument that the hydrogen-bond network–heme *a* system functions as the site for coupling the proton pumping process with the O_2 reduction.

As shown in Figure 4, a large conformational change is detectable in the X-ray structure of the water channel.[3] In the oxidized state, the hydroxyl group of Ser282 is hydrogenbonded with the OH group of the hydroxyfarnesylethyl group of heme *a*. Upon reduction, the hydrogen bond is broken and each of the two OH groups rotates by about 110°, accompanied by a significant migration of the three amino acid residues in helix X (Ser382, Leu381 and Val380). These conformational changes create a new water-containing space in the water channel. The water capacity change could promote effective proton collection by Arg38, by sucking up water molecules (or hydronium ions) upon reduction and by squeezing off hydroxide ions (or water molecules) upon oxidation. The effective pK_a of Arg38 must be significantly lower than the effective pK_a of a guanidinium group in aqueous solution since Arg38 is not exposed to the matrix space or to any cavity in which water molecules are trapped. The diameter of the water channel near the upper end is not sufficient to accommodate a water molecule and thus, the formyl group and Arg38 are only transiently exposed to bulk water molecules from the matrix space, depending on the dynamics of thermal protein motion. However, Arg38 remains likely to have strong affinity for protons in the reduced state of heme *a*. Thus, a hydronium ion or a water molecule, transiently exposed to the guanidino group would donate a proton to the guanidino group.

The redox-coupled structural changes near the hydroxyfarnesylethyl group are not detectable at 2.3/2.35 Å resolution in the oxidized/reduced states.[17] Consistently, no redox-coupled conformational changes at this site have been reported based on analysis of X-ray structures of bacterial cytochrome *c* oxidases at resolutions lower than 2.3/2.35 Å.[24,25,26] Thus, both configurations of the hydroxyfarnesylethyl group can accommodate the electron density observed at resolutions lower than 2.3/2.35 Å, in both oxidation states. No X-ray structure of bacterial cytochrome *c* oxidase with H-pathway has been solved at better than 2.3 Å resolution.

3.3 FTIR Analyses for Redox-coupled Conformational Changes in Bovine Heart Cytochrome *c* Oxidase

The difference spectrum of the fully oxidized against the fully reduced form of the bovine heart enzyme exhibits a peak at 1738 cm^{-1} and a trough at 1585 cm^{-1} which are assignable to COOH and COO$^-$, respectively.[3,27,28] These two bands are not detectable in the redox difference spectrum of the bacterial enzymes which lack Asp51.[27] Thus, it is reasonable to assign these bands to the Asp51 carboxyl group which independently confirms the conclusions drawn from the X-ray structural results. The two bands are detectable also in the difference spectrum of the mixed valence CO form against the fully reduced CO form, indicating that these bands are not due to heme a_3 or Cu_B.[3] Furthermore, reductive titration of the enzyme indicates that the absorbance changes in the two bands are controlled only by a single electron equivalent, necessarily at a single redox active metal site which could be either Cu_A or heme *a*.[3] Recent results show that two equivalents of protons are released from proteoliposomes, reconstituted with bovine heart cytochrome *c* oxidase in the right-side out state, upon complete oxidation of the fully reduced enzyme.[29] These results suggest that it is unlikely that Cu_A reduction induces the migration of Asp51 carboxyl group to the molecular

A

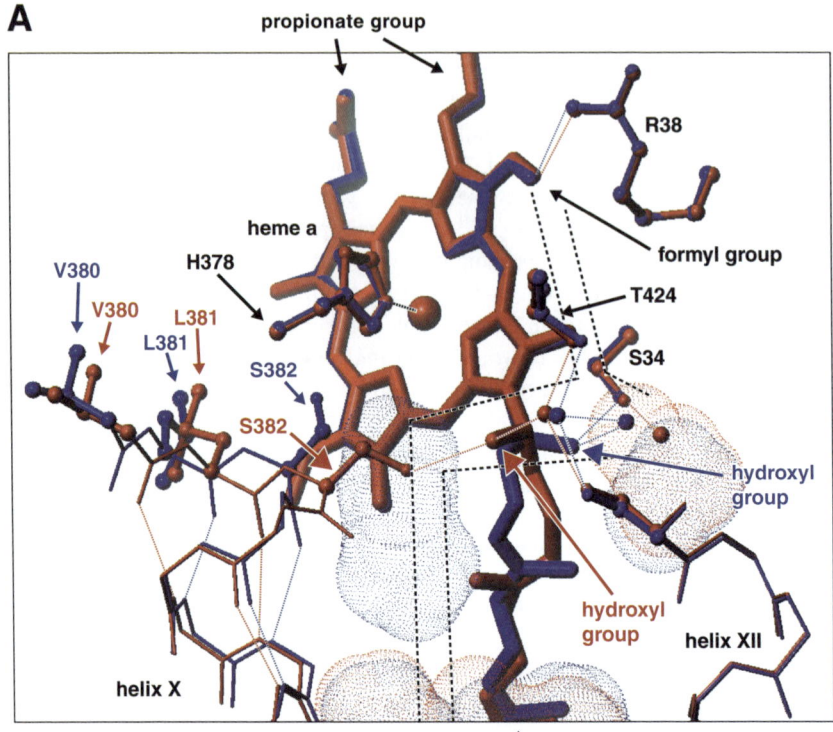

propionate group

heme a

H378

V380
V380

L381
L381

S382

S382

R38

formyl group

T424

S34

hydroxyl group

hydroxyl group

helix XII

helix X

hydroxyfarnesylethyl group

B

Oxidized   **Reduced**

propionate group   formyl group   heme a   R38   T424   O = C — S458   S461   S382   S34   hydroxyfarnesyl-ethyl group   H413   matrix

surface before proton dissociation, since reduction of Cu_A does not occur during the course of the oxidation of the fully reduced enzyme. It is highly likely that heme *a* controls the conformation of Asp51, although the observation of two equivalents of proton release as described above,[29] is not quantitatively consistent with the one equivalent of proton release estimated if heme *a* reduction controls the proton release.[3]

3.4 Mutation Analysis for Asp51

The above X-ray and FTIR results strongly suggest that H-pathway serves as the proton pumping system of cytochrome *c* oxidase. However, these results are not conclusive for the proton pumping function. For example, although the conformational changes of Asp51 and the FTIR results conclusively indicate the occurrence of a redox-coupled change in protonation state, the protons taken up by Asp51 upon oxidation of the enzyme could be obtained from the intermembrane phase *via* the short hydrogen-bond network connecting the molecular surface with Asp51 carboxyl group (Fig 2B), and not from the matrix phase through the H-pathway. In fact the large pK_a increase of the Asp51 carboxyl group that occurs upon oxidation is suggestive of the possibility that proton transfer is from the intermembrane phase.

One of the best methods for examining the functional role of an amino acid residue is its replacement with another residue by site-directed mutagenesis. However, the conventional site-directed mutagensis technique is impossible to apply for Asp51, since Asp51 is not conserved in bacterial enzymes. Therefore, a gene expression system for bovine enzyme subunit I (which contains Asp51) has been established in HeLa cells to prepare bovine human hybrid cytochrome *c* oxidase.[3]

The hybrid Asp51Asn mutant enzyme exhibits a rate of electron transfer 50% higher than that of the wild type hybrid enzyme. However, the proton pumping process is completely disabled, suggesting that Asp51 plays an important role in proton pumping.[3] This observation also indicates that proton uptake *via* the imidic acid intermediate is an absolute requirement for the overall proton pumping process, since Asn51 can form the hydrogen-bond structure in a manner identical to Asp51, but cannot receive a proton from the imidic acid intermediate. Furthermore, the result suggests that, in the oxidized state, the Ser441–OH group, which is located between the Asp51 carboxyl group and the bulk water molecules in the intermembrane side (Figure. 2B), effectively blocks reverse proton transfer from the intermembrane side to the deprotonated carboxyl group of Asp51 before receiving the proton from the imidic acid intermediate.

Figure 4 *X-ray structure of the water channel in the H-pathway. (A) Redox coupled conformational change observed in the upper part of the channel. The red and blue structures denote those in the oxidized and reduced states, respectively. Dotted and broken lines represent the hydrogen bonds and the location of the water channel, respectively. The dotted surfaces show the spaces in which water molecules are possibly located randomly. A large red ball is heme a iron. One of the histidine imidazoles bound to the iron is not shown. (B) A schematic representation of the redox-coupled conformational changes in the H-pathway. The area given in panel A is marked by a square. The blue oval and the blue ball given in the structure of the reduced form (the right panel) represent the water containing space and a fixed water which appear only in the reduced state, respectively*

It is possible that, upon oxidation of the enzyme, the conformational change of Asp51 comprising a movement of a negatively charged group (COO⁻) from the aqueous environment to an environment with a significantly lower effective dielectric constant (pK_a increase of about 5 pH units) limits the overall catalytic turnover. The conformational change must occur at a significantly faster rate in the Asp51Asn mutant enzyme, relative to the wild type enzyme because Asn51 does not accommodate a negative charge. Thus, the mutant enzyme exhibits a faster electron transfer rate as described above.[3] On the other hand, the conformational change of Asp51, which is expected to be an energetically unfavorable process, suggests that a positive charge on the imidic acid form of the peptide between Tyr440 and Ser441 serves to promote the conformational change of Asp51.

3.5   The Function of Heme *a*

The X-ray, infrared and mutagenesis strategies employed to probe the proton-pumping mechanism of bovine heart cytochrome *c* oxidase strongly suggest that heme *a* is the essential element that drives the proton-pumping process. Under anaerobic conditions, the redox potentials of the four redox-active metal sites are essentially identical.[30] However, once O_2 is bound to the heme a_3–Cu_B site, the redox potentials of these metal sites increase significantly in analogy to CO binding to the site, such that electron transfer from heme *a* to the O_2 reduction site is coupled with a large free energy change sufficient to drive the proton pumping process.[31] In addition, the high oxidation states of heme a_3 during the course of the O_2 reduction also provide large free energy changes upon heme *a* oxidation.[1] Therefore, the proton pumping process driven by heme *a* occurs with every electron transfer through heme *a*, in a mechanism that is consistent with the recent experimental results.[29,32] It should be noted that the H-pathway does not include the O_2 reduction site. Therefore, protons to be pumped cannot be used for water formation. This is one of the requirements for a proton-pumping system as described above.[19]

3.6   Comparison of the Proton-Pumping Mechanisms Proposed

As described above, Asp51 is not conserved in plant and bacterial enzymes.[3] This fact, however, does not provide conclusive evidence against its potential function as the proton-pumping site, since many different amino acid functional groups can transfer protons. In general, different sets of amino acids and prosthetic groups often perform identical physiological functions.[33] The superoxide dismutases are a notable example.[34] Furthermore, no amino acid residues in the D-pathway which is essential for O_2 reduction in cytochrome *c* oxidase are completely conserved.[33]

At present, X-ray structures of cytochrome *c* oxidases from three bacterial species other than bovine heart are available at resolutions better than 2.8 Å.[24,25,26,35] Two bacterial enzymes have H-pathways that are essentially identical to that of bovine enzyme, with the sole exception of Asp51. The X-ray structures of the two bacterial enzymes indicate the presence of glycine and a water molecule in the position occupied by Asp51 in the bovine enzyme.[24,25] The two bacterial enzymes with H-pathways seem to pump protons by essentially the same mechanism as that of the bovine enzyme, since the water molecule that takes the place of the carboxyl group of Asp51 relative to the bovine enzyme readily receives protons from the imidic acid intermediate which has a relatively low pK_a. The accessibility of the fixed water molecule to the bulk water phase must be blocked in order to prevent reverse proton transfer through the peptide bond. The redox-coupled conformational change required to make the hydronium ion accessible to the periplasmic phase may be too small to detect in the X-ray structures at the level of 2.8 Å resolution, at which the X-ray structures of the bacterial enzymes with H-pathway have been solved.

The proton-pumping mechanism of the third bacterial enzyme in which the heme *a* and the H-pathway are replaced with heme *b* and the Q-pathway is likely to be significantly different.[35] The Q-pathway across the enzyme molecule is not homogeneous to the H-pathway. However, the presence of heme *b* in the low spin state and the Q-pathway which does not include the O_2 reduction site suggests that protons are pumped through the Q-pathway, driven by heme *b*. The X-ray structure of the third bacterial enzyme has been solved at the highest resolution, but only in the oxidized state.[35] These X-ray structures of the bovine and three bacterial enzymes are strongly suggestive of significant variety in the proton-pumping mechanism amongst different species of organisms, while the role of the low-spin heme as the driving element and the absence of O_2 reduction site in the proton-pumping pathway are expected to be common. Extensive studies on the variety of proton-pumping mechanisms for the enzyme from various biological species would provide important insights for understanding the reaction mechanism of proton pumping.

Various amino acid residues in the H-pathway of the two bacterial enzymes described above have been probed extensively by site-directed mutagenesis.[36,37] Most of the mutant enzymes reported exhibit both O_2 reduction and proton-pumping activities equivalent to the native enzyme. These results are suggestive generally that the H-pathway of bacterial enzymes (as well as bovine enzyme) does not function as the proton-pumping conduit. However, X-ray structures of the bovine and bacterial enzymes indicate that this is not the case.[24,25,26,35] In order to minimize the secondary effects of mutations on the protein conformation, various amino acids located on the wall of the water channel were replaced with residues having less voluminous side chains. However, these mutations are inappropriate for examination of the water transportation function of the channel, since the channel would be expected to widen. Bacterial Tyr371 (bovine number) in the H-pathway hydrogen-bond network is replaced with phenylalanine without any significant effect on the enzyme activity.[36,37] This mutation is also inappropriate for examination of the function of the tyrosine residue in the proton transfer path since the mutation is expected to provide a space large enough to accommodate a water molecule which could serve as a proton conveyer much like tyrosine in the wild type enzyme. Arg38 (bovine number), another amino acid residue in the hydrogen-bond network, has been replaced with methionine, a mutation which causes a decrease in both O_2 reduction and proton-pumping activities.[38] However, the proton pumping efficiency as measured by the ratio of the two activities remains at unity as in the wild type enzyme. In this case the hydrogen bond between the formyl group and Arg38 in the wild type enzyme is abolished. However, the methionine thioether group is accessible to the upper end of the water channel since it is located at the position of the guanidino group in the wild type enzyme. The methionine thioether group forms two hydrogenbonds with other functional groups, each donating a hydrogen atom in the hydrogen bond formed.[39] Therefore, the thioether, which is located near the formyl group of heme *a*, would not block completely the proton transfer through the hydrogen bond network up to the peptide bond.

This enzyme has two possible proton transfer pathways (K and D) that connect the matrix surface with the O_2 reduction site.[31] The structure primarily suggests that each of the pathways serve as the proton transfer path for O_2 reduction. Some D-pathway mutations completely impair the proton-pump function, as well as the O_2 reduction activity.[40] Depending on the individual mutations, influence is exerted on the oxidation of the fully reduced enzyme with O_2 and the process is blocked at the P or F states.[40] From these results, it has been proposed that the D-pathway transfers protons both for the process of water formation and for the process of proton pumping. However, a blockage of proton transfer for water formation would block O_2 reduction and would also block proton pumping because of the loss of driving energy for the pumping process. Thus, it is possible that these mutations provide no significant perturbation of the structure of the proton pumping system which is isolated from the D-pathway. Recent studies of Asn98Asp (corresponding to the bovine enzyme number system), D-pathway mutants of bacterial cytochrome *c* oxidases, indicate that it has disabled proton pumping activity while exhibiting 100–200% of the O_2 reduction activity relative to the wild type enzyme.[41,42] The Asn98Asp mutation is not expected to have any significant effect on proton transfer through the D-pathway. The Asn98Asp mutation is expected to influence the conformation of, and thus impair, the proton-pumping system, without perturbing any proton transfer through the D-pathway. Thus, this mutation result also does not support transfer of pumping protons through D-pathway.

The Asn98Asp mutation exerts an influence on the pH dependence of the rate of transition from the P-form to the F-form during O_2 reduction, increasing the pK_a value of the pH dependence from 9.5 to 11.[43] The pH dependence has been thought to be due to the protonation state change of Glu242 (bovine number) located at the upper end of the D-pathway.[44,45] It has been proposed that Glu242 determines the direction of proton transfer toward the O_2 reduction site or toward the proton-pumping site which is thought to be located near the propionate groups of hemes *a* and a_3. In addition, based on the extensive kinetic analysis of the pH dependency of the O_2 reduction process, the proton affinity of the proposed proton-pumping site near the propionate groups has been proposed to be significantly lower than that of the O_2 reduction site.[44,45] In the wild type enzyme, the proton affinity of Glu242 is lower than those of both the O_2 reduction and proton pumping sites. However, the pK_a value of the Asn98Asp mutant is lower than that of the O_2 reduction site but higher than that of the proton-pumping site.[43] Therefore Glu242 can transfer protons only to the O_2 reduction site. This is a reasonable interpretation for the Asn98Asp mutation results. However, extensive experimental evaluations for the proposal are required since various alternative interpretations are possible especially for the pH dependency of the P–F transition rate. For example, it is equally possible that the pK_a change is due to Arg38 in the H-pathway. An increase in pK_a of the guanidino group of Arg38 would also inhibit the proton transfer through the hydrogen-bond network of the H-pathway.[3]

If the protons to be pumped are transferred through the D-pathway, which connects the O_2 reduction site with the matrix surface, the proton-pumping site must be at or near the O_2 reduction site. However, no functional group capable of a redox-coupled conformational change to produce significant changes in pK_a and accessibility to the enzyme molecular surface has been identified at or near the O_2 reduction site thus far in any X-ray structures. Although a small conformational perturbation near propionate groups of hemes *a* and a_3 of a bacterial enzyme, induced by a mutation in D-pathway, has been proposed,[26,45] the X-ray structural resolutions are not sufficiently high to conclude the conformational change. Furthermore, there is no experimental evidence for any functional group near the O_2 reduction site that would be capable of effectively sorting the protons to be pumped from the protons used for water formation. The possibility of redox-coupled proton pumping occurring at or near the O_2 reduction site has been proposed as a result of various theoretical calculations which have yet to be evaluated experimentally.[46,47,48]

Acknowledgements

This work was supported in part by the Grants-in Aid for 21st Century Center of Excellence Program and for Scientific Research on Priority Areas, Structures of Biological Macromolecular Assemblies from the Ministry of Education, Culture, Sports, Science and Technology, Japan. The author is a senior scientist in RIKEN Harima Institute and grateful for stimulating discussions with many colleagues on the subject of this review, particularly T. Tsukihara, K. Shinzawa-Itoh, H. Shimada, K. Muramoto, T. Ogura, T. Kitagawa, T. Sugimura and S. Sugano.

References

1. T. Kitagawa and T. Ogura, *Progr. Inorg. Chem.*, 1997, **45**, 431.
2. S. Yoshikawa, *Adv. Prot. Chem.*, 2002, **60**, 341.
3. T. Tsukihara, K. Shimokata, Y. Katayama, H. Shimada, K. Muramoto, H. Aoyama, M. Mochizuki, K. Shinzawa-Itoh, E. Yamashita, M. Yao, Y. Ishimura and S. Yoshikawa, *Proc. Nat. Acad. Sci. U.S.A.*, 2003, **100**, 15304.
4. W.S. Caughey, W.J. Wallace, J.A. Volpe and S. Yoshikawa in *The Enzymes*, P.D. Boyer (ed.), Academic Press, New York, 3rd ed., 1976, 299.
5. J.M. McCord, I. Fridovich, *J. Biol, Chem.*, 1969, **244**, 6049.
6. W.J. Wallace, J.C. Maxwell and W.S. Caughy, *Biochem. Biophys. Res. Comm.*, 1974, **57**, 1104.
7. B.F. Van Gelder and H. Beinert, *Biochim. Biophys. Acta.*, 1969, **189**, 1.
8. M. Mori and J.A. Weil, *J. Am. Chem. Soc.*, 1967, **89**, 3732.
9. S. Yoshikawa and W.S. Caughey, *J. Biol. Chem.*, 1990, **265**, 7945.
10. S. Yoshikawa, M. Mochizuki, X.-J. Zhao, W.S. Caughey, *J. Biol. Chem.*, 1995, **270**, 4270.
11. C. Varotsis, W.H. Woodruff and G.T. Babcock, *J. Am. Chem. Soc.*, 1990, **112**, 1297.
12. T. Ogura, S. Takahashi, K. Shinzawa-Itoh, S. Yoshikawa and T. Kitagawa, *J. Am. Chem. Soc.*, 1990, **112**, 5630.

13. S. Han, Y.-C. Ching and D.L. Rousseau, *Proc. Natl. Acad. Sci. U.S.A.*, 1990, **87**, 2491.

14. T. Ogura, S. Takahashi, S. Hirota, K. Shinzawa-Itoh, S. Yoshikawa and T. Kitagawa, *J. Am. Chem. Soc.*, 1993, **115**, 8527.

15. K. Oda, T. Ogura, E. H. Appelman and S. Yoshikawa, *FEBS Lett.*, 2004, **570**, 161.

16. D.A. Proshlyakov, T. Ogura, K. Shinzawa-Itoh, S. Yoshikawa, E.H. Appelman and T. Kitagawa, *Biochemistry*, 1996, **35**, 8580.

17. S. Yoshikawa, K. Shinzawa-Itoh, R. Nakashima, R. Yaono, E. Yamashita, N. Inoue, M. Yao, M. J. Fei, C. Peters Libeu, T. Mizushima, H. Yamaguchi, T. Tomizaki and T. Tsukihara, *Science*, 1998, **280**, 1723.

18. F.A. Cotton and G. Wilkinson, *Advanced Inorganic Chemistry: A Comprehensive Text*, John Wiley, New York, 4th ed., 1980.

19. R.J.P. Williams, *Nature*, 1995, **376**, 643.

20. N.S. Isaacs in *Physical Organic Chemistry* Longman, Essex, U.S.A., 2nd ed., 1995, 235.

21. C.L. Perrin, *Acc. Chem. Res.*, 1989, **22**, 268.

22. G.T. Bobcock and P.M. Callahan, *Biochemistry*, 1983, **22**, 2314.

23. M. Sassaroli, Y.-C. Ching and D.L. Rousseau, *Biochemistry*, 1989, **28**, 3128.

24. C. Ostermeier, A. Harrenga, U. Ermler and H. Michel, *Proc. Natl. Acad. Sci. U.S.A.*, 1997, **94**, 10547.

25. A. Harrenga and H. Michel, *J. Biol. Chem.*, 1999, **274**, 33296.

26. M. Svensson-Ek, J. Abramson, G. Larsson, S. Törnroth, P. Brzezinski and S. Iwata, *J. Mol. Biol.*, 2002, **321**, 329.

27. P. Hellwig, T. Soulimane, G. Buse and W. Maentele, *FEBS Lett.*, 1999, **458**, 83.

28. D. Okuno, T. Iwase, K. Shinzawa-Itoh, S. Yoshikawa and T. Kitagawa, *J. Am. Chem. Soc.*, 2003, **125**, 7209.

29. D. Bloch, I. Belevich, A. Jasaitis, C. Ribacka, A. Puustinen, M.I. Verkhovsky and M. Wikström, *Proc. Natl. Acad. Sci. U.S.A.*, 2004, **101**, 529.

30. M. Mochizuki, H. Aoyama, K. Shinzawa-Itoh, T. Usui, T. Tsukihara and S. Yoshikawa, *J. Biol. Chem.*, 1999, **274**, 33403.

31. S. Ferguson-Miller and G.T. Babcock, *Chem. Rev.*, 1996, **96**, 2889.

32. M. Ruitenberg, A. Kannt, E. Bamberg, K. Fendler, H. Michel, *Nature*, 2002, **417**, 99.

33. M.M. Pereira, M. Santana and M. Teixiera, *Biochim. Biophys. Acta.*, 2001, **1505**, 185.

34. M.E. Stroupe, M. DiDonato and J.A. Tainer in *Handbook of Metalloproteins* A. Messerschmidt, R. Huber, T. Poulos and K. Wieghardt (eds.), John Wiley, Chichester, 2001, 941.

35. T. Soulimane, G. Buse, G.P. Bourenkov, H.D. Bartunik, R. Huber and M.E. Than, *EMBO J.*, 2000, **19**, 1766.

36. U. Pfitzner, A. Odenwald, T. Ostermenn, L. Weingard, B. Ludwig and O.M. Richter, *J. Bioenerg. Biomembr.*, 1998, **30**, 89.

37. H.-M. Lee, T.K. Das, D.L. Rousseau, D. Mills, S. Ferguson-Miller and R.B. Gennis, *Biochemistry*, 2000, **39**, 2989.

38. A. Jasaitis, C. Backgren, J.E. Morgan, A. Puustinen, M.I. Verkhousky and M. Wikström, *Biochemistry*, 2001, **40**, 5269.

39. G. Modena, C. Paradisi and G. Scorrno in *Studies in Organic Chemistry* F. Bernardi, I.G. Csizmardia and A. Mangini (eds.), Elsevier Amsterdam, 1985, **19**, 568.

40. R.B. Gennis, *Biochim. Biophys. Acta.*, 1998, **1365**, 241.

41. U. Pfitzner, K. Hoffmeier, A. Harrenga, A. Kannt, H. Michel, E. Bamberg, O.-M. Richter and B. Ludwig, *Biochemistry*, 2000, **39**, 6756.

42. A.S. Pawate, J. Morgan, A. Namslauer, D. Mills, P. Brzezinski, S. Ferguson-Miller and R.B. Gennis, *Biochmistry*, 2002, **41**, 13417.

43. A. Namslauer, A.S. Pawate, R.B. Gennis and P. Brzezinski, *Proc. Natl. Acad. Sci. U.S.A.*, 2003, **100**, 15543.

44. A. Namslauer, A. Agaard, A. Katsonouri and P. Brzezinski, *Biochmistry*, 2003, **42**, 1488.

45. P. Brzezinski and G. Larsson, *Biochim. Biophys. Acta.*, 2003, **1605**, 1.

46. M. Wikström, M.I. Verkhousky and G. Hummer, *Biochim. Biophys. Acta.*, 2003, **1605**, 61.

47. P.E.M. Siegbahn, M.R.A. Blomberg and M.L. Blomberg, *J. Phys. Chem. B.*, 2003, **107**, 10946.

48. D.M. Popovic and A.A. Stuchebrukhov, *J. Am. Chem. Soc.*, 2004, **126**, 1858.

CHAPTER 4

Mechanisms of Redox-coupled Proton Pumping by Respiratory Oxidases

PETER BRZEZINSKI AND PIA ÄDDELROTH

Department of Biochemistry and Biophysics,
The Arrhenius Laboratories for Natural Sciences,
Stockholm University, SE-106 91 Stockholm, Sweden

Abbreviations

CcO, cytochrome c oxidase;

N- and P-sides, the (relatively) negatively and positively charged sides of the membrane, respectively;

The states of CcO are denoted with one-letter codes: (the superscripts denote the number of electrons added to the catalytic site in the different states). $O^{0(4)}$, oxidised state; E^1, a state in which Cu$_B$ at the catalytic site is reduced; R^2, both Cu$_B$ and haem a_3 are reduced; P^2, the peroxy intermediate formed upon reaction of CcO in the R^2 state with O$_2$; P^3, the peroxy intermediate with one more electron at the catalytic site than P^2 (formed upon reaction of R^2 with O$_2$ when also haem a is reduced); F^3, the ferryl intermediate formed upon addition of one electron and one proton to P^2 or one proton to P^3.

Unless indicated otherwise, the amino-acid residue numbering refers to the *Rhodobacter sphaeroides* cytochrome aa_3 sequence, and the residues are found in subunit I.

1 Introduction

The aerobic respiratory oxidases are integral membrane proteins that catalyse the four-electron reduction of O$_2$ to water. These enzymes are often referred to as *terminal oxidases* because they are the last components of the respiratory chains in eukaryotes and aerobic prokaryotes. One subgroup of the terminal oxidases is the *haem–copper oxidases*, in which the oxygen-reducing catalytic site consists of a haem group and a copper ion in close proximity, forming a binuclear centre. These enzymes are found in the inner mitochondrial membrane in eukaryotes or in the bacterial cytoplasmic membrane. Another subgroup of the terminal oxidases is the *alternative oxidases*, which are found, for example, in higher plants and fungi.[1] This subgroup is not further discussed here.

The haem–copper oxidase superfamily can be divided in two subgroups; the quinol (QH_2) and cytochrome (cyt.) c oxidases (CcOs), where the names refer to the identity of the electron donor. The chemical reactions catalysed by these two subgroups of the oxidases are:

cytochrome c oxidases: $4 \text{ cyt.}c_P^{2+} + O_2 + 4 H_N^+ \rightarrow 4 \text{ cyt.}c_P^{3+} + 2 H_2O$ (1a)

quinol oxidases: $2 QH_2 + O_2 + 4 H_N^+ \rightarrow 2Q + 2 H_2O + 4 \mathbf{H_P^+}$ (1b)

In CcO the reaction is arranged topographically such that the electrons are donated from one side of the membrane (the positive, *P*-side), while the protons are taken up from the other side (the negative, *N*-side) (see subscripts *P* and *N* in Eqs. 1a and b). The protons used for O_2 reduction will be referred to as 'substrate protons'. In the quinol oxidases the substrate quinol binds to the oxidase in the core of the membrane and upon oxidation to quinone (Q) the quinol protons are released to the *P*-side (H^+ in bold in Eq. 1b). As in CcOs, the substrate protons are taken up from the *N*-side. Thus, in both the CcOs and quinol oxidases the reduction of O_2 to water results in a charge separation that is equivalent to the transfer of one positive charge from the *N*- to the *P*-side, across the membrane, per electron transferred to O_2 (Figure 1a and b). In bacteria the periplasm (or outside) of the bacterium is the *P*-side, while in eukaryotes the *P*-side is the intermembrane space of the mitochondrion.

In mitochondria the proton electrochemical potential across the inner membrane is ~200 mV. Thus, moving one positive charge from the *N*- to the *P*-side requires a free energy of ~−200 meV. In the mitochondrial CcOs, the free energy difference for the electron transfer from cyt. c to O_2 at standard conditions is approximately −550 meV. Consequently, the transmembrane charge transfer upon the reaction shown in Eq. 1a results in conservation of less than half of the available free energy. In 1977 Mårten Wikström showed that in addition to the charge separation accomplished by taking electrons and protons for O_2-reduction from opposite sides of the membrane, CcO is also a proton pump,[2] which translocates one proton across the membrane for each electron transferred to O_2. Thus, the overall reaction catalysed by the CcOs results in translocation of two positive charges across the membrane per electron transferred to O_2. Because the transfer of two positive charges requires a free energy change of ~−400 meV, ~75% of the available free energy of ~−550 meV is conserved in the reaction catalysed by the CcOs.

In this review we focus on the mechanisms of O_2 reduction, electron and proton transfer, and proton pumping by the haem–copper oxidases. The discussion is primarily centred around a CcO from *Rhodobacter* (*R.*) *sphaeroides* (cytochrome aa_3, the nomenclature is explained below), which is a homologue of the eukaryotic enzymes[3] and one of the functionally and structurally most well-characterised haem–copper oxidases. The general functional mechanisms are believed to be the same for all CcOs.

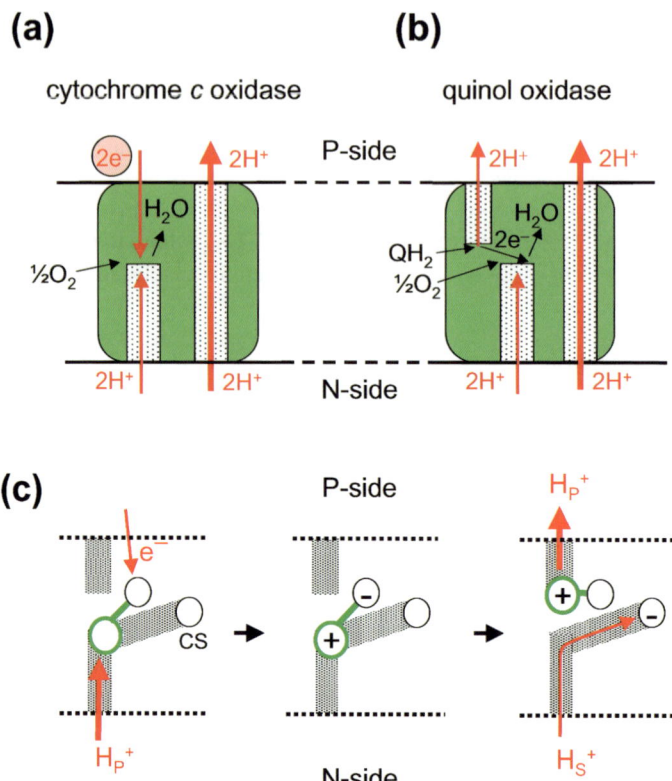

Figure 1 *The overall reaction catalysed by the haem–copper oxidases. (a) Cytochrome c (red circle), the electron donor to cytochrome c oxidase, binds on the positive (P) side of the membrane. Protons (substrate protons, thin red arrow) are transferred from the negative (N) side to oxygen, which binds at the catalytic site within the enzyme. The O_2-reduction reaction is energetically coupled to proton pumping across the membrane (thick red arrow). The overall reaction results in a charge separation across the membrane corresponding to the transfer of two positive charges from the N- to the P-side per electron transferred to O_2. Reactions shown with red arrows are associated with charge transfer perpendicular to the membrane surface (electrogenic reactions). (b) In the quinol oxidases the electrons are transferred from quinol (QH_2) and are then transferred to the catalytic site. Upon oxidation of QH_2 protons are released to the P-side, while the substrate and pumped protons are taken up from the N-side, as in cytochrome c oxidase. Thus, also the reaction catalysed by the quinol oxidases results in a charge separation corresponding to the transfer of two positive charges from the N- to the P-side per electron transferred to O_2. (c) A schematic general model of a redox-driven proton pump. The reduction of a redox site within the protein from the P-side is associated with proton uptake (of a proton that is 'pumped' H_P^+) by a protonatable site ('proton switch', shown in green) from the N-side. In the next step, the electron is transferred to the catalytic site (CS), associated with the uptake of a substrate proton, H_S^+. This proton uptake takes place only after the protonated 'switch' has moved towards the P-side, where the proton is released upon protonation of the catalytic site. The overall process results in the translocation of two positive charges from the N- to the P-side per electron transferred to the catalytic site. The dotted structures are intra-protein proton-transfer pathways (the model in (c) is modified from ref. 175)*

2 Redox-driven Proton Pumping – General Principles

As discussed above, C*c*O is a proton pump that translocates protons from one side of the membrane to the other side without using any (free) mobile proton carriers. In this section we consider some general design principles of proton pumps that have been discussed in the literature (for example[4–14]).

Protons are transferred across the membrane-spanning part of the protein through intraprotein proton-transfer pathways (see below for a specific description of those found in C*c*O). However, to avoid short-circuiting of the proton electrochemical gradient that is maintained across the membrane, it is crucial that there is never direct contact between the two membrane sides. Hence, the C*c*O must control the accessibility of protons to the two sides of the membrane. One possibility to control proton transfer is to introduce a *switch* (also referred to as *gate*) in the protein, which would provide *alternating access* (also referred to as *gating*) of protons to one or the other side of the membrane, but never both sides simultaneously. This switch could also act as a proton acceptor and donor during the pumping cycle (in this case it is referred to as a *pumping element*). One way to provide alternating access of protons to the two sides of the membrane is by modifying a segment of a proton-transfer pathway, *e.g.* by the breaking and forming of hydrogen bonds. An example of a possible pumping element is a protonatable amino-acid residue that can adopt two different conformations (see Figure 1c).

Independently of the specific proton-pumping mechanism, the changes in proton accessibility to the two sides of the membrane must be linked in time to specific reaction steps at the catalytic site. One possibility to achieve proton translocation is to use the free energy that is available from O_2 reduction to alter the pK_a value of a protonatable group in the proton-pumping pathway such that it has a high pK_a when in contact with the proton-input side and a low pK_a when in contact with the proton-output side. However, such pK_a changes are not strictly required.[7,15,16] In this context it should be mentioned that it has been argued that the 'pumped proton' must be transferred to the acceptor group before the substrate proton is transferred from solution to the catalytic site (see Figure 1c). If this were not the case the free energy available from the O_2-reduction reaction would be lost as heat. The significance of this requirement is discussed in more detail below in the framework of a specific pumping mechanism.

Alternatively, the continuous flow of electrons through the enzyme could drive the unidirectional flow of protons through a part of the C*c*O in which the electron and proton-transfer pathways overlap and interact strongly. This type of coupling would result in proton translocation without a gate (see for example[17,18]). However, the absence of a gate would result in a back-leak of protons when the electron flux decreases below a certain threshold, a problem that could be solved by introducing a 'valve' that closes when the electron flux decreases below a certain level.

3 Structure

The first crystal structures of C*c*Os were those of the enzymes isolated from the bacterium *Paracoccus (P.) denitrificans*[19,20] and from bovine heart[21–24] (first presented in 1995). Since then, three-dimensional structures of the ubiquinol oxidase cytochrome bo_3 from *Escherichia (E.) coli*,[25] and the C*c*Os cytochrome ba_3 from

Thermus (T.) thermophilus[26] and cytochrome aa_3 from *R. sphaeroides*[27] were determined (the X-ray structures of haem–copper oxidases that have been determined to date are summarised in Table 1).

The *R. sphaeroides* CcO consists of four subunits with a total molecular weight of ~130 kDa, while CcO from bovine heart mitochondria is much larger and composed of thirteen (all different) subunits with a total molecular weight of 220 kDa. The sequence similarity between subunits I–III of the mitochondrial CcO, which are encoded by the mitochondrial DNA, and the *R. sphaeroides* CcOs is very high,[28–31] and about 50% amino-acid residues are identical. The function of the nuclear-encoded subunits IV–XIII in the mitochondrial CcO is unclear, but they have been suggested to play regulatory roles, *e.g.* through binding of ATP.[32,33] These subunits probably also have stabilising roles which is reflected by the longer life-time of the eukaryotic CcO in the living cell of several days compared to hours for many bacterial enzymes.[5]

In subunit I of the haem–copper oxidases there are two haem groups and a copper ion. The haem groups may be of different types such as, for example, haem *a*, *b*, or *o*, where the two haems in subunit I may be of the same or of different types. Because many bacteria have several types of oxidases (expressed to various degrees depending on the environmental conditions such as, for example, the O_2 concentration), the haem–copper oxidases are often given a name that indicates the types of haem present, *e.g.* cytochrome ba_3. The CcO from *R. sphaeroides* discussed here, as well as the mitochondrial CcO, are both of the cytochrome aa_3 type.

The primary electron acceptor (from cytochrome *c*) in the *R. sphaeroides* CcO is a copper ion, Cu_A, which is located near the *P*-side surface in subunit II (Figure 2a and b). The secondary electron acceptor, which receives electrons from Cu_A, is a six-coordinated (two His axial ligands) haem group, haem *a*. From haem *a* electrons are transferred to the catalytic site, which is composed of a five-coordinated haem group (one His axial ligand), haem a_3, and a copper ion, Cu_B. The second axial ligand-binding position of haem a_3 binds O_2, as well as many other small ligands such as CO, OH^- and CN^-. In addition, a Tyr residue Y288 (Figure 2b) may be included as part of the catalytic site, because in most structures determined to date it is covalently bound to one of the His (H284) ligands of Cu_B and it is presumably involved

Table 1 *Structures of haem–copper oxidases from the Protein Data Bank (PDB) (found at http://www.rcsb.org)*

Haem–copper oxidase	Source	PDB code	Resolution (Å)	Comment
cytochrome aa_3	*P. denitrificans*	1AR1	2.7	two-subunit enzyme
		1QLE	3.0	four-subunit enzyme
cytochrome bo_3	*E. coli*	1FFT	3.5	
cytochrome ba_3	*T. thermophilus*	1EHK	2.4	
cytochrome aa_3	*R. sphaeroides*	1M56	2.3	wild-type enzyme
		1M57	3.0	E286Q mutant enzyme
cytochrome aa_3	bovine heart mitochondria	2OCC	2.3	oxidised state
		1OCR	2.35	reduced state
		1V54	1.8	oxidised state
		1V55	1.9	reduced state

in electron transfer to the catalytic site. In addition to the four redox-active metal sites, in many CcOs there is also a non-redox-active metal site containing a Mg^{2+} ion located 'above' the haem groups (Figure 2b and c).

Subunit III in cytochrome aa_3 from *R. sphaeroides* contains seven transmembrane helices, arranged in two bundles separated by a large V-shaped cleft. For CcOs from several species the minimal functional unit (*i.e.* that can reduce O_2 and pump protons) is the two-subunit (subunits I & II) CcO[34–37](see also[38]). However, without subunit III the timing of the proton-transfer reactions during specific reaction steps are dramatically altered, presumably because amino-acid residues of this subunit are found near the entrance to one of the proton-transfer pathways.[39–41] Because of this retardation of proton transfer, without subunit III the CcO is irreversibly inactivated ('suicide inactivation') after a number of turnovers, which indicates that subunit III is required for maintaining the structural integrity of the redox-active sites[42] (see also[36,39]).

Subunit IV of the *R. sphaeroides* cytochrome aa_3 consists of only one transmembrane helix, and its function is unknown.

3.1 Oxygen Channels

Analyses of the available X-ray crystal structures of the haem–copper oxidases, combined with results from theoretical and experimental studies, suggest that there are specific channels through which O_2 enters into the catalytic site. All the proposed channels start in the membrane phase and merge near the catalytic site. A hydrophobic channel, starting at the V-shaped cleft formed by subunit III, has been identified in the X-ray structures of CcOs from *P. denitrificans*[19] and *T. thermophilus*[26] and its function as a putative O_2 channel is supported from functional studies.[43,44] These channels overlap with one of the channels identified in the structure of the mitochondrial CcO.[23] On the basis of an analysis of the location of xenon atoms in the X-ray crystal structure of CcO from *R. sphaeroides*, a putative O_2 channel was proposed between helices II and III in subunit I,[27] supported by functional studies[45] (see Figure 2a). This channel converges with the channels starting in subunit III close to the catalytic site. Results from molecular-dynamics simulations[46] also indicate that O_2 is transported into the catalytic site along a specific trajectory. The relationship between the need for specific O_2 channels and other requirements for the function of a proton pump is discussed in a separate section below.

3.2 Proton-transfer Pathways

Two proton-transfer pathways, called the D-and K-pathways, leading from the *N*-side surface towards the catalytic site have been identified in the *R. sphaeroides* CcO as well as in many other CcOs (Figure 2b). A third pathway has also been proposed in the mitochondrial CcO and it is discussed briefly in this section.

3.2.1 D-pathway

The D-pathway is named after a highly conserved Asp residue, D132, located in a cavity forming the entry point of the pathway.[47–50] The pathway leads to another highly conserved residue, E286, at the end of the pathway, located ~24 Å from D132 and ~11 Å from Cu_B. The D132 and E286 residues are connected through a

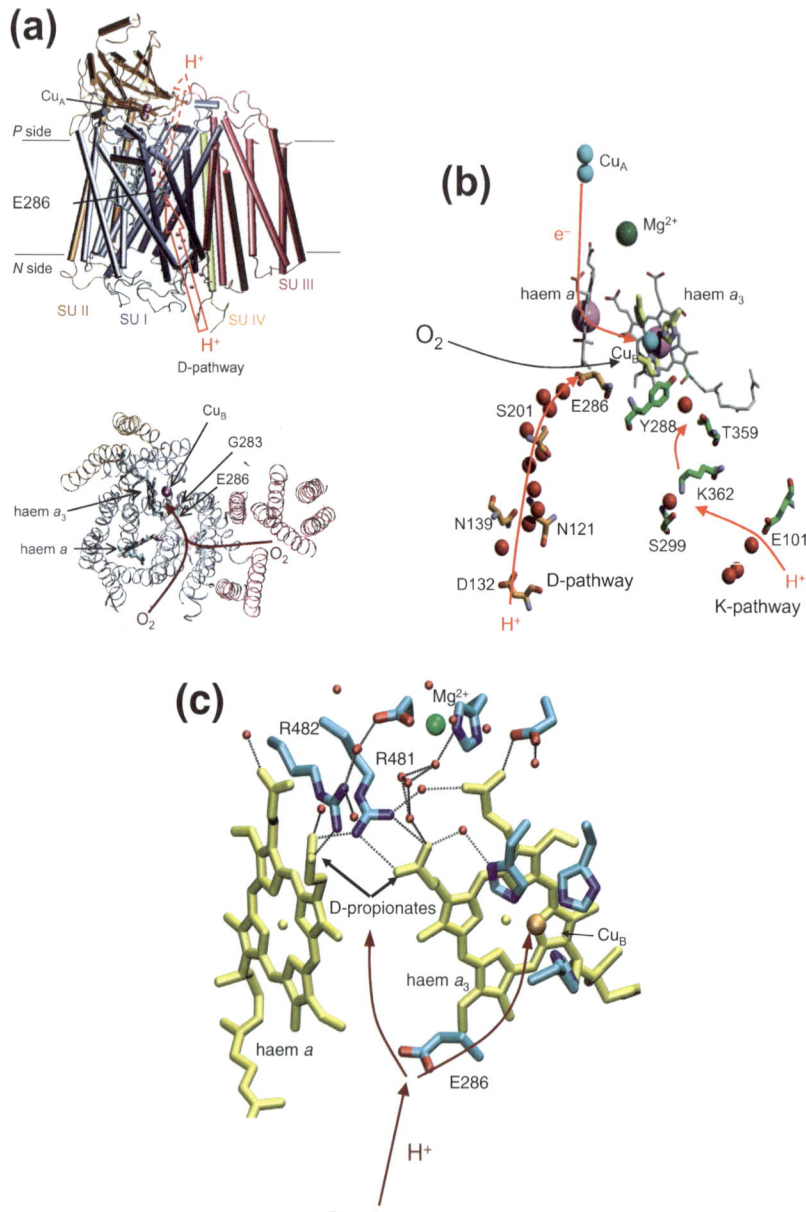

hydrogen-bonded chain of amino-acid residue side chains and water molecules resolved in the X-ray crystal structures (see Figure 2b) and predicted from theoretical studies.[46,51–54]

Mutant CcOs in which either the Glu or Asp are replaced by non-protonatable residues (*e.g.* Gln, Asn or Ala) were practically inactive because proton uptake through the D-pathway is blocked.[55–63]

3.2.2   K-pathway

A second proton pathway is called the K-pathway because it contains a highly conserved and essential lysine residue, K362, in the middle of the pathway.[64,65] Only a few water molecules are resolved in the X-ray crystal structures of CcOs within the K-pathway and it has been suggested that rapid proton transfer through the pathway requires rotation of the K362 side chain.[46] A thorough analysis of the water chain formation in the pathway was recently presented by Cukier.[66] On the basis of experimental studies with the *R. sphaeroides* cytochrome aa_3[67,68] and *E. coli* cytochrome bo_3,[69] supported by results from electrostatic calculations,[70,71] the entry point of the pathway was suggested to be located near E101 in subunit II (Figure 2b). However, different entry points may be found in CcOs from other species.[72] The pathway continues through the lysine, a threonine residue, T359, and the hydroxyl group of haem a_3 to Y288.

3.2.3   H-pathway

A third proton-transfer pathway, called the H-pathway, has been suggested on the basis of an analysis of the structure of CcO from bovine mitochondria and proposed to be used for proton pumping.[22,73] This hypothesis is supported by results from recent studies of a mutant CcO in which one of the key residues in the H-pathway, D51, (bovine CcO numbering) was modified.[24] However, the H-pathway is not fully conserved in the bacterial haem–copper oxidases and mutation of corresponding

---

Figure 2 *The structure of cytochrome c oxidase (cytochrome aa₃) from R. sphaeroides. (a) The overall structure of the enzyme (PDB number 1M56,²⁷). Subunits I, II, III and IV are shown in different colours as indicated. The approximate route of proton transfer through the D-pathway is indicated by the red arrow. The dashed arrow shows that 'pumped' protons are released towards the P-side (a specific proton-release pathway has not been identified). In the lower part (top view), the approximate location of two proposed O₂ channels are shown. (b) The D and K-proton-input pathways, together with the redox-active cofactors Cu_A, haem a, and haem a₃ and Cu_B (the catalytic site). All residues are found in subunit I, except E101, which is found in subunit II. Water molecules are shown as red spheres. The redox-inactive cofactor Mg²⁺ is also shown. (c) The detailed structure around the haem groups. The Arg groups (R481 and R482) form salt bridges with the haem propionates. Protons from E286, at the end of the D-pathway, are transferred either towards the output (P-side) or to the catalytic site (red arrows). In the relaxed state E286 is in rapid equilibrium with the N-side solution through the D-pathway. The pictures were prepared using the Visual molecular dynamics software¹⁸⁵ (c) is reproduced with permission from ref. 148 (Copyright 2005 Am. Chem. Soc.)*

residues in the bacterial C*c*Os did not show any effect on proton pumping,[59,74] therefore this pathway appears to be specific for the eukaryotic C*c*Os.

3.2.4   Proton Exit Pathway

Experiments and structural analyses have identified elements of the exit pathway for pumped protons in the region around the haem propionates (Figure 2c),[16,75–81] but the detailed path is still unknown. The presence of specific water-exit channels have been discussed in the literature[23,26,82] and the results from a recent study indicate that such a channel does connect the catalytic site with the Mg^{2+} site in the *R. sphaeroides* C*c*O.[83] Because the water molecules exit towards the membrane *P*-side, the water channel could also be used for proton transport out of C*c*O.[83]

3.3   Structural Requirements for Combining Rapid O_2 Delivery to the Catalytic Site with Gating of Protons

As discussed above, molecular O_2 is transferred to haem a_3 along a specific trajectory. Wikström and colleagues found that replacement of V279 (*P. denitrificans* amino-acid residue numbering), located in a putative O_2 channel near the catalytic site, by an Ile in the *P. denitrificans* C*c*O, resulted in a decrease of the O_2-binding rate to haem a_3 by a factor of ~30 (ref. 43, see also 44). This result indicates that introduction of steric hindrance in the O_2-channel results in slowed transfer of small gas molecules into the catalytic site. An even more dramatic effect was observed upon introduction of large residues at the G283 site (see Figure 2a), located in the O_2 channel ~8 Å from the haem a_3 iron and 6 Å (towards the catalytic site) from a narrow segment of the channel confined between residues F282 and W172. For example, in a mutant C*c*O in which G283 was replaced by a Val, binding of both O_2 and CO were slowed by several orders of magnitude.[45] Once the CO ligand was bound to haem a_3, it was trapped within the catalytic site. These results indicate that the C*c*O structure is highly rigid in a domain of the protein through which O_2 is transferred and that small gas molecules cannot be transferred unspecifically by way of structural fluctuations *via* the protein matrix.[45] The area where G283 is found overlaps with the 'gating' region of the C*c*O, which requires a rigid protein structure to prevent unspecific proton back-leaks. Consequently, because the proton-gating region is located in close proximity to the catalytic site, and this region must be rigid, O_2 diffusion cannot be supported by fluctuations of the protein matrix and the protein must provide a specific channel for the substrate O_2. In other words, the finding of specific O_2 channels in C*c*Os might be a consequence of the necessity to provide rapid access for the O_2 molecules while maintaining a high degree of rigidity to control the proton flux.[45]

4   Electron and Proton Transfer During C*c*O Turnover

4.1   Reduction of the Oxidised C*c*O in the Absence of O_2

The reduction of haem a_3 and Cu_B ($O^{0(4)} \rightarrow R^2$ in Figure 3) results in the uptake of two protons by the mitochondrial and *R. sphaeroides* C*c*Os,[84–87] but larger numbers have been reported for the *P. denitrificans* C*c*O.[88] The number of protons taken up upon

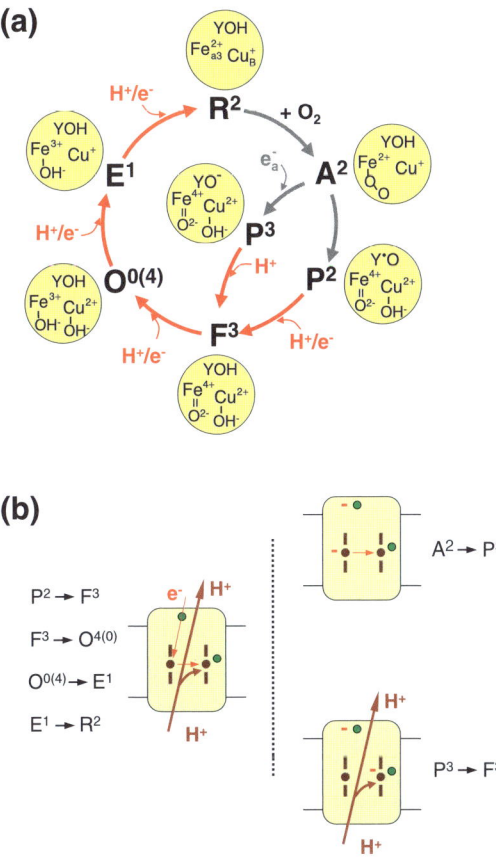

Figure 3 *The catalytic cycle of cytochrome c oxidase. (a) The intermediate states formed during the reaction are indicated by one-letter codes (see the 'abbreviations' section), where the superscript indicates the number of electrons transferred to the catalytic site. The reaction sequence along the outer circle is that observed during turnover when electrons are donated one at a time from cytochrome c. The reaction steps marked with red arrows are presumably associated with proton pumping (see text for details). The electrons (e⁻) marked in red are donated by cytochrome c. The reaction pathway within the cycle is that observed upon reaction of the fully reduced cytochrome c oxidase with O₂. In this case the third electron is transferred to the catalytic site from haem a (shown in grey) and the fourth electron is transferred from Cuₐ (not shown explicitly in the figure). The chemical structures of the reaction intermediates at the catalytic site are shown in the yellow circles. Two hydroxides are assumed to be bound at the catalytic site upon oxidation. One of these hydroxides presumably dissociates spontaneously if the enzyme is left oxidised for a prolonged time (see[133,186,127,187]). YOH is Y288. Water molecules that are released during the reaction are not shown. (b) In all transitions associated with proton pumping during turnover an electron is transferred through Cuₐ and haem a to the catalytic site (left side). During reaction of the fully reduced cytochrome c oxidase with O₂, formation of the P³ state is associated with electron transfer to the catalytic site. The following P³ → F³ transition involves only proton transfer to the catalytic site and this transition is associated with proton pumping*

reduction of haem a is less clear and values ranging from $\sim<0.5$ H^+ [84,89] to ~1 H^+ [78,85,88,90] have been reported. The midpoint potential of Cu_A is not pH-dependent,[91] and reduction of this site is not associated with proton uptake from the bulk solution[89] (but see[90]).

The reduction of the catalytic site is also associated with proton pumping, which is discussed in detail below.

Mutation of residues K362, S299, E101 (subunit II), T359 and T352 in or near the K-pathway result in a slowed reduction of the oxidised binuclear centre,[60,88,92–97] which was suggested to indicate that the proton uptake accompanying reduction of the site takes place through the K-pathway. The involvement of the K-pathway in proton transfer upon reduction of the binuclear centre is evident from experiments in which the reverse reaction, *i.e.* oxidation of haem a_3 was studied. After flash-induced dissociation of the CO ligand from the mixed-valence CcO there is first rapid electron transfer from haem a_3 to haem a (*i.e.* in the opposite direction compared to the normal turnover). This electron transfer is followed by slower proton release to the bulk solution on the ms timescale due to the oxidation of haem a_3. This proton release is slowed or impaired in mutant CcOs in which residues in the K-pathway have been modified.[67,93,98,99]

4.2 Reaction of the Reduced CcO with O_2

Understanding the mechanisms of O_2 reduction, electron and proton transfer, and proton pumping by CcO requires investigation of the reaction of the reduced CcO with O_2. Because most of the reaction steps in this reaction are faster than the mixing time of a conventional stopped-flow apparatus, a 'trick' that is called the 'flow-flash' technique must be used. In this method, the CcO is first reduced by two to four electrons in the absence of O_2, but in the presence of carbon monoxide, which binds to the reduced haem a_3. When the CcO–CO complex is mixed with an O_2-saturated solution, the rate of the reaction with O_2 is determined by the CO-off rate, which is relatively slow ($\tau \cong$ 30 s). Consequently, if the CO ligand is rapidly dissociated by means of a short light flash, allowing O_2 to bind, the partial reaction steps during reaction of the reduced CcO with O_2 can be resolved and followed in time using various spectroscopic techniques.

4.2.1 Reaction of the Two-electron Reduced CcO with O_2

When the catalytic site is reduced (state R^2) haem a_3 binds O_2 to form the haem a_3–O_2 complex (state A^2) with a time constant of ~30 μs (at 1 mM O_2). After binding, the O_2 molecule is partly reduced to form the so-called 'peroxy' state (P^2, also called P_M^{100}) with a time constant of ~300 μs.[101–105] The name 'peroxy' is used for historical reasons even though in this state the O–O bond is broken[106,107] and a ferryl ($Fe^{4+} = O^{2-}$) state is found at the catalytic site (see Figure 3).[108–112] Breaking of the O–O bond requires a total of four electrons, of which three originate from haem a_3 and Cu_B ($Fe_{a3}^{2+} \rightarrow Fe_{a3}^{4+}$, $Cu_B^+ \rightarrow Cu_B^{2+}$) and the fourth is presumably donated by the Y288 residue[47,112-114], which upon oxidation would form a radical (see also[115]). In other words, upon forming P^2 the two electrons transferred to the catalytic site when forming R^2 are used and two electrons are 'borrowed' from groups at the catalytic site. The $R^2 \rightarrow A^2 \rightarrow P^2$ transitions are not associated with any proton uptake from solution,[116] nor with any significant charge translocation across the membrane[117] and the reaction is only slightly exergonic.[118,119] As suggested by Babcock,[120] reduction of O_2 with four electrons in one step ($A^2 \rightarrow P^2$) assures that the release of potentially harmful partly reduced oxygen intermediates is prevented.

4.2.2 Reaction of the Four-electron Reduced C*c*O with O_2

When using the flow-flash technique with the fully reduced C*c*O, dissociation of the CO ligand is followed by binding of O_2 to the reduced haem a_3 with a time constant of ~ 10 μs (at 1 mM O_2). However, in contrast to the scenario when only haem a_3 and Cu_B are reduced (cf. the P^2 state discussed above), when starting from the fully reduced C*c*O, the O–O bond breakage is associated with simultaneous electron transfer from haem a to the catalytic site[87,103,121–127] with a time constant of ~50 μs. The state that is formed is also called 'peroxy', but in this case there is one more electron at the catalytic site as compared to P^2 and therefore we call this state P^3 (also called P_R^{100}). The $R^2 \rightarrow P^3$ transition is not associated with any significant charge displacement across the membrane,[128] nor is it associated with any proton uptake from solution.[87,129–132] Thus, because on the timescale of P^3 formation the negative charge transferred from haem a to the catalytic site is uncompensated, there is one more negative charge at the catalytic site in the P^3 state as compared to all other states that are discussed here (see Figure 3).

After formation of P^3, a proton is taken up from the bulk solution, neutralising the excess negative charge in the P^3 state and forming the F^3 (oxo-ferryl intermediate,[127,133]) state with a time constant of ~100 μs.[87,129–132] In the last reaction step the fourth electron is transferred to the catalytic site concomitantly with proton uptake from the bulk solution through the D-pathway forming state $O^{4(0)}$ with a time constant of ~1 ms. Proton uptake during both the $P^3 \rightarrow F^3$ and $F^3 \rightarrow O^{4(0)}$ transitions takes place through the D-pathway[60,61] and the transitions are associated with charge translocation across the membrane.[128,134,135] Below, we discuss the electron and proton-transfer reactions during these transitions in detail, in order to establish the basis for a discussion on the mechanism of proton pumping in the following sections.

4.3 The $P^3 \rightarrow F^3$ Transition

The $P^3 \rightarrow F^3$ transition (right side in Fig. 3b, $\tau \cong 100$ μs) is of special interest because in contrast to the other reaction steps discussed above (left side in Fig. 3b), it is not linked to electron transfer to the catalytic site, yet it is associated with proton pumping. Thus, even though the transition is associated with a *net* uptake of one proton, because it involves proton pumping on the same timescale, two protons are taken up and one is released on the 100 μs timescale. In this context one important question is which of the proton-transfer reactions is rate limiting for the overall reaction.

When proton transfer is blocked at the entry point of the D-pathway, for example by replacement of D132 by Asn, the F^3 state is still formed, but it is not associated with proton uptake from solution.[136–138] Because formation of F^3 requires proton transfer to the catalytic site, an internal group, presumably E286, has been suggested to act as a proton donor in this mutant C*c*O. Under these conditions the F^3 formation rate is the same as that in the wild-type CcO ($\tau \cong 100\ \mu s$), which indicates that the rate-determining step of the $P^3 \rightarrow F^3$ transition is the proton transfer from E286 to the catalytic site, *i.e.* the transfer of a substrate proton. This observation seemingly contradicts one of the basic requirements for the function of a proton pump (see Introduction), namely that a pumped proton should be taken up before the substrate proton. This problem is discussed in detail below.

The results discussed above indicate that proton transfer through the D-pathway takes place in two distinct steps: first a proton is transferred from E286 (or water molecules around E286) to the catalytic site with a time constant of ~100 µs ($k_H^0 \cong 10^4\ s^{-1}$), which determines the maximum rate constant of the transition. The proton transfer is then followed by rapid re-protonation of E286 from the bulk solution, through D132. The observed $P^3 \rightarrow F^3$ rate, k_{PF}, decreases with increasing pH above pH ~9, which most likely reflects the apparent pK_a of E286 (pK_{EH}). It has been shown that the observed transition rate is adequately modelled by the fraction of protonated E286, α_{EH}, times k_H^0:

$$k_{PF}(\text{pH}) = k_H^0 \cdot \alpha_{EH}(\text{pH}) \tag{2a}$$

$$\alpha_{EH}(\text{pH}) = \frac{1}{1 + 10^{\text{pH}-\text{p}K_{EH}}} \tag{2b}$$

where the pK_a of E286 has been determined to be 9.4 from the pH dependence of k_{PF}.[131] The proton-transfer rate from E286 to the catalytic site (k_H^0) is probably determined by the reorganisation of water molecules or a conformational change of the E286 side chain (see below). The observation of a high pK_a of E286 is consistent with results from experiments using FTIR spectroscopy, which show that the pK_a of the corresponding glutamate in haem–copper oxidases from different species is >9,[139–144] even though the results from some studies indicate that the pK_a may shift to lower values during turnover.[143,145]

With the same rate constant as the $P^3 \rightarrow F^3$ transition at the catalytic site, the fourth electron residing at Cu$_A$ equilibrates with haem *a*, which results in reduction of ~50% of haem *a*. The extent of this electron transfer varies between CcOs from different species, which reflects differences in the relative midpoint potentials of haem *a* and Cu$_A$.[87] In addition, the electron equilibrium between Cu$_A$ and haem *a* is sensitive to the protonation state of E286; if E286 is not re-protonated after the proton transfer to the catalytic site (*e.g.* in the D132N mutant CcO, see above), the electron is not transferred from Cu$_A$ to haem *a* on the timescale of the $P^3 \rightarrow F^3$ transition.[136,146] This absence of electron transfer is presumably due to electrostatic interactions between E286 and haem *a*, which result in lowering of the haem *a* midpoint potential when the Glu is deprotonated and hence negatively charged.

4.4 The $F^3 \rightarrow O^{4(0)}$ Transition

The $F^3 \rightarrow O^{4(0)}$ transition involves electron transfer from the haem *a*–Cu$_A$ equilibrium and proton transfer to the catalytic site. The rate of this transition, k_{FO}, has been modelled in terms of the fraction of the electron residing at the catalytic site ($F^{3(-)}$, the electron equilibrates between Cu_A, haem a and catalytic site), where the electron is trapped by proton transfer from E286:[87,147,148]

$$a^- F^3 \overset{K}{\longleftrightarrow} aF^{3(-)} \overset{k_H^0}{\longrightarrow} aO^4 \qquad (3)$$

where the equilibrium constant K reflects the electron equilibrium between haem a and the catalytic site in the F state (Cu_A is omitted for simplicity) and k_H^0 is the proton-transfer rate from E286 to the catalytic site (~10^4 s^{-1}, see above). The value of K is ~0.05, which means that the $F^{3(-)}$ state is populated to ~5%.[148]

As with the $P^3 \rightarrow F^3$ transition, the $F^3 \rightarrow O^{4(0)}$ rate decreases with increasing pH. However, in contrast to the $P^3 \rightarrow F^3$ transition the pH dependence in the range 6–11 cannot be fitted with a simple titration (Eq. 2) and the data are consistent with a titration curve displaying two pK_as of ~6.3 and ~8.9.[148,149] The higher pK_a (8.9) is similar to that seen in the pH dependence of the $P^3 \rightarrow F^3$ rate (9.4) and it is attributed to titration of E286. The lower pK_a (6.3), not observed in the pH dependence of the $P^3 \rightarrow F^3$ transition rate, is attributed to the electron equilibration component of the $F^3 \rightarrow O^{4(0)}$ rate, *i.e.* a pH dependence in the fraction reduced catalytic site (see Eq. 3).[148,149]

In a mutant CcO in which one of the Arg residues, R481, that forms a salt bridge with the haem D-propionates (Figure 2c), is replaced by a Lys (R481K), at pH >7, the pH-dependence of the rate constant for the $F^3 \rightarrow O^{4(0)}$ transition follows that of the wild-type CcO, while below pH 7 the rate is pH independent. These results have been interpreted to indicate that the protonation state of the R481–propionate cluster determines the $F^3 \rightarrow O^{4(0)}$ transition rate such that the rate increases upon protonation of a group within the cluster with an apparent pK_a of ~6.3. In addition, the low-pH titration is not observed with a mutant CcO in which proton pumping is uncoupled from the O_2-reduction reaction (N139D, Namslauer *et al.* unpublished), which suggests involvement of the haem D-propionates in proton pumping.

5 Proton Pumping

As discussed in the previous sections, when starting with the two-electron reduced CcO, the binding of O_2 to the reduced catalytic site (state A^2) and formation of P^2 are not associated with any proton uptake from solution, nor any voltage changes across the membrane. In other words, it is clear that the $R^2 \rightarrow A^2 \rightarrow P^2$ transitions are not associated with proton pumping. In addition, two protons are pumped upon oxidation of the reduced CcO, where one proton is pumped during each of the $P^2 \rightarrow F^3$ and $F^3 \rightarrow O^{4(0)}$ transitions. Thus, the remaining transitions (see Figure 3), $O^{0(4)} \rightarrow E^1 \rightarrow R^2$, are presumably linked to the pumping of one proton in each transition.[150–153] However, experimental studies of proton pumping in these reactions appear to be complicated by the presence of different states $O^{0(4)}$ (ref. 151). For example, it was noted about 30 years ago that there are functional differences between the 'as isolated' oxidised CcO and enzyme that has been reduced and then re-oxidised ('pulsed enzyme').[154,155] Results from more recent studies indicate that protons are pumped upon reduction of CcO, but only if the reduction reaction is immediately preceded by oxidation of the reduced CcO.[150,151] Consequently, it has been suggested that the $O^{4(0)}$ state can exist in either a 'relaxed' or a 'high-energy' form such that reduction of only the high-energy form results in proton pumping.[150,151] This issue is, however, controversial and Michel and colleagues recently presented results which suggest that the transfer of the second electron to the binuclear centre is associated with proton pumping without the need of a preceding turnover[153] (but see[152]).

Irrespective of the conditions under which proton pumping occurs upon reduction of the oxidised CcO, a common property of all the transitions associated with proton pumping, except the $P^3 \rightarrow F^3$ transition, is that the pumping occurs in reaction steps where an electron is transferred from cytochrome c to the catalytic site, through Cu_A and haem a, accompanied by a net uptake of one proton to the catalytic site (substrate proton) per electron (see Figure 3b), i.e. during the $O^0 \rightarrow E^1$, $E^1 \rightarrow R^2$, $P^2 \rightarrow F^3$ and $F^3 \rightarrow O^{4(0)}$ transitions. This scenario presumably means that a series of identical events, each associated with pumping of one proton, are repeated for each of the transitions, the only difference being the identity of the electron acceptor at the catalytic site in the different reaction steps (see[151]).

5.1   Proton Pumping During the $P^3 \rightarrow F^3$ Transition

It has already been mentioned above that the $P^3 \rightarrow F^3$ transition is unique in that it is associated with proton transfer, but not electron transfer to the catalytic site. Thus, the observation of proton pumping in this transition puts specific restrictions on the mechanism by which CcO pumps protons.

Early studies of proton ejection to the outside (P-side) of small unilamellar vesicles (SUVs) with reconstituted mitochondrial CcO showed that (pumped) protons were only released during the $F^3 \rightarrow O^{4(0)}$ transition ($\tau \cong 1$ ms) and no proton release was observed on the 100 µs timescale, i.e. that of the $P^3 \rightarrow F^3$ transition.[129] Later studies of charge translocation across the CcO–SUV membrane showed that the $P^3 \rightarrow F^3$ and $F^3 \rightarrow O^{4(0)}$ transitions are associated with equal charge transfer across the membrane, interpreted as proton pumping in each of the two transitions.[128,156] In a recent study we showed that 'pumped' protons are taken up from the inside solution of CcO–SUVs and released to the outside (P-side) bulk solution with the same time constant as that of the $P^3 \rightarrow F^3$ transition in *R. sphaeroides* CcO reconstituted in SUVs.[157]

In addition, proton pumping has been investigated in the M263L mutant CcO (Met-263 in subunit II is a ligand to Cu_A)[158,159] in which the Cu_A midpoint potential is increased significantly and, therefore, Cu_A stays reduced during the $P^3 \rightarrow F^3$ transition.[146] Thus, proton pumping on the timescale of the $P^3 \rightarrow F^3$ transition is not linked to any internal electron transfer. Even though the $P^3 \rightarrow F^3$ transition might be unique for the reaction of the fully reduced CcO with O_2, it is generally assumed that the proton-pumping mechanism is the same for each pumping event during turnover of CcO. Therefore, the observation of proton pumping in the absence of any electron transfer puts specific restrictions on possible mechanisms by which CcO, and most likely all proton-pumping haem–copper oxidases, pumps protons. In the last section of this review we discuss this observation in the framework of a molecular model for proton pumping.

5.2   Uncoupled Mutant Forms of Cytochrome *c* Oxidase

Studies of mutant CcOs which are capable of reducing O_2 to H_2O but do not pump protons, referred to as 'uncoupled' CcOs, have provided important mechanistic information. Understanding the origin of the uncoupling in these mutant CcOs at a molecular level is expected to provide important insights into the proton pumping machinery of CcO.

There are two main classes of uncoupled mutant CcOs: in one class the uptake of both substrate and pumped protons is inhibited to different degrees leading to slow catalytic turnover rates, whereas in the second class, only the uptake of pumped protons is impaired and the uptake of the substrate protons and the turnover rate is essentially unaffected.

One example of members of the first class is the mutant CcOs in which D132, at the entrance of the D-pathway, is replaced by a non-protonatable residue (*e.g.* by Asn or Ala[48,49,58,59]). The turnover activity of these mutant CcOs is substantially decreased and they do not pump protons. For example in the D132N mutant CcO, proton transfer into the D-pathway is dramatically slowed from <100 µs to ~1 s. The uncoupling is most likely a direct effect of the slowed proton uptake. In fact, in this mutant CcO the turnover activity increases upon increasing the membrane potential, which is interpreted in terms of proton uptake from the *P*-side of the membrane in the presence of a membrane potential.[75] Within the framework of a general design of a proton pump, this observation would indicate that the impaired proton uptake from the *N*-side leaves the pumping element in an unprotonated state during turnover such that there is time for its protonation from the *P*-side.

Another mutant CcO with similar behaviour, is one in which an Asp has been inserted in the vicinity of E286, at the position of S197, ~7Å from the carboxylate group of E286. In the S197D mutant CcO the steady-state activity is ~35% of that of the wild type CcO and the mutant CcO is not able to pump protons. Instead ~0.2 protons per electron are taken up from the *P*-side of the membrane during turnover in experiments in which the mutant CcO has been reconstituted in liposomes (Namslauer *et al.*, unpublished results).

The second class of uncoupled mutant CcOs, where turnover rates are unaffected, is perhaps the more interesting one from a mechanistic point of view. In CcO from *P. denitrificans* the introduction of an (acidic) Asp residue at the position of N131 (*P. denitrificans* CcO numbering) in the D-pathway results in a mutant CcO that is fully active (*i.e.* can reduce O_2 to H_2O with the same rate as the wild-type CcO), but it does not pump protons.[160] Essentially the same phenotype is observed with the corresponding mutant in the *R. sphaeroides* CcO even though in this case the N139D (*R. sphaeroides* CcO numbering) mutant CcO displays approximately two times higher steady-state activity than the wild-type enzyme.[161–163] A detailed analysis of the kinetic and thermodynamic properties of proton-transfer reactions through the D-pathway shows that the proton uptake rate from solution during the $P^3 \rightarrow F^3$ transition is the same as that of the wild-type CcO, *i.e.* the uncoupling is not due to inhibition of proton uptake through the D-pathway.[163] Instead the uncoupling has been rationalised by an increase in the pK_a of E286 by about two pH units.[163] This observation is discussed in detail below in the framework of a putative molecular mechanism for proton pumping.

5.3   The Role of E286

The results discussed in this review clearly show that the residue E286 plays a key role in proton transfer through the D-pathway. It is likely that the residue is a branch point from where protons are transferred either to the catalytic site (substrate protons) or towards an acceptor site located closer to the *P*-side of the enzyme.[19,46,77,164] In the X-ray crystal structure of the wild-type *R. sphaeroides* CcO at pH 6, E286 is protonated and the COOH group is hydrogen bonded to the M107 carbonyl oxygen[27] (see Figure 4). This hydrogen bond is broken and the conformation of the E286 side chain is altered in the E286Q mutant CcO. The same structural changes are observed also with the wild-type CcO at pH 10, where E286 is at least partly unprotonated. The structural changes around the E286 site appear to be linked to structural changes around the D-ring propionates of haems *a* and a_3. In the wild-type CcO there is a salt bridge between R481 and the D-ring propionate of haem a_3, which stabilises the propionate in its unprotonated form. In addition, R481 interacts with the D-ring propionate of haem *a*, which also forms a salt bridge with R482 (Figure 2c). In the E286Q mutant CcO and in the wild-type CcO at pH 10 the distance between the D-ring propionate of haem a_3 and R481 is increased, a water molecule enters between the two groups and the positions of other water molecules around the site change. These rearrangements result in breaking of the salt bridge between R481 and the D-ring propionate of haem a_3. Even though these structural changes are observed with the mutant CcO or with the wild-type CcO at pH 10, with the wild-type CcO at lower pH the E286 becomes transiently deprotonated every time a proton is transferred through the D-pathway to the catalytic site.

A structural change linked to the deprotonation of E286 is also supported by results from studies of the kinetic deuterium isotope effect of the $F^3 \rightarrow O^{4(0)}$ transition rate[147,165] and theoretical studies.[53,54] In addition, results from studies using FTIR spectroscopy indicate that there is a link between ligand binding at the catalytic site or redox reactions, and structural changes or changes in the protonation state of E286.[139–145,166–168]

Even though E286 is an essential residue in respiratory oxidases in which it is present, there are haem–copper oxidases that pump protons, but in which the Glu is not conserved (reviewed in[169]). Different combinations of other residues that can act as proton donors are found in these enzymes at the same approximate location as that of E286 in the *R. sphaeroides* CcO. In the *P. denitrificans* and *R. sphaeroides* enzymes, E286 has been replaced by combinations of residues mimicking the structural features of such alternative arrangements of residues, without loss of proton pumping.[82,170] Thus it is clear that other residues than E286 may act as transient proton donors to the catalytic site, and it is likely that independently of the identity of residues at the location of E286, the proton pumping mechanism is the same in all these haem–copper oxidases.

6   Molecular Mechanisms for Proton Pumping

Many molecular models describing proton pumping by CcO have been suggested, especially after publication of the high-resolution structures of the enzyme.[16,17,24,78,171–182] Because during CcO turnover electrons are transferred one by one from cytochrome *c*,

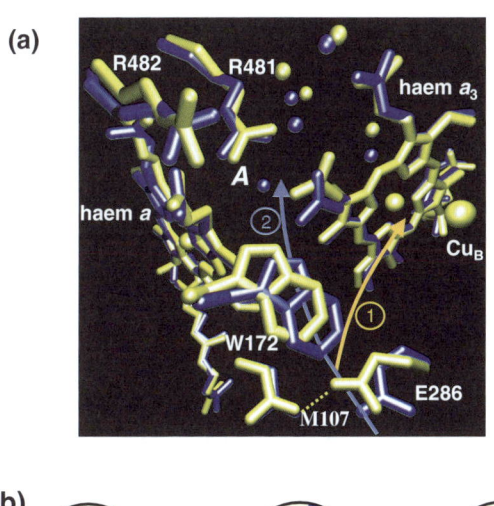

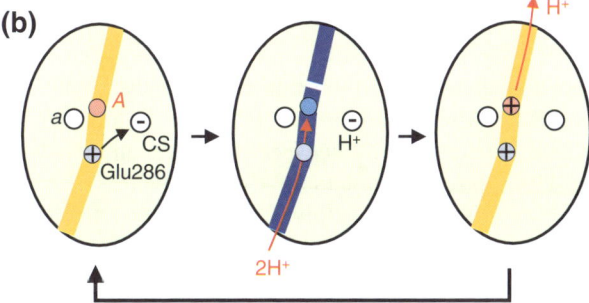

Figure 4 *A model for proton pumping by cytochrome c oxidase. The protein is initially in the relaxed state (yellow structure in (a)). After electron transfer to the catalytic site (CS) a proton is transferred from E286. Upon deprotonation of E286 the hydrogen bond between the E286 side chain and M107 is lost, which results in a local structural change around the Arg481/482–haem propionate cluster (proton acceptor, A) (blue structure in (a)). This structural change results in an increase in the pK$_a$ of A and rearrangement of water molecules around A such that A is now in slow equilibrium with the P-side (white line in the proton-transfer pathway in the middle picture in (b)). The increase in the pK$_a$ of A results in its protonation from the N-side, through the D-pathway, and reprotonation of E286. This reprotonation of E286 is followed by relaxation of the structure and proton release from A to the P-side. The free energy associated with the reaction at the catalytic site is conserved by way of changes in the pK$_a$ of A, coupled to deprotonation of E286. The numbers 1 and 2 in (a) indicate the sequence of the proton-transfer events. The picture in (a) is modified from ref 183 and that in (b) from ref. 14*

through Cu_A and haem *a*, to the catalytic site, many models involve the consecutive reduction and oxidation of one of the redox sites as a key step in the sequence of events that lead to proton translocation.

However, one specific reaction step that has been difficult to accommodate in such a scheme is the $P^3 \rightarrow F^3$ transition in which no internal electron transfer to the catalytic site takes place. As discussed in detail above, the transition is associated only with proton uptake through the D-pathway where both the substrate and pumped protons are taken up at the same rate. This rate is determined by the proton transfer from E286 to the catalytic site, *i.e.* an internal transfer of a substrate proton. To accommodate these observations in the framework of a model for proton pumping, we suggested recently that the series of events leading to proton pumping are triggered by the transfer of a substrate proton to the catalytic site rather than by electron transfer[14,183] (Figure 4). In this model, there is an acceptor for pumped protons, called *A*, assumed to be the R481/R482–haem propionate cluster.[19,46,51,75,77,78,164,175,176,184] Initially, *A* is in rapid equilibrium with the *P*-side, in a slow equilibrium with the *N*-side, and has a low pK_a (lower than the pH on the *P*-side). In the first step, electron transfer from haem *a* to the catalytic site (forming state P^3) is followed by proton transfer from E286, leaving E286 in a transiently deprotonated state. The deprotonation of E286 results in reorientation of the E286 side chain, which triggers a local structural change that propagates to the R481/R482–haem propionate cluster, *i.e.* the proton acceptor, *A* (cf. structures of the E286Q mutant CcO and of the wild-type CcO at high pH, discussed above). Upon the structural change, the pK_a of *A* increases, the connection to the *P*-side is broken and *A* is now in rapid equilibrium with the *N*-side. These structural changes result in protonation of *A* from the *N*-side and simultaneous re-protonation of E286. After re-protonation, E286 relaxes to the initial state, the pK_a of *A* decreases and the connectivity between the group and the *P*-side is re-established, which results in proton release to the *P*-side.

As is evident from this discussion, this scenario requires that the proton-transfer rates to and/or from *A* change between the two states, which may be accomplished through the rearrangement of the water structure observed around *A*. One such change is associated with the increase in the distance between R481 and the haem a_3 D-ring propionate, and the entering of a water molecule between these two groups.[183] This event may contribute to the increase in the pK_a of *A*; for example, if *A* is the propionate group, then an increase in the distance between the positively charged R481 and the propionate would stabilise the protonated state of the propionate. Another key requirement of the model is that *A* becomes protonated before the structural relaxation of E286, which occurs only after the re-protonation of E286. It should be noted that this requirement does not necessarily imply that E286 must be reprotonated after protonation of *A*. The sequence of the protonation events of E286 and *A* is irrelevant given that the structural relaxation is slower than both protonation events.

In one respect this model deviates fundamentally from the general principles underlying the function of redox-driven proton pumps (discussed in Section 2 above) because the substrate proton is transferred to the catalytic site before the uptake of the pumped proton. This deviation is however only apparent because the free energy is conserved by way of changes in the pK_a of *A* upon deprotonation of E286, which triggers the uptake of a pumped proton. Thus, this model can be accommodated within the framework of the general mechanism outlined above if E286 is included as part of the 'catalytic site'.

The model can be considered as a minimal scheme describing the key events. It is likely that during turnover the internal electron-transfer reactions are controlled by the protonation events such that the next electron is not transferred to the catalytic site unless the sequence of events associated with proton pumping of the preceding step is completed.[146,147] An experimental observation in support of this scenario is the electron equilibrium shift towards Cu_A when E286 is unprotonated, which would result in slowing of the electron transfer from cytochrome c (through Cu_A and haem a) to the catalytic site until re-protonation of E286 (*i.e.* the previous pumping event is completed). In addition, results from FTIR studies indicate that changes in the redox state of haem a may be linked to changes in the protonation state of E286.[145]

According to the model, the proton-pumping efficiency depends on the relative rates of proton transfer from the bulk solution to the pump site, A, and to E286, the reprotonation of which is associated with a structural relaxation. These rates depend, for example, on the relative pK_a values of E286 and the pump site in the 'proton-input conformation', as well as on the structural environment of E286. Thus, an increase in the pK_a of E286 or an alteration of the water structure around E286, as in the N139D mutant CcO, would result in an altered proton equilibrium between E286 and the pump site allowing reprotonation and relaxation of the E286 structure before the pumped proton is transferred to the pump site.

7   Final Remarks

The complexity of the reaction mechanism of CcO presents us with a great challenge when aiming at understanding the molecular mechanism of redox-driven proton pumping. Nevertheless, the use of site-directed mutagenesis and spectroscopy together with theoretical calculations have made it possible to identify structural elements involved in electron and proton transfer and the structures of the intermediate states formed at the catalytic site. In addition, it has been possible to identify specific reaction steps that are linked to proton pumping, and reactions that energetically drive proton translocation. In this review we have presented a proton pumping model and experimental data supporting this model. Clearly, however, many issues still remain to be resolved before the detailed catalytic mechanism and its coupling to proton pumping by CcO is fully understood.

Acknowledgement

We would like to thank Gisela Brändén and Lina Salomonsson for preparation of Figure 2.

References

1. A.L. Moore, M.S. Albury, P.G. Crichton and C. Affourtit, *Trends Plant Sci.* 2002, **7**, 478–81.
2. M.K.F. Wikström, *Nature,* 1977, **266**, 271–3.
3. J.A. García-Horsman, B. Barquera, J. Rumbley, J. Ma and R.B. Gennis, *J. Bacteriol.,* 1994, **176**, 5587–600.
4. M. Wikström and K. Krab, *Biochim. Biophys. Acta,* 1979, **549**, 177–22.
5. S. Ferguson-Miller and G.T. Babcock, *Chem. Rev.,* 1996, **96**, 2889–2907.

6. M. Wikström, *Biochim Biophys. Acta,* 1998, **1365**, 185–92.

7. B.G. Malmström, *Biochim. Biophys. Acta,* 1985, **811**, 1–12.

8. B.G. Malmström, *Acta. Physiol. Scand. Suppl.,* 1992, **607**, 209–11.

9. C. Tanford, *Proc. Natl. Acad. Sci. U.S.A.,* 1983, **80**, 3701–5.

10. D.F. Blair, J. Gelles and S.I. Chan, *Biophys. J.,* 1986, **50**, 713–33.

11. P.R. Rich, *Aust. J. Plant Physiol.,* 1995, **22**, 479–86.

12. D.A. Mills and S. Ferguson-Miller, *FEBS Letters,* 2003, **545**, 47–51.

13. M. Wikström, *Biochim. Biophys. Acta,* 2004, **1655**, 241–7.

14. P. Brzezinski, *Trends Biochem. Sci.,* 2004, **29**, 380–7.

15. M. Wikström, K. Krab and M. Saraste, *Cytochrome c oxidase–A synthesis* Academic Press, 1981.

16. M. Wikström, M.I. Verkhovsky and G. Hummer, *Biochim. Biophys. Acta,* 2003, **1604**, 61–5.

17. D.M. Popovic and A.A. Stuchebrukhov, *FEBS Letters,* 2004, **566**, 126–130.

18. F. Kamp, R.D. Astumian and H.V. Westerhoff, *Proc. Natl. Acad. Sci. U.S.A.,* 1988, **85**, 3792–96.

19. S. Iwata, C. Ostermeier, B. Ludwig and H. Michel, *Nature,* 1995, **376**, 660–9.

20. C. Ostermeier, A. Harrenga, U. Ermler and H. Michel, *Proc. Natl. Acad. Sci. U.S.A.,* 1997, **94**, 10547–53.

21. T. Tsukihara, H. Aoyama, E. Yamashita, T. Tomizaki, H. Yamaguchi, K. Shinzawa-Itoh, R. Nakashima, R. Yaono and S. Yoshikawa, *Science,* 1995, **269**, 1069–74.

22. S. Yoshikawa, K. Shinzawa-Itoh, R. Nakashima, R. Yaono, E. Yamashita, N. Inoue, M. Yao, M.J. Fei, C.P. Libeu, T. Mizushima, H. Yamaguchi, T. Tomizaki and T. Tsukihara, *Science,* 1998, **280**, 1723–29.

23. T. Tsukihara, H. Aoyama, E. Yamashita, T. Tomizaki, H. Yamaguchi, K. Shinzawa-Itoh, R. Nakashima, R. Yaono and S. Yoshikawa, *Science,* 1996, **272**, 1136–44.

24. T. Tsukihara, K. Shimokata, Y. Katayama, H. Shimada, K. Muramoto, H. Aoyama, M. Mochizuki, K. Shinzawa-Itoh, E. Yamashita, M. Yao, Y. Ishimura and S. Yoshikawa, *Proc. Natl. Acad. Sci. U.S.A.,* 2003, **100**, 15304–9.

25. J. Abramson, S. Riistama, G. Larsson, A. Jasaitis, M. Svensson-Ek, L. Laakkonen, A. Puustinen, S. Iwata and M. Wikström, *Nature Str. Biol.,* 2000, **7**, 910–7.

26. T. Soulimane, G. Buse, G.P. Bourenkov, H.D. Bartunik, R. Huber and M.E. Than, *EMBO. J.,* 2000, **19**, 1766–76.

27. M. Svensson-Ek, J. Abramson, G. Larsson, S. Törnroth, P. Brzezinski and S. Iwata, *J. Mol. Biol.,* 2002, **321**, 329–39.

28. J. Cao, J. Shapleigh, R. Gennis, A. Revzin and S. Ferguson-Miller, *Gene,* 1991, **101**, 133–7.

29. J. Cao, J. Hosler, J. Shapleigh, A. Revzin and S. Ferguson-Miller, *J. Biol. Chem.,* 1992, **267**, 24273–8.

30. J.P. Shapleigh and R.B. Gennis, *Mol. Microbiol.,* 1992, **6**, 635–42.

31. J.P. Hosler, J. Fetter, M.M. Tecklenburg, M. Espe, C. Lerma and S. Ferguson-Miller, *J. Biol. Chem.,* 1992, **267**, 24264–72.

32. B. Kadenbach, *J. Bioenerg. Biomembr.,* 1986, **18**, 39–54.

33. J.W. Taanman, P. Turina and R.A. Capaldi, *Biochemistry,* 1994, **33**, 11833–41.

34. T. Haltia, M. Saraste and M. Wikström, *EMBO J.,* 1991, **10**, 2015–21.

35. R.W. Hendler, K. Pardhasaradhi, B. Reynafarje and B. Ludwig, *Biophysical Journal*, 1991, **60**, 415–23.

36. M.R. Bratton, L. Hiser, W.E. Antholine, C. Hoganson and J.P. Hosler, *Biochemistry*, 2000, **39**, 12989–95.

37. F. Malatesta, G. Antonini, P. Sarti and M. Brunori, *Biochem. J.*, 1986, **234**, 569–72.

38. D.A.Thompson, L. Gregory and S. Ferguson-Miller, *J. Inorg. Biochem.*, 1985, **23**, 357–64.

39. J.P. Hosler, *Biochim. et Biophys. Acta*, 2004, 332–9.

40. G.Gilderson, L. Salomonsson, A. Aagaard, J. Gray, P. Brzezinski and J. Hosler, *Biochemistry*, 2003, **42**, 7400–9.

41. D.A. Mills, Z. Tan, S. Ferguson-Miller and J. Hosler, *Biochemistry*, 2003, **42**, 7410–7.

42. M.R. Bratton, M.A. Pressler and J.P. Hosler, *Biochemistry*, 1999, **38**, 16236–45.

43. S. Riistama, A. Puustinen, M.I. Verkhovsky, J.E. Morgan and M. Wikström, *Biochemistry*, 2000, **39**, 6365–72.

44. S. Riistama, A. Puustinen, A. Garcia-Horsman, S. Iwata, H. Michel and M. Wikström, *Biochim. Biophys. Acta*, 1996, **1275**, 1–4.

45. L. Salomonsson, A. Lee, R.B. Gennis and P. Brzezinski, *Proc. Natl. Acad. Sci. U.S.A.*, 2004, **101**, 11617–21.

46. I. Hofacker and K. Schulten, *Proteins*, 1998, **30**, 100–7.

47. R.B. Gennis, *Biochim. Biophys. Acta*, 1998, **1365**, 241–8.

48. J.W. Thomas, A. Puustinen, J.O. Alben, R.B. Gennis and M. Wikström, *Biochemistry*, 1993, **32**, 10923–8.

49. J.R. Fetter, J. Qian, J. Shapleigh, J.W. Thomas, A. García-Horsman, E. Schmidt, J. Hosler, G.T. Babcock, R.B. Gennis and S. Ferguson-Miller, *Proc. Natl. Acad. Sci. U.S.A.*, 1995, **92**, 1604–8.

50. J.A. García-Horsman, A. Puustinen, R.B. Gennis and M. Wikström, *Biochemistry*, 1995, **34**, 4428–33.

51. S. Riistama, G. Hummer, A. Puustinen, R.B. Dyer, W.H. Woodruff and M. Wikström, *FEBS Lett.*, 1997, **414**, 275–80.

52. X. Zheng, D.M. Medvedev, J. Swanson and A.A. Stuchebrukhov, *Biochim. Biophys. Acta*, 2003, **1557**, 99–107.

53. R.I. Cukier, *Biochim. Biophys. Acta*, 2004, 189–202.

54. E. Olkhova, H. Michel, M.C. Hutter, M.A. Lill and V. Helms, *Biophysical Journal*, 2004, 1873–89.

55. J.W. Thomas, M.W. Calhoun, L.J. Lemieux, A. Puustinen, M. Wikström, J.O. Alben and R.B. Gennis, *Biochemistry*, 1994, **33**, 13013–21.

56. D.M. Mitchell, R. Aasa, P. Ädelroth, P. Brzezinski, R.B. Gennis and B.G. Malmström, *FEBS Lett.*, 1995, **374**, 371–4.

57. M. Svensson-Ek, J.W. Thomas, R.B. Gennis, T. Nilsson and P. Brzezinski, *Biochemistry*, 1996, **35**, 13673–80.

58. M.L. Verkhovskaya, A. García-Horsman, A. Puustinen, J.L. Rigaud, J.E. Morgan, M.I. Verkhovsky and M. Wikström, *Proc. Natl. Acad. Sci. U.S.A.*, 1997, **94**, 10128–31.

59. U. Pfitzner, A. Odenwald, T. Ostermann, L. Weingard, B. Ludwig and O.M.H. Richter, *J. Bioenerg. Biomembr.,* 1998, **30**, 89–97.

60. A.A. Konstantinov, S. Siletsky, D. Mitchell, A. Kaulen and R.B. Gennis, *Proc. Natl. Acad. Sci. U.S.A.,* 1997, **94**, 9085–90.

61. P. Ädelroth, M. Svensson-Ek, D.M. Mitchell, R.B. Gennis and P. Brzezinski, *Biochemistry,* 1997, **36**, 13824–9.

62. N.J. Watmough, A. Katsonouri, R.H. Little, J.P. Osborne, E. Furlong-Nickels, R.B. Gennis, T. Brittain and C. Greenwood, *Biochemistry,* 1997, **36**, 13736–42.

63. S. Jünemann, B. Meunier, N. Fisher and P.R. Rich, *Biochemistry,* 1999, **38**, 5248–55.

64. J.P. Hosler, S. Ferguson-Miller, M.W. Calhoun, J.W. Thomas, J. Hill, L. Lemieux, J. Ma, C. Georgiou, J. Fetter, J. Shapleigh, M.M.J. Tecklenburg, G.T. Babcock and R.B. Gennis, *J. Bioenerg. Biomembr.,* 1993, **25**, 121–36.

65. J.P. Hosler, J.P. Shapleigh, D.M. Mitchell, Y. Kim, M.A. Pressler, C. Georgiou, G.T. Babcock, J.O. Alben, S. Ferguson-Miller and R.B. Gennis, *Biochemistry,* 1996, **35**, 10776–83.

66. R.I. Cukier, *Biochim. Biophys. Acta,* 2005, **1706**, 134–46.

67. M. Brändén, F. Tomson, R.B. Gennis and P. Brzezinski, *Biochemistry,* 2002, **41**, 10794–8.

68. F.L. Tomson, J.E. Morgan, G. Gu, B. Barquera, T.V. Vygodina and R.B. Gennis, *Biochemistry,* 2003, **42**, 1711–7.

69. J.X. Ma, P.H. Tsatsos, D. Zaslavsky, B. Barquera, J.W. Thomas, A. Katsonouri, A. Puustinen, M. Wikström, P. Brzezinski, J.O. Alben and R.B. Gennis, *Biochemistry,* 1999, **38**, 15150–6.

70. A. Kannt, C. Roy, D. Lancaster and H. Michel, *Biophys. J.,* 1998, **74**, 708–721.

71. A. Kannt, C.R. Lancaster and H. Michel, *J. Bioenerg. Biomembr.,* 1998, **30**, 81–7.

72. O.M. Richter, K.L. Durr, A. Kannt, B. Ludwig, F.M. Scandurra, A. Giuffre, P. Sarti and P. Hellwig, *FEBS J.,* 2005, **272**, 404–12.

73. S. Yoshikawa, K. Shinzawa-Itoh and T. Tsukihara, *J. Bioenerg. Biomembr.,* 1998, **30**, 7–14.

74. H.M. Lee, T.K. Das, D.L. Rousseau, D. Mills, S. Ferguson-Miller and R.B. Gennis, *Biochemistry,* 2000, **39**, 2989–96.

75. D.A. Mills, L. Florens, C. Hiser, J. Qian and S. Ferguson-Miller, *Biochim. Biophys. Acta,* 2000, **1458**, 180–7.

76. D.A. Mills and S. Ferguson-Miller, *Biochim. Biophys. Acta,* 1998, **1365**, 46–52.

77. A. Puustinen and M. Wikström, *Proc. Natl. Acad. Sci. U.S.A.,* 1999, **96**, 35–7.

78. H. Michel, *Proc. Natl. Acad. Sci. U.S.A.,* 1998, **95**, 12819–24.

79. L. Florens, B. Schmidt, J. McCracken and S. Ferguson-Miller, *Biochemistry,* 2001, **40**, 7491–7.

80. S. Ferguson-Miller, L. Florens, B. Schmidt, L. Qin and J. McCracken, (*Abstracts*), *American Chemical Society,* 2000, **220**, 88-PHYS.

81. J. Qian, W.J. Shi, M. Pressler, C. Hoganson, D. Mills, G.T. Babcock and S. Ferguson-Miller, *Biochemistry,* 1997, **36**, 2539–43.

82. C. Backgren, G. Hummer, M. Wikström and A. Puustinen, *Biochemistry,* 2000, **39**, 7863–7.

83. B. Schmidt, J. McCracken and S. Ferguson-Miller, *Proc. Natl. Acad. Sci., U.S.A.,* 2003, **100**, 15539–42.
84. R. Mitchell and P R. Rich, *Biochim. Biophys. Acta,* 1994, **1186**, 19–26.
85. N. Capitanio, T.V. Vygodina, G. Capitanio, A.A. Konstantinov, P. Nicholls and S. Papa, *Biochim Biophys Acta,* 1997, **1318**, 255–65.
86. E. Forte, M.C. Barone, M. Brunori, P. Sarti and A. Giuffrè, *Biochemistry,* 2002, **41**, 13046–52.
87. P. Ädelroth, M. Ek and P. Brzezinski, *Biochim. Biophys. Acta,* 1998, **1367**, 107–117.
88. E. Forte, F.M. Scandurra, O.M. Richter, E. D'Itri, P. Sarti, M. Brunori, B. Ludwig and A. Giuffrè, *Biochemistry,* 2004, **43**, 2957–63.
89. M.I. Verkhovsky, N. Belevich, J.E. Morgan and M. Wikström, *Biochim. Biophys. Acta.,* 1999, **1412**, 184–9.
90. N. Capitanio, G. Capitanio, D. Boffoli and S. Papa, *Biochemistry,* 2000, **39**, 15454–61.
91. M. Erecinska, B. Chance and D.F. Wilson, *FEBS Lett.,* 1971, **16**, 284–6.
92. S. Jünemann, B. Meunier, R.B. Gennis and P.R. Rich, *Biochemistry,* 1997, **36**, 14456–64.
93. P. Ädelroth, R.B. Gennis and P. Brzezinski, *Biochemistry,* 1998, **37**, 2470–6.
94. M. Wikström, A. Jasaitis, C. Backgren, A. Puustinen and M.I. Verkhovsky, *Biochim. Biophys. Acta.,* 2000, **1459**, 514–20.
95. T.V. Vygodina, C. Pecoraro, D. Mitchell, R. Gennis and A.A. Konstantinov, *Biochemistry,* 1998, **37**, 3053–61.
96. C. Pecoraro, R.B. Gennis, T.V. Vygodina and A Konstantinov, A. *Biochemistry,* 2001, **40**, 9695–708.
97. D. Zaslavsky and R.B. Gennis, *Biochemistry,* 1998, **37**, 3062–7.
98. P. Brzezinski and P. Ädelroth, *Acta Physiol. Scand.,* 1998, **163**, 7–16.
99. M. Brandèn, H. Sigurdson, A. Namslauer, R.B. Gennis, P. Ädelroth and P. Brzezinski, *Proc. Natl. Acad. Sci. U.S.A.,* 2001, **98**, 5013–8.
100. J.E. Morgan, M.I. Verkhovsky and M. Wikström, *Biochemistry,* 1996, **35**, 12235–40.
101. C. Greenwood, M.T. Wilson and M. Brunori, *Biochem. J.,* 1974, **137**, 205–15.
102. B.C. Hill and C. Greenwood, *Biochem. J.,* 1983, **215**, 659–67.
103. M. Oliveberg, P. Brzezinski and B.G. Malmström, *Biochim. Biophys. Acta,* 1989, **977**, 322–8.
104. M. Karpefors, P. Ädelroth, A. Namslauer, Y.J. Zhen and P. Brzezinski, *Biochemistry,* 2000, **39**, 14664–9.
105. J.E. Morgan, M.I. Verkhovsky, G. Palmer and M. Wikström, *Biochemistry,* 2001, **40**, 6882–92.
106. N.J. Watmough, M.R. Cheesman, C. Greenwood and A.J. Thomson, *Biochem. J.,* 1994, **300**, 469–75.
107. M. Fabian and G. Palmer, *Biochemistry,* 1995, **34**, 13802–10.
108. D.A. Proshlyakov, T. Ogura, K. Shinzawa-Itoh, S. Yoshikawa, E.H. Appelman and T. Kitagawa, *J. Biol. Chem.,* 1994, **269**, 29385–8.
109. D.A. Proshlyakov, T. Ogura, K. Shinzawa-Itoh, S. Yoshikawa and T. Kitagawa, *Biochemistry,* 1996, **35**, 8580–6.

110. D.A. Proshlyakov, T. Ogura, K. Shinzawa-Itoh, S. Yoshikawa and T. Kitagawa, *Biochemistry,* 1996, **35**, 76–82.
111. T. Kitagawa and T. Ogura, *J. Bioenerg. Biomembr.,* 1998, **30**, 71–9.
112. D.A. Proshlyakov, M.A. Pressler and G.T. Babcock, *Proc. Natl. Acad. Sci. U.S.A.,* 1998, **95**, 8020–5.
113. D.A. Proshlyakov, M.A. Pressler, C. DeMaso, J.F. Leykam, D.L. DeWitt and G.T. Babcock, *Science,* 2000, **290**, 1588–91.
114. M.R. Blomberg, P.E. Siegbahn and M. Wikström, *Inorg. Chem.,* 2003, **42**, 5231–43.
115. S. Lyubenova, F. MacMillan, O.-M.H. Richter, B. Ludwig, K. Budiman, A. Kannt and H. Michel, *Biochemistry,* 2004, 11709–16.
116. M. Karpefors, P. Ädelroth, A. Aagaard, I.A. Smirnova and P. Brzezinski, *Isr. J. Chem.,* 1999, **39**, 427–37.
117. A. Jasaitis, C. Backgren, J.E. Morgan, A. Puustinen, M.I. Verkhovsky and M. Wikström, *Biochemistry,* 2001, **40**, 5269–74.
118. M.R.A. Blomberg, P.E.M. Siegbahn, G.T. Babcock and M. Wikström, *J. Inorg. Biochem.,* 2000, **80**, 261–9.
119. G.T. Babcock and M. Wikström, *Nature,* 1992, **356**, 301–9.
120. Babcock, G. T. *Proc. Natl. Acad. Sci. U.S.A.,* 1999, **96**, 12971–3.
121. B.C. Hill and C. Greenwood, *Biochem. J.,* 1984, **218**, 913–21.
122. S.H. Han, Y.C. Ching and D.L. Rousseau, *Proc. Natl. Acad. Sci. U.S.A.,* 1990, **87**, 8408–12.
123. M.I. Verkhovsky, J.E. Morgan and M. Wikström, *Biochemistry,* 1994, **33**, 3079–86.
124. A. Sucheta, K.E. Georgiadis and Ó. Einarsdóttir, *Biochemistry,* 1997, **36**, 554–65.
125. B.C. Hill, *J. Bioenerg. Biomembr.,* 1993, **25**, 115–20.
126. S.W. Han, Y.C. Ching and D.L. Rousseau, *Proc. Natl. Acad. Sci. U.S.A.,* 1990, **87**, 2491–5.
127. S. Han, S. Takahashi and D.L. Rousseau, *J. Biol. Chem.,* 2000, **275**, 1910–9.
128. M.I. Verkhovsky, J.E. Morgan, M.L. Verkhovskaya and M. Wikström, *Biochim. Biophys. Acta,* 1997, **1318**, 6–10.
129. M. Oliveberg, S. Hallén and T. Nilsson, *Biochemistry,* 1991, **30**, 436–40.
130. S. Hallén and T. Nilsson, *Biochemistry,* 1992, **31**, 11853–9.
131. A. Namslauer, A. Aagaard, A. Katsonouri and P. Brzezinski, *Biochemsitry,* 2003, **42**, 1488–98.
132. S. Paula, A. Sucheta, I. Szundi and Ó. Einarsdóttir, *Biochemistry,* 1999, **38**, 3025–33.
133. S. Han, Y.C. Ching and D.L. Rousseau, *Nature,* 1990, **348**, 89–90.
134. D. Zaslavsky, A.D. Kaulen, I.A. Smirnova, T. Vygodina and A.A. Konstantinov, *FEBS Lett.,* 1993, **336**, 389–93.
135. S. Siletsky, A.D. Kaulen and A.A. Konstantinov, *Biochemistry,* 1999, **38**, 4853–61.
136. I.A. Smirnova, P. Ädelroth, R.B. Gennis and P. Brzezinski, *Biochemistry,* 1999, **38**, 6826–33.
137. A. Aagaard, A. Namslauer and P. Brzezinski, *Biochim. Biophys. Acta,* 2002, **1555**, 133–9.

138. A. Namslauer, M. Brändén and P. Brzezinski, *Biochemistry,* 2002, **41**, 10369–74.

139. P. Hellwig, J. Behr, C. Ostermeier,O.M. Richter, U. Pfitzner, A. Odenwald, B. Ludwig, H. Michel and W. Mäntele, *Biochemistry,* 1998, **37**, 7390–9.

140. M. Lübben, A. Prutsch, B. Mamat and K. Gerwert, *Biochemistry,* 1999, **38**, 2048–56.

141. A. Puustinen, J.A. Bailey, R.B. Dyer, S.L. Mecklenburg, M. Wikström and W.H. Woodruff, *Biochemistry,* 1997, **36**, 13195–200.

142. D. Heitbrink, H. Sigurdson, C. Bolwien, P. Brzezinski and J. Heberle, *Biophys. J.,* 2002, **82**, 1–10.

143. R.M. Nyquist, D. Heitbrink, C. Bolwien, R.B. Gennis and J. Heberle, *Proc. Natl. Acad. Sci. U.S.A.,* 2003, **100**, 8715–20.

144. R.M. Nyquist, D. Heitbrink, C. Bolwien, T.A. Wells, R.B. Gennis and J. Heberle, *FEBS Lett.,* 2001, **505**, 63–7.

145. B.H. McMahon, M. Fabian, F. Tomson, T.P. Causgrove, J.A. Bailey, F.N. Rein, R.B. Dyer, G. Palmer, R.B. Gennis and W.H. Woodruff, *Biochim. Biophys. Acta.,* 2004, **1655**, 321–31.

146. M. Karpefors, P. Ädelroth, Y. Zhen, S. Ferguson-Miller and P. Brzezinski, *Proc. Natl. Acad. Sci. U.S.A.,* 1998, **95**, 13606–11.

147. P. Ädelroth, M. Karpefors, G. Gilderson, F.L. Tomson, R.B. Gennis and P. Brzezinski, *Biochim. Biophys. Acta,* 2000, **1459**, 533–9.

148. G. Brändén, M. Brändén, B. Schmidt, D.A. Mills, S. Ferguson-Miller and P. Brzezinski, *Biochemistry,* 2005, **44**, 10466–74.

149. A. Namslauer and P. Brzezinski, *FEBS Lett.,* 2004, **567**, 103–10.

150. M.I. Verkhovsky, A. Jasaitis, M.L. Verkhovskaya, J.E. Morgan and M. Wikström, *Nature,* 1999, **400**, 480–3.

151. D. Bloch, I. Belevich, A. Jasaitis, C. Ribacka, A. Puustinen, M.I. Verkhovsky and M. Wikström, *Proc. Natl. Acad. Sci., U.S.A.,* 2004, **101**, 529–33.

152. M. Wikström and M.I. Verkhovsky, *Biochim. Biophy. Acta,* 2002, **1555**, 128–32.

153. M. Ruitenberg, A. Kannt, E. Bamberg, K. Fendler and H. Michel, *Nature,* 2002, **417**, 99–102.

154. E. Antonini, M. Brunori, A. Colosimo, C. Greenwood and M.T. Wilson, *Proc. Natl. Acad. Sci. U.S.A.,* 1977, **74**, 3128–32.

155. M. Brunori, G. Antonini, F. Malatesta, P. Sarti and M.T. Wilson, *Eur. J. Biochem.,* 1987, **169**, 1–8.

156. A. Jasaitis, M.I. Verkhovsky, J.E. Morgan, M.L. Verkhovskaya and M. Wikström, *Biochemistry,* 1999, **38**, 2697–706.

157. K. Faxén, G. Gilderson, P. Ädelroth and P. Brzezinski, *Nature*, In Press, 2005.

158. K.F. Wang, L. Geren, Y.J. Zhen, L. Ma, S. Ferguson-Miller, B. Durham and F. Millett, *Biochemistry,* 2002, **41**, 2298–304.

159. Y.J. Zhen, B. Schmidt, U.G. Kang, W. Antholine and S. Ferguson-Miller, *Biochemistry,* 2002, **41**, 2288–97.

160. U. Pfitzner, K. Hoffmeier, A. Harrenga, A. Kannt, H. Michel, E. Bamberg, O.M.H. Richter and B. Ludwig, *Biochemistry,* 2000, **39**, 6756–62.

161. A.S. Pawate, J. Morgan, A. Namslauer, D. Mills, P. Brzezinski, S. Ferguson-Miller and R.B. Gennis, *Biochemistry,* 2002, **41**, 13417–23.

162. S.A. Siletsky, A.S. Pawate, K. Weiss, R.B. Gennis and A.A. Konstantinov, *J. Biol. Chem.*, 2004, **279**, 52558–65.

163. A. Namslauer, A.S. Pawate, R.B. Gennis and P. Brzezinski, *Proc. Natl. Acad. Sci. U.S.A.*, 2003, **100**, 15543–7.

164. R. Pomès, G. Hummer and M. Wikström, *Biochim. Biophys. Acta*, 1998, **1365**, 255–60.

165. M. Karpefors, P. Ädelroth and P. Brzezinski, *Biochemistry*, 2000, **39**, 6850–6.

166. B. Rost, J. Behr, P. Hellwig, O.M.H. Richter, B. Ludwig, H. Michel and W. Mäntele, *Biochemistry*, 1999, **38**, 7565–71.

167. P. Hellwig, B. Rost, U. Kaiser, C. Ostermeier, H. Michel and W. Mäntele, *FEBS Lett.*, 1996, **385**, 53–7.

168. M. Iwaki, A. Puustinen, M. Wikström and P.R. Rich, *Biochemistry*, 2003, **42**, 8809–17.

169. M.M. Pereira, M. Santana and M. Teixeira, *Biochim. Biophys. Acta-Bioenerg.*, 2001, **1505**, 185–208.

170. A. Aagaard, G. Gilderson, D.A. Mills, S. Ferguson-Miller and P. Brzezinski, *Biochemistry*, 2000, **39**, 15847–50.

171. J.E. Morgan, M.I. Verkhovsky and M. Wikström, *J. Bioenerg. Biomembr.*, 1994, **26**, 599–608.

172. S. Papa, N. Capitanio and G. Villani, *FEBS Lett.*, 1998, **439**, 1–8.

173. A.V. Xavier, *FEBS Lett.*, 2002, **532**, 261–6.

174. R.B. Gennis, *Frontiers in Bioscience*, 2004, **9**, 581–91.

175. P.R. Rich, S. Jünemann and B. Meunier, *J. Bioenerg. Biomembr.*, 1998, **30**, 131–8.

176. D.A. Mills and S. Ferguson-Miller, *Biochim. Biophys. Acta*, 2002, **1555**, 96–100.

177. D. Zaslavsky and R.B. Gennis, *Biochim. Biophys. Acta*, 2000, **1458**, 164–79.

178. P.E.M. Siegbahn, M.R.A. Blomberg and M.L. Blomberg, *J. Phy. Chem. B*, 2003, **107**, 10946–55.

179. D.M. Popovic and A.A. Stuchebrukhov, *J. Am. Chem. Soc.*, 2004, **126**, 1858–71.

180. H. Michel, *Biochemistry*, 1999, **38**, 15129–40.

181. D.L. Rousseau, Y. Ching and J. Wang, *J. Bioenerg. Biomembr.*, 1993, **25**, 165–76.

182. N. Capitanio, G. Capitanio, M. Minuto, E. De Nitto, L.L. Palese, P. Nicholls and S. Papa, *Biochemistry*, 2000, **39**, 6373–9.

183. P. Brzezinski and G. Larsson, *Biochim. Biophys. Acta*, 2003, **1605**, 1–13.

184. J. Behr, H. Michel, W. Mäntele and P. Hellwig, *Biochemistry*, 2000, **39**, 1356–63.

185. W. Humphrey, A. Dalke and K. Schulten, *J. Molecular Graphics*, 1996, **14**, 33.

186. S.W. Han, Y.C. Ching and D.L. Rousseau, *J. Biol. Chem.*, 1989, **264**, 6604–7.

187. M. Brändén, A. Namslauer, Ö. Hansson, R. Aasa and P. Brzezinski, *Biochemistry*, 2003, **42**, 13178–84.

CHAPTER 5

Quantum Chemical Models of O_2 Bond Cleavage and Proton Pumping in Cytochrome Oxidase

PER E.M. SIEGBAHN AND MARGARETA R.A. BLOMBERG

Department of Physics,
Albanova University Center,
Stockholm University, S-106 91 Stockholm,
Sweden

1 Introduction

A few decades ago quantum chemical studies could only be applied to small molecules and the results were still of only moderate accuracy. During the 1980s a large improvement in the accuracy was obtained through the development of *ab initio* methods based on configuration interaction and coupled cluster theory, which were able to yield the major part of the correlation energy, provided very large basis sets were used. Still, the applications were restricted to small molecules with only a few atoms. The outlook for applications on larger systems was pessimistic due to the high scaling power with the size for these methods. The situation changed dramatically with the introduction of high accuracy density functional theory (DFT) methods in the beginning of the 1990s. At that time DFT methods, which scale much more favorably with the size of the system, had been available for two decades but the accuracy had not been sufficient. Through the introduction of terms in the functional that depend on the gradient of the density and a fraction of exact exchange [1], the accuracy obtainable became comparable to the most accurate *ab initio* methods. However, since these so called hybrid DFT methods were parametrized for small molecules with atoms from the first row of the periodic table, the hope that they would also perform well for transition metal complexes was low. An additional problem in this context was that it was hard to find out how high the accuracy actually was, since accurate experimental gas phase results for transition metal containing systems were essentially lacking, in particular for larger saturated complexes. This meant that comparisons would have to be done for systems in their actual chemical surroundings, and for this reason a careful development of adequate chemical models was required. This has been done over the last decade and it can be demonstrated that the hybrid DFT methods compare very favorably with experimental results, even for biochemical systems containing transition metals [2,3,4]. In this chapter, a quantum chemical study of one of the most interesting enzymes of this type, cytochrome oxidase, will be described. The description will rely on results published previously for O–O bond cleavage [5,6] and proton pumping [7], but new results will also be presented.

Cytochrome oxidase is the terminal enzyme in the respiratory chain. It catalyzes the reduction of dioxygen to water and couples this reaction to the translocation of protons across the mitochondrial (or bacterial) membrane. The reaction catalyzed is:

$$O_2 + 8H_{in}^+ + 4e^- \rightarrow 2H_2O + 4H_{out}^+$$

where H_{in}^+ are protons taken from the inside of the membrane and H_{out}^+ are protons translocated, or pumped, across the membrane from the inside to the outside [8]. The translocated protons drive the synthesis of ATP where the energy from food consumption and respiration is stored.

The X-ray structure of cytochrome oxidase has been solved for two types of bacteria [9,10] and also for a mammalian species [11]. There are four redox centers in the enzyme, Cu_A and heme a which function as electron transport cofactors, and Cu_B and heme a_3 which form the binuclear center (bnc), the active site for dioxygen reduction, see Figure 1. The first redox center that receives an electron is Cu_A, which gets the electron from cytochrome c on the outside of the membrane. From Cu_A the electron goes to heme a and then on to the binuclear center. Protons are transferred to the binuclear center from the inside of the membrane along two different pathways, the D-and the K-channels. It is believed that at most only one of the eight protons follows the K-pathway [12]. All four protons being pumped and at least three of the protons consumed in the dioxygen reduction thus follow the D-channel.

The main event in the catalytic cycle of cytochrome oxidase is the cleavage of dioxygen in the binuclear center. The first part of this chapter will describe the present status of the quantum chemical knowledge of this process. Experimentally, it is known that molecular oxygen coordinates to iron in the binuclear center forming the spectroscopically characterized compound A. In experiments on the fully reduced form of the enzyme, i.e. when all four metal sites are reduced, it is found that the life time of compound A is about 50 μs and the next observed intermediate is compound P_R in which the O–O bond is cleaved. Also the mixed valence form of the enzyme, *i.e.* the two electron reduced enzyme, cleaves the O–O bond, and in this case the product has a tyrosyl radical and is called compound P_M. In the mixed valence enzyme the life time of compound A is somewhat longer than in the fully reduced form, 200 μs [13], which corresponds to a free energy barrier of 12.5 kcal mol^{-1} using transition state theory. It is believed that the mixed valence form is the one that corresponds best to the working enzyme, and therefore P_M is one of the main intermediates of the catalytic cycle as shown in Figure 2.

After the O–O bond cleavage step, A to P_M, the catalytic cycle proceeds with four transitions, and in each transition one electron and one proton enters the binuclear center, see Figure 2. In each transition, at least one species has been spectroscopically

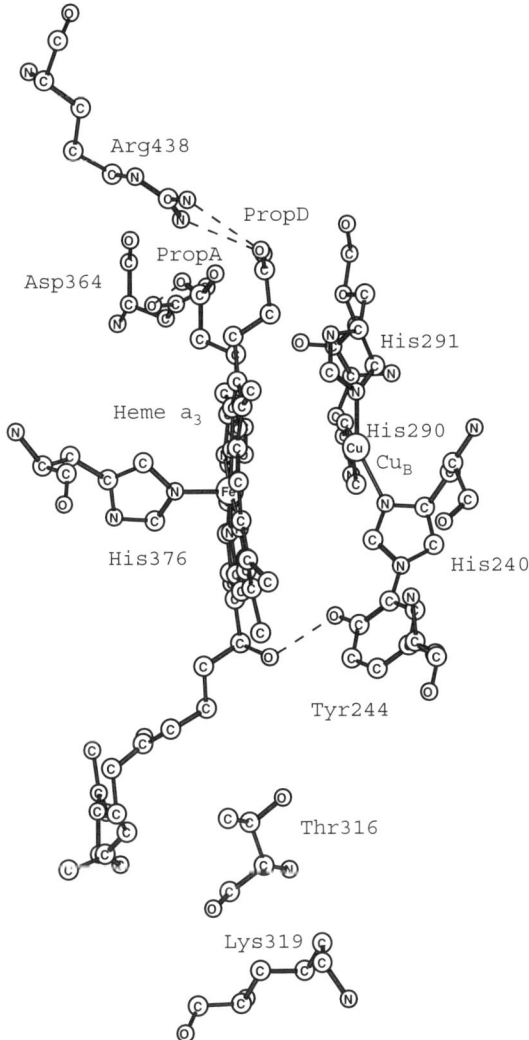

Figure 1 *X-ray structure of the binuclear center in cytochrome oxidase. The labels are from the bovine structure [11]*

characterized [14,15,16]. The P_M intermediate occurring immediately after the O–O bond cleavage has the oxidation states Fe(IV) and Cu(II), and also a tyrosine located in the binuclear center is oxidized. The species P_R mentioned above is the species observed when an electron has been received by species P_M. This electron is believed to go to the tyrosyl radical which then becomes a tyrosinate. The next observed intermediate in the cycle is F with oxidation states Fe(IV) and Cu(II), and after that the oxidized state O with Fe(III) and Cu(II) is observed. The state O is the resting state of the enzyme, and it has to be reduced by two electrons before a new O_2 activation can take place. The one-electron reduced state is labeled E with oxidation states

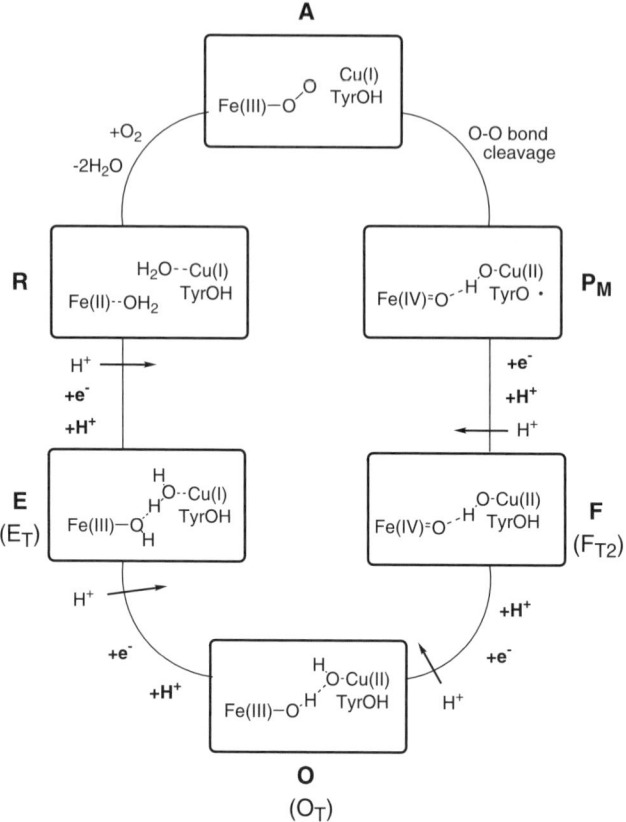

Figure 2 *The catalytic cycle of cytochrome oxidase. The arrows indicate proton translocation*

Fe(III) and Cu(I), and in the most reduced state, R, the oxidation states are Fe(II) and Cu(I). In Figure 2 the points where the electrons are transferred to the binuclear center are marked, as well as the positions where protons are taken up by the binuclear center from the inside, and where protons are being pumped across the membrane. As discussed below, the quantum chemical studies suggest a large number of additional states, some of them perhaps so short-lived that they might not be observable. In the suggested cycle, the appearance of 14 different species will be marked, most of them relating to the names of the 6 species shown in Figure 2, but will also contain subindices, as described below. Some of these, the ones that directly correspond to a species mentioned above, are shown in parenthesis in Figure 2.

The mechanism for proton pumping is one of the most important, and most difficult, mechanisms in biochemistry that still remains to be understood. Several suggestions have been made [17,18,19], each one probably containing some part of the true mechanism. The most difficult part to understand is how the protons can be allowed to move across the membrane from the inside to the outside against the gradient potential, but not be allowed to move the other way, which is thermodynamically easier. The general view has been that the same pathway has to be open sometimes and closed sometimes. This has been termed proton gating. An example of this which has been suggested is that when an electron is on heme *a* the proton translocation pathway is open and the pathway to the binuclear center is closed, while when the electron has moved to heme a_3, the opposite would apply [20]. The gating problem will be a main theme in the present review of the quantum chemically suggested mechanism. It will be suggested that a partial solution to this problem can be obtained without varying the barriers for the pathways over time. The way to accomplish this will be termed guiding rather than gating to emphasize the basic difference in character of the two approaches.

2 Methods and Models

The general approach adopted here is to use quantum mechanics to calculate essential parts of the potential energy surface, and to compare the results to available experimental data. For the dioxygen cleavage this means comparing mainly the activation energy obtained from the energy profile with the experimentally measured life time, applying transition state theory. For the proton translocation mechanism it means comparing the overall picture to experimental suggestions, but also judging from general principles if the calculated results can be fitted into a reasonable proton-pumping scheme. The calculations have to be performed on models of the active site of the enzyme, and the choice of model is therefore a very important part of any theoretical study. In the first place, the model cannot be too large, since determining reaction mechanisms requires a large number of calculations. At present, the quantum mechanically treated models can have up to about 120 atoms . At the same time the essential parts of the active site have to be included, *i.e.* those parts that might take part in the reaction. For metallo-enzymes this means that apart from the metal atoms, all amino acids and cofactors in the first coordination shell and sometimes a few of those in the second coordination shell should be included in the model. It also means that different models might be used for different aspects of the catalytic mechanism. As described below, different models of the binuclear center in cytochrome oxidase have been used to study the O–O bond cleavage mechanism on the one hand, and the proton-pumping mechanism on the other. To set up the models, X-ray structures were used, and for most of the models described below the bovine structure was used [11]. However, for the largest model used to study the proton pumping mechanism a bacterial structure with many resolved waters was used [10]. The protonation state and the number of electrons present in the particular state studied must be determined by calculations, since these are not elucidated by the crystal structure. However, reasonable guesses can be deduced from spectroscopic studies. In the calculations it may also be important to consider the effects of the part of the enzyme that surrounds the quantum mechanical model chosen. For metallo-enzymes it is found that effects coming from outside the metal complex are generally quite small, unless explicit charge separations are involved. These effects are therefore reasonably well treated by simple continuum methods for processes like the O–O bond cleavage. For the steps in the proton-pumping mechanism, where charge separations do occur, the results from the continuum methods have to be further parametrized to experiments, as described below.

To obtain a potential energy surface for a chemical reaction the relative energies of different intermediates and the activation energies for the transitions between them should ideally be calculated. The activation energies are calculated as the energy difference between the reactant and a transition-state structure. A full transition state optimization requires the calculation of the Hessian, *i.e.* the second derivatives of the energy with respect to the nuclear coordinates, which is very time consuming for large models. Therefore, in the present study of the O–O bond cleavage only approximate transition states have been determined by freezing the O–O bond distance at different values, optimizing all the other degrees of freedom, and determining the maximum on the potential energy surface. It has been demonstrated for smaller models that the activation energies normally change by less than 2 kcal-mol going from approximate transition states to the fully optimized ones. The calculated Hessians can also be used to estimate zero-point, thermal and entropy effects on the relative energies in the harmonic approximation, and in this way the free energy can be obtained. When the model used is too large for a Hessian calculation, these effects can be estimated from calculations on similar but smaller models.

For the active site model chosen several calculations were performed for each stationary point considered. The quantum mechanical method used was the hybrid density functional B3LYP [1]. Initially a geometry optimization was performed using a fairly small basis set (double zeta). In previous studies the geometry optimization was unconstrained, while in the presently reported calculations constrained optimizations were applied, where a few parameters were frozen to the corresponding values in the crystal structures. To evaluate relative energies, B3LYP calculations were further performed for the optimized geometries using larger basis sets (including diffuse and polarization functions). It has recently been shown that for transition metal systems, more accurate relative energies are sometimes obtained if the parameters of the B3LYP functional are modified in such a way that the exact exchange is given the weight 0.15 rather than the original value 0.20 [21]. The results reported for the O–O bond cleavage are those obtained with the modified functional, the main reason being that this has been found to better describe the particular case of dioxygen bonding and cleavage. On the other hand, the effect of such a modification of the function has been found to be small on electron and proton affinities, and therefore the results given for the proton pumping mechanism are those obtained using the original function. Another technical aspect of the calculations is that density functional theory (being a one configuration method) cannot correctly describe antiferromagnetic spin-coupling. Since unpaired electrons on the two metals of the binuclear center are known to be antiferromagnetically coupled, a spin correction [22] is introduced. Finally, dielectric effects are calculated in the optimized structures, also at the B3LYP level. The dielectric methods used employ cavities that follow the shapes of the molecular systems. The dielectric constant of the protein is the main empirical parameter of this model, and it is chosen to be equal to four in line with previous suggestions for proteins. The program used was the JAGUAR package [23].

The inherent accuracy of the B3LYP method has been estimated using the extended G3 benchmark set, consisting of enthalpies of formation, ionization potentials, electron affinities and proton affinities for first and second row molecules [24]. For a test set which excludes heats of formation for some large molecules the B3LYP functional has an error of 3.29 kcal-mol (301 entries). For transition metals there are no benchmarks due to the lack of accurate experimental gas phase numbers but indications from normal metal-ligand bond strengths are that the errors can be slightly larger, 3–5 kcal mol^{-1} [25]. With a limiting accuracy of 3–5 kcal mol^{-1}, it is clear that not all questions concerning reaction mechanisms can be meaningfully answered, and only the energetically most important aspects of the reaction mechanisms have therefore been addressed. When two different reaction mechanisms are compared, the preferred mechanism can only be safely determined when there is an energetic difference between them of more than 3–5 kcal mol^{-1}.

3 The A to P_M Step: Mechanism for O–O Bond Cleavage

The mechanism for the O–O bond cleavage in cytochrome oxidase is a difficult problem to attack with quantum chemical methods, one reason being that the active site is unusually large, with two metal centers and a porphyrin cofactor. On the other hand, the study of this process in the mixed valence form of the enzyme is simplified by the fact there are no electrons or protons entering from outside the binuclear center, and a direct comparison can be made with the measured lifetime of compound A. Molecular oxygen enters in the most reduced form of the binuclear center, R, with Cu(I) and Fe(II), and it coordinates to iron forming the experimentally observed compound A, see Figure 2. To cleave the O–O bond four electrons are needed and at least one proton, yielding Fe(IV)=O and Cu(II)OH, labeled compound P_M. Since the metals can only provide three of these electrons, one electron has to be taken from somewhere else. It has been suggested that the tyrosine cross-linked to one of the histidine ligands of copper (Tyr244, see Figure 1) provides this electron [26]. The same tyrosine has also been suggested to be the source of the proton needed to form the hydroxyl group on copper, giving a neutral tyrosyl radical as a component of P_M, see Figure 2.

The simplest model of the binuclear center used to study the O–O bond cleavage contains the heme a_3 unit, described by an essentially unsubstituted porphyrin with an imidazole modeling the axial histidine, Cu_B with three imidazoles modeling the ligated histidines, and a phenol modeling the cross-linked tyrosine, see Figure 3. To bridge the distance between tyrosine and O_2, making proton transfer possible, two water molecules have to be present in the binuclear center. Using a larger model including the valine residue (Val243) near Cu_B it was found that the binuclear center is able to accommodate these two water molecules without steric problems. Furthermore, the computed binding energy of these two water molecules in Compound A was found to be about the same as in bulk water. It was concluded that there is no energy cost to move these water molecules into the binuclear center at this point in the catalytic cycle, and they were therefore included in the model, see Figure 3.

For the type of model described above it has been shown previously that the barrier for the O–O bond cleavage step is higher than 25 kcal mol^{-1} [5,6], which is considerably more than the experimental value of 12.5 kcal mol^{-1}. New calculations have been performed using the same type of model, but with an improved computational

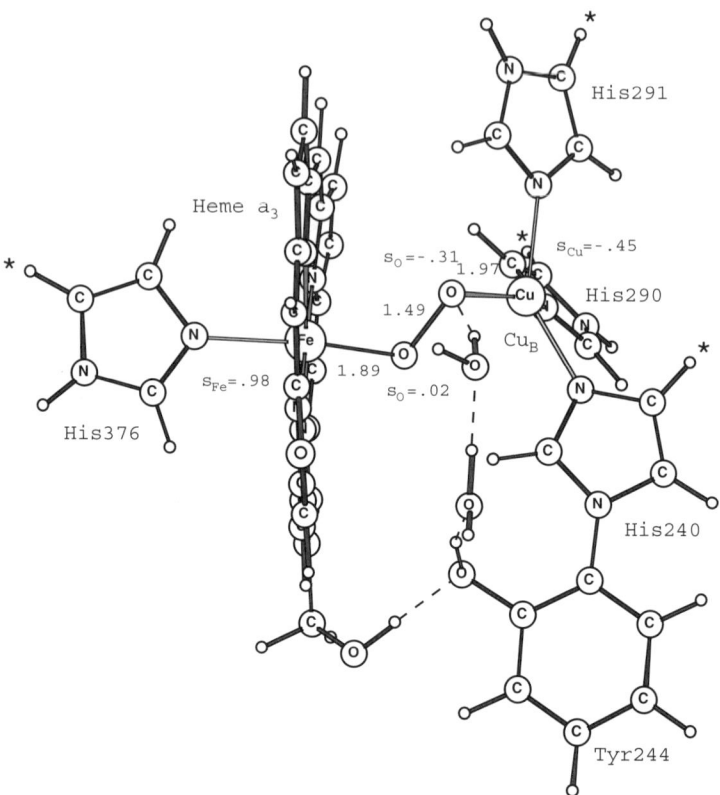

Figure 3 *Optimized geometry for the O-O bridge structure labeled compound B here, using the unprotonated model. Atoms marked with asterisks are frozen from the X-ray structure. The most important spin-populations are given*

procedure, and the results are reported as the dashed line in Figure 4. The improvements are concerned with the description of spin coupling and exact exchange in DFT, and are further described in Section 2. Both these improvements lower the barrier by a few kcal-mol. Another modification of the computational procedure is concerned with restrictions in the geometry optimization. In the previous study the model was fully optimized without constraints for all stationary points considered, while in the present study one point on each imidazole was kept frozen at its position in the X-ray structure. The frozen atoms are marked with asterisks in Figure 3. The purpose of this type of restriction in the geometry optimization is to try to simulate the steric effect of the surrounding protein, and as it turned out, this procedure also lowers the barrier relative to Compound A. The new value for the O–O bond cleavage barrier using the model in Figure 3 is 19.4 kcal mol^{-1}, which is still somewhat high compared to the experimental value of 12.5 kcal mol^{-1}.

Since the model shown in Figure 3 gives too high a barrier for the O–O bond cleavage, further modifications of the model that may lower the barrier were investigated. It was shown in previous studies that the addition of an extra proton to the binuclear

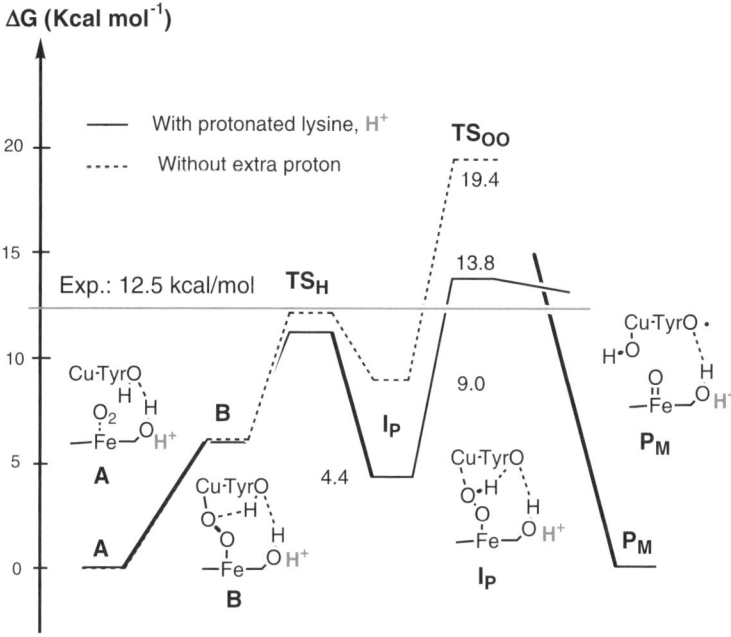

Figure 4 *Calculated energy profiles for the A to P$_M$ transition using different models. Note that the two water molecules in the binuclear center are omitted for clarity*

center decreases the barrier significantly [5,6]. It was argued that the farnesyl hydroxyl group on heme a_3 might be a possible site for this extra proton in compound A, due to its high proton affinity [5,6]. However, more recent calculations have shown that the proton affinity of that site is still not high enough to make protonation likely. Another possible modification of the model could be to find another source of the fourth electron. In the mixed-valence form of the enzyme there is no electron available on heme a, so there is no electron entering from outside the binuclear center, and a tyrosyl radical is most likely to be present in the final product of the O–O bond cleavage step, compound P$_M$. On the other hand, it is possible that the simultaneous delivery of a proton and an electron from tyrosine to oxygen during the O–O bond cleavage that has to occur in the model in Figure 3 causes the high barrier, and one alternative could be that the electron is temporarily delivered by another residue near the binuclear center, thereby decreasing the barrier [5,6]. This type of mechanism has previously been found plausible for ribonucleotide reductase (RNR) [27,28]. The most likely candidate for the electron donor was considered to be the conserved tryptophan Trp236 that is π-stacking with one of the histidine ligands on copper [6]. However, this type of mechanism could not be verified by calculations using a larger model including the tryptophan.

Therefore, the only proposal for how the O–O bond cleavage barrier might be lowered so far is the introduction of a positive charge in the binuclear center. In this context it can be noted that there are experimental results indicating that the highly conserved Lys319 in the K-channel (see Figure 1) is protonated at neutral pH, and that the K-channel is actually involved in the O_2 reduction, in contrast to what was previously assumed [29]. The experiments were performed on the fully reduced form of mutants in which the positive charge of Lys319 was either prevented or counterbalanced. Under these circumstances compound P_R could not be observed. These results could be interpreted to suggest that the protonated lysine is important for the O–O bond cleavage step. A new model was therefore constructed, including a protonated lysine in the K-channel, see Figure 5. The lysine was modeled by a protonated methylamine, and two water molecules were added to simulate the surroundings of the lysine. The effect of these water molecules was to stabilize the proton on the lysine, so that it was not too easily available for the binuclear center. Finally, to make the model practical for the calculations, the lysine was positioned next to the farnesyl hydroxyl group. In this way, the effect of a protonated Lys319

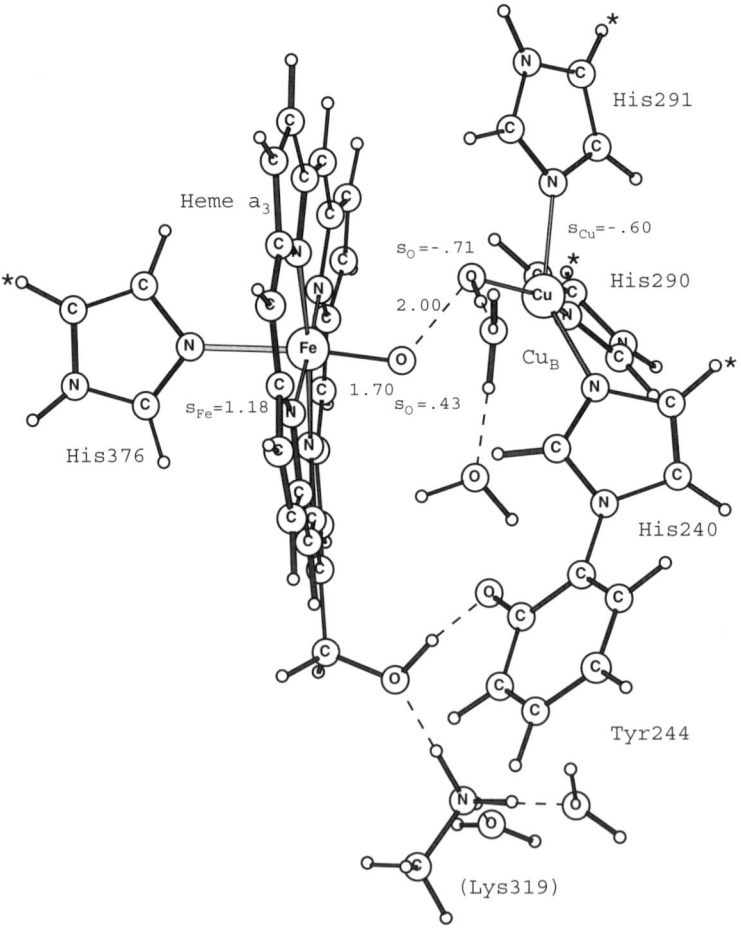

Figure 5 *Optimized approximate transition state for O–O bond cleavage using the model with a protonated residue in the K-channel. Atoms marked with asterisks are frozen from the X-ray structure. The most important spin-populations are given*

on the O–O bond cleavage step is admittedly exaggerated. On the other hand, it might also happen that during the bond cleavage step the proton actually moves in the K-channel closer to the binuclear center.

Using the model with a protonated lysine in the K-channel, the energy profile for the A to P_M step was calculated and the results are reported as the solid line in Figure 4. As can be seen from the figure, the agreement with experiment is quite good. The mechanism of the O–O bond cleavage will therefore be discussed in detail using this model. Calculations on compound A resulted in a mixture of the true compound A and a peroxide structure with oxygen bridging between the metals, see Figure 3 for the unprotonated model. The calculations thus indicated that the energies of compounds A and B (the bridging peroxide) are quite similar. On the other hand no intermediate of the peroxide type has been observed, and the explanation is that there is an entropy loss on going from compound A to compound B, making the free energy of B higher than that of A. An entropy effect of about 6 kcal mol^{-1} is observed experimentally on the O–O bond cleavage reaction [13], and also from calculations on smaller models, a similar entropy effect is obtained on the superoxo to peroxo transformation step [5,6]. Therefore, it is here assumed that the free energy of compound B is 6 kcal mol^{-1} higher than that of compound A, for both the unprotonated and the protonated model, see Figure 4.

The next step in the reaction is to transfer the proton from tyrosine to the oxygen atom coordinating to copper, forming a hydroperoxide which is bound to Fe(III) and weakly coordinating to Cu(II). This structure is here labeled I_P. The transition state for this step was not fully optimized, but preliminary calculations showed that for both models the barrier for this step was not rate limiting. For the protonated model this proton-transfer step is slightly exergonic from B, placing compound I_P 4.4 kcal mol^{-1} above compound A. This was the step where the protonated lysine makes a large difference, and for the unprotonated model the corresponding structure was at 9.0 kcal mol^{-1}, a difference of 4.6 kcal mol^{-1}. The proton transfer leads to a tyrosinate, and it is easy to understand that the formation of this negatively charged group is stabilized by a protonated group nearby (in the K-channel). It should be noted that the proton stays on the lysine during the whole O–O bond cleavage step, even though it was allowed to move in the optimization.

The barrier for the O–O bond cleavage relative to I_P was rather similar for the two models, only 1 kcal mol^{-1} higher for the unprotonated model, while the barrier relative to compound A differed by 5.6 kcal mol^{-1} due to the high energy of the I_P structure in the unprotonated model. The optimized transition state for the model with the protonated lysine is shown in Figure 5. The calculated barrier height relative to compound A for this model was 13.8 kcal mol^{-1}, very close to the experimental value of 12.5 kcal mol^{-1}. The O–O bond distance was 2.0 Å, and the spin populations indicated that only one electron had been transferred to the O–O antibonding orbital. As discussed previously [6], there is actually a very weak minimum on the potential energy surface closely after the transition state has been passed, with a similar electronic structure. As also discussed in the previous study [6], to make the second electron move to the antibonding orbital and complete the cleavage of the O–O bond, the hydrogen bonding pattern has to be changed. In the transition state, see Figure 5, the hydroxyl group on copper donates a hydrogen bond to one of the water molecules. If the hydrogen bonding pattern is changed, so that one of the water molecules donates a hydrogen bond to the hydroxyl oxygen, the electron affinity of this oxygen increases and the second electron is transferred from the tyrosinate, completely cleaving the O–O bond and yielding a neutral tyrosyl radical in the P_M product. The reaction step A to P_M is found to be close to thermoneutral for both models used. Taking into account the error bars of the calculations (3–5 kcal mol^{-1}) this is in fair agreement with the expected result that the reaction should be exothermic by a few kcal mol^{-1}.

In summary, for the O–O bond cleavage it was found that without a positive charge in the vicinity of the cross-linked tyrosine (Tyr244) the computed barrier is more than 5 kcal mol^{-1} higher than the experimental value. This indicated that there was something missing in that model, and it was shown that a protonated methylamine (modeling the highly conserved Lys319 in the K-channel) lowers the barrier by more than 5 kcal mol^{-1}, leading to much better agreement with experiment, see Figure 4. The effect of the extra proton is purely electrostatic, stabilizing the intermediately formed tyrosinate, since the proton remains on the model lysine throughout. It should be borne in mind that the model used might exaggerate the effect of the lysine, and it may also be possible that minor changes in the unprotonated model might lead to further decreases in the barrier.

Another electrostatic modification of the model tried was to introduce an unprotonated propionate on the heme, since the present model of heme a_3 would correspond to having the propionates protonated, and a negative charge might help to pull the proton from the tyrosine. However, no effect was found from this modification of the model.

4   The Catalytic Cycle: Mechanism for Proton Pumping

To address the mechanism of proton pumping, a chemical model has to be adopted for which it is possible to study both the proton pumping and the proton consumption. Proton consumption here refers to protons that are involved in the dioxygen reduction process to form water. Since the consumption of protons occurs either in the region between the two metals of the binuclear center, where oxygen binds, or at the histidine-linked tyrosine, it is clear that at least these two regions need to be included in the model. It is harder to define a model where the positions of protons being pumped are included, since the protons are not fully translocated until they have reached the outside of the membrane. In order not to make the chemical model too large to handle, a point reasonably close to the binuclear center therefore has to be defined, where it is possible to conclude that if a proton has reached this position it is on the translocation rather than on the consumption pathway. In the currently suggested mechanism for proton pumping, this point is considered to be propionate A of heme a_3, which is therefore also included in the chemical model.

With the above assumption it is possible to construct a model of the binuclear center that fulfills the fundamental criterion that it can be used to differentiate between protons being consumed and protons being pumped. The model of this type used in the main part of the calculations is shown in Figure 6. It consists of 120 atoms which is at the practical limit for the high number of exploratory calculations needed in this type of study. As discussed above, the propionate region is of special interest. Still

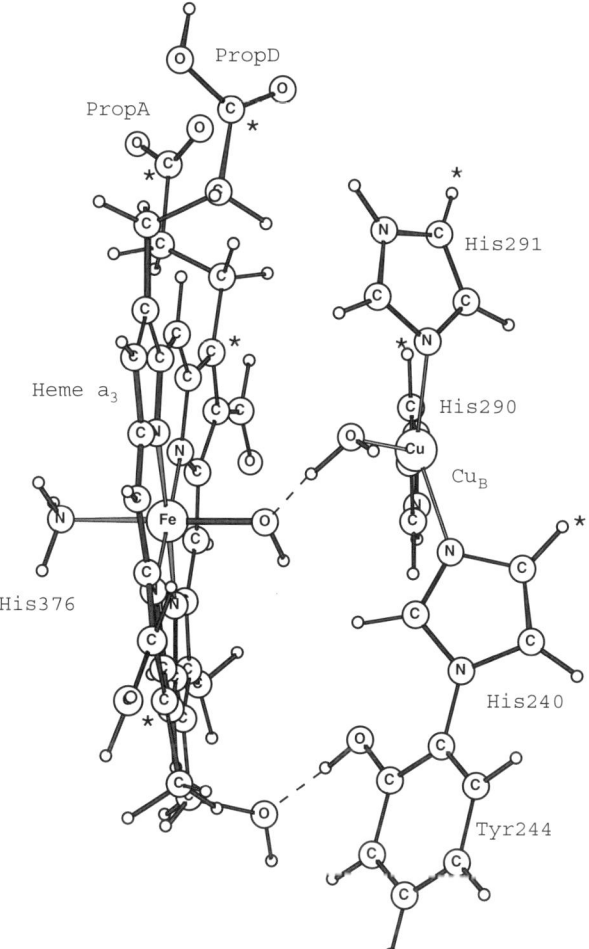

Figure 6 *Main model used in the calculations on the entire catalytic cycle. Atoms marked with asterisks are frozen from the X-ray structure*

this region has been quite simplified in the model to limit the size. In the X-ray structure both propionates are hydrogen bonded to other groups, propionate A to Asp364
and propionate D to Arg438. The hydrogen bonding noted in the X-ray structure
between the two negative groups, propionate A and Asp364, means that one of these
groups has to be protonated, for example Asp364. The effect of Asp364 is therefore
reduced to a normal, rather weak, hydrogen bond from a neutral residue, and it was
concluded that if a protonated Asp364 is left out of the model it should have only a
minor effect on the relative energies. In the case of the positively charged Arg438,
the effect was expected to be larger. The assumption made here is that the effect of
Arg438 is modeled by protonating propionate D. It is clear that these approximations
are not perfect, and a correction was therefore added as described below. Another
notable, but well tested, approximation is to use ammonia instead of imidazole for

the proximal histidine of the heme. The farnesyl chain of the heme has also been quite simplified, but all other substituents were kept. In order to keep the model used reasonably close to the X-ray structure, several positions were kept fixed as shown by asterisks in Figure 6.

To construct a catalytic cycle, electron affinities (redox potentials) and proton affinities (pK_a values) have to be calculated for different critical groups in the model. Obtaining absolute values for these quantities means comparing energies for models with different charge, which is known to be very difficult to do accurately using models of the present limited size. The dielectric effects on these values, for example, are quite large (of the order of 20 kcal mol^{-1}) and dependent on the rather arbitrary choice of dielectric constant. To circumvent the problem of calculating absolute values, the following procedure using only relative values was used. The redox potential of heme *a* is known experimentally to be about 0.40 V. The electron affinity for a model of heme *a*, as large as could be afforded, was then calculated, and the value obtained of 104 kcal mol^{-1} was set to correspond to 0.40 V. (Using an experimental standard normalization, 104 kcal mol^{-1} would correspond to 0.32 V, which is quite satisfactory but was not used when the energy diagram of the present study was constructed.) Since the model used for heme *a* was larger than the one used for heme a_3 in the binuclear model, a correction (-6.0 kcal mol^{-1}) was calculated by comparing the models, and this value was added to all the redox potentials calculated for heme a_3. This correction was not made in the previous study on the proton-pumping mechanism [7]. For the relative proton affinities a correction of -3.3 kcal mol^{-1} was added for the propionate due to the approximate treatment of this region, see above. To obtain relative pK_a values, the experimentally known driving force of 2.0 eV for the entire catalytic cycle was used. It turned out that a proton affinity value of 286 kcal mol^{-1} (corresponding to pH 7) leads to the experimental driving force. This value could only be obtained once all proton and electron affinities had been calculated. It should finally be emphasized that all relative redox potentials and pK_a values were directly taken from the calculations.

From the redox potentials and pK_a values, calculated as described above, a detailed energetic diagram could be set up for the entire catalytic cycle. This diagram, obtained without any gradient across the membrane, is shown in Figure 7, and contains the conventional labels from Figure 2 in bold, and the new labels assigned in the present study in parenthesis. It should be noted that this diagram is not exactly the same as the one published earlier [7], since a new correction of the redox potentials of heme a_3 has been applied, see above. The first observation that can be made is that the diagram fulfills two main criteria: the energy goes steadily downhill and there are no major thermodynamic barriers (kinetic barriers have not been calculated so far). It should be noted that the exergonicity of the entire cycle (45.9 kcal mol^{-1}) is a value that has been taken from experiments and used in the parametrization, but all the relative values in the individual steps have been obtained directly from the calculations.

The main order of events in each transition of the cycle can be described in essentially the same way. The resting state always has propionate A protonated. From the resting state, a proton first goes from the inside to the binuclear center (marked H^+_{bnc} in the figure). At the same time a proton is repelled to the outside from the propionate (marked H^+_{out}). This is overall an endergonic process and is only performed to allow the electron to reach the binuclear center, which happens in the next step. This

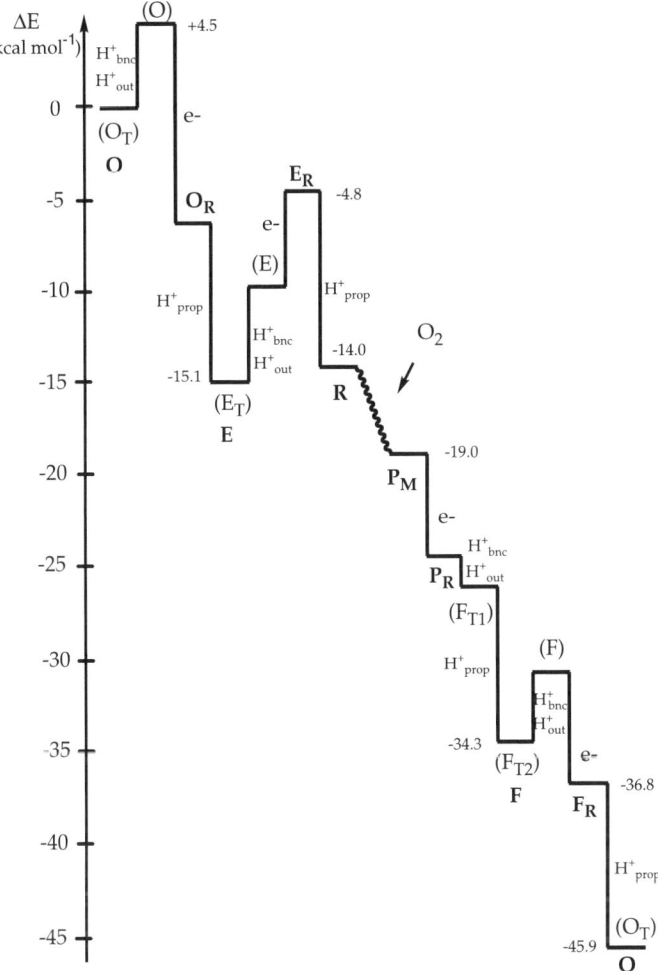

Figure 7 *Calculated energy profile for the entire catalytic cycle of cytochrome oxidase without gradient across the membrane*

electron transfer is usually exergonic. In the final exergonic step of each transition a proton moves from the inside to the propionate. No gate is needed to direct the protons to the propionate for translocation and prevent protons from going to the binuclear center for consumption, since the pK_a value for the propionate is higher than for the binuclear center. This result is in sharp contrast to previously suggested mechanisms, which have always required gating at this stage. In the presently suggested mechanism, the propionate region is thus an energetic sink for protons as compared to the outside, the inside and the binuclear center.

The result that the pumping of a proton from the inside all the way to the propionate region, close to the membrane outside, is quite exergonic in all steps of the cycle is an entirely new idea that comes out of the calculations and is quite surprising.

Energetically it might appear as if the proton pumping is driving the rest of the process, which is obviously not the case since the entire proton pumping is endergonic. The reason for this paradox is that the propionate region is an energy sink with an energy substantially below that of the inside both before and after an electron has reached the binuclear center (see below).

The order of the steps of the energy diagram described above is unconventional and surprising. The conventional view is that the step where the protons move from the inside to the binuclear site is the one that is energetically driving the cycle. It is clearly important to investigate this possibility using the same calculated values. When this is done the diagram shown in Figure 8 as the dashed line is obtained. In this figure the energetics from Figure 7 appear as the line (in order to fit both diagrams into the figure a different energy scale had to be used than the one in Figure 7). Again, as in the mechanism described above, each transition can be described in essentially the same

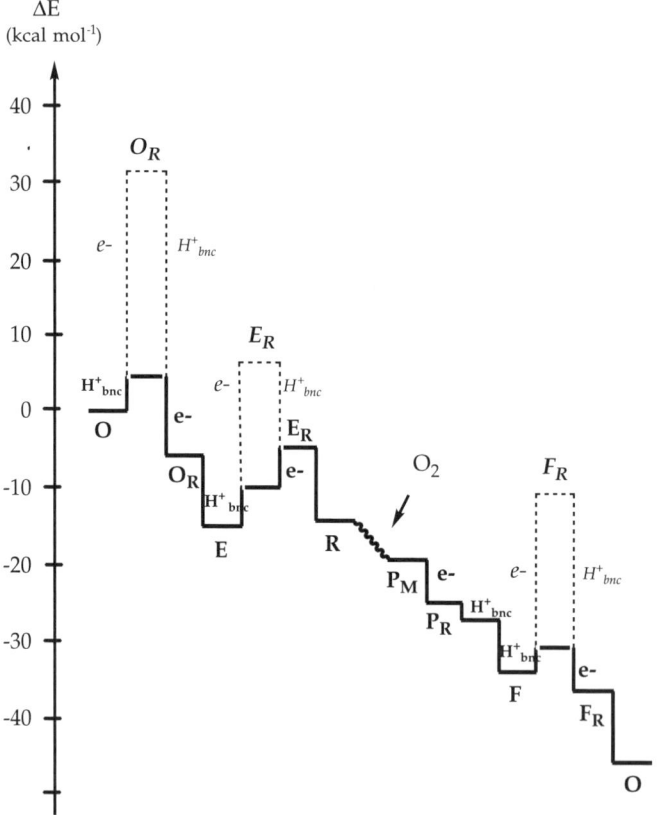

Figure 8 *Comparison between energy profiles for different mechanisms of the catalytic cycle. The dashed curve corresponds to a mechanism where the proton goes to the binuclear center after the electron has arrived, while the full curve corresponds to the mechanism suggested from the present calculations, where in most steps a proton arrives at the binuclear center before the electron comes in*

way. It starts by an endergonic electron transfer from heme *a* to the binuclear center followed by an exergonic proton transfer. The main problem here is that using the present results, this leads to a *very* endergonic electron transfer and a *very* exergonic proton transfer. Altogether, this leads to undesirable large thermodynamic barriers in the cycle which compares unfavorably with the rather smooth energetics with the alternative ordering of the mechanism, as described above. It should also be noted that in this conventional type of mechanism it is usually assumed that a gate is needed to direct the protons being translocated to the outside instead of to the binuclear center for consumption, which is not needed for the ordering in the presently suggested scheme as described above. It could be added that the water molecules in the binuclear center are quite important for preventing the thermodynamic barriers of the conventional ordering being even higher.

It is straightforward to get an idea of how a gradient across the membrane would affect the energy diagram. The experimentally known full membrane potential is 200mV (4.6 kcal mol⁻¹). Simply assuming a linear potential from the outside to the inside, and placing the binuclear center and the propionate at two–thirds the distance form the inside, leads to the energy diagram shown in Figure 9. This diagram appears in most respects reasonable with relatively small thermodynamic barriers. However, a few problems can be noted. First, the barrier from E to E_R of 16.4 kcal mol⁻¹ is too high by a few kcal mol⁻¹ for a process that should run in milliseconds. Second, the kinetic barrier for O–O bond cleavage would also be too high since the resting state is not R but E. These types of minor errors must be expected in this type of study, and give hints of what needs to be improved in the model.

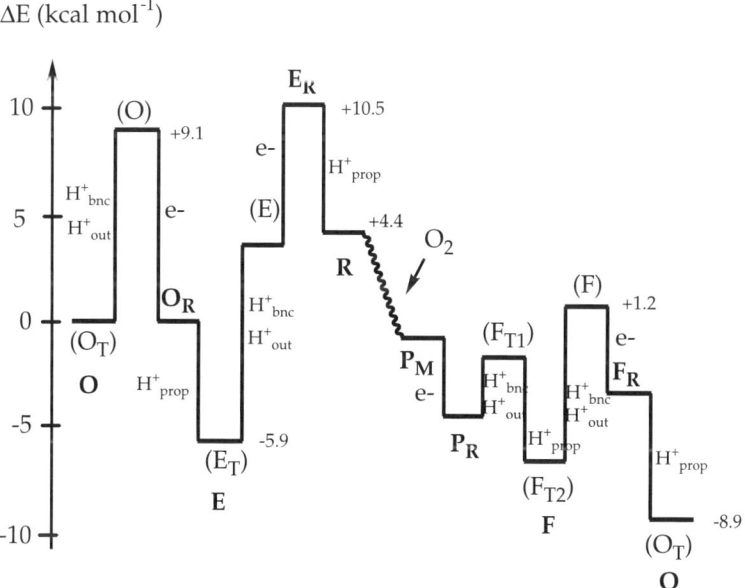

Figure 9 *Calculated energy profile for the entire catalytic cycle of cytochrome oxidase with a gradient applied across the membrane*

Finally, during the past year a much larger model containing 175 atoms has been used to test the adequacy of the previous model. This model is shown in Figure 10 and is primarily extended in the region around and outside the propionates. It is important to include water molecules and the coordinates were therefore taken from the spheroides structure [10] rather than from the bovine structure [11] as in the smaller model. The labeling in the figure is, however, taken from the bovine enzyme to fit with the rest of this chapter. The size of this model is at the limit of what is practical to use today and very long computation times were encountered. Not only was the size a problem, but also the appearance of very many different hydrogen-bonding

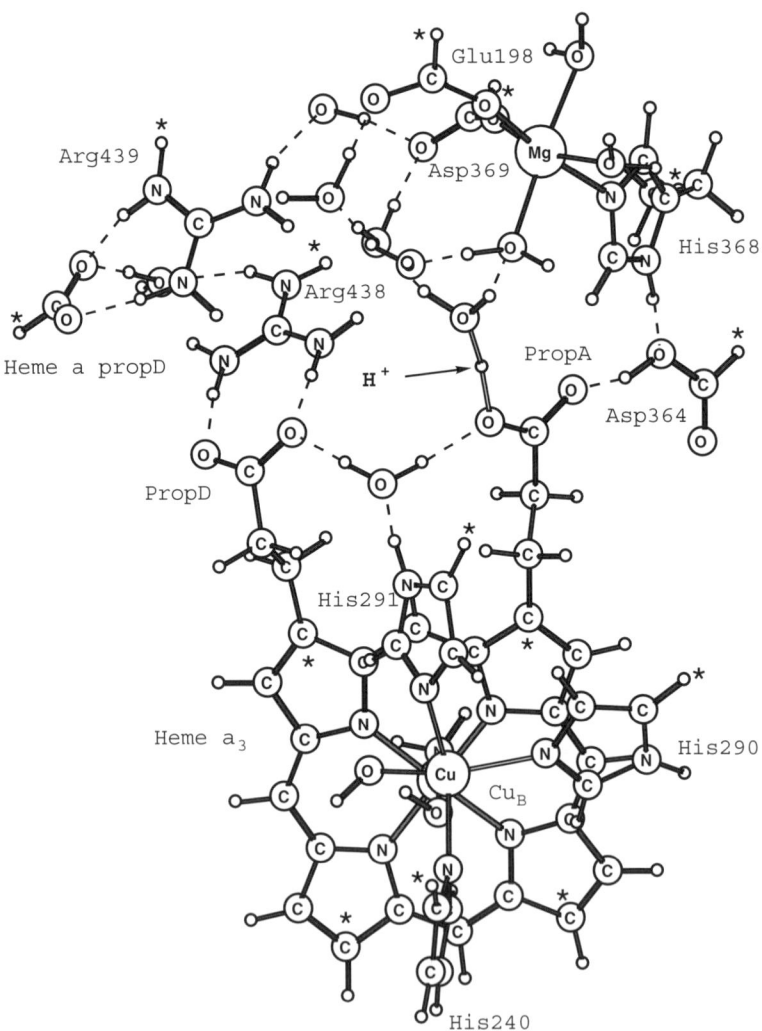

Figure 10 *Large model used to investigate the reliability of the smaller model. Atoms marked with asterisks are frozen from the X-ray structure*

minima, particularly in the region outside the propionates. Another technical problem was that one negatively charged group, Asp173, had to be left out since artificial spin appeared on this residue. This is a common artifact when negatively charged groups appear in the outskirts of a quantum chemical model. The effect of this group was instead estimated from pure electrostatics, which should be reasonably reliable in this case. Finally, rather stable results were obtained for the F to O transition. The most interesting result was the energy difference between the structures where either the propionate region or the binuclear center was protonated, after the electron had reached the binuclear center. For the large model this energy difference was 1.2 kcal mol^{-1} in favor of the propionate region compared to 4.5 kcal mol^{-1} for the small model. This means that the best scenario for the ordering of the events remains the same as described above for the small model. It should be noted that exact values cannot be obtained by the present techniques independently of how large the model is made. The main contribution of computational results like these is instead to suggest general mechanisms that fit the calculated energetics at a qualitative level, *i.e.* within 3–5 kcal mol^{-1} of the calculated results. The conclusion concerning the preferred mechanism would thus not change even if the large model gave a slightly negative value for the energy difference. It is the overall picture that counts. Another result of the large model should be mentioned, and this is that the proton placed on propionate A does not stay there but moves out among the water molecules between the propionates and the magnesium complex.

5   Proton Gating or Guiding

The most problematic part of any mechanism for proton pumping is explaining how protons are prevented from going from the outside to the inside, and yet are allowed to move in the other direction along the same pathway, even though it is energetically unfavorable. Also included in this problem is the question of how protons are prevented from reaching the binuclear center too early in the process since this would waste the energy.

The energy diagrams discussed above suggest a solution for the latter question, but the results need to be elaborated somewhat. A detailed energy diagram of what is happening directly after an electron has reached the binuclear center is sketched in Figure 11. Some values are taken directly from the calculations; for the other values the following rule has been used. The timescale of the process is of the order of milliseconds. This means that processes occurring with a barrier of 13 kcal mol^{-1} or less will be in thermodynamical equilibrium. Processes having a barrier of 16 kcal mol^{-1} or higher will not occur in this timescale. If two levels differ by 3 kcal mol^{-1}, the higher level will only have a population of 1% compared to the lower level. Besides the thermodynamic values, two barriers have also been put in the figure. One in the region of Arg438, which is on the pathway from the inside to the propionate region, but not on the pathway from the inside to the binuclear center, and the second barrier in between the propionate region and the outside.

After the electron has reached the binuclear center, the calculations indicate that it is exergonic to move a proton from the inside (H$_{in}^{+}$) both to the binuclear center (H$_{bnc}^{+}$) and to the propionate region (H$_{prop}^{+}$). An important result is that it is more exergonic

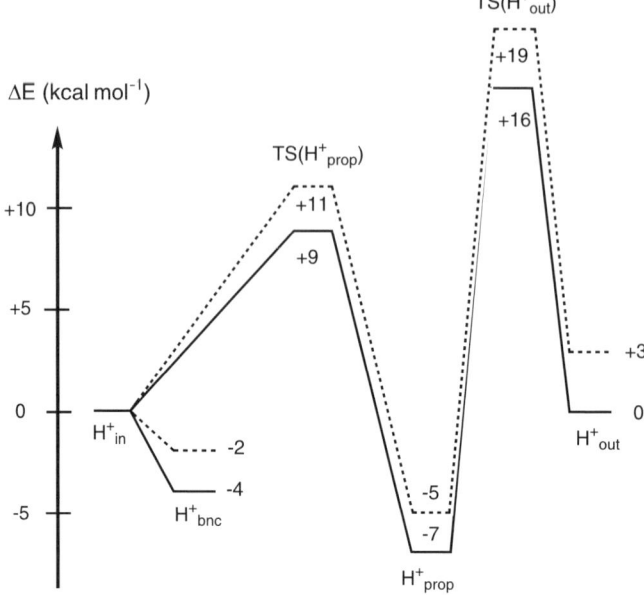

Figure 11 *Energy diagram for moving a proton from the inside, either to the outside over the propionate region, or to the binuclear center. The full line represents the case without a gradient across the membrane, while for the dashed line a part of the full gradient has been applied. Some energies are taken from the calculations, while some others are taken to be representative of what is reasonable, see text below*

to move the proton to the propionate; in the figure the energies are chosen as −7 and −4 kcal mol^{-1}, respectively, without membrane potential (−5 and −2 kcal mol^{-1} with part of the membrane potential). No gate is therefore needed at this stage to direct the protons to the propionate region. The barrier from the inside to the propionate is set to +9 kcal mol^{-1} (+11 kcal mol^{-1} with membrane potential). This means that a proton which happens to first go to the binuclear center will have a barrier of +13 kcal mol^{-1} to go to the propionate region, which is the desired place at this stage. The barrier for this of +13 (=4+9) kcal mol^{-1} is low enough that this will happen in the timescale of milliseconds (which is the reason the barrier was taken to be at +9 kcal mol^{-1}). In the other direction, from the propionate to the binuclear center, the barrier is +16 (=7+9) kcal mol^{-1} which leads to a too slow transfer. A further important point is that protons should not be allowed to reach the propionate region from the outside. A barrier of +16 kcal mol^{-1} was therefore placed on the pathway between these regions (to the right in the figure). In the opposite direction the barrier would be 23 (=7+16) kcal mol^{-1} and would not allow protons from the propionate region to reach the outside. The process sketched here for directing the protons for translocation is termed guiding, rather than gating. The figure indicates that the results do not change qualitatively if a gradient is applied across the membrane.

Figure 12 indicates what will happen when a second proton moves from the inside. At this stage the first proton is already at the propionate (H$_{in}^+$ H$_{prop}^+$) to the left in the figure, where the energy can again be set to zero, since it is the resting state from

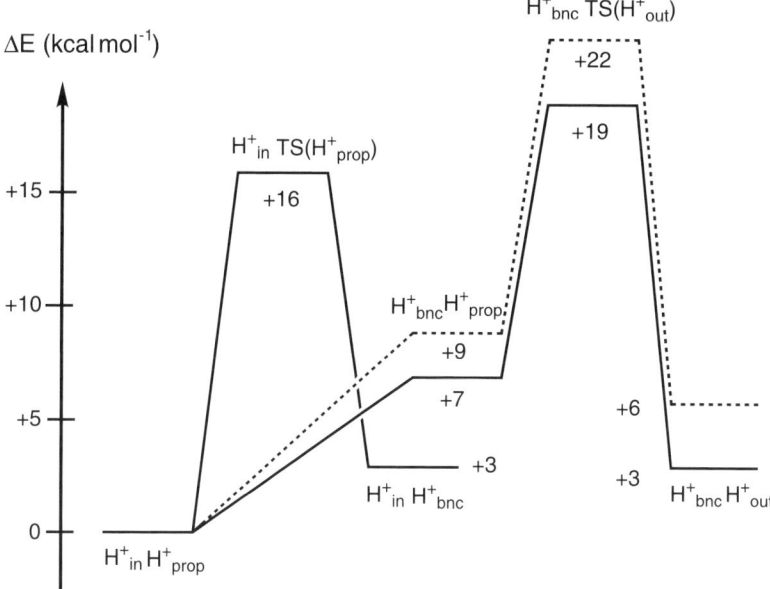

Figure 12 *Energy diagram involving two protons, where the propionate proton moves, either to the binuclear center, or to the outside with a simultaneous motion of the other proton from the inside to the binuclear center. The full line represents the case without a gradient across the membrane, while for the dashed line a part of the full gradient has been applied. Some energies are taken from the calculations, while some others are taken to be representative of what is reasonable, see text below*

Figure 11. The barrier for the propionate proton to go to the binuclear center is $+16$ kcal mol^{-1} as in Figure 11, preventing the proton from reaching the energy point H^+_{in} H^+_{prop} at $+3$ kcal mol^{-1}. It is important to prevent this, since if an electron reaches the binuclear center at that stage there will be no pumping. To continue the process, a proton must therefore move from the inside to the binuclear center, reaching an energy level of $+7$ kcal mol^{-1} ($+9$ kcal mol^{-1} with membrane potential) taken from the calculated values. When the propionate proton continues to move towards the outside it will reach the second barrier discussed above, and then go down in energy to $+3$ kcal mol^{-1} ($+6$ kcal mol^{-1} with membrane potential) for H^+_{in} H^+_{out}. The value of $+3$ kcal mol^{-1} is obtained by comparison to the starting point H^+_{in} H^+_{prop} and taking the energy difference between a proton at the propionate, -7 kcal mol^{-1}, and the binuclear center, -4 kcal mol^{-1}, where the energies are taken from Figure 11. With membrane potential the value changes to $+6$ kal mol^{-1}. The energy for the second transition state is then obtained as $+19$ kal mol^{-1} ($+22$ kal mol^{-1} with membrane potential) since the barrier from the outside was set to $+16$ kcal mol^{-1}. At the end-point in the diagram, at H^+_{bnc} H^+_{out}, the electron can be transferred from heme a to the binuclear site and the next transition can start.

The scenario described above would automatically guide protons to the point H^+_{bnc} H^+_{prop} and avoid too early a consumption in the binuclear center at the point H^+_{in} H^+_{bnc}.

The energy diagram also shows that the protons can never go from the outside into the propionate region since the barrier is too high at $+16$ kal mol^{-1}. However, the mechanism does not solve the gating problem entirely since the barrier at the point H^+_{bnc} TS(H^+_{out}) is also too high from the inside at $+19$ kal mol^{-1}, higher in fact than from the outside, which would lead to too slow a translocation. At the present stage, only a quite speculative suggestion can be formulated to solve this problem, and this is that a gate is present regulated by Cu$_A$, situated outside the Mg center in Figure 10. The assumption is that when there is no electron present on Cu$_A$ the barrier is high in both directions, while the barrier through the gate would be strongly reduced when an electron reaches Cu$_A$. Then it is assumed that there is no electron on Cu$_A$ when the process described in Figure 11 begins (H^+_{in} to the left in the figure). At the end of this process a proton is present in the propionate region (H^+_{prop} to the right in the figure). This proton will exert an attraction on an electron which is then assumed to move to Cu$_A$. This mutual attraction between the electron on Cu$_A$ and the proton in the propionate region (H^+_{prop}), will lead to a motion of the proton towards the outside, where the barrier is now lowered by this attraction. When the proton has moved sufficiently far from the binuclear center, a proton from the inside can move to the binuclear center. At this stage the binuclear center will be an energy sink for electrons and will attract the electron from Cu$_A$ over heme a. As the electron leaves Cu$_A$ the proton will be left on the outside. It should be added that a transfer from the outside, rather than from the inside, to Cu$_A$ will not lead the oxygen reduction process forwards and will therefore not occur even if this step is less energy demanding than taking the proton from the inside. There are clearly alternatives to this scenario, but it appears likely that some kind of gate is needed to allow protons to go in and out from the outside at the right moments, and Cu$_A$ is situated in a good position to handle this task.

A few additional comments should be added concerning the gating scheme described above. First, since there are some oxidases that do not have the Cu$_A$ center [30], the gating performed by Cu$_A$ should also be possible if it is substituted by a ubiquinone. This means that the gating mechanism has to be quite general, perhaps only requiring that a negative charge has to be present in this region at the right instant. Another comment is that all translocation schemes, not only the present one, will require a similar strategy for preventing protons from the outside getting in along the translocation pathway. At the very least a high barrier is required in the region between the propionate and the outside. Independent of the mechanism suggested, protons cannot be considered already translocated when they have reached the propionate region. The strongest argument that can be given here against a gate regulated by an electron on heme a, is that the computed redox potentials are far away from what is required for letting the electron in from heme a to the binuclear center at the right instant.

6 Summary

The present status of the quantum chemical modeling of cytochrome oxidase has been described. Two main questions have been addressed. The first one concerns the requirements for the cleavage of the O–O bond at the binuclear center. In line with experimental evidence, the O–O bond can be cleaved without electrons or protons entering from the outside to the binuclear region. The electrons required come from heme a_3 (two electrons), Cu_B (one) and the cross-linked tyrosine (one). The proton required comes from tyrosine. To transfer the proton from tyrosine to the bridging peroxide, two water molecules are required. These two water molecules have been found to fit well in the binuclear center, and to bind sufficiently strongly not to move away from this region. Still, without a second proton in the model, the barrier for O–O bond cleavage has been found to be too high. Based on experiments it has been concluded unlikely that a second proton would move a substantial distance in the direction of the membrane gradient. To fulfill this condition, a proton was added in the model in the K-channel rather close to the binuclear center. The suggested protonated residue is Lys319. In the step where the tyrosine proton moves to the peroxide (*via* the two waters), the K-channel proton moves closer to the binuclear center to stabilize the product tyrosinate. Calculations for this type of model do indeed lead to sufficiently low barriers.

The second question discussed in this paper is the mechanism for proton pumping. Calculations of the relevant redox potentials and pK_a values lead to an energy diagram for the pumping process with a few new features. The most important result is that the calculated energetics suggest that gating at the binuclear center may not be needed. As an electron reaches the binuclear center, the propionate region is found to function as an energetic sink, and the protons from the inside would therefore actually prefer to go to this region rather than to the binuclear center. Different possibilities of how the proton in the propionate region would continue to the outside, yet preventing proton flow from the outside have been sketched based on the calculated results and using a few assumptions. It has been shown that with the present results rather high barriers are needed both to prevent propionate protons going back to the binuclear center, and to prevent protons reaching the propionate region from the outside. If the latter barrier does not change with time, the barrier from the inside will be too high to allow proton translocation. Tentatively, it has been suggested that when an electron reaches Cu_A the negative charge should lower the barrier for proton transfer from both directions in this region and allow a proton from the inside to be translocated to the outside. As this happens, a proton could simultaneously move to the binuclear center from the inside leading to an electron transfer from Cu_A to the binuclear center over heme a. This transfer should close the gate in the Cu_A region again, preventing protons from entering from the outside. These are, of course, other possibilities.

References

1. A.D. Becke, *J. Chem. Phys.,* 1993, **98**, 5648.
2. P.E.M. Siegbahn and M.R.A. Blomberg, *Chem. Rev.,* 2000, **100**, 421.
3. P.E.M. Siegbahn and M.R.A. Blomberg, *J. Phys. Chem.* B, 2001, **105**, 9375.
4. P.E.M. Siegbahn, *Quart. Rev. Biophys.,* 2003, **36**, 91.
5. M.R.A. Blomberg, P.E.M. Siegbahn, G.T. Babcock and M.J. Wikström, *Am. Chem. Soc.,* 2000, **122**, 12848.
6. M.R.A. Blomberg, P.E.M. Siegbahn, and M. Wikström, *Inorg. Chem.,* 2003, **42**, 5231.

7. P.E.M. Siegbahn, M.R.A. Blomberg, *J. Phys. Chem. B*, 2003, **107**, 10946.

8. M. Wikström, *Nature*, 1997, **266**, 271.

9. C. Ostermeier, A. Harrenga, U. Ermler and H. Michel, *Proc. Natl. Acad. Sci., U.S.A.*, 1997, **94**, 10547.

10. M. Svensson-Ek, J. Abrahamson, G. Larsson, S. Tørnroth, P. Brzezinski and S. Iwata, *J. Mol. Biol.*, 2002, **321**, 329.

11. S. Yoshikawa, K. Shinzawa-Itoh, R. Nakashima, R. Yaono, E. Yamashita, N. Inoue, M. Yao, M.J. Fei, C.P. Libeu, T. Mizushima, H. Yamaguchi, T. Tomizaki and T. Tsukihara, *Science*, 1998, **280**, 1723.

12. M. Wikström, A. Jasaitis, C. Backgren, A. Puustinen, M.I. Verkhovsky, *Biochim. Biophy Acta*, 2000, **1459**, 514.

13. M. Karpefors, P. Ädelroth, A. Namslauer, Y. Zhen and P. Brzezinski, *Biochemistry*, 2000, **39**, 14664.

14. S. Ferguson-Miller and G. Babcock, *Chem. Rev.*, 1996, **96**, 2889.

15. H. Michel, J. Behr, A. Harrenga and A. Kannt, *Annu. Rev. Biophys. Biomol. Struct.*, 1998, **27**, 329.

16. J.E. Morgan, M.I. Verkhovsky, G. Palmer and M. Wikström, *Biochemistry*, 2001, **40**, 6882.

17. J.E. Morgan, M.I. Verkhovsky and M. Wikström, *J. Bioenerg. Biomemb.*, 1994, **26**, 599 M. Wikström, J.E. Morgan and M.I. Verkhovsky, *J. Bioenerg. Biomemb.*, 1998, **30**, 139; M. Wikström, *Biochim. Biophys. Acta*, 2000, **44827**, 1.

18. H. Michel, *Proc. Natl. Acad. Sci. U.S.A.*, 1998, **95**, 12819; H. Michel, *Biochemistry*, 1999, **38**, 15129; M. Ruitenberg, A. Kannt, E. Bamberg, K. Fendler and H. Michel, *Nature*, 2002, **417**, 99.

19. P. Brzezinski and G. Larsson, *Biochim. Biophys Acta*, 2003, **1604**, 61.

20. M. Wikström, M.I. Verkhovsky and G. Hummer, *Biochim. Biophys. Acta*, 2003, **160**, 61.

21. M. Reiher, O. Salomon and B.A. Hess, *Theor. Chem. Acc.*, 2001, **107**, 48.

22. L. Noodleman and D.A. Case, *Adv. Inorg. Chem.*, 1992, **38**, 423.

23. Jaguar 4.0, Schrödinger, Inc., Portland, OR, 1991–2000.

24. L.A. Curtiss, K. Raghavachari, R.C. Redfern and J.A. Pople, *J. Chem. Phys.*, 2000, **112**, 7374.

25. P.E.M. Siegbahn and M.R.A. Blomberg, *Ann. Rev. Phys. Chem.*, 1999, **50**, 221.

26. D.A. Proshlyakov, M.A, Pressler, C. DeMaso, J.F. Leykam, D.L. DeWitt and G.T. Babcock, *Science*, 2000, **290**, 1588.

27. P.E.M. Siegbahn, *Chem. Phys. Lett.*, 2002, **351**, 311.

28. J. Baldwin, C. Krebs, B.A. Ley, D.E. Edmondson, B.H. Huynh and J.M. Bollinger Jr., *J. Am. Chem. Soc.*, 2000, **122**, 12195.

29. M. Brändén, H. Sigurdson, A. Namslauer, R.B. Gennis, P. Ädelroth and P. Brzezinski, *PNAS*, 2001, **98**, 5013.

30. J. Abramson, S. Ristama, G. Larsson, A. Jasaitis, M. Svensson-Ek, L. Laakkonen, A. Puustinen, S. Iwata and M. Wikström, *Nature, Struct. Biol.*, 2000, **7**, 910.

CHAPTER 6

The bc_1 Complex: What is There Left to Argue About?

ANTONY R. CROFTS

Department of Biochemistry,
University of Illinois at Urbana-Champaign,
419 Roger Adams Lab,
600 S. Mathews Avenue,
Urbana,
IL 61801

1 Introduction

The cytochrome bc_1 complex and its relatives are the central enzymes of respiratory and photosynthetic chains, and catalyze the oxidation of hydroquinones in the membrane by small mobile aqueous redox proteins. The redox work is coupled to transfer of protons across the membrane. For the bc_1 complex, which oxidizes ubihydroquinone (quinol, QH_2), the reaction equations can be written to represent the overall reaction by its scalar and vectorial components:

$$QH_2 + 2 \text{ cyt } c^{3+} \leftrightarrows Q + 2 \text{ cyt } c^{2+} + 2H^+_{P(scalar)} \tag{1}$$

$$2H^+_N \leftrightarrows 2H^+_P \tag{2}$$

$$QH_2 + 2 \text{ cyt } c^{3+} + 2H^+_N \leftrightarrows Q + 2 \text{ cyt } c^{2+} + 2H^+_{P(scalar)} + 2H^+_P \tag{3}$$

The mechanism through which this overall reaction is accomplished has been the subject of extensive work from many labs, which has led to a consensus represented by a modified Q-cycle[1-3] (Figure 1, see[4] for a historical review). The mechanism appears at first sight to be so complicated as to discourage detailed physicochemical description. As the structures have shown,[5-8] the three catalytic subunits contain five catalytic interfaces, three of them concerned with processing external substrates, and these are connected through two separate-electron transfer chains. The complex is a homodimer, and much speculation about interaction between monomers has opened the possibility of even higher levels of complexity.[7, 9-14]

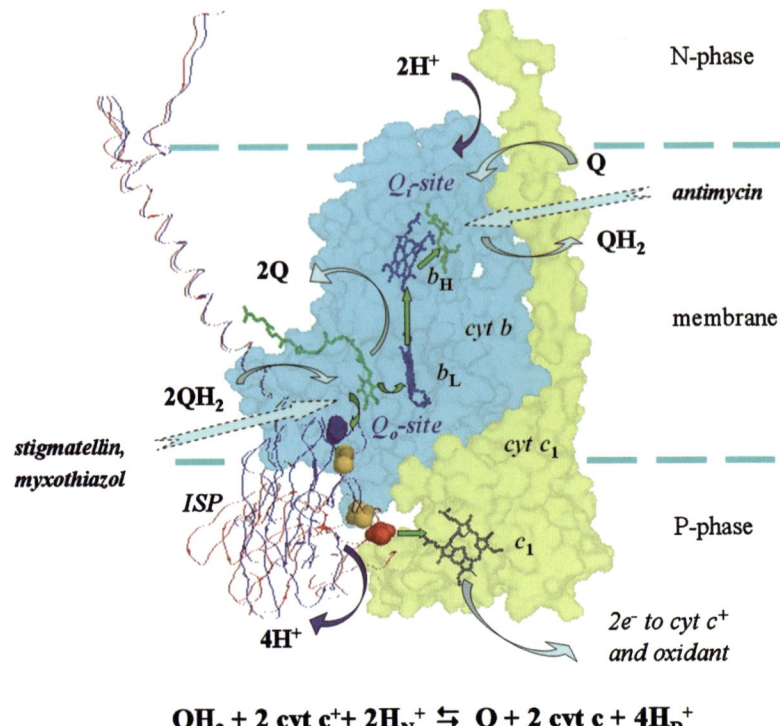

$$QH_2 + 2\ cyt\ c^+ + 2H_N^+ \leftrightarrows Q + 2\ cyt\ c + 4H_P^+$$

Figure 1 *The modified Q-cycle shown in the context the structure. The three catalytic subunits of a functional monomer of the bc_1 complex (PDB 2bcc) from chicken mitochondria are shown with a quinol modeled in the Q_o-site to replace stigmatellin, and quinone at the Q_i-site. The cyt b subunit is shown by its transparent cyan surface, and the cyt c_1 subunit by a transparent yellow surface. The Rieske iron sulfur protein is shown as a ribbon cartoon, docked at the b-interface (blue) or at the c-interface (red). The approximate position of the membrane is shown by the dashed lines. The b-hemes are shown in blue, heme c_1 in green. The [2Fe-2S] clusters are shown as spacefilling models, colored as for the protein. Two additional positions for the cluster are shown (in CPK colors) to indicate the trajectory of movement between b- and c-interfaces. Binding and unbinding of quinone species, and docking of cyt c, are shown by broad curved arrows (gray-blue). Electron transfers are shown by small green arrows, H^+ release and uptake are shown by curved blue arrows. The sites of action of inhibitors are shown by dotted gray arrows. For each turnover, two QH_2 molecules are oxidized to Q at the Q_o-site, and two successive electrons are passed down the pathways indicated. The two electrons going down the high-potential chain (ISP, cyt c_1, cyt c) are passed to a terminal oxidant. The two electrons going through the b-heme chain reduce Q to QH_2 at the Q_i-site through a two-electron gate, with storage of one electron as a SQ at the site. The modified Q-cycle in this form predated the structure by ~10 years, and the alignment here brings out the structural predictive power of the kinetic studies leading to the scheme*

In this review, I want to show that this complicated mechanism can be presented in terms of simple partial processes representing kinetic steps. For each of these, the parameters for a formal description (rate constants, equilibrium constant, activation barrier, *etc.*) can be measured, or constrained to well-defined ranges. For the critical rate-determining reaction, the bifurcated reaction leading to oxidation of QH$_2$ at the Q$_o$-site, we can offer a detailed description that explains the general properties, and several anomalous features that have been identified as problematic by other workers.

The Q-cycle mechanism appears to be essentially the same (with changes in parameters appropriate to species difference, *etc.*) whether observed in the isolated complex, in the mitochondrial respiratory chain, or in bacterial respiratory or photosynthetic chains. The photosynthetic bacteria offer many experimental advantages, and much of the work providing the formal descriptions presented here has been done in these systems, where turnover can be initiated *in situ* by activation of the photosynthetic chain. A brief saturating flash drives the photochemical reaction centers (RC) to generate the substrates for the *bc$_1$* complex in <10 μs. In both *Rhodobacter* species commonly used, the stoichiometry of the photosynthetic chain (~2 RC/*bc$_1$* complex) is such that, under uncoupled conditions, each *bc$_1$* complex turns over once in returning the system to the initial state. The kinetics of electron transfer associated with turnover can be followed spectrophotometrically by watching the elementary redox events associated with each of the three heme centers and RC, which have distinct spectra. In addition, electrochromic responses of carotenoid pigments of the light-harvesting complex make it possible to measure the electrogenic events associated with turnover, facilitating exploration of the relation between electron transfer and protonpumping (reviewed in[15]). Dissection of the partial processes has been facilitated by inhibitors that bind specifically at either of the two quinone-processing sites. These can be used to limit the set of partial reactions and simplify interpretation. In addition, the use of redox potentiometry[16] makes it possible to poise the system so that the effects of change in substrate concentration can be assayed over a wide range. Using these approaches, the modified Q-cycle was dissected into partial processes as summarized in Table 1. The rate and equilibrium constants shown are essentially those from work in the 1980s,[1, 17–25] updated where more complete data are now available.

A remarkable feature of the work from the 1980s was the demonstration that the equilibrium constants calculated from measurements of E_m values accounted for the dynamic distribution of electrons in the antimycin-inhibited complex (Fig. 2). The initial concentrations of substrate could be varied through reduction of the quinone pool, or by applying one or two saturating flashes, and the metastable state established 10–100 ms after each flash was then measured and the electron distribution compared to that expected from E_m values. In the context of measured stoichiometries and the linked electrogenic processes, a match was found, but only for a particular formulation of the basic Q-cycle[2, 3, 26–28] that had been largely ignored.[1, 4, 29, 30] The modified Q-cycle emerging from this work has stood the test of time,[31] and will be our basis for further discussion. Recent work from Osyczka *et al.*[32] has extended the equilibrium approach developed in this earlier work to analysis of the behavior in mutant strains in which the redox chains have been further truncated by mutagenesis.

One important conclusion from this extensive set of data is that the thermodynamic driving forces for the different partial processes are not markedly modified by hidden interactions; there was no evidence for long-range control of electron transfer by conformational or allosteric interactions.[18, 31, 32] A few qualifications are necessary:

(i) Direct measurement of electron exchange between hemes b_H and b_L in response to the transmembrane electric field generated by the photochemical reactions

Table 1 *Partial reactions of the bc₁ complex and their parameters*

#	Reaction	$\tau/\mu s$	$\Delta G^{o\prime}/meV$	K_{eq}	$\Delta E^{\#}/kJmol^{-1}$	Refs.
	Bifurcated reaction at Q_o-site					
a	$2(QH_2 + E.ISP_{ox}.b_L.b_H \leftrightarrows Q + E.ISPH.b_L^-.b_H + H_p^+)$	~770	−20	3	60–64	1, 15, 17–20, 4, 31, 166
	b-heme chain					
b	$2(E.b_L^-b_H \leftrightarrows E.b_L b_H^-)$	≤100	−130 (−60)	−160 (10)	<40	21–23, 33
	Q_i-site					
c	$E.b_L b_H^- + Q \leftrightarrows E.b_L b_H^-.Q_i$	<200	−94	40		23, 117
d	$E.b_L b_H^-.Q_i \leftrightarrows E.b_L b_H.SQ_i^-$	<200	−50	7		52, 118,
e	$E.b_L b_H^-.SQ_i^- + 2H_N^+ \leftrightarrows E.b_L b_H.Q_iH_2$	<200	−168	703		119
f	$E.b_L b_H.Q_iH_2 \leftrightarrows E.b_L b_H + QH_2$	<200	212	0.00025		
	sum $2\ E.b_L b_H^- + Q \leftrightarrows 2\ E.b_L b_H + QH_2$	≤200	−100	50	<40	
	High potential chain					
g	$2(E.ISPH.c_1^{2+} + cyt\ c^{3+} \leftrightarrows E.ISPH.c_1^{3+} + cyt\ c^{2+})$	150	−70	15.4	28	17, 20, 31, 93
h	$2(E.ISPH.c_1^{3+} \leftrightarrows E.ISP_{ox}.c_1^{2+} + H_p^+)$	10	20	0.46	18	116
i	sum $2(E.ISPH + cyt\ c^{3+} \leftrightarrows E.ISP_{ox} + cyt\ c^{2+} + H_p^+)$		−50	7		
	Overall reaction					
	$QH_2 + 2\ cyt\ c^{3+} + 2H_N^+ \leftrightarrows Q + 2\ cyt\ c^{2+} + 4H_p^+$		−500 (−2Δp)		60–64	

Notes:

Rates are given as lifetimes, τ, which for 1st-order processes are $1/k_{cat}$. The value for reaction a is from the rate measured at $E_{h,7} \sim 100$ mV (Q_{pool} 30% reduced), and represent about 80% V_{max}, obtained from extrapolation.[24] Values for $\Delta G^{o\prime}$ are given in meV ($\Delta G^{o\prime}/F = -z\Delta E^{o\prime}$), and calculated from measured $E^{o\prime}$ values, as follows: for cyt c_2, $E_{m,7} = 340$ mV; for cyt c_1, $E_{m,7} = 270$ mV; for ISP, $E_{m,7} = 290$ mV; for heme b_L, $E_{m,7} = -90$ mV; for heme b_H, $E_{m,7} = 40$ mV; for Q/QH₂(pool), $E_{m,7} = 90$ mV. [18, 28,] [120–122] Activation energies are taken from Arrhenius plots.[93]

Reaction a: $\Delta G^{o\prime} = -F((E^{o\prime}_{(ISP)} + E^{o\prime}_{(heme\ b_L)}) - 2\ E^{o\prime}_{(Q/QH2)})$.[1]

Reaction b: Values in parentheses reflect heme b_L, $E_{m,7} \sim -20$ when heme b_H charge is absent.[34]

Q_i-site (reactions c, d, e, f): The properties of the Q_i-site were modeled using the following values: $[Q_{total}] = 60$ mM; [bc_1 complex] = 1 mM; $K^a_{QH2} = 4000$ M⁻¹; $K^a_Q = 40$ M⁻¹; for bound Q/QH₂, $E_{m,7} = 149$ mV; for bound SQ/SQ, $E_{m,7} = 90$ mV; pK(heme b_H) = 7.8; pK(QH') = 12.5. Other parameters were calculated using the formalism in.[52, 123, 124] A Visual Basic program for modeling the site is available on request from the author.

enabled Shinkarev *et al.*[33] to explore the equilibrium constant under conditions in which only a single electron populated the chain, which was isolated from input or output by inhibitors. This worked confirmed an earlier suggestion,[34] that the measured redox potential of heme b_L reflected both an intrinsic E_m and an interaction with ferroheme b_H, likely coulombic in nature. The apparent E_m when heme b_H was oxidized, or removed by mutation of ligands, was ~60–80 mV higher than that measured by direct titration. Nevertheless, in kinetic experiments, heme b_H becomes reduced before b_L, and the E_m values appropriate to the metastable state after flash activation are the ones measured by redox titration.

(ii) Equilibration between the oxidized RC and the high potential chain occurs rapidly, but the apparent equilibrium constant is considerably lower than that expected from the E_m values.[35] Two mechanisms have been proposed, one based on supercomplexes ([35–38]), and the other on a heterogeneity in distribution of the components of the photosynthetic chain in the chromatophore vesicles.[39] Supporting evidence for supercomplexes is weak,[40–43] but the explanation in terms of heterogeneity seems to work quite naturally. The theoretical fits to the data depend on assumptions about the size of chromatophores and the stochastic distribution. The distribution function becomes much coarser for smaller numbers of entities per vesicle, and is therefore dependent on whether or not complexes are in dimeric association. Using the fact that both RC/LH1[38, 41, 42] and bc_1 complexes[44] are now known to be dimeric, application of the simple algorithm previously suggested[39] gives a quantitative fit to the data.

A second important conclusion, implicit in the equilibrium treatment, is that the partial processes are reversible in the ms range; recent attention has been drawn to this feature by Osyczka *et al.*,[32] and will be discussed in detail below.

A third conclusion from this treatment is also important. The bifurcated oxidation of QH_2 separates the two electrons, delivering one to the high potential chain of ISP, cyt c_1 and cyt c_2, and the other to the low potential *b*-heme chain. The high and low potential chains come rapidly to their own internal equilibria, and the two are in equilibrium, through the Q_o site, with the Q pool. However, the two chains do not otherwise exchange electrons in the ms range. This means that there are, on the timescale of normal turnover, no rapid bypass or short-circuit processes to lower the efficiency of the bifurcation or of the proton-pumping reactions that depend on it. Indeed, under static head conditions, the same equilibrium constants, modified only by the electrical backpressure along the low potential chain, were shown also to account for the redox poise of the photosynthetic chain.[45, 46] However, although the bifurcated reaction occurs with high efficiency, bypass or short-circuit reactions representing 1–3% of the maximal rate can be readily demonstrated in the presence of inhibitors.[47, 48] These are thought to involve the semiquinone intermediate formed at the Q_o site.[49–51]

Several recent publications have aimed to address the complexity of the bc_1 complex, [32, 52–55] in some cases by simplifying assumptions about the mechanism of the Q_o site. As noted above, Osyczka *et al.*[32] extended the equilibrium approach to show that the values for equilibrium constants calculated from redox measurements still

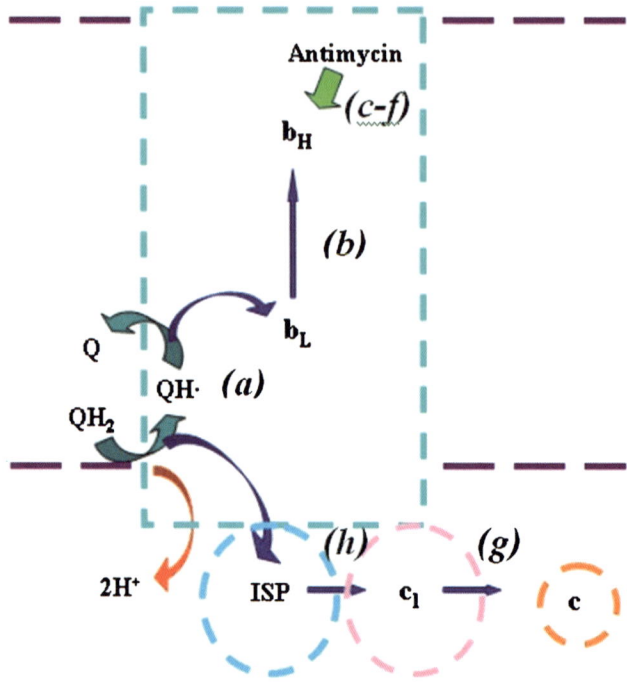

Figure 2 *The partial reactions of the Q-cycle in the presence of antimycin. By limiting the turnover at the Q_i site by the inhibitor antimycin, the reactions are constrained to the simpler scheme shown here, involving only the turnover of the Q_o site. The letters identify reactions shown in Table 1*

explained the distribution of electrons in chains truncated by mutagenesis. They emphasized the reversibility of the bifurcated reaction in the ms range, and discussed implications for mechanism in the context of rate constants expected from distances revealed by structures. They identified difficulties arising from attempts to account for forward, reverse, and bypass rates at calculated occupancies of the intermediate SQ, which has a putative role of in all these processes. To overcome these, they proposed a bold hypothesis that rejected all mechanisms involving SQ intermediates, and replaced them with two possibilities, a genuinely 'concerted' mechanism or a double-gated sequential mechanism in which SQ formed would be immediately consumed. However, in simplifying the problem, they overlooked complexities that allow alternative explanations. I will demonstrate that both of their mechanisms have problematic aspects, but that one of the rejected mechanisms involving a weakly populated SQ intermediate,[17, 52, 53, 56, 57] can account quite naturally for the experimental observations. A recent review by Rich [54] has a useful discussion of mechanism using a simple treatment through rate and equilibrium constants based on redox properties of the $Q/SQ/QH_2$ system, following earlier treatments involving an unstable SQ intermediate.[17, 54, 58] The reaction pathway discussed is similar to that previously proposed,[56] but the discussion also ignores complexities.

2 Control of Turnover by the Bifurcated Reaction at the Q$_o$ Site

Oxidation of QH$_2$ at the Q$_o$ site is the rate-limiting step, and tight control of the bifurcation of electrons is essential. With additional physicochemical data, and the availability of structures from crystallography starting about seven years ago,[5–7, 11, 13, 44, 55, 59–65] we now have important information that has allowed a deeper exploration of this central catalytic function. Several key features from structural analysis bear on the question of how the bifurcated reaction is controlled:

(i) The modified Q cycle maps naturally onto the structures.
(ii) The Q$_o$ site can be identified as the binding site for two main classes of inhibitors:
   (a) Those forming H-bonds to His-161 of ISP. They include stigmatellin, HHDBT, UHDBT, NQNO, and atovaquone (model structure). These necessarily occupy the domain of the Q$_o$ site near the *b*-interface, distal from heme b_L (distal domain). Biophysical evidence has shown that the reduced ISP (ISPH), in which His-161 is protonated, is the active form. The $>$C=O group of stigmatellin,[6] or a $>$C–O$^-$ group of HHDBT,[62] serves as H-bond *acceptor* from the imidazole $>$N$_\varepsilon$H of the *reduced* ISP. Stigmatellin is additionally H-bonded by Glu-272 (chicken numbering) from cyt *b*, with an −OH (across the chromone ring from the H-bond with His-161) acting as H-bond donor.[56]
   (b) Inhibitors binding in the domain proximal to heme b_L, and forming H-bonds with the peptide −NH of Glu-272.[57] These include myxothiazol, azoxystrobin, famoxazone, MOA–stilbene and related MOA inhibitors. With the exception of famoxazone,[65] all structures containing these inhibitors show the ISP displaced towards the *c*-interface.
(iii) The Q$_i$-site is identified through occupancy by inhibitors (antimycin, NQNO), or quinone. The binding pocket allows van der Waals contact with the heme b_H edge, and H-bonding of quinone or SQ that may be through sidechains (Asn-221, His-217, Asp-252 in *Rb. sphaeroides*) or water.[61,66–68]
(iv) Structures showing cyt *c* at a binding interface on cyt c_1 confirm the location of the third catalytic interface.[9,69,70]
(v) The extrinsic domain of the iron–sulfur subunit (ISP) is found in several different positions in different structures (at least eight to date). Berry and colleagues[6,71,72] suggested that the domain must act as a mobile shuttle of reducing equivalents between the Q$_o$ site and cyt c_1. This conclusion has been amply supported by biophysical, biochemical and molecular engineering studies[73–80]. The two interfaces for electron transfer are:
   (a) A concave external surface of cyt *b* (the *b*-interface) from which an access port allows N$_\varepsilon$ of His-161 of ISP (beef numbering) to form H-bonds with certain occupants (inhibitors,[6] and likely also Q and QH$_2$[56, 67, 81–84]) of the Q$_o$ site. His-161 is one of the ligands to the [2Fe-2S] cluster.
   (b) The *c*-interface on cyt c_1, at which N$_\varepsilon$ of His-161 forms an H-bond bridge to a heme c_1 propionate, with a distance appropriate for rapid electron transfer.[7]

(vi) Mutations giving rise to resistance to inhibitors for the most part map to the internal surfaces of the two inhibitor-binding domains.[11, 57, 69]

(vii) In the Q_o site, mutations giving rise to myxothiazol resistance also give rise, in the absence of inhibitor, to a slowing of electron transfer to the *b*-heme chain, suggesting that these mutations block occupancy of this proximal domain by some quinone species during the catalytic cycle, and thus inhibit turnover.[57]

(viii) None of the structures has shown any well-defined quinone occupant for the Q_o site. However, a recently deposited data set (PDB ID 1ntz) included coordinates for ubiquinone, albeit with B-factors that indicate a rather weak degree of confidence in assignment of electron density. The quinone was in the distal domain, but the distance from the quinone $>C=O$ to N_ϵ of His-161 (4.54 Å) was greater than expected for the H-bond suggested from EPR-studies.[67]

3   The ES Complex

The structural features summarized above have opened a flood of new experimentation, discussion, and speculation about mechanism, much of it reviewed earlier.[8, 10–12, 14, 52, 64, 69, 75] Rather than repeat this discussion, I will focus on aspects that remain controversial. Perhaps most pertinent to the present discussion is the suggestion that the ES complex, from which electron transfer proceeds, is formed at the distal end of the Q_o site, with the QH_2 forming two H-bonds. In one of these, –OH is a *donor* to the dissociated N_ϵ of His-161 from the *oxidized* ISP (ISP_{ox}), and in the other –OH is a donor to Glu-272 of cyt *b*.[53, 56, 85] A quinol can be modeled in place of stigmatellin without significant distortion of the structure, to provide distances appropriate to these H-bonds.[55, 56] Independent evidence in support of such a configuration comes from kinetic measurements to assay the binding forces involved in formation of the ES complex, and changes in these values found in mutant strains.[56, 86] The main enthalpic contributions are likely to come from the H-bonds above. Since QH_2 and His-161 of ISP form a mutual H-bond, it might be expected from simple mass-action considerations that formation of the ES complex would pull QH_2 'out of solution' in the lipid phase, and the dissociated active form of ISP_{ox} 'out of solution' from the mixed states of occupancy of the extrinsic domain in the P phase. Two effects quantified in a number of different labs can be associated with this mutual binding (reviewed in[53]). These are an apparent shift in the E_m of the Q/QH_2 couple involved in formation of the ES complex, and an apparent shift in the pK of ISP_{ox}, both assayed kinetically through [ES], measured by the rate of reduction of heme b_H in the presence of antimycin. These two displacements show similar driving forces.[53, 85]

The configuration of the ES complex suggested has important mechanistic implications. Indeed, the success with which the kinetic properties can be accounted for in terms of this starting configuration could be taken as additional justification for its validity. This configuration is shown in Figure 3 (top). The model provides a starting point for discussion of mechanism. Most importantly, because of the structural context, the discussion can be framed in light of the dependence of rate constant on distance demonstrated by Dutton's group.[87–90] In order to bring out the utility of this

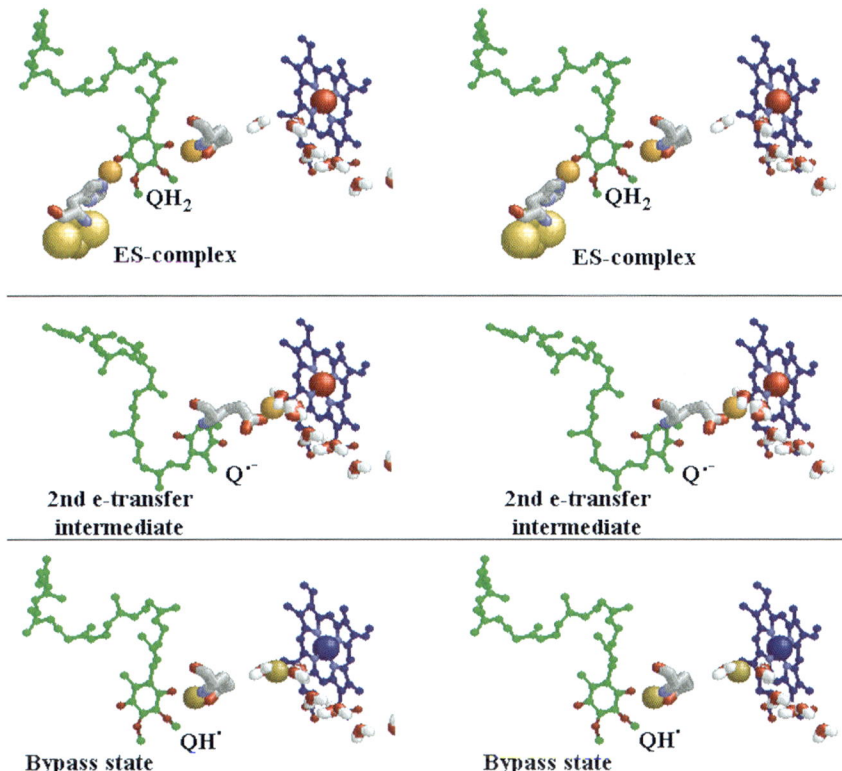

Figure 3 *Different states of occupancy of the Q_o site by ubihydroquinone and semiquinone. Top. The ES complex. The structure of the chicken complex in the presence of stigmatellin (from 2bcc) was modified as follows; the stigmatellin coordinates were stripped from the file. A ubiquinone molecule (ball-and-stick model with C-atoms in green) was then docked in the Q_o site as described in ref.[56] The model was allowed to relax with constrains on the position of ubiquinone through tethers to the two proposed ligands, to remove van der Waals clashes, followed by relaxation without constraints. Ligands are shown as tube models, with His-161 of ISP on the left, and Glu-272 of cyt b on the right. Heme b_L is shown as a ball-and-stick model, with C-atoms in blue. In all models, the protons associated with the mechanism of QH_2 oxidation are shown as van der Waals radius spheres, colored orange. Middle. The state from which the second electron transfer occurs. Heme b_L is oxidized (Fe colored red), and the Q^- intermediate can access the proximal domain. For this model, the quinone was relaxed in the context of a structure from which the myxothiazol occupant was replaced. Bottom. The bypass state. When heme b_L is reduced (Fe colored blue), oxidation of SQ cannot occur, and coulombic repulsion prevents formation of Q^- and coulombic attraction discourages H^+ exit down the water chain. Both effects contribute to keeping QH in the distal domain. The strong positive $\Delta G^{0'}$ keeps the occupancy of SQ at $< 10^{-6}$. Stereo pairs for crossed-eye viewing*

approach, note that electron transfer events in the pseudo-solid-state of the protein interior are first order, with rates dependent on fractional occupancy:

$$v = k_{cat}[\text{occupancy}] \tag{1}$$

This relationship is familiar from classical enzyme kinetics, where the fractional occupancy of the ES complex is assayed through the steady-state rate. The term [occupancy] here refers to the fraction of centers in the state from which reaction proceeds; for redox reactions, this will be determined by the product of donor and acceptor occupancies. In light of the above relationship, a distance from the structure can provide constraints, either if v or [occupancy] is known. A value for $\log_{10}k_o$ is calculated from the distance, R (see Equation 2), using the first two terms from the Moser–Dutton equation (with $\beta = 1.4$), and k_{cat} can then be estimated if appropriate values are available for the other parameters, ΔG^0 (standard free-energy change for the reaction, or driving force), and λ (Marcus reorganization energy), included in the right-hand term.

$$\log_{10} k_{cat} = 13 - \frac{\beta}{2.303}(R - 3.6) - \gamma\frac{(\Delta G^0 + \lambda)^2}{\lambda} \qquad (2)$$

The latter values are related to the activation energy though the Marcus expression, $\Delta G^{\#} = (\Delta G^0 + \lambda)^2/4\lambda$, suggesting several experimental routes for determination. Note that Equation 2 is an expanded form of the Arrhenius equation in $\log_{10}$ form, with the right-hand term replacing the Arrhenius activation barrier. The factor γ is either (F/(4 × 2.303 RT) ∼ 4.23 at 298 K) in the classical Marcus treatment,[91, 92] or a similar term modified by quantum mechanical contributions due to tunneling, with a value of 3.1 in the Moser–Dutton treatment. The Marcus value gives lower rate constants than the Moser–Dutton term, especially for high values of λ. The Arrhenius equation and its various derivatives also provide constraints on rate. Under conditions of substrate saturation, $k_{cat} = k_0\exp\{-\Delta G^{\#}/RT\} \approx k_0[ES^{\#}]$. This is because, in the standard treatment, $-\Delta G^{\#}/RT = \ln K^{\#}$, and $K^{\#} = [ES^{\#}]/[ES]$, so that $[ES^{\#}]$ becomes equal to the fractional occupancy of the activated state if $[ES^{\#}] \ll [E_{tot}]$.

4   Constraints from the ES-complex Model

Rate constants and activation barriers measured for the partial processes are shown in Table 1. In discussion of their results, Hong *et al.* [93] assumed that the bifurcated reaction occurred through two separate one-electron transfer reactions, with a finite, though low, occupancy of the intermediate bound SQ state $(QH^{\bullet}).E.b_L$.

$$QH_2 + ISP_{ox} + E.b_L \leftrightarrows ISP_{ox}.QH_2.E.b_L \leftrightarrows ISPH.(QH^{\bullet}).E.b_L$$

$$\leftrightarrows ISPH + (QH^{\bullet}).E.b_L \qquad (3)$$

$$(QH^{\bullet}).E.b_L \leftrightarrows Q + E.b_L^- + H^+ \qquad (4)$$

This assumption was based on a number of observations. A dependence of rate on driving force is predicted from the Marcus relationship. Although the E_m values of the SQ couples involved are unknown, on the assumption that these do not change, the driving forces for the two one-electron steps could be varied through changes in E_m values for their electron acceptors. The dependence of rate on the E_m of ISP, the acceptor in the high potential chain, has been studied in several labs through use of

ISP mutant strains with modified E_m.[53,82,86,94,95] The results showed that the rate of the overall reaction decreases with lowered driving force, as expected from Marcus theory if the first electron transfer is limiting. A remarkable uniformity in behavior across the bacterial–mitochondrial divide, and among different bacteria, provides a high level of confidence that the results are not due simply to non-specific effects. In contrast, no simple Marcus behavior has been reported on modification of the E_m of heme b_L, the acceptor in the low potential chain.[93] In all reported cases, the rate was lowered or unaffected when the E_m of heme b_L was raised by mutation. From this, it seems reasonable to conclude that transfer of the first electron from QH_2 to ISP determines the overall rate of the bifurcated reaction. In this sequential mechanism, SQ is an intermediate product, removed on transfer of the second electron to heme b_L.

An important lesson from these results is that they are in contradiction with any genuinely 'concerted' reaction.[53, 93] The two electron transfers to two distinct chains would have to occur from a common activated state with the same intrinsic rate constant and other Marcus parameters. If such a reaction did occur, both steps would be under the same constraint of driving force, and, apparently, they are not.

The alternative is some sort of sequential mechanism. What constraints are provided by the two hypotheses above, that the first electron transfer is rate limiting, and that the ES complex has the structural configuration proposed? A consequence of these two hypotheses is that the first (limiting) electron transfer occurs through the ~ 7 Å distance from the O atom of the quinol –OH to the nearest Fe atom of the cluster. The distance is bridged by the H-bond to His-161 and the histidine ring, providing a plausible though-bond path of this distance. With values for the Moser–Dutton variables in conventional ranges ($\beta \sim 1.4$, $\lambda \sim 0.7$ V, $\gamma - 3.1$), the expected rate constant would be in the sub-μs range. Even with the endergonic $\Delta G°$ appropriate for an unstable SQ, the expected rate is at least three orders of magnitude faster than measured. The problem is therefore how to explain the slow rate observed. The rate can be accounted for only if a value for $\lambda > 2.0$ V is used. This is consistent with the high activation barrier for the overall reaction ($\Delta G^{\#} \sim 550–650$ mV), supporting the hypothesis that this barrier is in the first electron transfer.[17,93] However, this is not altogether satisfactory, because there are no obvious features of the structure that might explain the high value for reorganization energy, λ, and the linked high $\Delta G^{\#}$.

5   Proton-coupled Electron Transfer

In order to proceed further, we now have to consider several complicating factors. The first of these is the involvement of a proton in the first electron transfer. In order to form the proposed H-bond, either His-161 of ISP_{ox} or the quinol –OH would have to lose a proton. It now seems well established that the pK of 7.6 on ISP_{ox} is due to dissociation of one of the histidine ligands to the cluster, likely His-161.[96–99] If so, the probabilities for dissociation are easily calculated from the pKs, and they favor by a factor of ~ 1000 the dissociation of His-161. However, on reduction, the pK of ISP is shifted from 7.6 to 12.4, and this would lead to uptake of H^+. At the same time, the pK of the quinol would shift on oxidation from > 11.3 to the acid range for $QH^{\cdot +}$, favoring release of a H^+ to yield $QH^{\cdot}$. The net result would be transfer of both the electron and a proton to ISP. What does this entail mechanistically? If the configuration of the protein interface in the ES-complex state is similar to that in the stigmatellin complex, then the interface around the H-bond is anhydrous and apolar; there are no groups immediately available for proton exchange. The electron and the proton must therefore both be transferred across the H-bond joining the two groups. Similar reactions in chemistry are discussed as proton-coupled electron transfer reactions.[100, 101] We then have two obvious possibilities for the sequence of the partial processes: electron first or proton first.[85, 99] Since the proton transfer probability is determined by the pKs of donor and acceptor groups (QH_2, ISP_{ox}) through a Brønsted relationship, the different probabilities can be calculated from the pKs using a thermodynamic square and the overall free-energy change.[85] For the endergonic first electron transfer expected, transfer of the proton first is strongly favored, despite the fact that this step is itself endergonic ($\Delta G^0 = 2.303RT(pK_D-pK_A) > 21-27$ kJ mol^{-1}, depending on choice of bound or free pK values for ISP_{ox}). The slow rate can then be readily explained by the fact that electron transfer occurs from this weakly occupied intermediate state. The electron transfer rate constant, k_{ET}, has the value expected from the distance and a conventional value for λ, but the observed rate constant, given by $k_{app} = k_{ET}K_{proton}$, where K_{proton} is between 10^{-4} and 10^{-6}, is lowered by the weak occupancy. The first electron transfer can now be dissected into the partial processes shown in Figure 4, with approximate $\Delta G^{0\prime}$ values (in electrical units), and rate constants, as shown in the legend. Expanding the relationship for k_{app} into the Moser–Dutton equation gives:

$$\log_{10} k_{app} = 13 - \frac{\beta}{2.303}(R - 3.6) - \gamma\frac{(\Delta G_{ET}^0 + \lambda)^2}{\lambda} - (pK_D - pK_A)$$

In addition to explaining the paradoxical properties of the first electron transfer, this equation also accounts for the previously unexplained kinetic behavior of an ISP mutant strain in which the pK was changed from 7.6 to 8.5.[53, 86]

A third possibility for the sequence of partial processes is simultaneous transfer of H^+ and e^-, essentially along the diagonal of the thermodynamic square.[99, 100] Although not excluded, such mechanisms have the disadvantage that the diagonal pathway would necessarily be of lower energy than the alternatives, removing the explanation for the slow rate provided by the proton-first mechanism.

6   The Second Electron Transfer, From SQ to Heme b_L

Consideration of the second electron transfer brings up another complicating factor. Attempts to measure the intermediate SQ state through EPR have failed to detect any species with the expected properties.[17, 58] It might be argued that this is not surprising. In order to reduce heme b_L, the Q/SQ couple would have to have an $E_m < -90$ mV. Substitution of values in this range into the standard equations (using an E_m of 90 mV for the quinone pool), give values for the SQ/QH$_2$ couple appropriate for reduction of ISP at an E_m of 300. A stability constant for the disproportionation reaction, $K_S \leq 10^{-6}$, can be calculated from these E_m values, giving an equilibrium occupancy in the range of $10^{-4} - 10^{-3}$ monomer^{-1}, which would be undetectable. Osyczka *et al*[32] discussed the mechanistic occupancy at the Q$_o$ site in terms of K_S. However, the stability constant is clearly not appropriate for calculation of mechanistic occupancies. The reaction

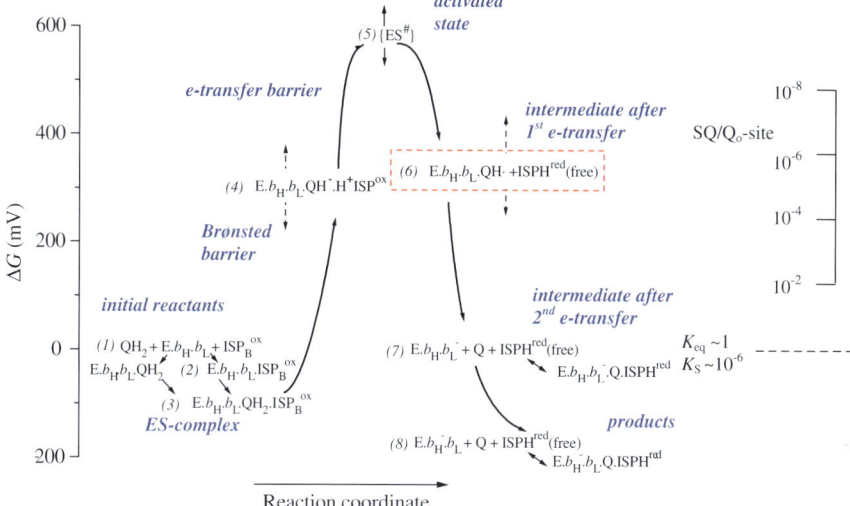

Figure 4 *An energy landscape for the Q_o-site reactions. In order to keep track of the thermodynamic interplay in this complex set of reactions, a computer program to model the Q_0-site reactions was written that balances the terms within appropriate physicochemical constraints, and allows the user to explore how each of the parameters discussed in the text effects the rate. Variables include driving forces for both electron transfer steps, relative binding constants for Q and QH₂, stability of the SQ, λ for each electron transfer reaction, contributions of the Brønsted term, distances, activation barriers, pK values, etc. The Visual Basic program in executable and source form is available for downloading.[125] The data are presented as free-energy differences, but it should be obvious that conventional conversions allow alternative presentation in terms of equilibrium constants and rate constant, using the data from Table 1 and below. The inserted scale to the right shows occupancies for the SQ state, based on equilibrium constants for the one-electron transfer reactions, in units of [SQ] [bc₁ monomer]⁻¹. Partial reactions for the first electron transfer are:*

Reaction	Process	$\Delta G^{0'}$	$k_{forward}$	$k_{reverse}$
$QH_2 + ISP_{ox} + E.b_L \leftrightarrows ISP_{ox}.QH_2. E.b_L$	(1–3)	−80*	1.4×10^5 M⁻¹s⁻¹	~10⁴ s⁻¹
$ISP_{ox}.QH_2. E.b_L \leftrightarrows ISPH^+.QH^-. E.b_L$	(4)	360	10^5 s⁻¹	10^{11} s⁻¹
$ISPH^+.QH^-. E.b_L \leftrightarrows \{ISPH^+QH^-\}^\#.E.b_L$	(5)	300	10^6 s⁻¹	10^{11} s⁻¹
$\{ISPH^+QH^-\}^\#.E.b_L \leftrightarrows ISPH + (QH^\bullet).E.b_L$	(6)	−300	10^{11} s⁻¹	10^6 s⁻¹
$QH_2 + ISP_{ox} + E.b_L \leftrightarrows ISPH + (QH^\bullet).E.b_L$	(1st e⁻)	280	1.3×10^3 s⁻¹	6×10^7 s⁻¹

In these equations heme b_H has been omitted for clarity, $ISP_{ox}.QH_2. E.b_L$ is the ES-complex, and $\{ISPH^+QH^-\}^\#.E.b_L$ is $\{ES^\#\}$ in the Figure. Equation identifiers refer to numbers in the Figure. Values are approximate within the ranges shown by arrows on the scheme (see [93] for details).

*Note. The value in electrical units is derived from apparent binding constants and is not normalized to concentration.[53]

under consideration is not the equilibrium disproportionation of Q and QH$_2$, but the bifurcated electron transfer to two separate acceptors. This does not occur at redox equilibrium, but only under the metastable conditions in which the quinone pool is reduced and ISP oxidized. For the sequence of reactions shown by Equations 3 and 4, the equilibrium constants could in principle be calculated separately, but with the constraint that $K^3_{eq} \cdot K^4_{eq} = K_{Q0}$, the equilibrium constant for the bifurcated reaction. The value for K_{Q0} is between 1 and 20 (depending on species, and choice between bound and free Q/QH$_2$). If the E_m values were as above, K_{eq} would be ~1 for each process in *Rb. sphaeroides*, and a SQ occupancy of ~0.33 monomer^{-1} would be expected, well within the range of detectability.

Nevertheless, a much lower occupancy is expected for a different reason. The bc_1 complex, when operating in the presence of antimycin, or under other conditions (high proton motive force, mutation in the Q$_i$ site) leading to an inhibition of flux out of the Q$_0$ site, generates superoxide, a precursor to the reactive oxygen species that lead to DNA damage and cellular aging.[48, 49] The reductant for O$_2$ is thought to be the SQ generated at the Q$_0$ site.[49] In order to minimize this deleterious reaction, it is likely that evolution has designed the site to keep the SQ at as low a concentration as possible compatible with rapid forward electron transfer, and to insulate it from reaction with O$_2$.[57]

7   Kinetic Estimation of SQ Occupancy

In order to find limiting values for the lowest kinetically compatible SQ occupancy, Hong *et al.*[93] explored a range of scenarios for the second electron transfer, based on use of Equation 1 in consideration of the parameters noted above, the constraints on rate constant from distances shown by the structures, and the occupancy. They considered three possibilities:

(i) That the SQ remained in the location at the distal end of the site where it was formed.

(ii) That the SQ could move from this site to the proximal domain close to heme b_L.

(iii) That a second quinone could bind in the domain proximal to heme bL and facilitate electron transfer between the Q/SQ generated in the distal domain and heme b_L.

The reaction configuration summarized under (i) would strongly constrain the kinetic values because of the long distance for transfer of the second electron from SQ to heme b_L. For example, with $\lambda = 0.75$ V, $\Delta G^o = -.38$ V, and a distance of 12.4 Å, k_{cat} is ~ 1.22×10^7 s^{-1}. As a consequence, any SQ occupancy lower than ~10^{-4} would give a rate lower than that measured, and was therefore considered to be unrealistic in this scenario.

Mechanisms in which the intermediate state is constrained to the distal domain include variants of that proposed by Link,[102] and discussed by Hong *et al.*[93] Berry and Huang[55] have also discussed a version of this mechanism, in which the ISP$_{ox}$.QH$_2$ ⇋ ISPH.QH· equilibrium mix acts as the electron transfer state from which heme b_L is reduced. The fractional occupancy of the right-hand state would be the critical term, and could be determined by an equilibrium constant higher than allowed in models in which dissociation occurs. Kinetic measurements provide constraints on the highest possible occupancy of intermediate states.[52] The ~120 μs lag in reduction of heme b_H following flash activation of turn-over in *Rb. sphaeroides* chromatophores must be accounted for in terms of all delays in delivery of substrates to the Q_o site following the flash. With the quinone pool partly reduced, these delays are in the high-potential chain, and in the population of intermediate states in the bifurcated reaction. If, as we have demonstrated, the intermediate SQ state is after the activation barrier, it must be formed at the limiting rate. Since lags in the high potential chain account for ~100 μs, maximal occupancy is given approximately by (unaccounted lag)/(limiting lifetime), or 20/770 ~ 0.025 (lag and lifetime values are in μs). Thus, mechanisms in this class are constrained to occupancies between ~10^{-2} and 10^{-4} SQ/monomer.

7.1 Estimation of SQ Occupancy from Bypass Rates

From the importance of occupancy of SQ in this discussion, it is clear that an alternative method for estimation of this value would be useful. A number of potential bypass (or short-circuit) reactions have been previously discussed, all of which involve SQ as donor or acceptor of electrons.[51] In view of this, Osyczka *et al.*[32] suggested an approach based on the rate of these reactions. The general idea follows the method already discussed in [93] and above. Using Equation 1, occupancy can be calculated from a rate constant estimated from the structures, and from the measured short-circuit rate. The four pathways discussed are shown in Table 2. In estimation of rate constants, Osyczka *et al.*[32] assumed that the SQ was positioned in the domain occupied by stigmatellin (the distal domain), and that the appropriate distances were either 6.8 Å to ISP for oxidation of SQ, or 12.4 Å for reduction by heme b_L^-. The first reaction discussed was the reduction of ISP_{ox} by the SQ formed after the first electron transfer (Table 2 (a)). From the limited information presented, it seems likely that they used a distance of ~7 Å, a value for λ ~1.0 V, and an assumed $\Delta G^{0\prime}$~0, to give k ~7×10^7 s^{-1}. Using the rate they measured (0.3 s^{-1}), this gives a SQ occupancy of ~4×10^{-9} monomer^{-1}, from which K_S ~ 10^{-16}, the value they gave. However, as noted above, the equilibrium constant, K_S, although a convenient shorthand for the E_m values of the one electron couples, is not related to the actual equilibrium constant for the electron transfer reactions, and its use could be misleading. A second complication is that the short-circuit rates they measured were an order of magnitude slower than those reported in the literature (*cf.*[48, 51] and see Table 3). Using a more realistic rate of 3 s^{-1}, and otherwise similar assumptions, occupancy would be ~4×10^{-8} monomer^{-1}. Thirdly, the reaction equations all involve electron transfer between donor and acceptor pairs; the calculated occupancy of SQ would have to be greater if the occupancy of its reaction partner were < 1. With these caveats, we can use the occupancy of SQ and the arguments above to calculate a rate for the second electron transfer required for QH_2 oxidation. The rate using a rate constant of ~1.22×10^7 s^{-1} (calculated as above) is then incompatible with the observed overall rate of turnover; the maximal rate would be <1 s^{-1} instead of 1.3 $\times 10^3$ s^{-1}. In light of the argument leading to this rate constant, and as previously

Table 2   *Reactions proposed for bypass or short-circuit of the bifurcated reaction*

Partial reactions leading to bypass	Occupancy	k_{prox} or (k_{dist})	Rate (s^{-1})	ID and refs.
$QH_2 + ISP_{ox} + E.b_L^- \leftrightarrows ISPH + (QH^{\bullet}).E.b_L^-$ $ISPH + cyt\ c_2^+ \leftrightarrows ISP_{ox} + cyt\ c_2 + H^+$ $(QH^{\bullet}).E.b_L^- + ISP \leftrightarrows Q + E.b_L^- + ISPH$ $ISPH + cyt\ c_2^+ \leftrightarrows ISP_{ox} + cyt\ c_2 + H^+$	4×10^{-8}	7×10^7	2.8	(a) {c} 32, 51
$QH_2 + ISP_{ox} + E.b_L^- \leftrightarrows ISPH + (QH^{\bullet}).E.b_L^-$ $ISPH + cyt\ c_2^+ \leftrightarrows ISP_{ox} + cyt\ c_2 + H^+$ $(QH^{\bullet}).E.b_L^- + O_2 \leftrightarrows Q + E.b_L^- + O_2^-$ $O_2^- + cyt\ c_2^+ \leftrightarrows O_2 + cyt\ c_2 + H^+$	4×10^{-8}	7×10^7	2.8	(a') 32, 45, 51
$E.b_L^- b_H^- + Q + H^+ \leftrightarrows (QH^{\bullet}).E.b_L b_H^-$ $(QH^{\bullet}).E.b_L b_H^- \leftrightarrows (QH^{\bullet}).E.b_L^- b_H$ $(QH^{\bullet}).E.b_L^- b_H + H^+ \leftrightarrows QH_2 + E.b_L b_H$ $QH_2 + E.b_L b_H + cyt\ c_2^+ \leftrightarrows Q + E.b_L b_H^- + cyt\ c_2 + 2H^+$ $QH_2 + E.b_L b_H^- + cyt\ c_2^+ \leftrightarrows Q + E.b_L^- b_H^- + cyt\ c_2 + 2H^+$	0.2 4×10^{-11}	1.2×10^3 $<1 \times 10^{11}$	240 <4	(b) {d} 32
$QH_2 + ISP_{ox} + E.b_L^- \leftrightarrows ISPH + (QH^{\bullet}).E.b_L^-$ $(QH^{\bullet}).E.b_L^- + H^+ \leftrightarrows QH_2 + E.b_L$ $QH_2 + ISP_{ox} + E.b_L \leftrightarrows Q + ISPH + E.b_L^- + H^+$ $2(ISPH + cyt\ c_2^+ \leftrightarrows ISP_{ox} + cyt\ c_2 + H^+)$	4×10^{-8}	1×10^{11} or (5×10^7)	4000 or 2	(c) {e} 32, 45, 51
$E.b_L^- b_H^- + Q + H^+ \leftrightarrows (QH^{\bullet}).E.b_L b_H^-$ $(QH^{\bullet}).E.b_L b_H^- + ISP_{ox} \leftrightarrows Q + E.b_L b_H^- + ISPH$ $ISPH + cyt\ c_2^+ \leftrightarrows ISP_{ox} + cyt\ c_2 + H^+$ $QH_2 + E.b_L b_H^- + cyt\ c_2^+ \leftrightarrows Q + E.b_L^- b_H^- + cyt\ c_2 + 2H^+$	0.2 4.10^{-8}	1.2×10^3 7.10^7	240 2.8	(d) {f} 32

Notes:

For each of the above sets of partial processes, the overall reaction is the scalar part of the reaction catalyzed by the uninhibited Q_o site, $QH_2 + 2\ cyt\ c^{3+} \leftrightarrows Q + 2\ cyt\ c^{2+} + 2H^+$. The letters in the right-hand column identify either (in parentheses) reactions referred to in the text, or (in curly brackets) the equivalent reaction in Figure. 6 of.[32] Values for occupancy, rate constants and rate are calculated as described in the text.

Table 3 *Rates of bypass reactions in wild type and mutant strains*

Reaction used to measure bypass	Inhibitor	Bypass Rate (e/bc$_1$/s)	QH$_2$ oxidation (s^{-1})	QH$_2$ oxidation measured through:	Bypass as % QH$_2$ oxidation	Ref.
Heme b_L reoxidation at pH 7	H212N (heme b_H removed)	0.32	1200	redn. heme b_H (in WT)	0.026	32
Cyt c reduction, yeast bc$_1$ complex[a]	antimycin	3.25	160	cyt c redn. with no inhibitor	2.03	51
Flux into high pot. chain, Q-pool red[b]	antimycin, anaerobic	4–8	600	reduction heme b_H	~1.0	45
RC reduction[c]						
WT	antimycin, aerobic	1.35	237	Reduction of heme b_H with antimycin, or (RC redn. with no inhibitor)	0.6	See note[d]
E295W		1.5	2.0		75	
E295L		2.8	7.18		39	
E295G		1.46	2.7		54	
E295Q		1.76	9.8		18	
E295D		1.56	14 (23)		11 (7)	
E295K		1.25	3.2 (6.7)		39 (19)	

[a] The bypass rate due to O$_2^-$ production was measured as 2.26 s^{-1} electrons/Q$_o$ site; that due to reduction of semiquinone by heme b_L was 0.99 s^{-1}.
[b] Myxothiazol sensitive, antimycin insensitive rate of RC reduction, anaerobic conditions, E_h 120 mV.
[c] Aerobic conditions with 1 mM Na-ascorbate added as reductant (~1 QH$_2$/bc$_1$), 2 mM KCN to block oxidases.
[d] Lhee, S., Crofts, S.B., Li, J.-Y., Rose, S. and Crofts, A.R., in preparation.

noted,[93] we can effectively eliminate from consideration hypotheses in which a reaction intermediate is constrained to the distal domain, and has an occupancy in this range. A similar constraint applies to 'concerted' reactions, for which an occupancy for $[ES^\#] \sim 10^{-10}$ monomer^{-1} is determined by $\Delta G^\#$.

Osyczka *et al.* [32] framed their main discussion around the lower occupancy they calculated, the difficulties that even this low occupancy gave rise to in understanding how short-circuits are avoided, and how the rapid rate for back reactions needed to explain the reversibility could be accounted for. Indeed, if the rate constant for reduction of Q by heme b_L^- is calculated assuming occupancy of the distal position, and the highly endergonic reaction implied by the value of E_m Q/SQ derived by calculation from their K_S value, the back-reaction rates would certainly be much too slow to explain reversibility. However, Berry and Huang[55] made the interesting point that their equilibrium mix could be a potential substrate for the backreaction from heme b_L^-. The (ISP$_{ox}$.QH$_2 \leftrightarrows$ ISPH.QH·) state differs from the enzyme-product complex represented by the $g_x = 1.800$ complex (ISPH-Q) by one reducing equivalent. The latter is stabilized in the distal domain by the binding constant involved in its formation, with occupancy ~ 0.5 under physiological conditions.[31,103,104] Any value for $\Delta G^{0\prime}$ for formation of ISPH.QH· <0.3 V (with $\lambda \sim 0.7$ V) would give a rapid enough back reaction. However, since this complex is constrained to the distal domain, the arguments above place limits on the occupancy of this SQ state for viable forward rates.

8 Mobility in the Q_o site

The second possibility considered by Hong *et al.*[93] ((ii) above), opens up a wider range of possibilities, but requires particular characteristics of the reaction mechanism. They suggested that, after formation of SQ, the intermediate EP complex dissociates to liberate ISPH and SQ. A mechanism in which ISPH dissociates from the intermediate ISPH.SQ complex also seems to be required to account for bypass reactions.[45,51] Dissociation to liberate SQ in the Q_o site allows movement close to heme b_L (Figure 3, middle), and this can change the rate constant dramatically. Taking a distance of <6.3 Å (the distance from myxothiazol to heme b_L), but otherwise the same parameters as above, k_{cat} has a value $> 6.5\ 10^{10}$ s^{-1}. The calculated rate using the occupancy above is then in the range observed experimentally.

Arguments similar to those developed above can be used in consideration of the back reaction. The electron acceptor from heme b_L^- is Q, which would also likely be mobile in the Q_o site, and therefore able occupy the proximal domain. Assuming an endergonic $\Delta G^{0\prime}$ of <0.48 V, the reverse rate constant, $k_{-2} \geq 1.2 \times 10^3$ s^{-1}, would certainly be adequate to explain reversibility, even with an occupancy of Q < 1.

From this discussion, it is clear that the scope for plausible mechanisms is greatly increased by allowing consideration of mobility for SQ and Q in the Q_o site.

The values used in outlining the arguments above are approximations. In particular, we have used the Moser–Dutton term in all calculations, rather than the classical Marcus term, and contributions of protolytic processes (except for the first electron transfer) have not been separated out. The calculations should therefore be seen as illustrative rather exact. Hong *et al.*[93] discussed a wider range of scenarios, and the limitations on the approximations they used. I will not elaborate further here, except to reiterate that values for K_{eq} for the reactions involving SQ are more appropriately calculated from the E_m of the bound rather than free couples (E_m (Q/QH$_2$-bound) ~130 mV rather than 90 mV for the pool). We have used the same equation for both exergonic and endergonic processes, since, if a value for γ of 4.23 (appropriate to the Marcus treatment) is used, the same result is obtained with equations recommended for exergonic[88] or endergonic[90] reactions. However, the endergonic treatment recommended by Page *et al.*[90] gives a change in slope at the exergonic to endergonic transition, because the quantum mechanical correction leading to $\gamma = 3.1$ is applied to only a part of the endergonic activation barrier (see Appendix 1).

The energy landscape of Figure 4 summarizes the results of an exploration of the Q$_o$ site reactions compatible with low SQ occupancy, using the computer program described in the legend. The free-energy differences between different intermediate states of the bifurcated reaction are shown plotted against an arbitrary reaction coordinate. The dependence of all rates on occupancy, and the interdependence of the bound Q, SQ, QH$_2$ system parameters, introduce some interesting play-offs between driving force and occupancy that have to be considered in the context of the different choices for distance and standard thermodynamic constraints, as discussed at length by Hong *et al.*[93] Since values for several parameters are unknown except within limits, some energy levels are shown by ranges. These represent values that allow forward and reverse rates compatible with the data as summarized in Table 1, assuming the mobility of quinone species in the Q$_o$ site discussed above.

It seems clear from the above that some of the problems hinted at by Osyczka *et al.* [32] depend on choice of constraints. If the reactions of the quinone species are all limited by occupancy of the distal domain, rates involving the intermediate SQ state are problematic because of conflicting values needed to match measured rates of forward, reverse and bypass reactions.

9 Other Problematic Short Circuits and Their Prevention

Osyczka *et al.*[32] concluded that a value for $K_S < 10^{-16}$ (for bypass reactions (a) and (b), Table 2), or $K_S < 10^{-22}$ (for (c) and (d)) would be required to slow the short-circuits to the seconds range. They provided no detailed justification for these values, but they correspond to SQ occupancies of $<5 \times 10^{-9}$ and $<5 \times 10^{-12}$, respectively, and reflect the slow rate of bypass they measured, as discussed above. Because we have used short-circuit (a) as the basis for calculation of SQ occupancy, this provides a reference point rather than a problem. However, using the occupancy of 4×10^{-8}, some of the other short-circuits resulting from reduction of Q or SQ by heme b_L^- ((b)–(d)) do appear to provide difficulties. Two of these, (b) and (d), involve the reduction of Q by heme b_L^- to form SQ. These would occur in competition with the normal reverse reaction. The reaction would occur only under metastable conditions, would be strongly endergonic in the first step, and would lead to a low occupancy of the SQ state. In (b), the second electron transfer through heme b_L would occur from a low occupancy of the b_L^- state due to the large equilibrium constant with heme b_H^-. The product of occupancies is in the range

10^{-11}, compensating for even the highest rate constant expected (see below), to give a low net rate. In short-circuit (d), leading to the subsequent oxidation of SQ by ISP_{ox}, the latter rate would be limited by the constraints, already considered above, arising from the low occupancy of the SQ state. Reaction (c) involves the reduction of SQ, generated in the first electron transfer of the forward reaction, by heme b_L^-. This was previously considered by Chen[45] and by Muller et al.[51] as a bypass reaction leading to net oxidation of QH_2, as shown by the partial reactions in Table 2. The strongly exergonic reduction of SQ by heme b_L^- leads to a calculated rate constant much higher than that for reduction of Q. This would certainly be problematic if the SQ intermediate could be reduced in the proximal domain. For example, with SQ in the proximal domain, a value for $k_{cat} \sim 1 \times 10^{11}$ is calculated using $\Delta G^{0\prime} \sim -0.5$ V, and $\lambda \sim .75$ V, and this would allow rates in the 10^3 s^{-1} range even at the low occupancies suggested above. A similar situation would arise on reduction of Q through heme b_L (reaction (b)) if electron transfer between the b_L hemes across the dimer interface could occur. For comparison, the experimental rate, determined from the fraction of bypass that could be attributed to reduction of SQ by heme b_L^-, was about 30% of the total bypass in the presence of antimycin,[51] giving a rate of ~ 1 e$^-$ monomer^{-1} s^{-1}. Clearly, the thousand-fold discrepancy between the calculated and observed rates needs to be addressed.

9.1   Double-gating

As an alternative to their 'concerted' reaction, Osyczka et al.[32] considered solutions in which the bifurcated reaction operated with a SQ intermediate, but with double-gating to ensure that no reaction generating the SQ could take place unless the complementary reaction removing it also occurred. This would ensure that no build-up of SQ could occur, neatly explaining the low rates for short-circuit reactions they had observed, while allowing a higher occupancy of intermediate states than the 'concerted' mechanism. Criticism of this solution is difficult because the authors provided no specific suggestion as to how it might work; the justification provided was the low rate of bypass reactions. Since the short-circuit rates they measured were 10-fold slower than those measured by other groups, this justification is somewhat weakened. The relatively rapid rates of generation of superoxide by the bc_1 complex (1–3% maximal rate) have been the subject of an extensive literature, in which SQ at the Q_o site has been considered as the reductant for O_2^{49} (see Table 2, reaction a′). Similar rates are observed under anaerobic conditions, when the second electron is transferred to cyt c rather than to $O_2^{45,51}$ (reaction a). It might therefore be fruitful to discuss mechanisms in which the SQ is generated as a natural component of the reaction mechanism, even under inhibited conditions, rather than to reject all such mechanisms. In order to decide between these options, we must look more carefully at possible constraints on short-circuit rates.

9.2   Coulombic Gating in a Sequential Mechanism

If the mechanism is sequential, and involves SQ in the distal domain, the value of k_{cat} calculated for reduction of SQ by heme b_L^- is $\sim 5 \times 10^7$, which would match rates in the experimental range if occupancy was $\sim 10^{-8}$ SQ monomer^{-1}. We should note however that if the SQ in the distal domain were at an occupancy in the $>10^{-4}$ range (required for rapid electron transfer in the forward direction from this domain), this short-circuit would be in the range $>10^3$ s^{-1}. That such bypass rates are not observed means that mechanisms with high SQ occupancy fail this critical test. It seems therefore that an explanation for the slow bypass rates observed in the context of the models involving SQ must be sought in terms of constrains that prevent the SQ from accessing the proximal domain *under circumstances in which heme* b$_L^-$ *might be present*, while maintaining an occupancy $\leqslant 10^{-6}$ monomer^{-1} in the distal domain.

The sequential mechanism previously suggested,[53, 56, 93] involving a mobile SQ, provides a natural explanation for these otherwise paradoxical properties. To show how, we need to consider the fate of the second H$^+$, and the interplay of charges in the site. The structures show that Glu-272 can occupy two different configurations. When involved in H-bonding with the distal domain occupant (stigmatellin in known structures, QH$_2$ initially in our model of the ES-complex), it points away from heme b$_L$. When not so involved, the side-chain rotates $>120°$ so that the carboxylate group forms a H-bonding contact with a water molecule at the terminus of a water chain leading along the heme edge past one of the propionates to the P-phase water.[56, 60, 105] A natural pathway for proton exit from the site involves transfer of the H$^+$ from QH· (generated after the first electron transfer) to the Glu-272 carboxylate through the H-bond, to liberate the SQ anion, Q$^{\bullet-}$. This would be followed by rotation of the Glu-272 side chain (now in the acid form) to the configuration in contact with the water chain, followed by release of H$^+$ to allow exit to the P-phase.[56,85] In the configuration before rotation, Glu-272 and two H-bonded waters occupy most of the Q$_o$-site volume proximal to heme b$_L$. Rotation would open up the domain to occupancy by movement of Q$^{\bullet-}$ close to heme b$_L$. Rotation of Glu-272 and mobilization of Q$^{\bullet-}$, with the potential for movement to the proximal domain, would therefore be linked processes:

$$ISP_{ox}.Q_dH_2\cdots\overline{OOC}\text{-glu} \leftrightarrows ISPH_b + Q_d{}^{\bullet}H\cdots OOC^-\text{-glu} \leftrightarrows ISPH_c$$
$$+ Q^{\bullet-} + HOOC\text{-glu} \tag{5}$$

$$Q^{\bullet-} + HOOC\text{-glu} + \text{heme } b_L + \{\text{water chain}\} \leftrightarrows Q_p{}^{\bullet-}.\text{heme } b_L + {}^-OOC\text{-glu}$$
$$\{\text{water chain} + H^+\} \tag{6}$$

(where subscripts b and c to ISPH indicate location at the b- and c-interfaces, respectively, and subscript d and p to SQ species indicate location in the distal and proximal domains of the Q$_o$ site, respectively). In considering how this plays out in relation to bypass reactions, we need to take account of two effects:

(i) The location of Q$^{\bullet-}$ would likely depend on the state of heme b_L. With the heme reduced, coulombic repulsion of Q$^{\bullet-}$ by heme b$_L^-$ would favor location in the distal domain, and ensure that the Q$^{\bullet-}$ remained there unless the *oxidized* heme was available as an acceptor. With heme b$_L$ oxidized, movement of Q$^{\bullet-}$ to the proximal domain to facilitate rapid electron transfer would be unconstrained, but a new set of coulombic interactions would couple the transfer of the electron on to heme b_H to release of the H$^+$ from Glu-272 to the water chain.

(ii) In order to reduce Q *via* heme b_L, reversal of the H^+ transfer would be necessary. Although the neutral Q could access the domain proximal to heme b_L^- without coulombic consequence, generation of $Q^{\cdot-}$ on reduction of Q by heme b_L^- would have obvious consequences on re-reduction of the heme. In addition, reduction by a second electron would not be favorable until protonation could occur (this sequence has been well characterized in the Q_B site of bacterial reaction centers[106]), and this would require movement to the distal domain.

In general, the choreography of the second electron transfer in both directions will reflect the coulombic interactions between $Q^{\cdot-}$, the glutamate carboxylate, the H^+ in the water chain, and the electron on heme b_L. The two effects above provide the constraints necessary to prevent excessive bypass rates, and with this natural extension of the hypothesis, the kinetic problems seem to be satisfactorily explained. Gating reflects both the coulombic effects, and the proton exchanges involving Glu-272. The gating is not to prevent formation of SQ, but to ensure that it is constrained to the distal domain unless heme b_L is oxidized (Figure 3, bottom).

9.3 Double Occupancy

Recent reports in favor of a double occupancy of the Q_o site by quinone have been reviewed elsewhere.[52,84,107,108] Although this issue was not addressed by Osyczka et al.,[32] a second quinone in the Q_o site could introduce rate constants that in principle would overcome the objections, developed above, to mechanisms in which the intermediate state from which the second electron transfers is constrained to the distal domain. By introducing an additional redox center in the otherwise too long path, rate constants in both directions could be raised by $>10^3$. However, the kinetic arguments [32] leading to rapid short-circuits would then apply, giving bypass rates much greater than those observed. The double-occupancy solution therefore seems untenable for reactions involving a SQ intermediate in the distal domain.

9.4 Location of the ES Complex or Activated State at Some Alternative Position

As noted above, the difficulties in accounting for the rates observed arise from the postulated location of the ES-complex at the distal domain, constrained by its H-bonds. Hong et al. [93] overcame these problems by allowing the SQ to move. An alternative is to suggest that the ES complex is located elsewhere and that the Q_o site reaction occurs through a 'concerted' process. In the reaction proposed,[32] both electron transfers would have to occur from a state with no SQ property, at the top of the activation barrier, and with both electrons leaving within a single vibrational frequency. This would preclude nuclear movement, so both electrons would have to transfer from the same position. By locating the intermediate state halfway between ISP_{ox} and heme b_L ($R = 9.6$ Å, giving $k_o \sim 2.25 \times 10^9$), and assuming that $\Delta G^{0\prime} \sim 0$ for both electron transfers, and that $\lambda \sim 4\Delta G^{\#} \geqslant 2.0$ V, the $k_{cat} \leqslant 1.5 \times 10^3$ s^{-1} calculated (using the Moser–Dutton term) provides the rate observed. There are five problems with this scenario. The first is that such a location has no structural justification. The second is the difficulty of accounting for the linked H$^+$ transfers that must occur. The third is the abundant evidence that, in contrast to the expectations of this mechanism, short-circuits do occur. The fourth is the experimental evidence showing that the first electron transfer, but not the second, is rate limiting, as discussed above. Finally, Kramer and colleagues,[109] in a detailed study of the yeast bc$_1$ complex, have shown that electron transfer in the absence (normal forward QH$_2$ oxidation) and presence (bypass reaction) of antimycin showed essentially the same activation barriers and kinetic isotope effects. The characteristics strongly suggested that both involved the same mechanism, generation of a SQ intermediate at low occupancy.

10 Studies Using Glu-272 Mutants to Explore the Role of This Residue in Control of the Q$_o$ site Reaction and Protection Against Excess ROS Production

The critical role we have suggested for Glu-272 in the mechanism of the Q$_o$ site reaction can be tested by mutagenesis. Effects predicted from the above hypothesis are: (i) an involvement in the binding of stigmatellin, (ii) an involvement in binding of QH$_2$ (but not Q), and (iii) a critical role in transfer of the second electron. We have previously reported experimental evidence in favor of some of these effects on mutation of Glu-295, the equivalent residue in *Rb. sphaeroides*, to glutamine, aspartate or glycine.[56] The overall electron transfer was slowed, the binding constant for QH$_2$ was lowered (K_m was increased), and the site became weakly resistant to stigmatellin. Moreover, in the two most inhibited strains (E295G, E295Q) a normal $g_x = 1.80$ line was seen in the EPR spectrum of ISPH, indicating that its interaction with Q at the distal domain could still occur. From these results, we concluded that in the forward reaction, it might still be possible for the ES complex to form in these mutant strains, but that further progress would be blocked by inhibition of the second electron transfer, and by the strongly endergonic nature of the first electron transfer reaction. We have more recently constructed additional mutants at this site, and assayed these strains to explore the role of this residue further (see Figure 5 for example traces). In these experiments, chromatophores were suspended under aerobic conditions in a reaction medium containing 1 mM sodium ascorbate as reductant, and 1 mM potassium cyanide to block cyt *c* oxidase activity. Redox mediators were omitted to avoid their contribution to electron transfer. Valinomycin and nigericin were added so as to minimize the development of any proton gradient or contribution to absorbance changes from electrochromic effects. In a series of experiments with each strain, electron transfer kinetics of cyt c_1 plus c_2, hemes b_H and b_L, and the reaction center (BChl)$_2^+$ (RC) were followed after one or six saturating actinic flashes. The experiments were performed in the absence of inhibitors (left column), in the presence of antimycin (middle column), or in the presence of both antimycin and myxothiazol (right column). In the presence of antimycin, activation by six flashes fully oxidized the high potential chain, and introduced several additional QH$_2$ into the pool, to set up conditions for maximal 'oxidant-induced reduction of cyt *b*'.[110] These are also the conditions for maximal bypass activity at the Q$_o$ site, assayed through the rate of re-reduction of RC. In *Rb. sphaeroides*, myxothiazol eliminates >90% of this activity,

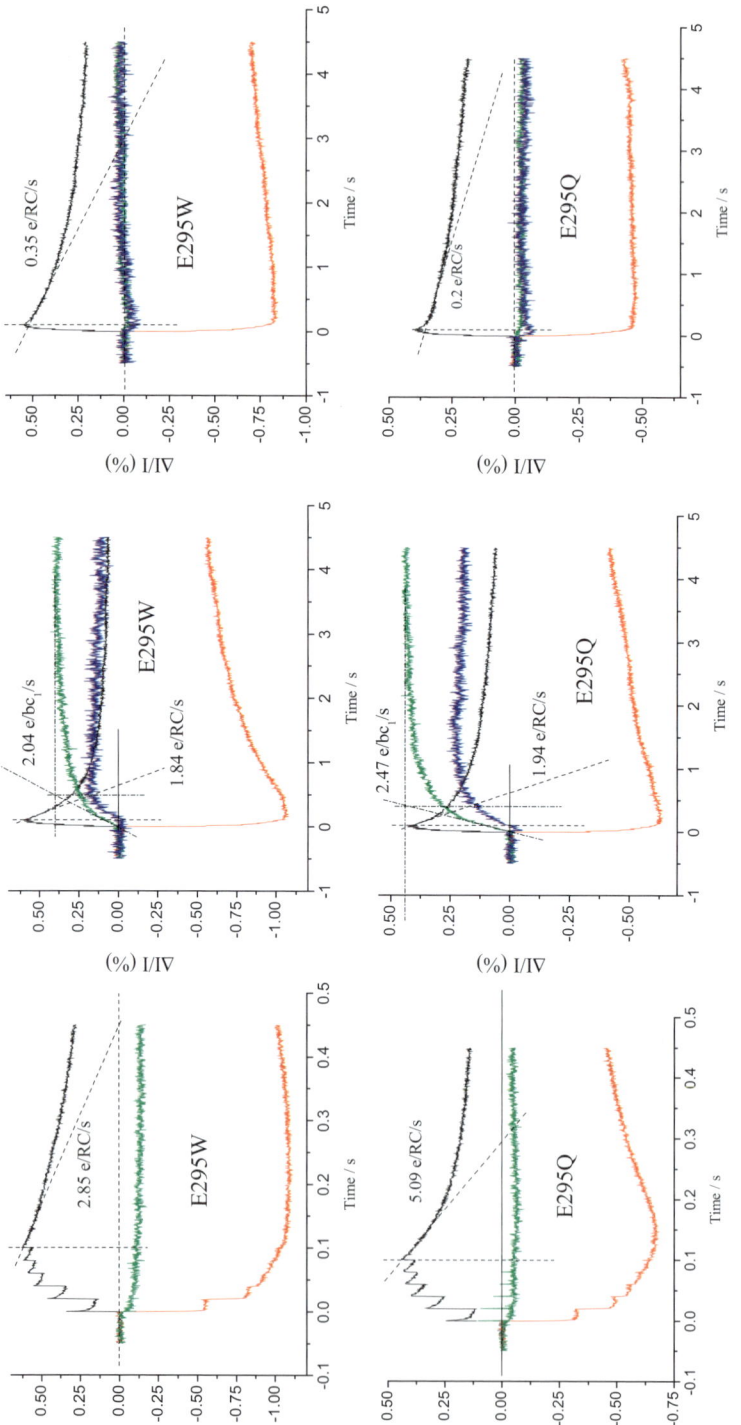

Figure 5 *Kinetic traces showing measurement of bypass reactions in Glu-295 mutant strains. Top, E295W strain; bottom, E295Q strain. The green traces show kinetics measured at 561–569 nm, to assay heme b_H reduction (upward deflection), with minor contributions from cyt c_1 plus c_2 subtracted; the black traces were measured at 542 nm to assay RC oxidation (upward deflection) and reduction (downward deflection) and reduction of cyt c_1 plus c_2; the blue traces show heme b_L reduction (upward deflection), measured at 566–575, with contributions from heme b_H, RC, and cyt c subtracted. The left hand column shows electron transfer in the absence of inhibitor; the middle column in the presence of antimycin (10 μM), and the right column with antimycin (10 μM) and myxothiazol (5 μM) were present throughout to allow rapid equilibration of the proton gradient, and traces were meas- ured in a chromatophore suspension containing 0.5 μM RC in 50 mM MOPS, 100 mM KCl at pH 7.0*

so the right-hand traces show this myxothiazol-insensitive component. Much of this residual activity can be attributed to back reactions of RC. All mutant strains showed inhibited electron transfer through the Q$_o$ site, as assayed either by re-reduction of RC in the absence of inhibitor, or by reduction of heme b_H in the presence of antimycin. However, all strains also showed rates for bypass reactions that accounted for a substantial fraction of the residual turnover of the site (Table 3). These were in some cases in excess of bypass rates in the wild type. These results show clearly that the first electron transfer could still occur to generate SQ, but that the second electron transfer is inhibited when Glu-295 is mutated. Furthermore, the slowed reduction of heme b_H in the mutant strains shows that SQ formed in the distal end of the Q$_o$ site cannot transfer electrons to heme b_L at a normal rate. Since the distance is that from the distal domain discussed above, it seems clear that at this occupancy, the actual rate constant is too low. However, the rate constant estimated from the distance (as discussed above) is of about the right magnitude to account for the slow rates observed. It is noteworthy in the kinetic traces, but especially apparent in the presence of antimycin, that the mass action effects expected from the equilibrium constants operating within the separate chains, and between these and the Q pool catalyzed through the bifurcated reaction, are always in evidence. In the highly inhibited mutants, the consequence is that the quasi-equilibrium condition is not yet attained until some time after the last of the six flashes, so that the *b*-hemes continue to be reduced in the dark. In the seconds ensuing after maximal reduction is attained, the relaxation shows the pattern expected from the equilibria discussed in the introductory sections above; the reversible bifurcated reaction is still at play, but with much reduced rate constants. This is nicely explained by the hypothesis above; the SQ is now constrained to the distal domain, and movement of SQ (or Q) in the site has been lost, and with that, so have the rapid rates in both forward and reverse directions.

11   Conclusions

The critical problems identified by Osyczka *et al.*[32] certainly apply to mechanisms that invoke a SQ intermediate and restrict it to the distal domain of the Q$_o$ site. However, one of the solutions they propose, the 'concerted' reaction, seems quite improbable, and the doubly gated mechanism that prevents any SQ from accumulating seems tenable only with additional *ad hoc* hypotheses to explain the bypass reactions. On the other hand, the sequential mechanism we have proposed,[56] in which an intermediate semiquinone is transiently formed during ubihydroquinone oxidation, but can move within the Q$_o$ site, can provide a satisfactory explanation for the properties observed. The gating functions discussed above represent a natural extension of the original hypothesis. Conclusions as to the demise of such hypotheses therefore seem premature. Notwithstanding the obvious utility of the Moser–Dutton approach,[88–90] and the elegant simplification of electron transfer kinetics in the context of distance, our understanding of the Q$_o$ site reactions has advanced through recognition of paradoxes revealed when application of the equations have showed an apparent contradiction with observed rates. A paradox has been defined as 'Truth standing on her head to get attention',[111] and this seems an appropriate take-home lesson from the above. The movement of the ISP extrinsic domain, the slowness of the rate-limiting step in relation to the distance for electron transfer, and the difficulty in explaining rates pointed out by Osyczka *et al.*[32] all appear paradoxical, but can be explained by mobility or recognition of reaction complexity.[53, 112] The roadmaps provided by successful paradigms are the very stuff of science, but new understanding sometimes evolves when we are forced to look again.

Although the mechanism of the Q_o site seems well explained, several other areas remain controversial. The H-bonding of quinone at the Q_i site varies between structures.[61, 66, 68] Different crystal-packing forces could possibly stabilize configurations involved in different stages of the catalytic cycle.[113] The liganding of the relatively stable SQ at this site has been explored by high-resolution pulsed EPR and ENDOR, but again with differences, this time between species ([68, 113] and F. MacMillan, personal communication). Since an understanding of the liganding is needed before a detailed mechanism can be justified, this controversy is still unresolved. Another critical question is that of electron transfer between monomers. Application of the Moser–Dutton equation to the dimeric structures suggests that electron transfer between the b_L hemes should be rapid.[32, 69] This would allow both Q_o sites to communicate with both Q_i sites, so that in the presence of excess antimycin, titration of the Q_o site activity through inhibition of heme b_H reduction using myxothiazol or stigmatellin would give rise to strongly convex titration curves. In our hands, although the convex titration curves expected could be obtained when delivery of substrates was limiting, the titrations were linear when QH_2 was saturating. The convexity under limiting conditions seems better explained by a model of the sort suggested by Kroger and Klingenberg.[114] Perhaps this is another paradox. Other kinetic experiments on mitochondrial complexes have been interpreted as showing dramatic interactions both across the dimer interface, and between the Q_o and Q_i sites,[14,70,115] which seem in contradiction to the experimental data in the bacterial systems[1,4,31,32] and the simple model discussed above. Further work will be needed to sort out this apparent discrepancy between mitochondrial and bacterial complexes.

Acknowledgements

Although the views expressed here are my own, they have been formed in the context of useful discussions with many colleagues, in particular Les Dutton, David Kramer, Peter Rich, Colin Wraight, Vlad Shinkarev, Ed Berry, Sergei Dikanov, Fraser MacMillan, and Fevzi Daldal. Support for research on the bc_1 complex from grant RO1 GM035438 from NIH is also gratefully acknowledged.

References

1. A.R. Crofts, S.W. Meinhardt, K.R. Jones, *et al.*, Biochim. Biophys. Acta, 1983, **723**, 202.
2. P. Mitchell, *FEBS Lett.,* 1975, **56**, 1.
3. P. Mitchell, *J. Theor. Biol.,* 1976, **62**, 327.
4. A.R. Crofts, *Photosynth. Res.,* 2003, **80**, 223.
5. D. Xia, C.-A. Yu, H. Kim, *et al.*, *Science*, 1997, **277**, 60.
6. Z. Zhang, L.-S. Huang, V.M. Shulmeister, *et al.*, *Nature* (Lond.), 1998, **392**, 677.

7. S. Iwata, J.W. Lee, K. Okada, *et al.*, *Science*, 1998, **281**, 64.
8. E. Berry, M. Guergova-Kuras, L.-S. Huang, *et al.*, *Annu. Rev. Biochem.* 2000, **69**, 1007.
9. C. Hunte, S. Solmaz and C. Lange, *Biochim. Biophys. Acta* 2002, **1555**, 21.
10. C. Hunte, H. Palsdottir and B.L. Trumpower, *FEBS Lett.* 2003, **545**, 39.
11. H. Kim, D. Xia, C.A. Yu, *et al.*, *Proc. Natl. Acad. Sci. (U.S.)* 1998, **95**, 8026.
12. B.L. Trumpower, (2002).
13. K. Xiao, G. Engstrom, S. Rajagukguk, *et al.*, *J. Biol. Chem.* 2003, **278**, 11419.
14. C.-A. Yu, X. Wen, K. Xiao, *et al.*, *Biochim. Biophys. Acta* 2002, **1555**, 65.
15. A.R. Crofts and C.A. Wraight, *Biochim. Biophys. Acta* 1983, **726**, 149.
16. P.L. Dutton and D.M. Wilson, *Methods Enzymol.* 1976, **54**, 411.
17. A.R. Crofts and Z. Wang, *Photosynth. Res.* 1989, **22**, 69.
18. A.R. Crofts, in *The Enzymes of Biological Membranes*, Vol. 4, A.N. Martonosi(ed.) Plenum Publ. Corp., New York., 1985, 347.
19. S.W. Meinhardt and A.R. Crofts, *FEBS Lett.* 1982, **149**, 217.
20. S.W. Meinhardt and A.R. Crofts, *FEBS Lett.* 1982, **149**, 223.
21. S.W. Meinhardt and A.R. Crofts, *Biochim. Biophys. Acta* 1983, **723**, 219.
22. E.G. Glaser and A.R. Crofts, *Biochim. Biophys. Acta* 1984, **766**, 322.
23. E.G. Glaser, S.W. Meinhardt, and A.R. Crofts, *FEBS Lett.* 1984, **178**, 336.
24. M. Snozzi and A.R. Crofts, *Biochim. Biophys. Acta* 1984, **766**, 451.
25. M. Snozzi and A.R. Crofts, *Biochim. Biophys. Acta* 1985, **809**, 260.
26. P.R. Rich, *Biochim. Biophys. Acta* 1984, **768**, 53.
27. B.L. Trumpower, *Biochim Biophys Acta* 1981, **639**, 129.
28. J.R. Bowyer and A.R. Crofts, *Biochim. Biophys. Acta* 1981, **636**, 218.
29. P.B. Garland, R.A. Clegg, D. Boxer, *et al.*, in *Electron Transfer Chains and Oxidative Phosphorylation*, E. Quagliariello, S. Papa, F. Palmieri, E. C. Slater and N. Siliprandi (eds.), North-Holland Publishing Co., Amsterdam, The Netherlands., 1975, 351.
30. A.R. Crofts and S.W. Meinhardt, *Biochem. Soc. Trans.* 1982, **10**, 201.
31. A.R. Crofts, V.P. Shinkarev, D.R.J. Kolling, *et al.*, *J. Biol. Chem.* 2003, **278**, 36191.
32. A. Osyczka, C.C. Moser, Daldal, F. and P.L. Dutton, *Nature* 2004, **427**, 607.
33. V.P. Shinkarev, A.R. Crofts, and C.A. Wraight, *Biochemistry* 2001, **40**, 12584.
34. C.-H. Yun, A.R. Crofts and R.B. Gennis, *Biochemistry* 1991, **30**, 6747.
35. P. Joliot, A. Vermeglio, and A. Joliot, *Biochim. Biophys. Acta* 1989, **975**, 336.
36. J. Lavergne, P. Joliot, and A. Vermeglio, *Biochim. Biophys. Acta* 1989, **975**, 347.
37. A.T.a.J. Verméglio, P., *Trends in Microbiol.* 1999, **7**, 435.
38. C. Jungas, J.-L. Ranck, J.-L. Rigaud, P. Joliot and A. Verméglio, *EMBO J.* 1999, **18**, 534–542.
39. A.R. Crofts, M. Guergova-Kuras and S. Hong, *Photosynth. Res.* 1998, **55**, 357.
40. A.R. Crofts, *Trends in Microbiol.* 2000, **8**, 105.
41. C. Siebert, P. Qian, D. Fotiadis, A. Engel, C.N. Hunter and P.A. Bullough, *EMBO J.* 2004, **23**, 690.
42. S. Scheuring, F. Francia, J. Busselez, B.A. Melandri, J.-L. Rigaud, and D. Lévy, *J. Biol. Chem.* 2004, **279**, 3620.
43. F. Francia, J. Wang, G. Venturoli, B.A. Melandri, W.P. Barz and D. Oesterhelt, *Biochemistry* 1999, **38**, 6834.

44. E.A. Berry, L.-S. Huang, L.K. Saechao, N.G. Pon, M. Valkova-Valchanova and F. Daldal, *Photosynth. Res.* 2004, **81**, 251.

45. Y. Chen, in *Ph.D. Thesis in Biophysics*, University of Illinois at Urbana-Champaign, 1989, 143.

46. Y. Chen and A.R. Crofts, in *Current Research in Photosynthesis*, Vol. III, M. Baltscheffsky (ed.), Kluwer Academic Publishers, Dordrecht/Boston/London., 1990, 287.

47. A. Boveris, *Methods Enzymol.* 1984, **105**, 429.

48. F. Muller, *J. Am. Aging Assoc.* 2000, **23**, 227.

49. J.F. Turrens, A. Alexandre, and A.L. Lehninger, *Arch. Biochem. Biophys.* 1985, **237**, 408.

50. F.L. Muller, A.G. Roberts, M.K. Bowman, *et al.*, *Biochemistry* 2003, **42**, 6493.

51. F. Muller, A.R. Crofts, and D.M. Kramer, *Biochemistry* 2002, **41**, 7866.

52. A.R. Crofts, *Annu. Rev. Physiol.*, 2004, **66**, 689.

53. A.R. Crofts, *Biochim. Biophys. Acta* 2004, **1655**, 77.

54. P.R. Rich, *Bioçhim. Biophys. Acta* 2004, **1658**, 165.

55. E.A. Berry and L.S. Huang, *FEBS Lett.* 2003, **555**, 13.

56. A.R. Crofts, S.J. Hong, N. Ugulava, *et al.*, *Proc. Natl. Acad. Sci. U.S.A.*, 1999, **96**, 10021.

57. A.R. Crofts, B. Barquera, R.B. Gennis, *et al.*, *Biochemistry* 1999, **38**, 15807.

58. S. Junemann, P. Heathcote and P.R. Rich, *J. Biol. Chem.* 1998, **273**, 21603.

59. Z.-L. Zhang, E.A. Berry, L.-S. Huang, *et al.*, *Subcell. Biochem.* 2000, **35**, 541.

60. C. Hunte, J. Koepke, C. Lange, *et al.*, *Structure*, 2000, **8**, 669.

61. X. Gao, X. Wen, L. Esser, *et al.*, *Biochemistry*, 2003, **42**, 9067.

62. H. Palsdottir, C.G. Lojero, B.L. Trumpower, *et al.*, *J. Biol. Chem.* 2003, **278**, 31303.

63. J.J. Kessl, B.B. Lange, T. Merbitz-Zahradnik, *et al.*, *J. Biol. Chem.* 2003, **278**, 31312.

64. C.-A. Yu, D. Xia, H. Kim, *et al.*, *Biochim. Biophys. Acta* 1998, **1365**, 151.

65. X. Gao, X. Wen, C.-A. Yu, *et al.*, *Biochemistry* 2002, **41**, 11692.

66. C. Lange, J.H. Nett, B.L. Trumpower, *et al.*, *EMBO J.* 2001, **23**, 6591.

67. R.I. Samoilova, D. Kolling, T. Uzawa, *et al.*, *J. Biol. Chem.* 2002, **277**, 4605.

68. S.A. Dikanov, R.I. Samoilova, D.R.J. Kolling, *et al.*, *J. Biol. Chem.* 2004, **279**, 15814.

69. A.R. Crofts and E.A. Berry, *Curr. Opinions in Struc. Biol.*, 1998, **8**, 501.

70. C. Lange and C. Hunte, *Proc. Natl. Acad. Sci. U.S.A.*, 2002, **99**, 2800.

71. A.R. Crofts, M. Guergova-Kuras, L.-S. Huang, *et al.*, *Biochemistry* 1999, **38**, 15791.

72. A.R. Crofts, S. Hong, Z. Zhang, *et al.*, *Biochemistry* 1999, **38**, 15827.

73. A.R. Crofts, E.A. Berry, R. Kuras, *et al.*, in *Photosynthesis: Mechanisms and Effects*, Vol. III, G. Garab (ed.) Kluwer Academic Publ., Dordrecht/Boston/London., 1998, 1481.

74. E. Darrouzet and F. Daldal, *Biochemistry* 2003, **42**, 1499.

75. E. Darrouzet, C.C. Moser, P.L. Dutton, *et al.*, *TIBS* 2001, **26**, 445.

76. K. Xiao, L. Yu and C.A. Yu, *J. Biol. Chem.* 2000, **275**, 38597.

77. C.H. Snyder, E.B. Gutierrez-Cirlos and B.L. Trumpower, *J. Biol. Chem.* 2000, **275**, 13535.
78. H. Tian, L. Yu, M. W. Mather, *et al.*, *J. Biol. Chem.* 1998, **273**, 27953.
79. H. Tian, S. White, L. Yu, *et al.*, *J. Biol. Chem.* 1999, **274**, 7146.
80. J.H. Nett, C. Hunte and B.L. Trumpower, *Eur. J. Biochem.* 2000, **267**, 5777.
81. D.E. Robertson, F. Daldal and P.L. Dutton, *Biochemistry* 1990, **29**, 11249.
82. C. Snyder and B.L. Trumpower, *Biochim. Biophys. Acta* 1998, **1365**, 125.
83. N.B. Ugulava and A.R. Crofts, *FEBS Lett.* 1998, **440**, 409.
84. H. Ding, D.E. Robertson, F. Daldal, *et al.*, *Biochemistry* 1992, **31**, 3144.
85. A.R. Crofts, M. Guergova-Kuras, R. Kuras, *et al.*, *Biochim. Biophys. Acta* 2000, **1459**, 456.
86. M. Guergova-Kuras, R. Kuras, N. Ugulava, *et al.*, *Biochemistry* 2000, **39**, 7436.
87. C.C. Moser, J.M. Keske, K. Warncke, R.S. Farid and P.L. Dutton, *Nature* 1992, **355**, 796.
88. C.C. Moser, C.C. Page, R. Farid and P.L. Dutton, *J. Bioenerg. Biomembranes* 1995, **27**, 263.
89. C.C. Moser, C.C. Page, X.X. Chen and P.L. Dutton, *J. Biol. Inorg. Chem.* 1997, **2**, 393.
90. C.C. Page, C.C. Moser, X.X. Chen, *et al.*, *Nature* 1999, **402**, 47.
91. R.A. Marcus and N. Sutin, *Biochim. Biophys. Acta* 1985, **811**, 265.
92. D. DeVault, *Q. Rev. Biophys.* 1980, **13**, 387.
93. S.J. Hong, N. Ugulava, M. Guergova-Kuras, *et al.*, *J. Biol. Chem.* 1999, **274**, 33931.
94. E. Denke, T. Merbitzzahradnik, O.M. Hatzfeld, *et al.*, *J. Biol. Chem.* 1998, **273**, 9085.
95. T. Schröter, O.M. Hatzfeld, S. Gemeinhardt, *et al.*, *Eur. J. Biochem.* 1998, **255**, 100.
96. T.A. Link, *Adv. Inorg. Chem.* 1999, **47**, 83.
97. C. Colbert, M.M.-J. Couture, L.D. Eltis, *et al.*, *Structure* 2000, **8**, 1267.
98. G.M. Ullmann, L. Noodleman and D.A. Case, *J. Biol. Inorg. Chem.* 2002, **7**, 632.
99. Y. Zu, M.M.-J., M.M.-J. Couture, *et al.*, *Biochemistry* 2003, **42**, 12400.
100. R.I. Cukier, *Biochim. Biophys. Acta* 2004, **1655**, 37.
101. C.J. Chang, M.C.Y. Chang, N.H. Damrauer and D.G. Nocera, *Biochim. Biophys. Acta* 2004, **1655**, 13.
102. T.A. Link, *FEBS Lett.* 1997, **412**, 257.
103. A.R. Crofts, V.P. Shinkarev, S.A. Dikanov, *et al.*, *Biochim. Biophys. Acta* 2002, **1555**, 48.
104. V.P. Shinkarev, D.R.J. Kolling, T.J. Miller, *et al.*, *Biochemistry* 2002, **41**, 14372.
105. S. Izrailev, A.R. Crofts, E.A. Berry, *et al.*, *Biophys. J.* 1999, **77**, 1753.
106. C.A. Wraight, *Frontiers in Bioscience* 2004, **9**, 309.
107. S. Bartoschek, M. Johansson, B.H. Geierstanger, *et al.*, *J. Biol. Chem.* 2001, **276**, 35231.
108. H. Ding, C.C. Moser, D.E. Robertson, *et al.*, *Biochemistry* 1995, **34**, 15979.

109. D.M. Kramer, J.L. Cape, I. Forquer, F. Muller and M.K. Bowman, submitted (2005).

110. E.C. Slater, in *Chemiosmotic Proton Circuits in Biological membranes*, V.P. Skulachev and P.C. Hinkle (ed.), Addison-Wesley Publ. Co., Reading, Mass., 1981, 69.

111. G.K. Chesterton, in *The Paradoxes of Mr. Pond* (Dover Books, 1937).

112. V.L. Davidson, *Biochemistry* 1996, **35**, 14036.

113. D.R.J. Kolling, R.I. Samoilova, J.T. Holland, *et al.*, *J. Biol. Chem.* 2003, **278**, 39747.

114. A. Kroger and M. Klingenberg, *Eur. J. Biochem.* 1973, **34**, 358.

115. R. Covian, E.B. Gutierrez-Cirlos and B.L. Trumpower, *J. Biol. Chem.* 2004, **279**, 15040.

116. G. Engstrom, K. Xiao, C.-A. Yu, *et al.*, *J. Biol.Chem.* 2002, **277**, 31072.

117. B. Hacker, B. Barquera, A.R. Crofts, *et al.*, *Biochemistry* 1993, **32**, 4403.

118. A.R. Crofts, B. Barquera, G. Bechmann, *et al.*, in *Photosynthesis: From Light to Biosphere.*, Vol. II, P. Mathis (ed.), Kluwer Academic Publ., Dordrecht., 1995, 493.

119. D.E. Robertson, R.C. Prince, J.R. Bowyer, *et al.*, *J. Biol. Chem.* 1984, **259**, 1758.

120. P.L. Dutton and J.B. Jackson, *Eur. J. Biochem.* 1972, **30**, 495.

121. J.R. Bowyer, S.W. Meinhardt, G.V. Tierney, *et al.*, *Biochim. Biophys. Acta* 1981, **635**, 167.

122. J.R. Bowyer, P.L. Dutton, R.C. Prince, *et al.*, *Biochim. Biophys. Acta* 1980, **592**, 445.

123. B. Hacker, B. Barquera, R.B. Gennis, *et al.*, *Biochemistry* 1994, **33**, 13022.

124. A.R. Crofts, B. Barquera, G. Bechmann, *et al.*, in *Photosynthesis: From Light to Biosphere.*, Vol. II, P. Mathis (ed.), Kluwer Academic Publ., Dordrecht., 1995, 493.

125. A.R. Crofts, http://www.life.uiuc.edu/crofts/Marcus_Bronsted/, 2004.

Appendix 1

Marcus Treatment of Endergonic Reactions

An alternative form of the Marcus equation has been suggested for treatment of endergonic processes [1], using the equation:

$$\log_{10} k_{et}^{end} = 13.0 - 0.6(R - 3.6) - 3.1(-\Delta G + \lambda)^2/\lambda - \Delta G/0.06 \quad (A.1)$$

In addition to what appears to be a conventional Marcus term, this equation seems to contain an extra Boltzmann term contributing to the energy barrier (the right-most term). This representation is misleading and unnecessary. The 'additional term' arises when estimating the activation barrier and rate-constant in the endergonic direction (which cannot be measured easily) using the reaction properties measured in the exergonic direction. It can be derived in its simplest form as follows (using electrical units for ΔG):

$$K_{end} = k_f^{end}/k_b^{end} = k_f^{end}/k_f^{ex} \quad (A.2)$$

$$k_f^{end} = k_f^{ex} K_{end} \quad (A.3)$$

$$\log_{10} k_f^{end} = \log_{10}k_f^{ex} + \log_{10}K_{end} = \log_{10}k_f^{ex} - \Delta G_{end}^0 \, F/2.303RT \quad (A.4)$$

This expression can be extended using classical Marcus treatment [4, 5], and substitution using the Arrhenius expression, $k_f^{ex} - k_0 exp(-\Delta G^\#.F/RT)$, the Marcus term, $\Delta G^\# = (\Delta G^o + \lambda)^2/4\lambda$, and the Moser *et al.* [2] expression for k_0 in terms of distance, R.

$$\log_{10} k_f^{end} = 13.0 - 0.6(R - 3.6) - \gamma(\Delta G_{ex}^0 + \lambda_{ex})^2/\lambda_{ex} - \Delta G_{end}^0 \, F/2.303RT \quad (A.5)$$

The Moser–Dutton [1, 2] formulation, in which quantum-mechanical corrections modify the pre-exponential factor, has $\gamma = 3.1$ instead of 4.23. The Page et al. [1] version is then obtained by substituting $-\Delta G_{end}^o$ for ΔG_{ex}^o. The λ used here is actually that for the exergonic reaction; the value is the same in both directions if the parabolas have the same shape (the same 'spring constant'), but not otherwise.

An alternative treatment can be framed in terms of a set of Marcus parabolas [4, 5]. The endergonic energy barrier, $\Delta G_{end}^\#$, is calculated by adding $\Delta G_{ex}^\#$, the energy barrier in the exergonic direction, to the energy difference in the endergonic direction (the Boltzmann term in question). $\Delta G_{ex}^\#$ is estimated using the Marcus expression, with ΔG_{ex}^o for the exergonic reaction direction, and a value for λ, λ_{ex}, obtained in principle from measurement of the exergonic rate. The tricky factor is again the replacement of ΔG_{ex}^o by $-\Delta G_{end}^o$, its numerical equivalent. The underlying rate equation in the endergonic direction has the same conventional Arrhenius form as that in the exergonic direction.

$$k_{cat} = k_0 \exp(-\Delta G^{\#}_{end}/RT) = k_0\exp(-(\Delta G^{\#}_{ex} + \Delta G^{0}_{end})/RT) \qquad (A.6)$$

Derivation of the Marcus term for $\Delta G^{\#}$ from Hookes' Law, using parabolas drawn to represent either the exergonic or endergonic direction, results in the same equation.

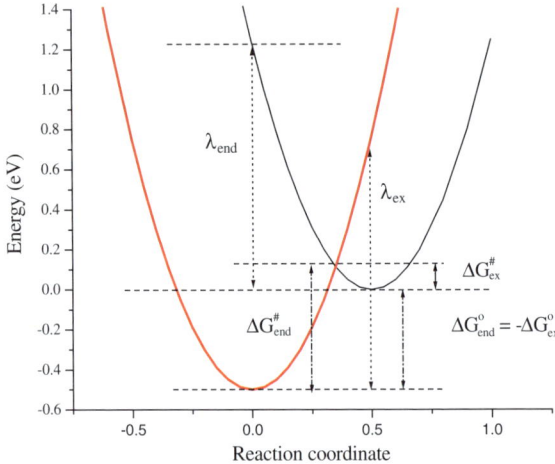

Figure A1 *Marcus parabolas for an endergonic process, labeled to show the terms discussed in the text.*

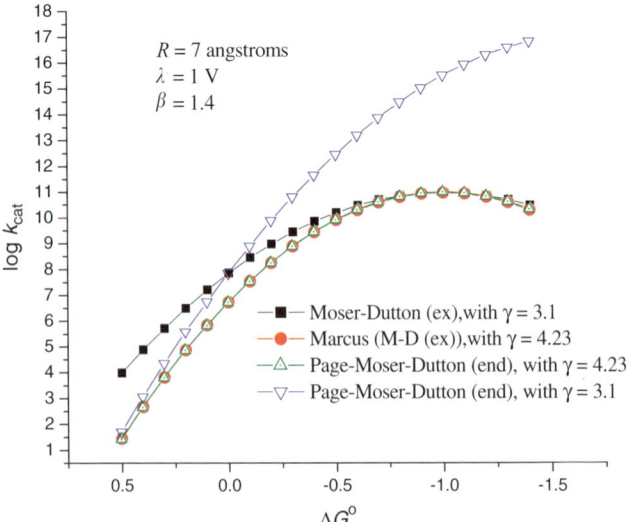

Figure A2 *Curves showing the dependence of rate constant on driving force using different versions of the Moser-Dutton equation*

The "additional Boltzmann term" of the Page *et al.* [1] equation is not an extra term contributing to the barrier, and the treatment used in (3) and this review, in which the endergonic region is treated using the standard Marcus term, is perfectly appropriate.

$$\log_{10} k_f^{end} = 13 - \frac{\beta}{2.303}(R - 3.6) - \gamma\frac{(\Delta G_{end}^0 + \lambda_{end})^2}{\lambda_{end}} \tag{A.7}$$

The identity of the two equations can most easily be demonstrated by numerical substitution, as in Figure A2. Identical Marcus curves are generated using the exergonic or endergonic forms of the Page–Moser–Dutton equation, as long as a simple Marcus treatment (giving γ of 4.23 at 298 K) is used. However, it should be noted that with the Page *et al.* [1] equation, although $\Delta G_{end}^{\#}$ and $(\Delta G_{ex}^{\#} + \Delta G_{end}^0)$ have the same numerical value, the quantum mechanical correction implicit in $\gamma = 3.1$ is applied only to the part of this value corresponding to $\Delta G_{ex}^{\#}$. As a consequence, different values are generated for k_f^{end} when using the two equations, and the Marcus curve resulting from the treatment recommended in [1] (in which the 'endergonic' equation is used only for the endergonic part) shows a discontinuity in slope at the exergonic to endergonic transition, which is clearly unnatural.

References

1. C.C. Page, C.C. Moser, X. Chen, and P.L. Putton *Nature,* 1999 **402**, 47.
2. C.C. Moser, J.M. Keske, K. Warncke, R.S. Farid and P.L. Dutton, *Nature,* 1992. **355**, 796.
3. A.R. Crofts, *Biochim. Biophys. Acta,* 2004 **1655**, 77.
4. R. Marcus, *J. Chem. Phys.,* 1965 **43**, 679.
5. D. DeVault, *Q. Rev. Biophys.,* 1980 **13**, 387.

CHAPTER 7

Insights into the Mechanism of Mitochondrial Complex I from its Distant Relatives, the [NiFe] Hydrogenases

STEFAN KERSCHER, VOLKER ZICKERMANN, KLAUS ZWICKER AND ULRICH BRANDT

Universität Frankfurt,
Fachbereich Medizin,
Gustav Embden Zentrum der Biologischen Chemie,
Frankfurt am Main, Germany

1 Introduction

At about 1.000 kDa, up to 46 different subunits, and at least 10 prosthetic groups, proton-translocating NADH:ubiquinone oxidoreductase (complex I) from mitochondria is among the largest and most complicated integral membrane protein complexes known.[1,2] This holds also for the bacterial counterpart, although its total mass is only about half and the number of different subunits is only one-third of that of the mitochondrial enzyme.[3] The 14 subunits found in both eukaryotic and prokaryotic complex I (Table 1) are the central subunits that carry all redox prosthetic groups and perform all bioenergetic functions of complex I. The numerous accessory subunits of mitochondrial complex I are not discussed here.

In comparison to the size and complexity of mitochondrial complex I, the redox reaction it catalyzes that is linked to the vectorial translocation of 4 $H^+/2e^{4,5}$ across the inner membrane seems to be a rather simple task. In fact, so-called alternative NADH-dehydrogenases that are found in fungal and plant mitochondria, as well as in many bacteria, catalyze the same redox reaction with a single subunit, have a mass of only 50–70 kDa and contain merely one flavine as a prosthetic group.[6] However, these enzymes do not pump protons across the membrane suggesting that the more

Table 1 *Evolutionary relationships between complex I and other enzyme complexes*

	Subunit symbol				Homologous subunits in related enzymes					
Functional module	Bos taurus	Yarrowia lipolytica	Escherichia coli	Redox prosthetic groups	NAD+-reducing hydrogenase Alcaligenes eutrophus	Water soluble [NiFe] hydrogenase e.g. Desulfovibrio fructosovorans	Membrane-bound [NiFe] hydrogenase Methanosarcina barkeri	Hydrogenase 3 (FHL-1) Escherichia coli	Hydrogenase 4 (FHL-2) Escherichia coli	Antiporter e.g. Bacillus subtilis
N	75 kDa	NUAM	NuoG	N1b N5 N4, N1c/N7	HoxU[a]	—	—	—	—	—
	51 kDa	NUBM	NuoF	FMN, N3	HoxF	—	—	—	—	—
	24 kDa	NUHM	NuoE	N1a	HoxF	—	—	—	—	—
Q	49 kDa	NUCM	NuoD[b]	—	—	Large subunit	EchE	HycE	HyfG	—
	30 kDa	NUGM	NuoC[b]	—	—		EchD	HycE	HyfG	—
	TYKY	NUIM	NuoI	N6a,b	—		EchF	HycF	HyfH	—
	PSST	NUKM	NuoB	N2	—	Small subunit	EchC	HycG	HyfI	—
P	ND1	ND1	NuoH	—	—		EchB	HycD	HyfC	—
	ND2	ND2	NuoN	—	—		EchA$_c$	HycC[c]	HyfB,D,F[c]	MrpD[c]
	ND3	ND3	NuoA	—	—		—	—	—	—
	ND4	ND4	NuoM	—	—		EchA$_c$	HycC[c]	HyfB,D,F[c] HyfEβ	MrpD[c]
	ND4L	ND4L	NuoK	—	—		—	—	—	—
	ND5	ND5	NuoL	—	—		EchA$_c$	HycC[c]	HyfB,D,F[c]	MrpA[c]
	ND6	ND6	NuoJ	—	—		—	—	—	—

[a] only the first 200 residues of the 75 kDa subunit are homologous to HoxU.
[b] In *E.coli* both subunits are fused (NuoCD).
[c] The ND2, 4 and 5 subunits are weakly homologous to each other, an assignment of the individual subunits to other proteins is ambiguous.

complicated design and larger size of complex I may be related to this process of energy conversion. The thermodynamic efficiency of complex I is illustrated by the full reversibility of its mechanism.[7] Under certain metabolic regimes in some microorganisms and probably also in mitochondria, proton-motive-force-driven reduction of NAD$^+$ is the physiological mode of operation for complex I.[8]

Eventually, high-resolution structural data that are still not available (with the exception of a small accessory subunit of human complex I[9]) will be required to understand how the structure of complex I relates to its function. However, from insights into the evolution and assembly of complex I, low-resolution structural data, as well as biochemical and mutagenesis studies, a general picture is emerging of how the mechanism of complex I is reflected in its structural organization. Electron microscopic studies indicate that mitochondrial complex I is an L-shaped particle with a hydrophobic membrane arm and a peripheral arm that protrudes into the matrix space.[10–12] The three phases of the catalytic mechanism are implemented as three separate modules (Figure 1). In the forward mode, NADH oxidation is catalyzed by the N-module, ubiquinone reduction by the Q-module and protons are pumped by the P-module. Electrons are transferred through a series of iron–sulfur clusters in the N- and the Q-modules. Recent findings suggest that redox energy is converted into conformational energy by the Q-module and then transferred to the P-module, where it is used to drive vectorial proton translocation. Here we present evidence supporting this model of complex I. The discussion focuses on the energy-converting Q-module and implications for the molecular mechanism will be discussed.

2   The Three Modules of Complex I

So far, insight into the overall architecture of complex I and its three functional modules has been provided by the identification of building blocks based on their homology to other enzymes,[13–16] by biochemical dissection of the holoenzyme into subcomplexes[17–20] and by electron microscopic analysis.[21–23]

2.1   The N-module

The N-module has evolved from NAD$^+$-reducing hydrogenases (Table 1). The 51 kDa and 24 kDa subunits are homologous to the α subunit of the NAD$^+$- reducing hydrogenase encoded by *hoxF* from *Alcaligenes eutrophus* while the N-terminal part of the 75 kDa subunit is homologous to the γ subunit encoded by *hoxU*.[24] This evolutionary relationship is matched very well by subcomplexes that can be split off from complex I by chaotropes: the so-called flavoprotein (FP) from bovine complex I contains the 51 and 24 kDa subunits[17,25] and the NADH dehydrogenase fragment prepared from the *Escherichia coli* complex contains the 75, 51 and 24 kDa subunits.[19] According to electron-microscopic single-particle analysis of the *E. coli* complex,[22] the N-module resides on the inner side of the peripheral arm (Figure 1). During the forward reaction, the N-module operates as an electron input device. Hydride is transferred from NADH onto FMN within the 51 kDa subunit. From there single electrons flow through a 'wire' made of iron–sulfur clusters towards the P-module. This is illustrated by the fact that the isolated FP fragment is able to transfer electrons from

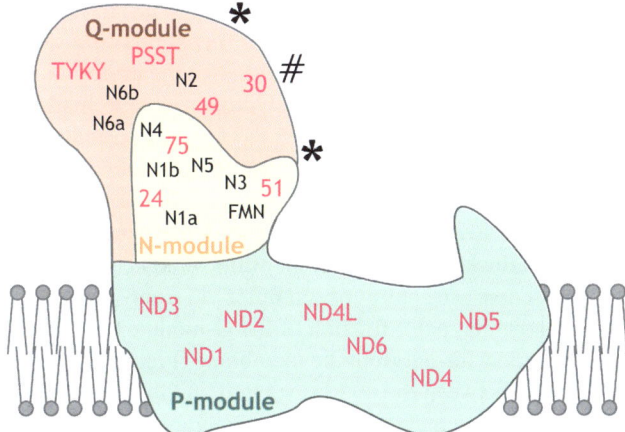

Figure 1 *Possible arrangement of the three modules of complex I. Hypothetical positions of the seven central, mitochondrially coded and the seven central, nuclear-coded subunits of complex I are given together with the redox prosthetic groups they contain. The approximate locations of two monoclonal antibody epitopes in the 49 kDa subunit (*) and of the hexa-histidine tag attached to the C-terminus of the 30 kDa subunit (#), as derived by electron microscopic single particle analysis of* Y. lipolytica *complex I[23] are marked*

NADH to artificial electron acceptors.[26–28] The tetranuclear cluster N3 is bound to the 51 kDa subunit and the binuclear cluster N1a is bound to the 24 kDa subunit. The other iron–sulfur clusters of the N-module, the tetranuclear clusters N4 and N5 and the binuclear cluster N1b are bound to the 75 kDa subunit.[29] In *E. coli* and some other bacteria a third cluster called N1c is found that is most likely bound to the part of the 75 kDa subunit not homologous to the γ subunit of NAD$^+$-reducing hydrogenases.[30] Recent evidence suggests that this cluster is likely to be a tetranuclear cluster and it has been proposed that it should therefore be renamed cluster N7.[31] The exact sequence how the iron–sulfur clusters are arranged within the N-module is unknown. It seems that in some enzymes that are related to complex I but use other electron donors the 51 and 24 kDa subunits are missing. One example is the complex I-like enyzme of *Helicobacter pylori* and *Campylobacter jejuni*.[32,33] Another somewhat different situation has been described for the F420H2:quinone reductase from the archaeon *Archaeoglobus fulgidus*[34] and complex I-like complexes in chloroplasts and blue algae, where in addition the 75 kDa subunit is missing.[35]

2.2 The Q-module

The Q-module has evolved from [NiFe] hydrogenases.[13,14] The large and small subunits of water soluble [NiFe] hydrogenases are homologous to the 49 kDa and part of the PSST subunits, respectively (Table 1). In *E. coli* the 30 kDa subunit that is found as a separate subunit in most organisms is fused to the N-terminus of the 49 kDa subunit.[30] A so-called connecting fragment can be prepared biochemically from complex I that in addition to these subunits contains the TYKY subunit.[19] The sequence of the TYKY subunit identifies it as a typical ferredoxin-type protein carrying two tetranuclear iron–sulfur clusters. It seems that this subunit has taken over the function of two of the three iron–sulfur cluster-carrying domains found in the small subunit of water soluble [NiFe] hydrogenases. Remarkably, the same rearrangement, *i.e.* a shortened small subunit homolog and an additional ferredoxin-like subunit homologous to TYKY, is found in membrane-bound type 3 hydrogenase from *E. coli*[13] and Ech-hydrogenase from *Methanosarcina barkeri*[36] that will be discussed in more detail below. Decorating complex I particles with specific monoclonal antibodies has allowed the localization of the 49 kDa and 30 kDa subunits in the electron microscopic structure of complex I from *Yarrowia lipolytica*.[23] According to this study, the major fraction of the Q-module corresponds to the part of the peripheral arm most distant from the membrane (Figure 1). On the other hand, chemical cross-linking between the 49 kDa subunit and the membrane integral subunit ND3 suggests a direct connection to the membrane arm.[37] Different pieces of information provide clues to how the Q-module uses electrons that are delivered from the N-module to reduce ubiquinone: all remaining (tetranuclear) iron–sulfur clusters reside in the Q-module. Clusters N6a and N6b[38] are likely to accept the electrons from the N-module. Cluster N2 in the PSST subunit[39,40] has been shown to reside near an ubisemiquinone species formed during turnover.[41] A photoreactive derivative of pyridaben, an inhibitor known to act at the level of ubiquinone reduction, has been found to label the PSST subunit.[42] Mutants exhibiting resistance towards various ubiquinone-site inhibitors have been found identified in the 49 kDa subunit.[43,44]

Together with the N-module, the Q-module can be split off from fungal and mammalian complex I as a hydrophilic subcomplex called 1λ in the case of the bovine heart enzyme.[20,45] A similar subcomplex can be released from complex I from *Y. lipolytica*.[46] Remarkably, these functional modules have been combined also in NAD$^+$-reducing hydrogenase: subunits of this bacterial enzyme exhibit sequence similarities to subunits of the N-module (Table 1) and homology has also been reported between the 49 kDa and PSST subunits of complex I, and the β subunit encoded by *hoxH* and the δ subunit encoded by *hoxY*, respectively.[14] Considering that the similarities between the two enzymes are rather weak, one may speculate that the two functional modules have been combined independently in complex I and NAD$^+$-reducing hydrogenase.

It can be concluded for complex I that the entire electron transfer chain from NADH to ubiquinone is contained in the peripheral arm of complex I as none of the subunits assigned to the N- and Q-modules is predicted to contain a transmembrane domain. It follows that proton pumping requires a third, membrane-integral module.

2.3   The P-module

Critical parts of the P-module have evolved from Na$^+$/H$^+$ antiporters.[47,48] The *mrp* (multiple resistance and pH adaptation) operon in *Bacillus subtilis* encodes two polypeptides that show weak similarity to the complex I subunits ND2, ND4 and ND5 (Table 1). The transmembrane topology of the three putative transporter proteins has been studied.[48] The MrpA and MrpD proteins both act as Na$^+$/H$^+$ antiporters and have a function in Na$^+$ resistance and pH homeostasis. Related genes have been found in other bacteria. It is straightforward to assume that the transporter function may have been retained by the related membrane integral subunits of complex I. It should be noted at this point that there is a current controversy about whether bacterial complex I pumps protons, sodium ions or both.[49,50] However, this question is of secondary importance for the concepts discussed here and will not be further addressed. To define the components of the P-module one has to know which of the transporter-derived subunits actually function as ion pumps in complex I. It has been shown that amiloride derivatives, compounds known to inhibit Na^+/H^+ antiporters, inhibit complex I activity most likely by acting on the ND5 subunit.[51,52] However, nothing is known about the ion-translocating properties of the other two transporter-like subunits.

The 'minimal form' of complex I[53] comprises four more membrane integral subunits (ND1, ND3, ND4L and ND6) that also have to be considered as parts of the P-module (Figure 1). There is very little known about their function, but mutations that cause mitochondrial disorders have been found[54,55] and characterized[56,57] in all hydrophobic subunits that are encoded by the mitochondrial genome in eukaryotes. In addition to the subunits found in subcomplex Iλ, subcomplex Iα contains ND1 and ND2 suggesting that these subunits form part of the junction between the peripheral and the membrane arm. Comparison of 2D crystals of the *Neurospora crassa* membrane arm[58] and a bovine subcomplex suggest that subunit ND5 is located at the distal end of the membrane arm.[59] The finding that ND4 and ND5 can be split off together in subcomplex Iβ places ND4 in the same region of complex I[18,20] (Figure 1).

This rather crude picture of the P-module gains some support from another surprisingly close evolutionary relationship between complex I and membrane bound type 3 [NiFe] hydrogenases present, for example, in *E. coli*[47] and the archaeon *M. barkeri*:[60] in addition to the already discussed subunits of the Q-module these enzymes contain an ND1 homolog and one subunit of the Na^+/H^+ antiporter family.[61] Based on sequence comparisons the transporter subunit has been proposed to be a homolog of ND5,[47] but it has been argued that this assignment is ambiguous.[48] The localization of ND2 in subcomplex Iγ and the distal position of ND5 rather suggests that the transporter homolog found in type 3 hydrogenases is actually related to ND2. Three additional subunits that are homolog to complex I subunits have been identified in the hydrogenase part (Hyd-4, *hyf* operon) of the formate hydrogen lyase complex FHL-2 from *E. coli*,[62] namely two more transporter proteins and a ND4L homolog.[16] Another enzyme family that has a subunit composition related to membrane bound [NiFe] hydrogenases are the CO dehydrogenases, *e.g.* from *Rhodospirillum rubrum*.[61]

It can be concluded from this intricate evolutionary relationship that membrane-bound [NiFe] hydrogenase has already acquired part of the P-module of complex I. In the absence of direct evidence, it remains unclear, however, whether these hydrogenases are capable of translocating protons or sodium ions across a membrane or whether this capacity is specific to complex I. At any rate, comparison of the homologous subunits with [NiFe] hydrogenases may serve as a helpful guide to understand the function of the Q-module and gain insight how it may be mechanistically linked to the P-module to drive proton pumping.

3   [NiFe] Hydrogenases, a Model for the Q-Module of Complex I

3.1   Conserved Structural Elements

X-ray structures are available for water-soluble [NiFe] hydrogenases from several *Desulfovibrio* species[63–65] which may serve as structural models for the 49 kDa and PSST subunits of complex I. The first step towards construction of a hydrogenase model for the 'ubiquinone reducing catalytic core' within the Q-module of complex I, is to search for corresponding regions between the subunits of the two distant relatives using sequence comparisons and insights from secondary structure predictions.[66]

In spite of low overall sequence similarity and identity scores, 'ortholog hopping', from water-soluble [NiFe] hydrogenases, to membrane-bound [NiFe] hydrogenases from eu- and archaebacterial sources, and finally to complex I has provided significant and robust alignments for critical segments from both the 49 kDa and PSST subunits of complex I with corresponding regions in the large and small subunits of water soluble [NiFe] hydrogenases (Figure 2). In the hydrogenase structure these conserved segments are combined to from recognizable domains. In these domains, a remarkable number of amino acids are strictly conserved within each enzyme family as well as between hydrogenase and complex I. Significantly, most representatives of this group are amino acids with positively or negatively charged sidechains or prolines suggesting important structural roles or glycines that typically mark boundaries of conserved secondary structure elements (Figure 2). The conservation of an RGXE motif among [NiFe] hydrogenases and complex I was recognised many years ago.[13,14]

Similarities between [NiFe] hydrogenases and complex I become much clearer when conservation of secondary structure elements is taken into consideration as well (Figure 2). The large subunit of water-soluble [NiFe] hydrogenases contains two large β-sheets critically important for its tertiary structure (Figure 3A). Both are composed of three anti-parallel strands and situated very close to the N-terminus and quite close to the C-terminus of the mature subunit, respectively. Both are conserved in the 49 kDa subunit of complex I.

As has been described in detail before, the structural folds of the *Desulfovibrio fructosovorans* protein[65] that make close contact with the [NiFe] active site are conserved in the 49 kDa subunit where they are involved in forming the ubiquinone-reducing catalytic core of complex I.[44,67] Four structural elements in the hydrogenase large subunit contribute to this conserved fold (Figure 4): element A is the helical domain around the two N-terminal ligands of the [NiFe] site (C72, C75), element B is the loop containing H228, element C is formed by the two β-strands around P475, and element D is the domain around the two C-terminal ligands of the [NiFe] site (C543, C546).

Significantly, C72 and C546 of the *D. fructosovorans* sequence both correspond to strictly conserved acidic residues in complex I (D143 and E463 in *Y. lipolytica*, Figure 2). C543 corresponds to a strictly conserved valine (V460 in *Y. lipolytica*), while the position corresponding to C75 can be occupied by various small amino

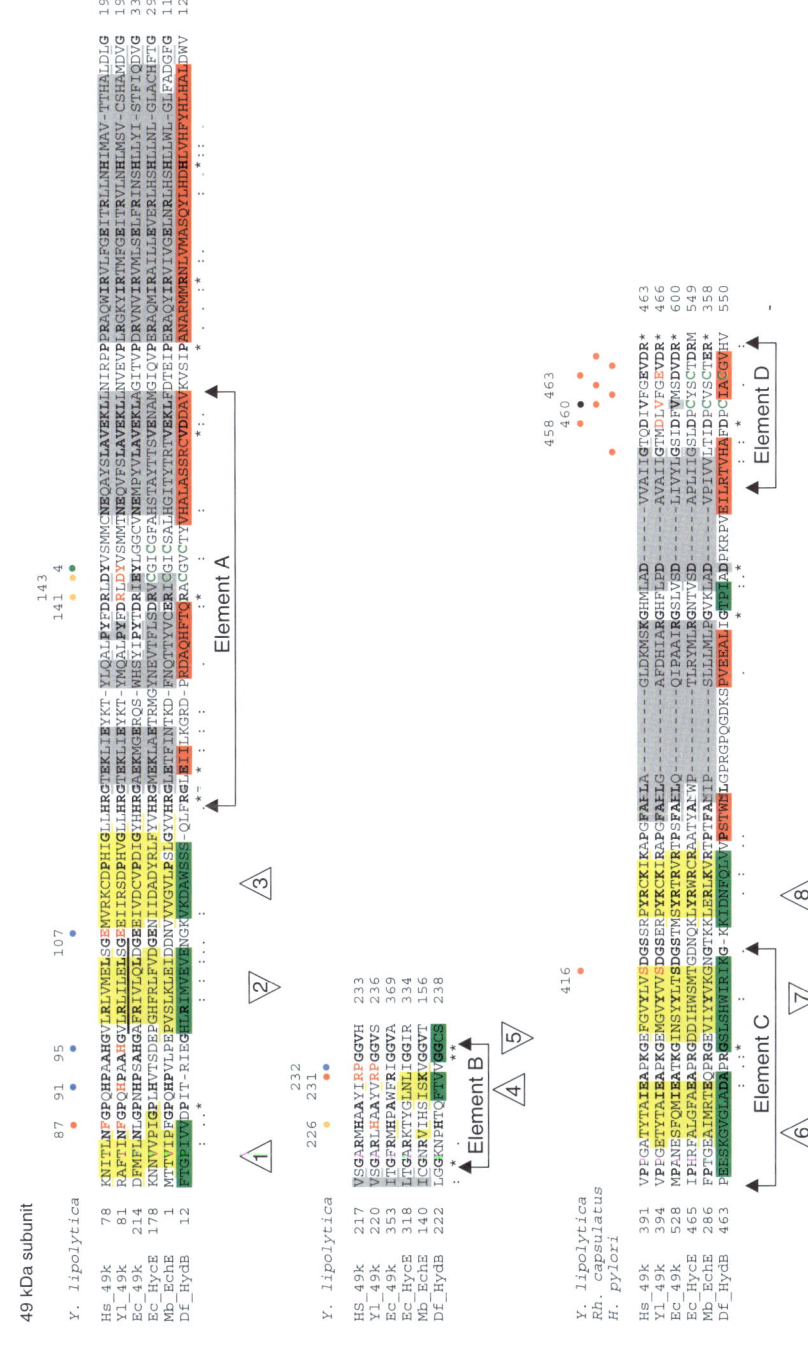

Figure 2 *see p. 165 for full caption*

B

PSST subunit

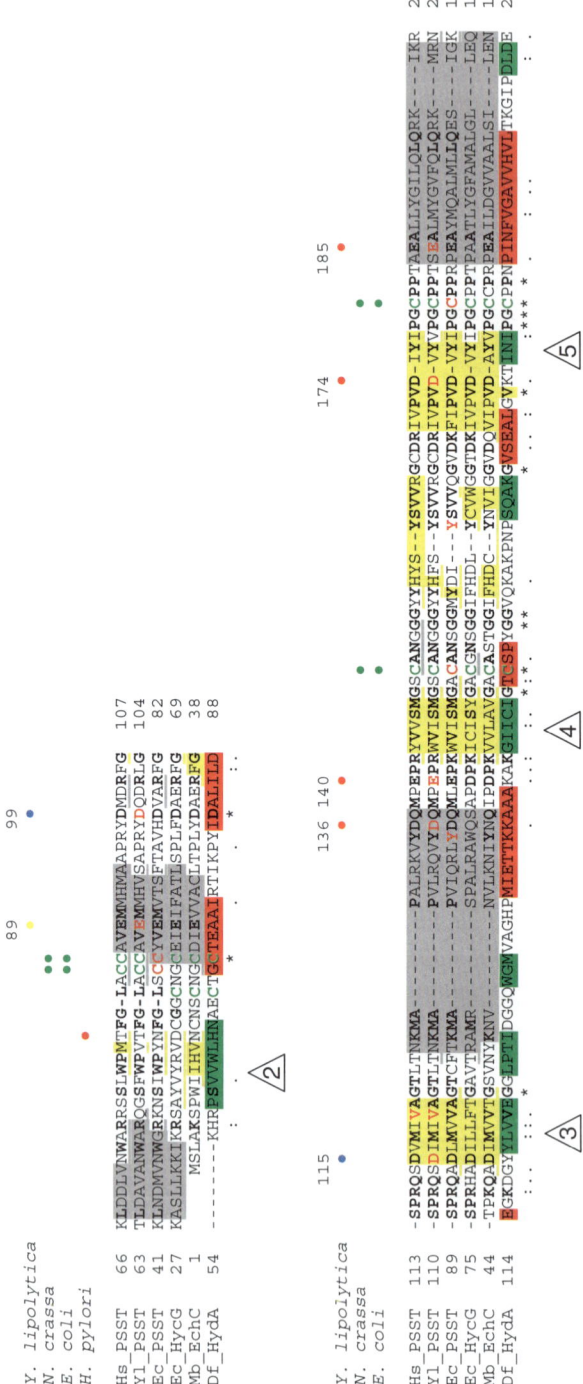

Figure 2 *Sequence alignment between complex I and [NiFe] hydrogenase subunits. Alignment of corresponding sequence segments from complex I, membrane-bound [NiFe] hydrogenases (Hyc and Ech) and water-soluble [NiFe] hydrogenases (Hyd).* **A**, *49 kDa subunits;* **B**, *PSST subunits of complex I and their homologs from [NiFe] hydrogenases. Hs, Homo sapiens (mammals); Yl, Yarrowia lipolytica (fungi); Ec, Escherichia coli (α-proteobacteria); Mb, Methanosarcina barkeri (archaebacteria); Df, Desulfovibrio fructosovorans (δ-proteobacteria). Alignments were created using ClustalW at http://www.ebi.ac.uk/clustalw/index.html with minor manual editing. Secondary structures were predicted using the PROF algorithm[66] at http://www.aber.ac.uk/~phiwww/prof/. Observed α-helices and β-strands in the D. fructosovorans [NiFe] hydrogenase are highlighted in red and green, respectively; Predicted α-helices and β-strands in the other sequences are highlighted in yellow and grey, respectively and underlined in yellow and grey, respectively; if pH or pE is between 0.33 and 0.5. Amino acid positions are printed in bold face if conserved in all complex I sequences; (D/E) and (K/R) are treated like identities. Identities (*), strong (:) and weak (.) similarities are also indicated below the alignment. Numbered triangles indicate the orientations of β-strands. The segment underlined in black in the Y. lipolytica 49 kDa subunit represents the monoclonal antibody epitope located at the membrane-distal end of the peripheral arm of Y. lipolytica complex I[23] (Figure 1). Coloured dots indicate the positions where complex I mutations in the respective model organisms were found to display the following effects: yellow, effects on cluster N2 (shifts in the field position of at least one of the cluster N2 EPR signals, g_x and/or g_{xy} or strong reduction of signal intensities); red, effects on binding of quinone substrates or quinone-like inhibitors (resistance of colonies carrying the respective mutation, or elevated K_m for nonyl-ubiquinone or decyl- ubiquinone, or at least 1.5 fold increase in I_{50} for DQA or rotenone); blue, loss of catalytic activity (less than 10% of the original rate; P232Q [17%] was included as it causes Leigh's syndrome in man); orange, combination of yellow and red; green, combination of yellow and blue; black, combination of yellow, red and blue. In contrast to Table 2, the colour of some dots is a cumulative representation of the phenotypes displayed by several mutations at the same position. Cysteine ligands of the [NiFe] center and the proximal cluster in the hydrogenase and of cluster N2 in the complex I sequences are printed in bold green. Red bold face letters in the alignments indicate that the respective mutations were either found in human Leigh syndrome or created in Y. lipolytica or E. coli. The positions of mutations created in Y. lipolytica are also indicated above the alignments*

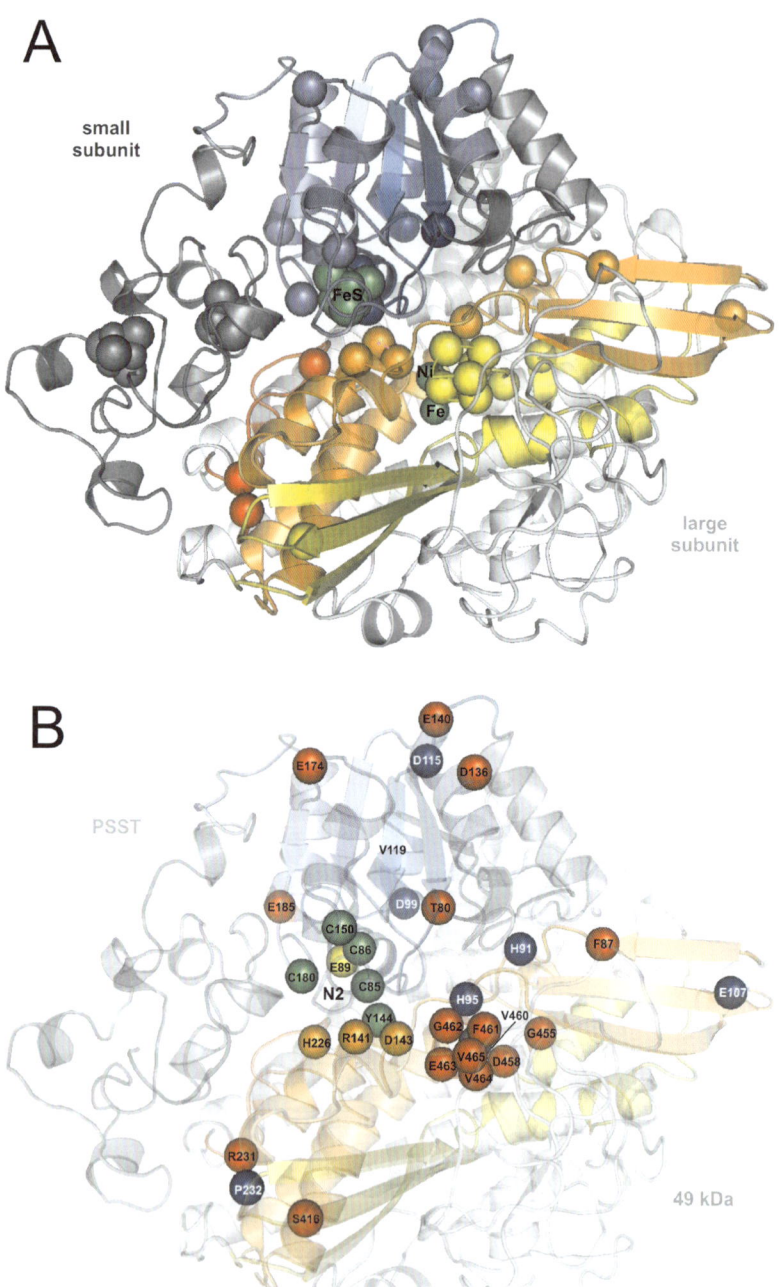

acids, like serine or glycine (S146 in *Y. lipolytica*). Histidine 228 at the tip of element B in the *D. fructosovorans* sequence is very close to the [NiFe] active site (Figure 4) and forms a hydrogen bond to the proximal iron–sulfur cluster in the neighbouring small subunit.[63] This histidine corresponds to a strictly conserved histidine in complex I (H226 in *Y. lipolytica*). Element C contains a proline residue (P475) that is conserved in all hydrogenase and complex I sequences, with the notable exception of *E. coli* complex I (Figure 2). It is involved in forming the turn that connects the first and second β-strands in the C-terminal β-sheet in the large hydrogenase subunit.

The small subunit of water-soluble [NiFe] hydrogenase contains a large β-sheet composed of five parallel strands (Figure 3A). Three to four of these β-strands appear to be conserved in the PSST subunit of complex I. Strand #1 (Figure 2) and the region between β-strands #1 and #2 is certainly not conserved in complex I, and strand #2 is only weakly predicted by the PROF algorithm.[66] Apart from these β-strands, all segments that are conserved between the small [NiFe] hydrogenase subunit and the PSST subunit of complex I are involved in the coordination of the so-called proximal iron–sulfur cluster of [NiFe] hydrogenases or of cluster N2 of complex I, respectively (Figure 3A). As mentioned before, two iron–sulfur clusters and the domains that ligate them are absent in complex I. The twin cysteine motif that provides two of the ligands for cluster N2 (see below) is situated at the beginning of a conserved α-helix. Another α-helix is conserved immediately upstream of β-strand #4 and the third cysteine ligand of cluster N2. While the segment that contains this cysteine is not predicted as α-helical by the PROF algorithm in the PSST subunits of complex I, it is bounded by one and two strictly conserved glycines, respectively. These may represent the boundaries of a short α-helical segment, flanked by β-strands. The short β-strand #5 appears to be present in all enzymes included in the alignment. It is situated immediately upstream of the perfectly conserved PGC(P/C)P motif that contains the fourth cysteine ligand of cluster N2 (Figure 2).

In summary, major structural elements from the large and small subunits of water-soluble [NiFe] hydrogenases seem to be conserved in the 49 kDa and PSST subunits of complex I. These include β-sheets that are critically important for the tertiary structures of these subunits and the structural elements that provide the ligands for the active site [NiFe] cluster and the proximal cluster of water-soluble [NiFe] hydrogenases. As evidenced by the effects of natural and site-directed mutations, these

---

Figure 3 *Structural domains conserved between [NiFe] hydrogenase and complex I. A, segments of the large and small subunits that correspond to segments in the 49 kDa and PSST subunits of complex I (see Figure 2) are highlighted (orange, red, and yellow: large subunit; and blue: small subunit) in the X-ray structure of the [NiFe] hydrogenase from* D. fructosovorans,[65] *using PDB data file 1FRF and Pymol version 0.96, www.pymol.org. Mutations are marked by representing the Cα-atom of respective amino acid position in space fill. The nickel and iron ions in the large and the proximal iron–sulfur cluster in the small subunit are shown in green spacefill representation. B, map positions and effects of mutations are given by putting residues of complex I (*Y. lipolytica *numbering) into the corresponding positions of the hydrogenase structure. Effects of mutations are marked as in A and colour coded as in Figure 2*

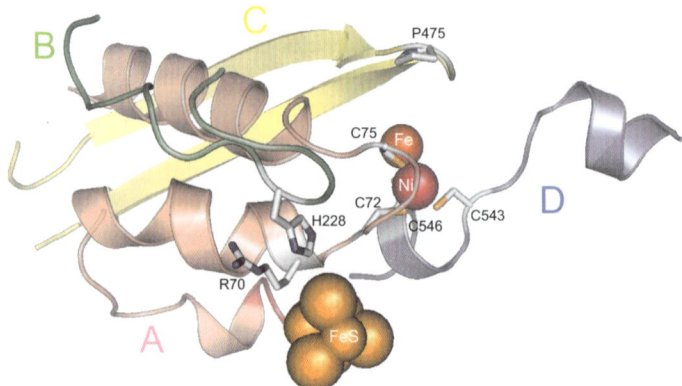

Figure 4 *Structural elements forming the hydrogenase-derived fold at the PSST/49 kDa subunit interface of complex I. The four structural elements in the large subunit of the D. fructosovorans [NiFe] hydrogenase X-ray structure[65] (using PDB data file 1FRF and Pymol version 0.96, www.pymol.org) that make close contact to the [NiFe] center (see Figure 2). Element A is shaded in salmon pink with the slightly darker segment corresponding to the RGXE sequence motif, element B in green, element C in yellow and element D in blue. Sidechains of residues R70, C72 and C75 (all in element A), H228 (element B), P475 (element C), C543 and C546 (element D) are colour coded according to CPK conventions. Nickel (pink) and iron (orange) ions in the large subunit and the proximal iron–sulfur cluster (orange) in the small subunit are also shown*

conserved structural elements play important roles in complex I by forming a significant part of the ubiquinone-reducing catalytic core in the 49 kDa subunit and by harbouring iron–sulfur cluster N2 in the PSST subunit, respectively (Figure 3A).

3.2 Natural and Site-directed Mutations in Complex I

Natural mutations in the genes for subunits of complex I, which may arise 'spontaneously', become recognizable when they result in a specific phenotype. Typical examples are the partial loss-of-function alleles that underlie human genetic disease syndromes and the gain-of-function alleles that give rise to resistance against certain complex I inhibitors. Site-directed mutagenesis has been applied to complex I from several model organisms to tackle diverse questions. An important prerequisite for a useful model organism is that it has to be able to survive in the partial or total absence of complex I activity.

The bacterium *Rhodobacter capsulatus*[8] is capable of photosynthetic growth, which is independent of complex I, while heterotrophic growth in the dark does depend on complex I. Complex I independent growth of many bacteria, plants and fungi like *N. crassa* or *Aspergillus spec.* is made possible by the presence of so-called alternative NADH dehydrogenases. In eucaryotes, these enzymes can be classified into external and internal forms, with their active site oriented towards the outer or inner face of the inner mitochondrial membrane.[6] Introduction of the *E. coli* gene for alternative NADH dehydrogenase into the genome of the bacterium

Paracoccus denitrificans allowed the generation of complex I mutant strains.[68] In the obligate aerobic yeast *Y. lipolytica*, alternative NADH dehydrogenase activity is conferred by a single, external enzyme, called NDH2.[69] Only after NDH2 had been redirected to the internal face of the mitochondrial inner membrane, did it become possible to generate deletions of genes encoding complex I subunits and other loss-of-function alleles in complex I.[70] For *Y. lipolytica*, complex I is essential for survival and it seems that therefore complex I expression is not downregulated when its activity is compromised. Reductions in complex I content of isolated mitochondrial membranes that can be estimated as NADH:hexammineruthenium oxidoreductase activity,[71] thus essentially reflect assembly defects or reduced stability of the enzyme.

Historically, it had been commonly assumed that the mitochondrially encoded, membrane-integral ND subunits of eucaryotic complex I (or their bacterial homologs) form the binding site for the hydrophobic substrate ubiquinone. However, a screen of chemically induced mutants of *Rh. capsulatus* for resistance to the classical complex I inhibitor piericidin A provided the first hint that domains within the hydrophilic 49 kDa subunit of complex I related to the hydrogenase [NiFe] center structural fold make substantial contributions to the ubiquinone-reducing catalytic core. In this screen, a single mutation in the 49 kDa subunit (V407M) was retrieved several times independently and exhibited cross resistance to other complex I inhibitors.[43] Significantly, the affected residue corresponds to one of the cysteine ligands (C543) for the hydrogenase [NiFe] center within element D (Figure 2). Two site-directed mutations of *Rh. capsulatus* created in the vicinity of the original V407M resistance mutation (G409A, D412E) also displayed resistance towards various complex I inhibitors[72] (Table 2).

A screen for benzimidazole resistance in *H. pylori*[73] also yielded three substitutions in the 49 kDa subunit (G398S, F404S, V407M in *H. pylori*, corresponding to G455, F461 and V464 in *Y. lipolytica*) and one in the PSST subunit (T27A, corresponding to T80 in *Y. lipolytica* and N58 in *E. coli*). The most obvious explanation for this result is that complex I is the previously unknown target of benzimidazole compounds. Remarkably, all three 49 kDa subunit mutants identified in this study also exhibited alterations in affinity to the complex I inhibitor rotenone.[73]

Reconstruction of the *Rh. capsulatus* V407M mutation at the corresponding position in *Y. lipolytica* complex I (V460M) also resulted in inhibitor resistance, which was more pronounced against DQA than against rotenone.[44] Much higher resistance was observed when a nearby aspartate residue (D458 in *Y. lipolytica*), which is strictly conserved among hydrogenase and complex I sequences, was mutated.[44] In several studies, site-directed mutations of conserved acidic and basic amino acids that align with residues that reside within the active site domain of the large subunit of [NiFe] hydrogenases, or with residues that reside in the vicinity of the proximal iron–sulfur cluster in the small subunit of [NiFe] hydrogenases were created in the 49 kDa and PSST subunits of complex I from *Y. lipolytica*.[74–76] Some of these residues are even invariant among [NiFe] hydrogenases and complex I. The results of this approach are fully consistent with the concept that the active site structural fold of [NiFe] hydrogenases has been conserved in complex I and evolved to form a significant part of the ubiquinone-reducing catalytic core, and that the proximal

Table 2   *Selected mutations in the catalytic core of complex I*

Effects [a] of mutations in the 49 kDa subunit of Y. lipolytica complex I

Strain	Content[b] (%)	Activity[c] (%)	K_m for DBQ (µM)	I_{50} (nM) DQA	I_{50} (nM) rot.	N2 by EPR	Reference
Parental[d]	100	100	11	13	600	normal	44
F87L •	130	59	21	50	600	n.d.	94
H91A •	120	<5	—	—	—	normal	76
H91M •	80	<5	—	—	—	normal	76
H91R •	80	<5	—	—	—	normal	76
H95A •	130	<5	—	—	—	normal	76
H95M •	120	<5	—	—	—	normal	76
H95R •	100	<5	—	—	—	normal	76
E107A •	140	<5	—	—	—	n.d.	94
R141K •	140	45	13	55	1500	decreased	76
R141M •	110	40	11	11	500	not detectable	76
R141A •	130	17	10	21	570	not detectable	76
D143E •	170	23	n.d.	38	630	unstable	44
Y144H •	70	<5	n.d.	—	—	altered	44
H226A •	100	20	n.d.	18	770	not detectable	44
H226Q	110	56	9	13	600	decreased	76
H226C	120	43	5	13	600	decreased	76
H226M •	120	80	9	9	300	altered	76
R231Q •	84	101	16	17	900	n.d.	94
R231E •	112	74	29	20	850	n.d.	94
P232Q	38	17	—	—	—	n.d.	94
P232G	67	56	7	23	500	n.d.	94
S416P •	90	111	17	17	1100	n.d.	94
S416A •	109	86	39	19	1500	n.d.	94
D458A •	93	28	n.d.	520	5200	normal	44
V460A •	90	<5	—	—	760	not detectable	44
V460M •	100	28	n.d.	53	760	decreased	44
E463Q •	60	20	n.d.	34	780	decreased	44

[a] illustrated by coloured dots: yellow, effects on cluster N2; red, effects on binding of quinone substrates or quinone-like   inhibitors; blue, loss of catalytic activity; orange, combination of yellow and red; green, combination of yellow and blue.
[b] expressed as relative specific NADH: hexammineruthenium oxidoreductase activity. 100% = 1 µmol min^{-1} mg^{-1}.
[c] normalized for complex I content. 100% = 0.3 µmol min^{-1} mg^{-1}.
[d] plasmid complemented *nucm::URA3* deletion strain.

Table 2 *(Continued)*

Effects [a] of mutations in the PSST subunit of Y. lipolytica complex I

Strain	Content[b] (%)	Activity[c] (%)	K_m for DBQ (μM)	I_{50} (nM) DQA	I_{50} (nM) rot.	N2 by EPR	Reference
Parental [d]	100	100	22	20	500	normal	75
E89A	160	108	19	13	590	altered	74
E89C	160	98	18	14	630	altered	74
E89Q	110	128	9	11	620	altered	74
D99N	90	<5	—	—	—	decreased	75
D99E	90	<5	—	—	—	n.d.	75
D99G	90	<5	—	—	—	n.d.	75
D115N	105	6	—	—	—	decreased	75
D115E	101	5	—	—	—	n.d.	75
D115G	104	6	—	—	—	n.d.	75
V119M	110	55	12	10	700	decreased	93
D136N	110	25	18	48	100	n.d.	74
E140Q	110	30	15	45	150	n.d.	74
D174N	93	100	35	20	500	n.d.	75
E185Q	110	35	50	20	500	normal	75

[a] illustrated by coloured dots: yellow, effects on cluster N2; red, effects on binding of quinone substrates or quinone-like inhibitors; blue, loss of catalytic activity.
[b] expressed as relative specific NADH: hexammineruthenium oxidoreductase activity. 100% = 1 μmol min^{-1} mg^{-1}.
[c] normalized for complex I content. 100% = 0.4 μmol min^{-1} mg^{-1} in Ref. 93 and 0.3 μmol min^{-1} mg^{-1} in all other Ref.
[d] plasmid complemented *nukm::LEU2* deletion strain.

Effects[a] of mutations in the 49 kDa subunit of Rh. capsulatus complex I

Strain	Y. lipolytica equivalent	Content[b] (%)	Activity[c] (%)	I_{50} (nM) piericidin	rotenone	pyridaben	Reference[d]
Parental	—	100	100	0.6	100	30	72
V407M	V460	220	50	5	600	50	72
G409A	G462	160	15	1.5	100	55	72
D412E	D465	210	40	3.6	400	75	72

[a] red dots indicate resistance to various quinone-like inhibitors.
[b] expressed as relative specific NADH: hexammineruthenium oxidoreductase activity. 100% = 0.6 μmol min^{-1} mg^{-1}.
[c] normalized for complex I content. 100% = 0.3 μmol min^{-1} mg^{-1}.
[d] Format adapted to match the *Y. lipolytica* data.

iron–sulfur cluster of [NiFe] hydrogenases has been transformed into cluster N2 of complex I (Figure 3B). Basically, the mutants display three types of effect: (i) alterations of the N2 cluster environment, evidenced by reduced or shifted EPR signals, (ii) effects on ubiquinone binding, evidenced by an increase in the K_m for decyl- or nonylubiquinone and/or resistance (or, in rare cases hypersensitivity) towards quinone-like hydrophobic inhibitors of complex I and (iii) loss of catalytic activity (Table 2).

When the effects of mutations from various model systems (*Y. lipolytica*, *N. crassa*, *Rh. capsulatus*, *E. coli*, *H. pylori*) are related to their map positions in the hydrogenase model for the 49 kDa and PSST subunits of complex I, a striking picture emerges (Figure 3). Mutations of the cysteine ligands, generated in *N. crassa* or *E. coli* lead to inactivity and the complete absence of the cluster N2 EPR signals. These and other mutations that influence the EPR signature of cluster N2 will be discussed in detail below. Most of the resistance mutations in the 49 kDa subunit, including D458A, cluster around the former [NiFe] site. Resistance and hypersensitivity mutations in the PSST subunit are found at various positions, some of which are quite distant from cluster N2 and from the interface with the 49 kDa subunit (Figure 3B). It is likely that such mutations exert their effects by altering the secondary structure of the PSST subunit and perhaps even the neighbouring 49 kDa subunit. Mutations that cause inactivity can also be found at some distance from the former [NiFe] center. Among these, mutations of two conserved aspartates in the PSST subunit (D99 and D115) are especially interesting. Their individual replacement by glutamate leads to the same complete loss of activity as replacement by glycine. This may indicate that the exact position of their side chain carboxylate groups is critical for catalytic activity. It is tempting to speculate that they may be critical parts of proton pathways for chemical protons required for ubiquinone reduction.[75]

Significantly, positions in the 49 kDa subunit of *Y. lipolytica* complex I where mutations affect the EPR signature of cluster N2 and also produce additional effects like inactivity (Y144), resistance (R141, D143) or hypersensitivity (H226) map close with the interface with the PSST subunit, in the vicinity of cluster N2. Strikingly, at position V460 in the 49 kDa subunit of *Y. lipolytica*, corresponding to the hydrogenase [NiFe] center ligand C543, mutations could be generated that displayed all three types of effect (Figure 3B).

3.3 Human Pathogenic Mutations

Considering the large number of subunits present in mammalian complex I, it is not surprising that numerous human pathogenic mutations have been described in both mitochondrial and nuclear genes coding for subunits of complex I. While complex I patients are generally heterogeneous with respect to age of onset and disease symptoms, the most common clinical picture is Leigh syndrome, or subacute necrotizing encephalomyopathy, with or without cardiomyopathy.[77–79] Mutations causing complex I deficiency have been discovered in all seven mitochondrial ND genes[80–82] and in the following nuclear genes: *NDUFS1* (75 kDa), *NDUFS2* (49 kDa),

NDUFS3 (30 kDa), *NDUFS4* (AQDQ), *NDUFS6* (13 kDa), *NDUFS7* (PSST), *NDUFS8* (TYKY), *NDUFV1* (51 kDa) and *NDUFV2* (24 kDa).[83–92] It is likely that many more mutations, especially in genes for accessory subunits remain to be discovered.

Several human point mutations, including those located in the 49 kDa and PSST subunits, have been reconstructed in complex I from the obligate aerobic yeast *Y. lipolytica*[93,94] and/or the fungus *N. crassa*.[95] As expected, analysis of these reconstructed pathogenic complex I mutations demonstrated that in most cases they reduce, but not completely eliminate, complex I activity. Interestingly, in the hydrogenase structural model for the ubiquinone reducing catalytic core of complex I, these mutations do not reside in the immediate vicinity of the former [NiFe] site, but rather at the boundaries of conserved structural elements.[94] Two mutations in the human 49 kDa subunit, namely F84L and E104A, corresponding to F87L and E107A in *Y. lipolytica*, map to the N-terminal three-stranded antiparallel β-sheet (Figure 3B). F87 resides in the loop connecting β-strands #1 and #2, and E107 in the sharp turn connecting β-strands #2 and #3 (Figure 2). Human pathogenic mutations R228Q and P229Q[88] (corresponding to R231Q and P232Q in *Y. lipolytica*) are both close to the region corresponding to element B in the 49 kDa subunit primary sequence (Figure 3B). However, since the corresponding residues in the hydrogenase structure are part of a small, antiparallel β-sheet (β-strands #4 and #5 in Figure 2) at the base of the His228 loop that is furthest away from the [NiFe] center, this segment of the 49 kDa subunit can be expected to be relatively far away from the actual ubiquinone-binding pocket. Human mutation S413P,[88] corresponding to S416P in *Y. lipolytica*, is situated in the turn connecting β-strands #7 and #8 of the C-terminal β-sheet (Figure 3B).

Interpretation of the effects of these mutations, as observed after reconstruction in *Y. lipolytica*,[94] is not always straightforward due to the inherent limitations of the homology-based structural model for the ubiquinone-reducing catalytic core of complex I. Mutations R231Q resulted in a 1.5 fold, and mutation S416P in a 2.0 fold increase in the I_{50} value for rotenone but left activity and the K_m for ubiquinone almost unaffected. On the other hand, mutations F87L and P232Q, where this effect was even more pronounced, resulted in unexpectedly large reductions in complex I activity. Mutation P232Q also resulted in a drastic reduction in complex I content. This effect was alleviated in a P232G mutant. Complex I from the E107A mutant was completely inactive and since the corresponding human mutation was observed in a compound heterozygous Leigh-syndrome patient, so far a complete loss-of function phenotype in man cannot be excluded for this mutation.

The V122M Leigh syndrome mutation[85] in the human PSST subunit (which corresponds to V119M in *Y. lipolytica*) resides within β-strand #3 of the predicted four to five stranded parallel β-sheet. Since cluster N2 is sandwiched between the 49 kDa subunit and the parallel β-sheet in the PSST subunit, the position of the V119M mutation is on the far side of cluster N2 when viewed from the site corresponding to the [NiFe] center. Consistent with this position, the V119M mutation had no effect on the K_m for ubiquinone[93] but resulted in a reduction of the N2 EPR signal amplitude of about 50% and a similar reduction in catalytic activity.[94] Similarly, the same mutation in *N. crassa* only caused a slight destabilization of complex I but left its EPR signature and catalytic activity virtually unaffected.[95]

4   Addressing the Function of Cluster N2 by EPR Spectroscopy and Site-Directed Mutagenesis

Iron–sulfur cluster N2 resides within the Q-module of complex I and is considered to play a central functional role in the coupling of electron-and proton-transport processes. This hypothesis is supported by three important experimental findings: (i) N2 has the most positive redox midpoint potential of all iron–sulfur centers in complex I,[96–98] (ii) the midpoint potential is pH-dependent,[98] and (iii) in EPR spectroscopy, cluster N2 shows a magnetic interaction with semiquinone radicals.[99] These three results make N2 the most likely candidate to be the endpoint of the intrinsic electron-transferring chain of complex I directly reducing ubiquinone. The pH-dependent redox behaviour or 'redox-Bohr effect' would potentially allow a direct coupling of electron transfer and proton translocation.

4.1   Subunit Assignment of Iron–Sulfur Cluster N2

Despite the assumedly prominent role of N2 in the catalytic mechanism, the assignment of this cluster to an individual complex I subunit covered a long period of controversial discussions. By overexpression of subunits and reconstitution of iron–sulfur clusters, site-directed mutagenesis studies, and EPR spectroscopy it was possible to assign clusters N1a, N1b, N3, N4, N5 (and N7/N1c of *E. coli*) to individual subunits of the N-module of complex I.[31,100–104] Deduced from potential iron–sulfur cluster binding motifs and the homology of the complex I Q-module to [NiFe] hydrogenases two subunits, TYKY[105] and PSST,[29] were candidates for harbouring cluster N2.

TYKY contains two iron–sulfur cluster-binding motifs in its primary sequence and it was proposed that the EPR spectrum of iron–sulfur cluster N2 resulted from two very similar tetranuclear clusters residing in this subunit.[106] Based on UV-VIS spectroscopy of complex I from *E. coli* and *N. crassa*, it was suggested that these binding motifs contain two tetranuclear iron–sulfur clusters in a ferredoxin-like arrangement. So far these clusters could not be identified clearly by EPR spectroscopy, but their redox midpoint potential was determined for *N. crassa* using optical spectroscopy.[38,107] The redox potential was pH-insensitive, associated with transfer of two electrons and much more negative (–270 mV) than that obtained from redox titrations of cluster N2 EPR signals in complex I from other sources ($E_{m,7} > -200$ mV, pH dependent). From these results it was concluded that subunit TYKY contains two previously uncharacterised iron–sulfur clusters that were named N6a and N6b.

The electron-transfer step between cluster N2 and quinone is specifically blocked by a number of structurally diverse inhibitors.[108,109] Labelling of mitochondrial and bacterial membranes with a photosensitive pyridaben derivative supported the hypothesis that the PSST subunit is involved in ubiquinone redox reactions with cluster N2 as the endpoint of the iron sulfur–cluster wire.[110] In contrast to the homologous small subunit of [NiFe] hydrogenase, PSST contains just one potential iron–sulfur cluster-binding motif but with an unusual arrangement of four conserved cysteines, two of which are immediately adjacent. Site-directed mutagenesis of both cysteines in complex I from *N. crassa* prevented assembly of cluster N2.[39] Individual mutations of all four cysteines in subunit NuoB (PSST) and all eight cysteines in NuoI (TYKY) of *E. coli* complex I followed by EPR spectroscopic analysis of mutants suggest that cluster N2 is bound by the PSST subunit.[39,40] This assignment is strongly supported by the mutagenesis results summarized in Table 2 and illustrated in Figure 3. According to the homology model of the Q-module, mutants of both the 49 kDa and PSST subunit that affect the properties of iron–sulfur cluster N2 are predominantly found in the vicinity of the conserved iron–sulfur cluster (see above).

4.2 The Fourth Ligand of Iron–Sulfur Cluster N2

As already mentioned, one of the cysteines ligating the proximal iron–sulfur cluster in hydrogenase is not conserved in subunit PSST but is replaced by another cysteine immediately adjacent to one of the conserved ligands (Figure 2). This raises the question whether the tetranuclear cluster N2 is ligated by the unusual four-cysteine motif or whether there is another amino acid serving as the fourth ligand like the conserved glutamate (E89 in the *Y. lipolytica* PSST subunit), three residues downstream of the adjacent cysteines.[111] This possibility could be ruled out by site-directed mutagenesis in *Y. lipolytica*. EPR analysis demonstrated that cluster N2 was still present even when glutamate was exchanged with alanine.[74] A more elaborate study mutating all eight highly conserved acidic residues in subunit PSST of *Y. lipolytica* revealed that none of these amino acids is involved in the ligation of iron–sulfur cluster N2.[75] Based on the structural model for the Q-module we proposed that the fourth ligand might also be provided by the 49 kDa subunit as cluster N2 is predicted to be located at the interface of this subunit with the PSST subunit[44] (Figure 4). Several invariant or highly conserved amino acids reside around the [NiFe] binding site and the proximal iron–sulfur cluster of [NiFe] hydrogenases. The significance of corresponding residues in the 49 kDa subunit was analysed by functional and spectroscopic characterization of site-directed mutants in *Y. lipolytica*.[44,76] Despite the negative finding that none of the mutated residues in the 49 kDa subunit could be identified as the fourth ligand of cluster N2, the results strongly supported the hypothesis that the Q-binding site of complex I has evolved from the [NiFe] site of hydrogenases and that the proximal cluster of hydrogenases corresponds to cluster N2 in complex I (see above).

The analysis of isolated *Y. lipolytica* complex I by relaxation-filtered hyperfine spectroscopy (REFINE), a technique allowing spectroscopic characterization of individual paramagnetic centers in a protein containing several paramagnetic groups[112,113] yielded no nitrogen modulations in the ESEEM spectrum of cluster N2, excluding a direct coordination by a histidine, another amino acid with nitrogen in the side chain, or a backbone nitrogen.

Recently, the structure of the PSST-homologous small subunit of the *Desulfovibrio gigas* [NiFe] hydrogenase was used to create an *in silicio* mutant with two consecutive cysteines. Molecular dynamics calculations showed a slight main-chain conformational change would accommodate coordination of a tetranuclear cluster.[114] So far, coordination by two adjacent cysteines has only been reported for a special binuclear iron–sulfur cluster in the Fhuf protein of *E. coli*.[115] A similar sequence motif can be found in the small subunit of hydrogenase III from *Aquifex aeolicus* but it remains unclear whether both consecutive cysteines are involved in binding the proximal tetranuclear cluster in this case.[116] Although a clear experimental proof for this kind of coordination is still missing, the hypothesis that cluster N2 is ligated by the four cysteines of PSST seems most likely.

4.3   Redox-dependent Protonations/Deprotonations

As discussed above, iron–sulfur cluster N2 exhibits a marked pH-dependence of its redox midpoint potential. This has prompted several studies to identify protonable amino-acid residues that are responsible for this so-called redox-Bohr effect. On the basis of electrochemically induced FT-IR double difference spectra it was suggested that an aspartate or glutamate sidechain gets protonated upon oxidation of cluster N2,[117] while one or more tyrosine residues are protonated upon reduction of cluster N2.[118] Using site-directed mutagenesis of the *nuoB* gene from *E. coli*, two strictly conserved tyrosine residues in the PSST subunit (Tyr114 and Tyr139) were proposed to be responsible for the latter effect, while mutagenesis of Tyr154 did not markedly influence the FT-IR spectra.[118]

For several reasons, it seems questionable whether these interpretations are valid. It may be argued that the observed FT-IR signals could also indicate changes in the local environment of residues that are protonated in both the oxidised and reduced states of cluster N2. Moreover, as the spectra were obtained by subtraction of oxidised-minus-reduced spectra of the *E. coli* NADH dehydrogenase fragment from oxidised-minus-reduced spectra of complete complex I, assembly or preparation artefacts cannot be excluded: EPR spectra of complex I prepared from tyrosine single mutants exhibit reduced g_z signal intensities for iron–sulfur clusters N1a and N1b and these signals are virtually absent in complex I from the double mutant. For the same reason, it is unclear whether minor changes in the position of the cluster N2 g_z signal reflect specific effects of these mutations. All three conserved tyrosine residues discussed here correspond to residues in hydrogenase that reside at a Cα distance of 14–19 Å from the proximal iron–sulfur cluster. The hydrogenase model therefore also argues against their involvement in the redox-Bohr group associated with cluster N2. Even when one accepts the authors' conclusions that the FT-IR data are indicative of specific protonation/deprotonation events, this cannot be taken as evidence for a direct role of cluster N2 in proton translocation. Proton channelling to the active site must occur in both hydrogenases and in complex I to deliver the protons that are consumed during hydrogen production and ubiquinone reduction, respectively. However, since proton channels have not been identified in water-soluble hydrogenases so far, it is impossible to speculate about potential conservation in complex I.

H226 in element B of the catalytic core corresponds to H228 in the large subunit of [NiFe] hydrogenase (Figure 4), where it forms a hydrogen bond to the proximal iron–sulfur cluster[63] (Figure 4). This suggests that it may undergo redox-dependent protonation changes due to the close proximity to iron sulfur cluster N2. Depending on the kind of H226 mutation, complete loss of the cluster N2 EPR spectrum (H226A) or decreased and shifted N2 signals (H226M) were observed in *Y. lipolytica* complex I (Table 2). More strikingly, as a result of this mutation the redox midpoint potential of iron–sulfur cluster N2 lost its characteristic pH-dependence and was markedly shifted to the negative, to a value (< -200 mV) very close to the other 'isopotential' clusters (Zwicker *et al.* in preparation). Thus, H226, although it is not a direct ligand of cluster N2, is the redox-Bohr group that is responsible for the rather positive and pH-dependent midpoint potential of cluster N2. Complex I from mutant H226M exhibited essentially wild-type ubiquinone-reductase activity and its capacity to pump protons was not affected (Zwicker *et al.* in preparation). This excludes a direct involvement of cluster N2 and its redox-Bohr group in the proton-pumping mechanism of complex I.

4.4 Ubiquinone Reduction

As mentioned above a prominent role of cluster N2 in catalysis has been implied from its interaction with ubisemiquinone radicals.[41,119,120] Up to now, semiquinone radicals could only be identified in preparations from bovine heart mitochondria. Three species (SQ_{Nf}, SQ_{Ns}, SQ_{Nx}) were detected by EPR spectroscopy.[99] These differ primarily in power saturation behavior of the EPR signal and dependence on $\Delta\mu_{H+}$. The fast-relaxing SQ_{Nf} is dependent on $\Delta\mu_{H+}$, both slow-relaxing species, SQ_{Ns} and SQ_{Nx} are not. Hence the latter two semiquinones should allow characterization with samples of isolated complex I as well.[121] In isolated complex I from bovine heart, a semiquinone EPR signal can be detected that originates only from one paramagnetic species. Redox midpoint potentials were estimated from plotting the signal intensity as a function of ambient redox potential at $E_{m1,7.8}$ (Q/SQ) = -45 mV and $E_{m2,7.8}$ (SQ/QH$_2$) = -63 mV. The power saturation behavior of the semiquinone signal was found to be dependent on the reduction state of iron–sulfur cluster N2 suggesting an interaction with this redox center. From this interaction the distance N2–SQ_{Ns} can be estimated at ~30 Å. Since this property is only known from SQ_{Ns}, it was concluded that the SQ_{Nx} species does not occur in the isolated enzyme and might be an artefact of the three-component resolution of the power-saturation curve of SQ species detected in submitochondrial particles.

Under conditions of steady state NADH oxidase or NADH:ubiquinone-1 oxidoreductase activity, tightly coupled bovine heart SMP allows detection of the fast-relaxing semiquinone SQ_{Nf} that exhibits a pronounced pH dependence.[122] The spin-spin interaction of this radical with reduced cluster N2 resulted in splitting of both a component of the cluster N2 EPR spectrum (g_z–N2) and the g = 2.004 radical signal. The center to center distance between the two paramagnets was estimated to be 12 Å. The SQ-signal linewidths of both SQ_{Ns} and SQ_{Nf} suggest that both semiquinones are in their anionic form.

Mutagenesis of R141 of the 49 kDa subunit in *Y. lipolytica* provided some additional clues on the role of iron–sulfur cluster N2 in the ubiquinone reduction reaction of complex I. R141 corresponds to R70 in the large subunit of the *D. fructosovorans* hydrogenase, located in element A of the catalytic core, immediately upstream of C72 and C75, two ligands of the [NiFe] center and about 6 Å away from the proximal iron–sulfur cluster in the small subunit (Figure 4). In all mutants at this position analyzed so far (R141A, R141K, R141M) the typical cluster N2 EPR signature was lost but catalytic activity was still present ranging up to 45% of the parental enzyme[76] (Table 2). Obviously, the absence of an EPR signal cannot prove the absence of a redox center, but even by thorough analysis, no spectroscopic evidence for the presence or the involvement of cluster N2 in the enzymatic reaction was found in these mutants. A possible, although speculative, explanation for this surprising result might be offered by electron transfer theory: if one assumes that the distance between redox groups in complex I is similar (~5 Å) as in [NiFe] hydrogenases[63,65] or other iron–sulfur proteins, electron transfer rates can be estimated by applying the 'Moser–Dutton ruler'[123] – increasing the distance from 5 to 10 Å by removing one iron–sulfur cluster in a chain is predicted to result in a slow down of the electron transfer rate by 4 to 5 orders of magnitude from 10^8–10^9 s^{-1} which would still clearly be sufficient to sustain the steady-state turnover rate of complex I (~10^2 s^{-1}).

5   Implications for the Mechanism of Proton Pumping by Complex I

The mechanism of how mitochondrial complex I uses redox energy to pump protons across the inner membrane remains elusive. However, recent progress in understanding the structural organization of complex I provides some clues as to how such a mechanism could operate, and even more importantly, excludes a number of possible mechanisms that have been discussed in the past. Several fundamental implications from the structural organization of complex I pose significant restraints on any hypothesis of its catalytic mechanism.

As electron transfer and proton translocation occur in distinct structural modules, a direct-coupling mechanism driving proton movement by charge-compensation processes is unlikely. Such a mechanism would require close and immediate electrostatic coupling between at least one of the redox prosthetic groups or substrate ubiquinone and proton pathways in the P-module. For an efficient design of a pumping device of this type, one would expect the critical redox events to occur in close proximity to the membrane-integral proton-transducing elements. However, none of the known redox prosthetic groups of complex I is found in the membrane-integral part of complex I and there is good evidence that most of the Q-module and therefore the ubiquinone-reducing catalytic core of complex I is located well away from the headgroup region of the membrane in the peripheral arm.[23] Moreover, two of the three subunits (subunits ND4 and ND5) that are likely candidates for acting as proton transporters in complex I, seem not to even reside in the part of the P-module that is in direct contact with the peripheral arm.[59] On the other hand, as discussed above in detail, there is good evidence that iron–sulfur cluster N2 is the immediate electron donor for ubiquinone that resides in the Q-module, but is not directly involved in proton pumping.

Taken together, evidence seems to exclude a direct mechanism of coupling between electron transfer and proton pumping and as a consequence, transmission of energy between the Q- and the P-module must occur by conformational coupling. This type of mechanism was proposed a long time ago[124] and more recently several studies provided direct evidence for redox-dependent conformational changes of complex I.[125–128] Considering that the model for the structural organization of complex I discussed here places the likely proton translocators ND4 and ND5, and the site of ubiquinone reduction at opposite ends of the large complex I particle, one has to envision long-range conformational changes that would have to transmit energy over distances well over 100 Å. In this respect, the mechanism of complex I would resemble those of ATP-Synthase[129] or Ca^{2+}-ATPase.[130]

Within this concept of a long-range conformational proton pump for complex I, there are two central questions that are not even close to being answered conclusively by available evidence. The first question deals with the problem how ubiquinone with its long and extremely hydrophobic side chain reaches the active site that is predicted to be located far up in the peripheral arm. As a solution to this problem, we have proposed that the pathway for ubiquinone may be provided by a hydrophobic or amphipathic ramp that allows its headgroup to diffuse up the peripheral arm.[23,128] The fact that inhibitor binding studies have shown that probably all of the numerous hydrophobic inhibitors of complex I bind to one large binding domain[109,131] and that, in contrast to the cytochrome *bc₁* complex, hydrophobicity is not the primary determinant for ubiquinone derivatives to be good substrates of complex I, provides some support for this idea. Some hydrophilic subcomplexes of complex I are still capable of reducing hydrophilic quinones, but this reaction is not sensitive towards the hydrophobic inhibitor rotenone.[45] This dissociation of catalytic activity and inhibitor sensitivity is also observed frequently, when intact complex I is treated harshly, *e.g.* with high concentrations of detergents. This is also in line with the picture of a 'ubiquinone-ramp' in complex I, where the active site for the ubiquinone headgroup and major parts of the binding domain for the hydrophobic tail of ubiquinone and the hydrophobic inhibitors are spatially distinct moieties. The second question is how the redox energy that has to be transmitted to the P-module is converted into conformational energy and which redox active components are immediately involved in this process. Very little information is available on this issue so far. Iron–sulfur cluster N2, which has been discussed as one of the possible major components of the proton pump in recent years, seems not to be directly involved (Zwicker *et al.* in preparation) as removing its pH-dependence and shifting its midpoint potential to the negative does not impair proton pumping. Therefore, it seems very likely that the primary role of the redox prosthetic groups of complex I is to deliver electrons to ubiquinone and that processes directly associated with the two-step redox reaction of the substrate itself are generating the conformational strain then used for pumping protons.

Acknowledgements

We thank Drs S. Dröse, A. Galkin and H. Schägger for critically reading the manuscript and for helpful discussions. This work was supported by the Deutsche Forschungsgemeinschaft (SFB 472 - Molekulare Bioenergetik, Project P2).

References

1. J. Carroll, I.M. Fearnley, R.J. Shannon, J. Hirst and J.E. Walker, *Mol. Cell. Proteomics*, 2003, **2**, 117.
2. J. Hirst, J. Carroll, I.M. Fearnley, R.J. Shannon and J.E. Walker, *Biochim. Biophys. Acta*, 2003, **1604**, 135.
3. T. Yagi, T. Yano, S. Di Bernardo and A. Matsuno-Yagi, *Biochim. Biophys. Acta*, 1998, **1364**, 125.
4. K. Krab, J. Soos and M.K.F. Wikström, *FEBS Lett.*, 1984, **178**, 187.
5. A.S. Galkin, V.G. Grivennikova and A.D. Vinogradov, *Biochemistry-Moscow*, 2001, **66**, 435.
6. S. Kerscher, *Biochim. Biophys. Acta*, 2000, **1459**, 274.
7. A.D. Vinogradov, *Biochim. Biophys. Acta*, 1998, **1364**, 169.
8. A. Dupuis, M. Chevallet, E. Darrouzet, H. Duborjal, J. Lunardi and J.P. Issartel, *Biochim. Biophys. Acta*, 1998, **1364**, 147.
9. C. Brockmann, A. Diehl, K. Rehbein, H. Strauss, B. Korn, R. Kuhne and H. Oschkinat, *Structure*, 2004, **12**, 1645.
10. V. Guenebaut, R. Vincentelli, D. Mills, H. Weiss and K. Leonard, *J. Mol. Biol.*, 1997, **265**, 409.
11. N. Grigorieff, *J. Mol. Biol.*, 1998, **277**, 1033.
12. R. Djafarzadeh, S. Kerscher, K. Zwicker, M. Radermacher, M. Lindahl, H. Schägger and U. Brandt, *Biochim. Biophys. Acta*, 2000, **1459**, 230.
13. R. Böhm, M. Sauter and A. Böck, *Mol. Microbiol.*, 1990, **4**, 231.
14. S.P.J. Albracht, *Biochim. Biophys. Acta*, 1993, **1144**, 221.
15. T. Friedrich and H. Weiss, *J. Theor. Biol.*, 1997, **187**, 529.
16. M. Finel, *Biochim. Biophys. Acta*, 1998, **1364**, 112.
17. C. I. Ragan, Y. M. Galante, Y. Hatefi and T. Ohnishi, *Biochemistry*, 1982, **21**, 590.
18. M. Finel, J.M. Skehel, S.P.J.Albracht, I.M. Fearnley and J.E. Walker, *Biochemistry*, 1992, **31**, 11425.
19. H. Leif, V.D. Sled, T. Ohnishi, H. Weiss and T. Friedrich, *Eur. J. Biochem.*, 1995, **230**, 538.
20. L.A. Sazanov, S.Y. Peak-Chew, I.M. Fearnley and J. E. Walker, *Biochemistry*, 2000, **39**, 7229.
21. K. Leonard, G. Hofhaus and H. Weiss, *Biol. Chem. Hoppe-Seyler*, 1991, **372**, 551.
22. B. Böttcher, D. Scheide, M. Hesterberg, L. Nagel-Steger and T. Friedrich, *J. Biol. Chem.*, 2002, **277**, 17970.
23. V. Zickermann, M. Bostina, C. Hunte, T. Ruiz, M. Radermacher and U. Brandt, *J. Biol. Chem.*, 2003, **278**, 29072.
24. S.J. Pilkington, J.M. Skehel, R.B. Gennis and J.E. Walker, *Biochemistry*, 1991, **30**, 2166.
25. Y.M. Galante and Y. Hatefi, *Methods Enzymol.* 1978, **53**, 15.
26. G. Dooijewaard and E.C. Slater, *Biochim. Biophys. Acta*, 1976, **440**, 1.
27. Y.M. Galante and Y. Hatefi, *Arch. Biochem. Biophys.*, 1979, **192**, 559.
28. E.V. Gavrikova, V.G. Grivennikova, V.D. Sled, T. Ohnishi and A.D. Vinogradov, *Biochim. Biophys. Acta*, 1995, **1230**, 23.
29. T. Ohnishi, *Biochim. Biophys. Acta*, 1998, **1364**, 186.

30. T. Friedrich, *Biochim. Biophys. Acta*, 1998, **1364**, 134.

31. E. Nakamaru-Ogiso, T. Yano and T. Ohnishi, *J. Biol. Chem.*, 2005, **280**, 301.

32. M. Finel, *Trends Biochem. Sci.*, 1998, **23**, 412.

33. M.A. Smith, M. Finel, V. Korolik and G.L. Mendz, *Arch. Microbiol.*, 2000, **174**, 1.

34. H. Brüggemann, F. Falinski and U. Deppenmeier, *Eur. J. Biochem.*, 2000, **267**, 5810.

35. P. Cardol, F. Vanrobaeys, B. Devreese, J. van Beeumen, R.F. Matagne and C. Remacle, *Biochim. Biophys. Acta*, 2004, **1658**, 212.

36. J. Meuer, S. Bartoschek, J. Koch, A. Künkel and R. Hedderich, *Eur. J. Biochem.*, 1999, **265**, 325.

37. M.C. Kao, A. Matsuno-Yagi and T. Yagi, *Biochemistry*, 2004, **43**, 3750.

38. T. Rasmussen, D. Scheide, B. Brors, L. Kintscher, H. Weiss and T. Friedrich, *Biochemistry*, 2001, **40**, 6124.

39. M. Duarte, H. Populo, A. Videira, T. Friedrich and U. Schulte, *Biochem. J.*, 2002, **364**, 833.

40. D. Flemming, A. Schlitt, V. Spehr, T. Bischof and T. Friedrich, *J. Biol. Chem.*, 2003, **278**, 47602.

41. A.D. Vinogradov, V.D. Sled, D.S. Burbaev, V.G.X. Grivennikova, I.A. Moroz and T. Ohnishi, *FEBS Lett.*, 1995, **370**, 83.

42. F. Schuler, T. Yano, S. Di Bernardo, T. Yagi, V. Yankovskaya, T.P. Singer and J.E. Casida, *Proc. Natl. Acad. Sci. U.S.A.*, 1999, **96**, 4149.

43. E. Darrouzet, J.P. Issartel, J. Lunardi and A. Dupuis, *FEBS Lett.*, 1998, **431**, 34.

44. N. Kashani-Poor, K. Zwicker, S. Kerscher and U. Brandt, *J. Biol. Chem.*, 2001, **276**, 24082.

45. M. Finel, A.S. Majander, J. Tyynelä, A.M.P. de Jong, S.P.J. Albracht and M.K.F. Wikström, *Eur. J. Biochem.*, 1994, **226**, 237.

46. A. Abdrakhmanova, V. Zickermann, M. Bostina, M. Radermacher, H. Schägger, S. Kerscher and U. Brandt, *Biochim. Biophys. Acta*, 2004, **1658**, 148

47. I.M. Fearnley and J.E. Walker, *Biochim. Biophys. Acta*, 1992, **1140**, 105.

48. C. Mathiesen and C. Hägerhäll, *Biochim. Biophys. Acta*, 2002, **1556**, 121.

49. A.C. Gemperli, P. Dimroth and J. Steuber, *Proc. Natl. Acad. Sci. USA*, 2003, **100**, 839.

50. Y.V. Bertsova, A.V. Bogachev, *FEBS Lett.*, 2004, **563**, 207.

51. E. Nakamaru-Ogiso, B.B. Seo, T. Yagi and A. Matsuno-Yagi, *FEBS Lett.*, 2003, **14**, 43.

52. J. Steuber, *J. Biol. Chem.*, 2003, **278**, 26817.

53. T. Friedrich and D. Scheide, *FEBS Lett.*, 2000, **479**, 1.

54. R.H. Triepels, L.P. van den Heuvel, J.M. Trijbels and J.A. Smeitink, *Am. J. Med. Genet.*, 2001, **106**, 37.

55. M. Zeviani and S. DiDonato, *Brain*, 2004, **127**, 2153.

56. A. Majander, M. Finel, M.L. Savontaus, E. Nikoskelainen and M.K.F. Wikström, *Eur. J. Biochem.*, 1996, **239**, 201.

57. V. Zickermann, B. Barquera, M.K.F. Wikström and M. Finel, *Biochemistry*, 1998, **37**, 11792.

58. G. Hofhaus, H. Weiss and K. Leonard, *J. Mol. Biol.*, 1991, **221**, 1027.

59. L.A. Sazanov and J.E. Walker, *J. Mol. Biol.*, 2000, **392**, 455.

60. A. Künkel, J.A. Vorholt, R.K. Thauer and R. Hedderich, *Eur. J. Biochem.*, 1998, **252**, 467.

61. R. Hedderich, *J. Bioenerg. Biomembr.*, 2004, **36**, 65.

62. S.C. Andrews, B.C. Berks and J. McClay, A. Ambler, M.A. Quail, P. Golby, J.R. Guest, *Microbiology*, 1997, **143**, 3633.

63. A. Volbeda, M.H. Charon, C. Piras, E.C. Hatchikian, M. Frey and J.C. Fontecilla-Camps, *Nature*, 1995, **373**, 580.

64. Y. Higuchi, T. Yagi and N. Yasuoka, *Structure*, 1997, **5**, 1671.

65. Y. Montet, P. Amara, A. Volbeda, X. Vernede, E.C. Hatchikian, M.J. Field, M. Frey and J.C. Fontecilla-Camps, *Nature Struct. Biol.*, 1997, **4**, 523.

66. M. Ouali and R.D. King, *Protein Sci.*, 2000, **9**, 1162.

67. S. Kerscher, N. Kashani-Poor, K. Zwicker, V. Zickermann and U. Brandt, *J. Bioenerg. Biomembr.*, 2001, **33**, 187.

68. M. Finel, *FEBS Lett.*, 1996, **393**, 81.

69. S. Kerscher, J.G. Okun and U. Brandt, *J. Cell Sci.*, 1999, **112**, 2347.

70. S. Kerscher, A. Eschemann, P.M. Okun and U. Brandt, *J. Cell Sci.*, 2001, **114**, 3915.

71. K. Matsushita, T. Ohnishi and H.R. Kaback, *Biochemistry*, 1987, **26**, 7732.

72. I. Prieur, J. Lunardi and A. Dupuis, *Biochim. Biophys. Acta*, 2001, **1504**, 173.

73. S.D. Mills, W. Yang and K. McCormack, *Antimicrob. Agents Chemother.*, 2004, **48**, 2524.

74. P. Ahlers, K. Zwicker, S. Kerscher and U. Brandt, *J. Biol. Chem.*, 2000, **275**, 23577.

75. A. Garofano, K. Zwicker, S. Kerscher, P. Okun and U. Brandt, *J. Biol. Chem.*, 2003, **278**, 42435.

76. L. Grgic, K. Zwicker, N. Kashani-Poor, S. Kerscher and U. Brandt, *J. Biol. Chem.*, 2004, **279**, 21193.

77. B.H. Robinson, *Biochim. Biophys. Acta*, 1998, **1364**, 271.

78. J.L.C.M. Loeffen, J.A.M. Smeitink, J.M.F. Trijbels, A.J.M. Janssen, R.H. Triepels, L.P. Sengers and L. P. van den Heuvel, *Human Mutat.*, 2000, **15**, 123.

79. R.H. Triepels, L.P. van den Heuvel, J.M. Trijbels and J.A. Smeitink, *Am. J. Med. Genet.*, 2001, **106**, 37.

80. D.M. Kirby, M. Crawford, M.A. Cleary, H.H. Dahl, X. Dennett and D.R. Thorburn, *Neurology*, 1999, **52**, 1255.

81. S. Lebon, M. Chol, P. Bénit, C. Mugnier, D. Chretien, I. Giurgea, I. Kern, E. Girardin, L. Hertz-Pannier, P. De Lonlay, A. Rötig, P. Rustin and A. Munnich, *J. Med. Genet.*, 2003, **40**, 896.

82. R. McFarland, D.M. Kirby, K.J. Fowler, A. Ohtake, M.T. Ryan, D.J. Amor, J.M. Fletcher, J.W. Dixon, F.A. Collins, D.M. Turnbull, R.W. Taylor and D.R. Thorburn, *Ann. Neurol.*, 2004, **55**, 58.

83. J. Loeffen, J. Smeitink, R. Triepels, R. Smeets, M. Schuelke, R. Sengers, F. Trijbels, B. Hamel, R. Mullaart and L. van den Heuvel, *Am. J. Hum. Genet.*, 1998, **63**, 1598.

84. L. van den Heuvel, W. Ruitenbeek, R. Smeets, Z. Gelman-Kohan, O. Elpeleg, J. Loeffen, F. Trijbels, E. Mariman, D. de Bruijn and J. Smeitink, *Am. J. Hum. Genet.*, 1998, **62**, 262.

85. R. Triepels, L.P. van den Heuvel, J.L. Loeffen, C.A. Buskens, R.J.P. Smeers, M.E. Rubio-Gozalbo, S.M.S. Budde, E.C.M. Mariman, F.A. Wijburg, P.G. Barth, J.M. Trijbels and J.A. Smeitink, *Ann. Neurol.*, 1999, **45**, 787.
86. M. Schuelke, J. Smeitink, E. Mariman, J. Loeffen, B. Plecko, F. Trijbels, S. Stöckler-Ipsiroglu and L. van den Heuvel, *Nature Genetics*, 1999, **21**, 260.
87. P. Bénit, D. Chretien, N. Kadhom, P. de Lonlay-Debeney, V. Cormier-Daire, A. Cabral, S. Peudenier, P. Rustin, A. Munnich and A. Rötig, *Am. J. Hum. Genet.*, 2001, **68**, 1344.
88. J. Loeffen, O. Elpeleg, J. Smeitink, R. Smeets, S. Stöckler-Ipsiroglu, H. Mandel, R. Sengers, F. Trijbels and L. van den Heuvel, *Ann. Neurol.*, 2001, **49**, 195.
89. P. Bénit, R. Beugnot, D. Chretien, I. Giurgea, P. de Lonlay-Debeney, J.P. Issartel, S. Kerscher, P. Rustin, A. Rötig and A. Munnich, *Human Mutat.*, 2003, **21**, 582.
90. P. Bénit, J. Steffann, S. Lebon, D. Chretien, N. Kadhom, P. De Lonlay, A. Goldenberg, Y. Dumez, M. Dommergues, P. Rustin, A. Munnich and A. Rötig, *Hum. Genet.*, 2003, **112**, 563.
91. P. Bénit, A. Slama, F. Cartault, I. Giurgea, D. Chretien, S. Lebon, C. Marsac, A. Munnich, A. Rötig and P. Rustin, *J. Med. Genet.*, 2004, **41**, 14.
92. D.M. Kirby, R. Salemi, C. Sugiana, A. Ohtake, L. Parry, K.M. Bell, E.P. Kirk, A. Boneh, R.W. Taylor, H.H.M. Dahl and D.R. Thorburn, *J. Clin. Invest.*, 2004, **114**, 837.
93. P. Ahlers, A. Garofano, S. Kerscher and U. Brandt, *Biochim. Biophys. Acta*, 2000, **1459**, 258.
94. S. Kerscher, L. Grgic, A. Garofano and U. Brandt, *Biochim. Biophys. Acta*, 2004, **1659**, 197.
95. A. Videira, M. Duarte and A. Ushakova, *Biochim. Biophys. Acta*, 2004, **1657**, Supplement 27.
96. T. Ohnishi, *Biochim. Biophys. Acta* 1975, **387**, 475.
97. T. Ohnishi, *Eur. J. Biochem.* 1976, **64**, 91.
98. W.J. Ingledew and T. Ohnishi, *Biochem. J.*, 1980, **186**, 111.
99. S. Magnitsky, L. Toulokhonova, T. Yano, V.D. Sled, C. Hagerhall, V.G. Grivennikova, D.S. Burbaev, A.D. Vinogradov and T. Ohnishi, *J. Bioenerg. Biomembr.*, 2002, **34**, 193.
100. T. Yano, T. Yagi, V. D. Sled and T. Ohnishi, *J. Biol. Chem.*, 1995, **270**, 18264.
101. T. Yano, V.D. Sled, T. Ohnishi and T. Yagi, *FEBS Lett.*, 1994, **354**, 160.
102. T. Yano, V.D. Sled, T. Ohnishi and T. Yagi, *Biochemistry*, 1994, **33**, 494.
103. T. Yano, V.D. Sled, T. Ohnishi and T. Yagi, *J.Biol.Chem.*, 1996, **271**, 5907.
104. T. Yano, J. Sklar, E. Nakamaru-Ogiso, T. Yagi and T. Ohnishi, *J. Biol. Chem.*, 2003, **278**, 15514.
105. A. Dupuis, J.M. Skehel and J.E. Walker, *Biochemistry*, 1991, **30**, 2954.
106. M. Chevallet, A. Dupuis, J.P. Issartel, J.L. Lunardi, R. van Belzen and S.P.J. Albracht, *Biochim. Biophys. Acta*, 2003, **1557**, 51.
107. T. Friedrich, B. Brors, P. Hellwig, L. Kintscher, T. Rasmussen, D. Scheide, U. Schulte, W. Mäntele and H. Weiss, *Biochim. Biophys. Acta*, 2000, **1459**, 305.
108. M. Degli Esposti, *Biochim. Biophys. Acta*, 1998, **1364**, 222.
109. J.G. Okun, P. Lümmen and U. Brandt, *J. Biol. Chem.*, 1999, **274**, 2625.

110. F. Schuler, T. Yano, S. Di Bernardo, T. Yagi, V. Yankovskaya, T.P. Singer and J.E. Casida, *Proc. Natl. Acad. Sci. U.S.A.*, 1999, **96**, 4149.

111. T. Ohnishi, *J. Bioenerg. Biomembr.*, 1993, **25**, 325.

112. T. Maly and T.F. Prisner, *J. Magnetic Res.*, 2004, **170**, 88.

113. T. Maly, F. MacMillan, K. Zwicker, N. Kashani-Poor, U. Brandt and T.F. Prisner, *Biochemistry*, 2004, **43**, 3969.

114. M. Gurrath and T. Friedrich, *Proteins*, 2004, **56**, 556.

115. K. Müller, B.F. Matzanke, V. Schünemann, A.X. Trautwein and K. Hantke, *Eur. J. Biochem.*, 1998, **258**, 1001.

116. M. Brugna-Guiral, P. Tron, W. Nitschke, K.O. Stetter, B. Burlat, B. Guigliarelli, M. Bruschi and M.T. Giudici-Orticoni, *Extremophiles*, 2003, **7**, 145.

117. P. Hellwig, D. Scheide, S. Bungert, W. Mäntele and T. Friedrich, *Biochemistry*, 2000, **39**, 10884.

118. D. Flemming, P. Hellwig and T. Friedrich, *J. Biol. Chem.*, 2003, **278**, 3055.

119. D.S. Burbaev, I.A. Moroz, A.B. Kotlyar, V.D. Sled and A.D. Vinogradov, *FEBS Lett.*, 1989, **254**, 47.

120. A.M.P. de Jong, A.B. Kotlyar and S.P.J. Albracht, *Biochim. Biophys. Acta*, 1994, **1186**, 163.

121. T. Ohnishi, J.E. Johnson, T. Yano, R. LoBrutto and W.R. Widger, *FEBS Lett.*, 2005, **579**, 500.

122. T. Yano, W.R. Dunham and T. Ohnishi, *Biochemistry*, 2005, in press.

123. C.C. Page, C.C. Moser, X. Chen and P.L. Dutton, *Nature*, 1999, **402**, 47.

124. G. Belogrudov and Y. Hatefi, *Biochemistry*, 1994, **33**, 4571.

125. M. Yamaguchi, G. Belogrudov and Y. Hatefi, *J. Biol. Chem.*, 1998, **273**, 8094.

126. A.A. Mamedova, P.J. Holt, J. Carroll and L.A. Sazanov, *J. Biol. Chem.*, 2004, **279**, 23830.

127. P. Hellwig, S. Stolpe and T. Friedrich, *Biopolymers*, 2004, **74**, 69.

128. U. Brandt, S. Kerscher, S. Dröse, K. Zwicker and V. Zickermann, *FEBS Lett.*, 2003, **545**, 9.

129. J.P. Abrahams, A.G.W. Leslie, R. Lutter and J.E. Walker, *Nature*, 1994, **370**, 621.

130. C. Toyoshima, M. Nakasako, H. Nomura and H. Ogawa, *Nature*, 2000, **405**, 647.

131. J. G. Okun, V. Zickermann and U. Brandt, *Biochem. Soc. Trans.*, 1999, **27**, 596.

CHAPTER 8

Current Knowledge About the Mechanism of Energy Transduction by Respiratory Complex I

JUDY HIRST

Medical Research Council Dunn Human Nutrition Unit,
Wellcome Trust/MRC Building, Hills Road,
Cambridge, CB2 2XY, United Kingdom

1 Introduction

Complex I is a complicated, membrane-bound, multi-subunit enzyme. No high resolution structural model is available, and no substantial mechanism has been proposed for energy transduction by complex I. However, the subunit compositions of both eukaryotic and prokaryotic complexes I have been defined, the primary sequences of their subunits are known, and a consistent picture of the cofactor composition has been developed. Structural and functional studies are being undertaken on a variety of experimental systems, and they are rapidly producing new (and sometimes contradictory) results.[*] The aim of this chapter is to examine, critically, current data relating to the mechanism of complex I, and to construct a framework for the discussion of possible conclusions and future mechanistic proposals.

2 Complex I in Energy Transduction

Complex I (NADH-ubiquinone oxidoreductase) is the first enzyme of the respiratory electron transport chain in mitochondria. The electron-transport chain transfers two

---

[*]Between submission and publication of this manuscript the first detailed structural information on complex I has been reported (P. Hinchliffe and L.A. Sazanov, *Science*, 2005, **309**, 771). The positions of the nine iron-sulphur clusters present in the hydrophilic domain of the complex I from *Thermus thermophilus* have been reported and found an 84 Å long chain through the protein.

electrons from NADH to oxygen, and uses the energy which is available from each of three sequential steps to transport protons across the inner-mitochondrial membrane. Therefore, the three 'proton-pumping' enzymes of the respiratory chain produce and sustain the proton-motive force which supports the synthesis of ATP.[1,2] Complex I oxidises NADH, produced by the tricarboxylic acid cycle, glycolysis and the β-oxidation of fatty acids, to NAD^+, maintaining the $NAD^+/NADH$ ratio in the mitochondrial matrix (Figure 1). The two electrons from NADH oxidation are used to reduce ubiquinone (Q) to ubiquinol (QH_2) in the inner-mitochondrial membrane, and the ubiquinol is subsequently reoxidised by complex III (ubiquinol-cytochrome *c* oxidoreductase, the cytochrome bc_1 complex). The reduction of ubiquinone by NADH (eq. 1)

$$NAD^+ + 2H^+ + 2e^- \Leftrightarrow NADH + H^+ \; E_{pH7} = -0.320 \text{ V } (20°C)$$

$$Q + 2H^+ + 2e^- \Leftrightarrow QH_2 \; E_{pH7} = +0.110 \text{ V}^§ (20°C) \tag{1}$$

is thermodynamically very favourable ($\Delta E_{pH7} = 0.43$ V under standard conditions), although under physiological conditions the NAD^+:NADH and Q:QH_2 ratios are not unity and so $\Delta E_{(NADH/QH2)}$ is modulated by the Nernst equation, as well as by the solution and membrane conditions. For example, isolated mitochondria respiring on succinate or sulphite, at close to equilibrium, gave $\Delta E_{(NADH/QH2)} \approx 0.36$ V.[3] For complex

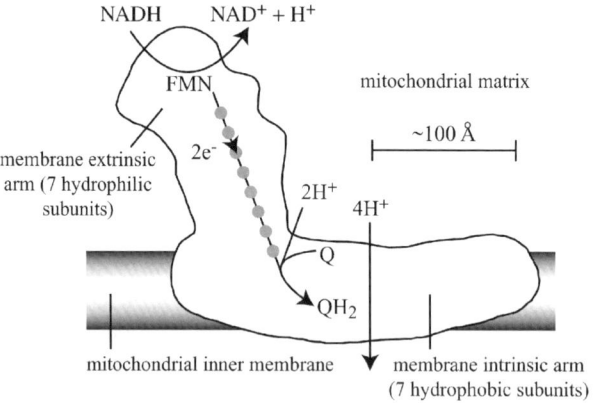

Figure 1 *Schematic diagram showing the reactions catalysed by complex I in the inner-mitochondrial membrane. The enzyme is L-shaped, with the membrane-extrinsic arm protruding into the mitochondrial matrix. The outline is taken from single-particle electron microscopy studies of complex I from E. coli,[5] to illustrate the approximate dimensions of a complex consisting of only the 14 core subunits. In E. coli, complex I is located in the cytosolic membrane. NADH is oxidised by a non-covalently bound FMN, then the two electrons are transferred to bound quinone by a linear chain of FeS clusters (cluster-to-cluster distance ~14 Å)*

---

§A range of values has been reported for the Q/QH_2 reduction potential, depending on, for example, the hydrophobicity, ionic strength, polarity, and protic/aprotic nature of the environment.

I in the inner-mitochondrial membrane, the reaction between NADH and ubiquinone is coupled to proton translocation across the membrane. This conserves the redox energy by transforming it into a proton-motive force: proton transport against the proton-motive force is energetically unfavourable and opposes the favourable redox reaction. It is generally accepted that, in mitochondria, complex I translocates four protons for each NADH oxidised (Figure 1).[4] Therefore, under standard conditions, the maximum proton-motive force that complex I can support is 0.22 V (0.43 V (2 e$^-$), 4 H$^+$ translocated). A typical value for the proton-motive force across the inner membrane, under steady-state conditions, is 0.2 V, indicating that complex I is catalysing a process that is close to equilibrium, and thus it is operating efficiently.

3 The Location of the Cofactors and Substrate Binding Sites in Complex I

3.1 The L-shaped Structure of Complex I

Complex I from bovine heart mitochondria contains 14 core subunits and 32 supernumerary subunits.[6] Only the core subunits are considered here. They are conserved in all complexes I, are sufficient for energy transduction, and constitute the 'minimal' prokaryotic complex I. The core subunits are readily divided into two groups: seven are predominantly hydrophilic and encoded in the nucleus, and seven are highly hydrophobic, predicted to be dominated by transmembrane helices, and encoded by the mitochondrial genome. In bovine complex I the seven hydrophilic subunits are the 75, 51, 49, 30 and 24 kDa subunits, TYKY and PSST. The seven hydrophobic subunits are ND1-6 and ND4L.[6,7]

Electron microscopy has shown that the complexes I from both prokaryotic and eukaryotic organisms are 'L-shaped', with one arm in the plane of the membrane, and the other protruding into the mitochondrial matrix (Figure 1).[8,9] The two arms are considered to correspond to the two sets of core subunits: the seven hydrophilic subunits constitute the membrane-extrinsic arm, the seven hydrophobic subunits constitute the membrane-intrinsic arm. This picture is supported by the resolution of complex I from bovine mitochondria into subcomplexes Iα, Iβ and Iλ.[6] Subcomplexes of the two arms of *Neurospora crassa* complex I have been characterised by single-particle electron microscopy,[8] and subcomplexes comprising the hydrophilic and hydrophobic core subunits of complex I from *Escherichia coli* have been produced.[5,10]

3.2 The Location of the Cofactors in Complex I

The cofactors of complex I are a non-covalently bound flavin mononucleotide (FMN, the primary oxidant of NADH), and a number of iron–sulphur (FeS) clusters. They are all bound by the seven hydrophilic core subunits, as summarised in Table 1. Although the information in Table 1 is consistent with most experimental data, an unambiguous picture is unlikely to emerge until a high-resolution structural model is available. No cofactors are known conclusively to be bound by the hydrophobic core subunits (see below for a discussion of quinone binding), although UV-visible spectra have been

Table 1 Current consensus on the number and location of the cofactors bound by the seven hydrophilic core subunits of complex I

Subunit	Cofactors	EPR signal	Proposed ligation	Supporting evidence
75 kDa*	[4Fe-4S] [4Fe-4S] [2Fe-2S]	N4 N5 N1b	$Cys-X_2-Cys-X_2-Cys-X_{43}-Cys-Pro$ $His-X_3-Cys-X_2-Cys-X_5-Cys$ (?) $Cys-X_{10}-Cys-X_2-Cys-X_{13}-Cys$ (?)	Over-expressed subunit from P. denitrificans and homology to Fe-only hydrogenase;[15] N1b and N4 observed in E. coli NADH dehydrogenase fragment.[10]
51 kDa	FMN [4Fe-4S]	N3	$Cys-X_2-Cys-X_2-Cys-X_{39}-Cys$	FMN and N3 present in Fp subcomplex;[16] arylazido-β-alanyl NAD$^+$ labels the 51 kDa subunit;[17] N3 interacts with FMN radical.[18]
24 kDa	[2Fe-2S]	N1a	$Cys-X_4-Cys-X_{35}-Cys-X_3-Cys$	Over-expressed subunit;[19,20] homologous to known ferredoxins from Clostridium pasteurianum and Aquifex aeolicus.
49 kDa	—	—	Ligand to N2 ?	Mutations in 49 kDa subunit affect cluster N2.[21,22]
30 kDa	—	—	—	—
PSST	[4Fe-4S]	N2 ?	$Cys-Cys-X_{63}-Cys-X_{29}-Cys-Pro$ motif; cluster ligands not identified.	Homologous to small subunit of NiFe hydrogenases;[23] present in 'connecting fragment' of E. coli enzyme;[10] site-directed mutations in PSST disrupt cluster N2.[24,25]
TYKY	[4Fe-4S] [4Fe-4S]	N6a, b?	$Cys^*-X_2-Cys^*-X_2-Cys^*-X_3-Cys-Pro-X_{27}-Cys-X_2-Cys-X_2-Cys-X_3-Cys^*-Pro$	Canonical 8 Fe-ferredoxin ligation motif.

*In the complexes I from E. Coli and Thermus thermophilus an additional [4Fe-4S] cluster morif is present in the 75 kDa subunit.

used to suggest the presence of a novel cofactor, perhaps a quinoid group, in the membrane domains of the complexes I from *E. coli* and *N. crassa*.[11] All the clusters observed in bovine complex I by EPR spectroscopy were observed in subcomplex Iλ also, although the signal from cluster N2 was modified.[12] All the EPR-detectable clusters of *E. coli* complex I were detected either in the 'NADH-dehydrogenase' fragment (75, 51 and 24 kDa subunits) or the 'connecting' fragment (cluster N2; 49 and 30 kDa, PSST, and TYKY subunits).[10] In *N. crassa*, the membrane-extrinsic subcomplex contained all the EPR-observable clusters except cluster N2;[13] cluster N2 co-localised with the membrane arm.[14] Therefore, although all the known cofactors of complex I are bound by the seven hydrophilic subunits, it is likely that cluster N2 is close to the interface with the membrane-bound subunits. The implication that all the cofactors are in the membrane-extrinsic domain (Figure 1) is important in the formulation of mechanistic hypotheses for both the redox and proton-pumping reactions.

The 75, 51 and 24 kDa subunits of complex I are closely associated: they form the 'NADH-dehydrogenase' fragment of complex I from *E. coli*,[10] and they are homologous to the α and γ subunits of NAD$^+$-reducing hydrogenases.[26] The cofactors in the 51 and 24 kDa subunits (Table 1) are clearly defined, but the assignment of two 4Fe–4S clusters to the 75 kDa subunit, and of the ligands to the clusters in this subunit, remain unconfirmed.

The 49 and 30 kDa, PSST, and TYKY subunits form a second fragment of the membrane-extrinsic domain of complex I from *E. coli*,[10] and they are thought to lie between the NADH-dehydrogenase subcomplex and the membrane domain. The membrane association of PSST and TYKY is supported by differential extraction of the subunits of *Paracoccus denitrificans* complex I by chaotropic agents,[27] and by cross-linking between PSST and ND3.[28] The two clusters in TYKY are not readily observed by EPR spectroscopy, though signals observed in the connecting fragment of *E. coli* complex I have been assigned to them.[29] It is possible that: i) their reduction potentials are too low for reduction by NADH or commonly used mediators, they are strongly interacting, or they have an unusual spin state, or ii) the assignments of clusters to subunits in Table 1 is incorrect (see, for example, ref. 30). Clearly, resolving this issue is crucial for understanding the mechanism of complex I. The assignment of cluster N2 to PSST is based on mutagenesis studies,[24] on the homology of PSST and the 49 kDa subunit to the small and large subunits of NiFe hydrogenases,[23] and on electron transfer from N2 to quinone being affected by inhibitors which bind to PSST.[31] The four conserved cysteines in PSST (including two adjacent residues) may coordinate cluster N2 (Table 1),[24] but this motif has not been identified elsewhere, so a non-cysteine ligand may be involved, or the 49 kDa subunit may supply a ligand.[21] Mutations of candidate residues have so far failed to confirm either suggestion.[22,25] Cluster N2 is likely to be the direct electron donor to ubiquinone because it interacts magnetically with semiquinone radicals which are formed during turnover[32] (see below).

3.3 The Quinone Binding Site(s)

There are several reasonable constraints on the position of the quinone binding site(s) (Q-sites) in complex I. First, ubiquinone reduction requires proton uptake. These protons are likely to bind from the matrix (augmenting, not dissipating, the proton-motive force), consistent with electron transfer from cluster N2 (Figure 1). Second, it is very likely that the Q-site(s) are in the membrane domain, to accommodate the hydrophobic isoprenoid chain (Figure 2). However, the relatively hydrophilic headgroup forms hydrogen bonds and must be connected to an aqueous phase for protonation. This suggests that the quinone headgroup(s) may be located in the interfacial region/level with the phospholipid headgroups, as observed in many other quinone oxidoreductases, for example, fumarate reductase and succinate dehydrogenase,[33] formate dehydrogenase,[34] the cytochrome bc_1 complex,[35] and the photosynthetic reaction centre.[36] The quinone headgroup may interact with the hydrophilic subunits; for example, the soluble domain of the Rieske protein contributes to the Q_O-site in the cytochrome bc_1 complex,[35] and the iron–sulphur protein contributes to the Q-site in succinate dehydrogenase and fumarate reductase.[33] Third, any redox-active quinones must be within reasonable distance ($\sim$14 Å[37]) of the electron donor (for example, cluster N2). Consequently, multiple (unknown) cofactors would be required to access quinone at the far end of the membrane domain (Figure 1).

Direct evidence about the number and location of the Q-sites in complex I is limited. Only substoichiometric amounts of ubiquinone were retained in highly active preparations of complex I from *Yarrowia lipolytica*[40] and bovine heart mitochondria,[41] indicating that no strongly-bound, non-exchangeable quinones (comparable to Q_A in the photosynthetic reaction centre) are required for activity. Subunit ND4 of *E. coli* complex I was labelled by an azido-ubiquinone derivative.[42] Semiquinones

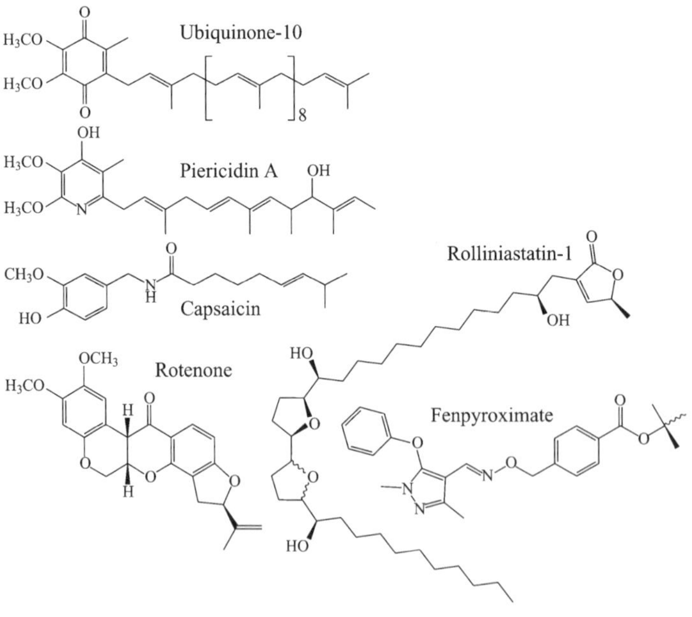

Figure 2 *The structures of ubiquinone-10 and some of the inhibitors of complex I*[38,39]

have been detected in complex I reduced by NADH, and three different species (SQ_{Nf}, SQ_{Ns}, and SQ_{Nx}) have been identified by their different relaxation rates and responses to pH, membrane potential, and inhibitors.[32] They are all piericidin-A sensitive. SQ_{Nf} is fast-relaxing, sensitive to uncouplers and rotenone, and proposed to interact paramagnetically with cluster N2 (from 8–11 Å). SQ_{Ns} and SQ_{Nx} are slow-relaxing, do not interact significantly with other paramagnetic species within complex I (indicating they are distant from the FeS clusters), and are uncoupler and rotenone insensitive. However, none are observed stoichiometrically, and SQ_{Ns} and SQ_{Nx}, in particular, may be formed at the same site.

Many diverse hydrophobic compounds inhibit complex I[38,39] (Figure 2) and have been proposed to bind at, or close to, the Q-site(s), in analogy with structural models of inhibitors bound to the cytochrome bc_1 complex and the bacterial reaction centre, and because only NADH:ubiquinone (not NADH:ferricyanide) oxidoreduction is inhibited. However, they may also trap the enzyme in a state in which the Q-site is occluded, or inhibit a different step in the catalytic cycle (for example, electron transfer, proton translocation, or a conformational change). Furthermore, although some inhibitors clearly resemble the quinone substrate, for others the relationship is not so obvious (Figure 2). Complex I inhibitors have been divided into three classes, consistent with the existence of more than one Q-site.[38] Alternatively, competitive binding experiments (class B inhibitors (rotenone) compete with class A (piericidin A) and class C (capsaicin) inhibitors, but class A and C inhibitors do not compete) have been used to propose the existence of just one large site with overlapping domains.[43] Note, however, that different inhibitors may bind to different catalytic states, and so compete 'functionally', not spatially. Radiolabelling experiments have localised bound inhibitors to ND1 (rotenone,[44] DCCD,[45] and pyridaben[31]), PSST (pyridaben[31]) and ND5 (fenpyroximate[46]). Fenpyroximate (Figure 2) is an instructive example: it may inhibit proton translocation rather than quinone binding (it is displaced by amiloride derivatives, which inhibit Na^+/H^+ antiporters), and it is reported to be displaced by rotenone, piericidin A, *and* capsaicin.[46]

Mutations in the 49 kDa subunit imposed inhibitor resistance on complex I from *Rhodobacter capsulatus*, and decreased the rate of quinone reduction. Therefore, the 49 kDa subunit was proposed to bind the quinone headgroup.[47] Further mutations in the 49 kDa subunit of *Y. lipolytica* suggested that the 49 kDa and PSST subunits (homologous to the NiFe hydrogenases) form a 'functional core' of complex I, with bound quinone replacing the NiFe cofactor.[21] However, it is also possible, for example, that alterations in the 49 kDa subunit propagate to a distant Q-site, perhaps *via* a conformational change. Recently, electron microscopy was used to localise the 49 kDa subunit to the top of the peripheral arm of the complex, inferring that the Q-site is 70–80 Å from the membrane.[48] This contradicts the idea that the hydrophobic isoprenoid chain is retained in the membrane, and it discounts the FeS clusters as a method of long-range energy transfer between the hydrophilic and hydrophobic domains.

In summary, although defining the quinone binding sites in complex I is fundamental for our elucidation of the mechanism of the enzyme, both their number and location remain controversial.

4   The Redox Reaction: Oxidation of NADH and Reduction of Quinone

Figure 3 presents a basis for discussion of the electron-transfer reactions catalysed by complex I, in the absence of a proton-motive force. The reduction of ubiquinone by NADH is thermodynamically very favourable, and so the reverse reaction is negligible under these conditions. NADH is oxidised by non-covalently bound FMN, then the two electrons are transferred, one at a time, along a chain of FeS clusters to ubiquinone. Ubiquinone reduction requires two protons and two electrons, and so the uncoupled redox reaction is electroneutral, and does not involve charge transfer across the membrane. In addition, complex I is an important source of reactive oxygen species (ROS) in mitochondria (a small percentage of the electrons 'escape' to oxygen, instead of being transferred to quinone), although a consensus has not yet been reached on the cofactor(s) involved directly in this reaction.

4.1   The Role of the FMN: Oxidation of NADH and Electron Transfer

The NADH radical is thermodynamically very unstable, so it is likely that the reaction between NADH and FMN is a hydride (obligatory two–electron) transfer (Figure 3). In complex I the FMN has a two-electron potential of -0.38 V (pH 7.5),[18] close to the potential of NADH. Therefore, the reaction between NADH and FMN is thermodynamically reversible, as demonstrated by electrochemical experiments in which no 'overpotential' (extra driving force) was required for the interconversion of NADH and NAD^+.[53] Following the reaction with NADH, $FMNH^-$ (or $FMNH_2$) is reoxidised by the FeS clusters. Crucially, the FMN has a thermodynamically accessible (but not thermodynamically favoured) semiquinone state,[18] allowing $FMNH^-$ to be oxidised in two one-electron steps (Figure 3). The flavin semiquinone radical interacts magnetically with cluster N3, indicating that N3 is the primary electron acceptor.[18]

4.2   Electron Transfer via the Iron–Sulphur Clusters

The oxidation of NADH and the reduction of quinone are spatially separated in complex I, so electrons must be transferred efficiently between the two. It is very likely that this occurs *via* a 'chain' of FeS clusters (Figures 1 and 3), as exemplified by the structures of succinate:quinol oxidoreductase,[33] quinol:nitrate oxidoreductase,[54] and formate:quinone oxidoreductase.[34] Typically, the intercluster distances are 11–13 Å, consistent with a physiological rate of electron transfer (<14 Å).[37] Therefore, the eight clusters in complex I are able to span ~120 Å, commensurate with the height of the membrane-extrinsic domain (Figure 1). Furthermore, providing that the intercluster distances are short enough, the exact values of the individual reduction potentials are not critical, and significantly out-of-line reduction potentials can be tolerated.[37] Figure 4 presents the corresponding profile for succinate dehydrogenase (complex II), in which the out-of-line cluster is known, from the structure, to participate in electron transfer.[33] Consequently, in complex I the relatively low potential of the [2Fe-2S] cluster N1a[20] does not rule it out of the electron-transfer chain. In addition, the relatively

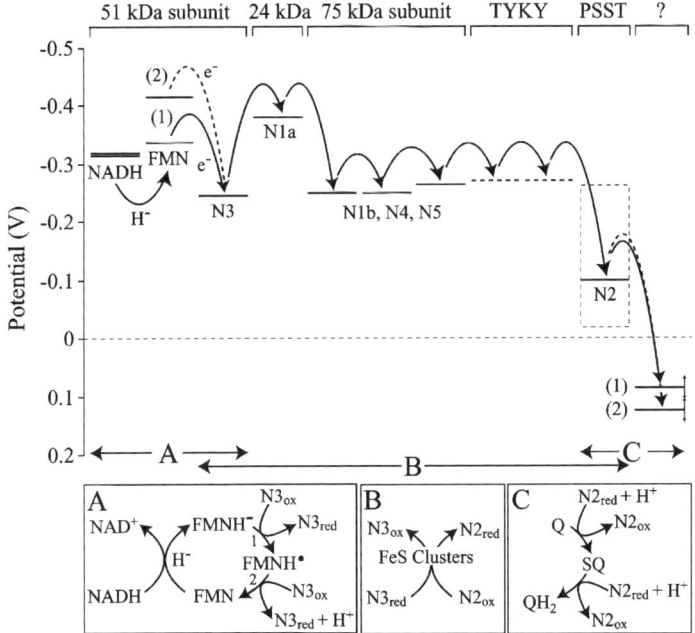

Figure 3 *Schematic representation of the movement of electrons through complex I, in the absence of a proton-motive force. Reduction potentials are presented on the y-axis, inverted to represent free-energy changes, and are taken from studies of complex I from bovine heart mitochondria;[18,49–51] the two one-electron potentials of bound quinone are not known. NADH transfers a hydride (two electrons) to FMN; the FMNH$^-$ is reoxidised, in a stepwise manner, by cluster N3 (box A). The FMN potentials are 'crossed'[18] so that the first oxidation (1) is less favourable than the second (2). The electrons are passed, one at a time, along the chain of FeS clusters, given here in the order N3-N1a-N1b-N4-N5-(TYKY clusters)-N2 (box B). The potentials of the TYKY clusters[29] have not been defined unambiguously, and a wide range of different values have been reported for cluster N2.[50,52] Finally, the two electrons are donated, sequentially, to bound quinone (box C)*

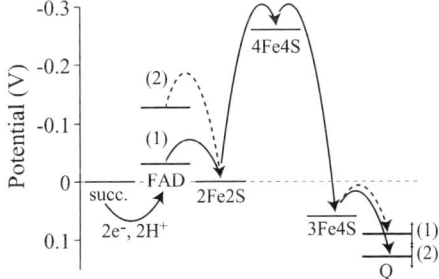

Figure 4 *Schematic representation of the movement of electrons through succinate: ubiquinone oxidoreductase (complex II), presented for comparison with Figure 3. Structural models show unambiguously that the [4Fe4S] cluster forms part of the electron transfer chain in this enzyme, despite its 'out-of-line' potential[33]*

high potential of cluster N2 is not sufficient to define it as the last cluster in the series, or as the direct electron donor to quinone.

4.3 Quinone Reduction

Cluster N2 is proposed to be the FeS cluster that donates electrons directly to bound quinone because it is thought to interact magnetically with bound semiquinone (see above). Figure 3 shows how cluster N2 may donate two electrons, in a stepwise manner, to reduce the bound quinone. The (one-electron) reduction potential of cluster N2 is significantly lower than the (two-electron) potential of ubiquinone (Figure 3), apparently favouring electron transfer from N2 to ubiquinone. However, the two one-electron potentials of the bound ubiquinone are unknown, and so how one (or two) confined electrons would distribute between cluster N2 and bound quinone species cannot be determined. Electron delivery to cluster N2, from NADH *via* the FeS chain, is expected to be fast (non rate determining). Therefore, a semiquinone formed by the addition of one electron should be quickly reduced by a second. A bound semiquinone could be stabilised by: (i) separation of the one-electron potentials ($E_{Q/SQ} > E_{SQ/QH2}$), (ii) alteration of the relative potentials of N2 and quinone, (iii) kinetically. Consequently, it is interesting that the fast-relaxing semiquinone SQ_{Nf}, proposed to be spin-coupled to cluster N2, is only observed in the presence of a membrane potential,[32] suggesting possible coupled proton translocation at this point in the mechanism. Note that bound semiquinone is a possible source of ROS in mitochondrial complex I, and that ROS production has been observed to depend on the proton-motive force.[55] Finally, mutants of complex I which apparently lack cluster N2, but which catalyse NADH:ubiquinone oxidoreduction at a reasonable rate, have been generated recently in *Y. lipolytica*.[22] The inference that cluster N2 is not required for catalysis is obviously inconsistent with the model presented above, and requires further investigation.

5 Mechanisms of Proton Pumping

The mechanism of proton pumping in complex I must fulfil a number of criteria. First, it must be reversible – although the mechanism cannot be microscopically reversible, it must be viable for reverse catalysis (for example, succinate-driven reverse electron transport in the presence of a membrane potential). Second, it must account for four protons translocated vectorially across the membrane per NADH oxidised. Third, it must be efficient (short-circuits are avoided[56]): every oxidation of NADH is coupled to proton transfer, and the proton-motive force cannot push protons backwards without NAD^+ reduction. This is true even though the uncoupled redox reaction and proton 'leak' are favoured thermodynamically, and it demonstrates the importance of kinetic coupling.

At present there is little direct evidence for any mechanism of energy transduction in complex I, so mechanistic models are inspired largely by other membrane-bound energy-transducing enzymes:

(i) Q-cycle mechanisms use quinol as a mobile proton/electron (hydrogen) carrier, and are based on analogy with the cytochrome bc_1 complex. Earlier proposals which employed $FMNH_2$ instead are inconsistent with the location of the FMN, the conservation of potential energy during electron transfer along the FcS chain (Figure 3), and with the fact that NADH oxidation, supported by artificial electron acceptors, does not pump protons;

(ii) In directly-coupled mechanisms, proton transfer across a hydrophobic barrier is controlled *directly* by a gating reaction, such as a coupled electron-transfer reaction, positioned at (or close to) the barrier. Cytochrome *c* oxidase is an example of a directly-coupled proton-pump;

(iii) In indirectly-coupled mechanisms the driving reaction is spatially separated from proton transfer across the hydrophobic barrier, and the two are coupled *indirectly* by interactions with the protein structure. Indirectly coupled mechanisms are exemplified by the rotary mechanism of F-type ATPases,[57] and the conformational changes observed in Ca^{2+}-ATPase.[58] They are not associated typically with redox enzymes, in which electron transfer provides an obvious 'in-place' mechanism for long-range energy transfer.

5.1 The Q-cycle Mechanism

Dutton and coworkers proposed that the mechanism of complex I is equivalent to that of the cytochrome bc_1 complex operating 'in reverse' (Figure 5), with the FMN and the FeS clusters acting only to supply low-potential electrons.[59] In support of this attractive proposition, semiquinones have been observed during turnover of complex I, and inhibitor-binding studies can be rationalised in terms of more than one Q-site (see above). However, a number of observations are not consistent with a Q-cycle mechanism. First, no cofactors have been found in the membrane domain, to mediate electron transfer between the two Q-sites (analogous to the *b*-haems in the cytochrome bc_1 complex). It is possible that a strongly-bound quinone-cofactor

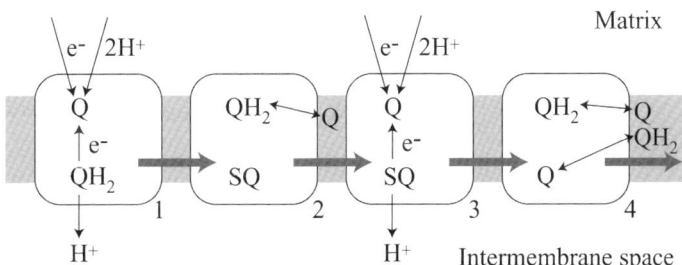

Figure 5 *The reverse Q-cycle mechanism proposed for complex I. 1. Ubiquinone (Q) is bound at the matrix face of the membrane, and accepts one electron from the FeS clusters (from NADH oxidation) together with one electron from ubiquinol (QH_2) bound at the cytoplasmic side. This is the reverse of the reaction at the Q_o site in the bc_1 complex, and it generates semiquinone at the cytoplasmic site (SQ). 2. The QH_2 formed is replaced by a second Q from the membrane. 3. Step 1 is repeated, generating Q at the cytoplasmic side. 4. The QH_2 and Q formed are replaced, to regenerate the starting species. In total, two Q are reduced, one QH_2 is oxidised, and two protons are translocated, for each NADH oxidised*

fulfils this role,[59] or the quinone headgroups may be located close to the middle of the membrane, and connected to the aqueous phase by proton channels. Second, because the Q-cycle can only account for two protons pumped per NADH oxidised, a second proton-pumping component is required; Dutton and coworkers suggested that the putative quinone-cofactor couples proton transfer to electron transfer between the Q-sites.[59] Third, no semiquinone radical has been detected at the Q_O-site of the cytochrome bc_1 complex, perhaps because quinol oxidation is a concerted process.[56] Consequently, a semiquinone close to cluster N2 may actually be evidence against a Q-cycle mechanism. Finally, the complexes I from *E. coli* and *Klebsiella pneumoniae* have been observed to pump sodium ions,[60] in clear contradiction of a Q-cycle mechanism. However, this observation has been disputed by other researchers, and is currently an unresolved issue. Further variations on Q-cycle mechanisms have also been proposed,[61] but they are less obviously related to a known mechanism in another enzyme, and they remain speculative.

5.2 Coupled Reactions

Figure 3 suggests two possibilities for the reaction which is coupled to proton translocation: the oxidation and reduction of cluster N2, and the conversion of ubiquinone into ubiquinol. At these two points the electron loses a significant amount of potential energy, whereas 'upstream' of cluster N2 its potential energy is essentially conserved. Figures 6A and 6B show examples of how it is possible to modify Figure 3, to include the thermodynamic contribution of the proton-motive force. In Figure 6A, proton transfer is coupled to a redox event, in this case at cluster N2. An electron is transferred rapidly to N2, forming a high-energy state ($N2^*_{red}$), perhaps by altering the protein configuration around the active site, and conserving potential energy. The high-energy state is able to do work, and its relaxation to the 'normal' state ($N2_{red}$) is coupled to the translocation of two protons, requiring 0.4 V ($2 \times$ PMF). Consequently, the normal state, $N2_{red}$, has an apparent reduction potential of $+0.06$ V, and is still able to reduce bound (semi)quinone, completing the redox reaction. The reverse reaction is possible because protons moving back across the membrane 'lift' $N2_{red}$ into $N2^*_{red}$, thus allowing reduction of the upstream FeS clusters. Similar coupled electron–proton transfer reactions could be proposed also for the oxidation and reduction of a semiquinone, for example. Alternatively, Figure 6B suggests that proton transfer could be coupled to substrate binding or produce dissociation. In Figure 6B, the potential of bound quinone is much lower than the potential of free quinone (quinone is bound more strongly than quinol) and so electron transfer to bound quinone conserves the potential energy. The potential energy is released, and coupled to the translocation of four protons, upon release of QH_2 or binding of Q. Reverse catalysis can occur as the proton-motive force acts to overcome the (otherwise unfavourable) binding or dissociation step, lifting Q_{Memb} into Q_{CI}.

Importantly, Figures 6A and 6B do not contain any information about the *molecular* mechanism of coupled proton translocation. The two possibilities discussed below are supported by the limited amount of experimental data which is currently available; they are certainly not the only viable possibilities.

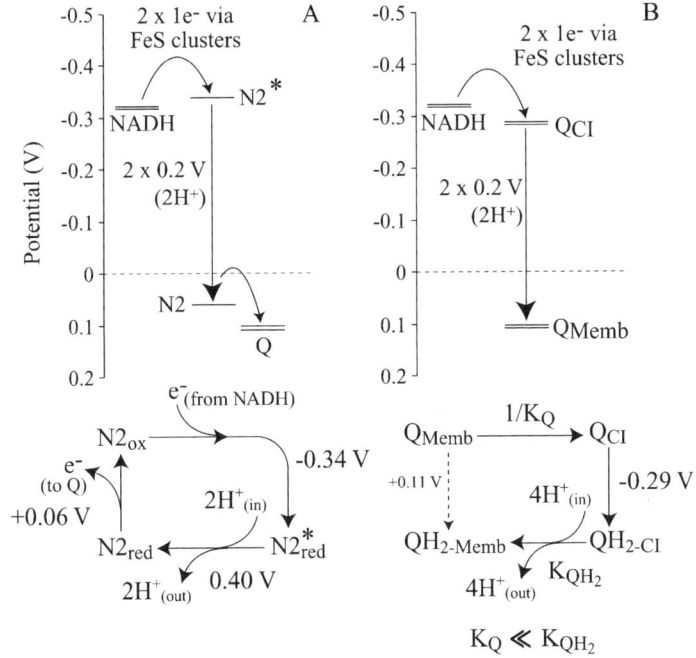

Figure 6 *Possible schemes for the coupling of proton and electron transfer in complex I. A. The translocation of two protons is coupled to the reduction and oxidation of cluster N2. A low-potential electron is delivered to N2 (from NADH), energising the protein and forming N2*$_{red}$. Relaxation of N2*$_{red}$ to N2$_{red}$ is coupled to proton translocation, resulting in the stabilisation of N2 by 0.4 V (2 × PMF). Consequently, the reoxidation of N2$_{red}$ by bound quinone occurs at high potential. B. Proton translocation is coupled to quinone reduction, reflected in the different binding constants for Q (K_Q) and QH$_2$ (K_{QH2}), and in the different two-electron reduction potentials for free (0.11 V) and bound (−0.29 V) quinone. In Figure 6B proton translocation is shown coupled to quinol dissociation*

5.2.1 Evidence for Directly-Coupled Proton–Electron Transfer at Cluster N2

Cluster N2 has long been a candidate for part of the proton-pumping mechanism of complex I. First, its reduction potential is pH dependent,[49,50] indicating that either the cluster, or a nearby residue, is protonated only when the cluster is reduced. Proton transfer across a hydrophobic barrier, in response to the change in oxidation state of an FeS cluster, has been characterised in detail in *Azotobacter vinelandii* ferredoxin I,[62] providing insights into how a proton pump might be constructed at cluster N2. High efficiency is achieved only if pK_{ox} and pK_{red} (the pK values in the two redox states) are sufficiently separated: for >95% of the system to change its protonation state upon oxidation or reduction (over a range of one pH unit) $pK_{red} - pK_{ox}$ must be >3.6. The two pK values have not been defined in bovine complex I, but in *Y. lipolytica* a separation of only ~one pH unit has been reported.[63] This questions the relevance of the pH-dependence of N2, but it is important to note that a redox titration (at equilibrium) may not be relevant to (cyclic) catalytic turnover. Second, the reduction potential of cluster N2 responds to the membrane potential,[50,64] consistent with Figure 6A; it is possible that this effect contributes to the variation in reported values of the reduction potential of N2.[50,52] Third, cluster N2 is proposed to be the direct electron donor to bound quinone, and it is likely to be in an appropriate position for controlling proton transfer across a hydrophobic barrier (in, or close to, the membrane domain). Finally, FTIR studies on *E. coli* complex I have recently suggested that a number of residues protonate or deprotonate upon the oxidation or reduction of cluster N2, and that a redox-coupled conformational change takes place.[65,66] Taken together, these observations provide evidence which is consistent with proton pumping by cluster N2, but they are, at present, insufficient to justify the formulation of specific mechanistic hypotheses.

5.2.2   Evidence for an Indirectly Coupled Proton Transfer Mechanism

Typically, indirectly coupled mechanisms are used to couple a chemical transformation (for example ATP synthesis) to ion transport across a membrane, and they involve long-range energy transfer between a hydrophilic and a hydrophobic protein domain.[57,58] Although it is not clear why complex I should adopt indirect coupling as a method of long-range energy transfer, perhaps linking the redox reaction of cluster N2 or the chemical conversion of ubiquinone to proton pumping, conformational changes have been reported, and used to suggest the operation of an indirectly coupled mechanism. Note that a conformational change also takes place in the cytochrome bc_1 complex.[35] However, it is not part of an indirectly coupled proton-transfer mechanism, but part of the mechanism for the bifurcation of electron transfer at the Q_O site: a conformational change does not necessarily imply indirect coupling. A 'horseshoe' conformation has been reported for complex I, as an alternative to the more usual L-shaped structure,[67] but has been questioned by other researchers.[68] Complex I from *E. coli* has been reported to adopt a more 'open' conformation upon its reaction with NADH, although the structural change shown is not specific to any particular part of the structure.[69] Differences in cross-linking and trypsinolysis of the hydrophilic subunits of bovine and *E. coli* complex I upon reaction with NADH substrates have been reported.[69,70] However, it is clear that further data are required before a functional correlation may be drawn between these preliminary observations and the proton-pumping mechanism of complex I.

References

1. B.E. Schultz and S.I. Chan, *Annu. Rev. Biophys. Biomol. Struct.*, 2001, **30**, 23.
2. M. Saraste, *Science*, 1999, **283**, 1488.
3. G.C. Brown and M.D. Brand, *Biochem. J.*, 1988, **252**, 473.
4. M. Wikström, *FEBS Lett.*, 1984, **169**, 300.
5. P.J. Holt, D.J. Morgan and L.A. Sazanov, *J. Biol. Chem.*, 2003, **278**, 43114.

6. J. Hirst, J. Carroll, I.M. Fearnley, R.J. Shannon and J.E. Walker, *Biochim. Biophys. Acta*, 2003, **1604**, 135.

7. J.E. Walker, *Q. Rev. Biophys.*, 1992, **25**, 253.

8. G. Hofhaus, H. Weiss and K. Leonard, *J. Mol. Biol.*, 1991, **221**, 1027.

9. V. Guénebaut, A. Schlitt, H. Weiss, K. Leonard and T. Friedrich, T., *J. Mol. Biol.*, 1998, **276**, 105.

10. H. Leif, V.D. Sled, T. Ohnishi, H. Weiss and T. Friedrich, *Eur. J. Biochem.*, 1995, **230**, 538.

11. T. Friedrich, B. Brors, P. Hellwig *et al.*, *Biochim. Biophys. Acta*, 2000, **1459**, 305.

12. M. Finel, A.S. Majander, J. Tyynelä *et al.*, *Eur. J. Biochem.*, 1994, **226**, 237.

13. D.-C. Wang, S.W. Meinhardt, U. Sackmann, H. Weiss and T. Ohnishi, *Eur. J. Biochem.*, 1991, **197**, 257.

14. M. Schmidt, T. Friedrich, J. Wallrath, T. Ohnishi and H. Weiss, *FEBS Lett.*, 1992, **313**, 8.

15. T. Yano, J. Sklar, E. Nakamaru-Ogiso *et al.*, *J. Biol. Chem.*, 2003, **278**, 15514.

16. C.I. Ragan, Y.M. Galante, Y. Hatefi and T. Ohnishi, *Biochemistry*, 1982, **21**, 590.

17. S. Chen and R.J. Guillory, *J. Biol. Chem.*, 1981, **256**, 8318.

18. V.D. Sled, N.I. Rudnitzky, Y. Hatefi, and T. Ohnishi, *Biochemistry*, 1994, **33**, 10069.

19. T. Yano, V.D. Sled, T. Ohnishi and T. Yagi, *FEBS Lett.*, 1994, **354**, 160.

20. Y. Zu, S. Di Bernardo, T. Yagi and J. Hirst, *Biochemistry*, 2002, **41**, 10056.

21. N. Kashani-Poor, K. Zwicker, S. Kerscher and U. Brandt, *J. Biol. Chem.*, 2001, **276**, 24082.

22. L. Grgic, K. Zwicker, N. Kashani-Poor, S. Kerscher and U. Brandt, *J. Biol. Chem.*, 2004, **279**, 21193.

23. S.P.J. Albracht, *Biochim. Biophys. Acta*, 1993, **1144**, 221.

24. D. Flemming, A. Schlitt, V. Spehr, T. Bischof and T. Friedrich, *J. Biol. Chem.*, 2003, **278**, 47602.

25. A. Garofano, K. Zwicker, S. Kerscher, P. Okun and U. Brandt, *J. Biol. Chem.*, 2003, **278**, 42435.

26. S.J. Pilkington, J.M. Skehel, R.B. Gennis and J.E. Walker, *Biochemistry*, 1991, **30**, 2166.

27. T. Yano and T. Yagi, *J. Biol. Chem.*, 1999, **274**, 28606.

28. S. Di Bernardo and T. Yagi, *FEBS Lett.*, 2001, **508**, 385.

29. T. Rasmussen, D. Scheide, B. Brors *et al.*, *Biochemistry*, 2001, **40**, 6124.

30. M. Chevallet, A. Dupuis, J.-P. Issartel *et al.*, *Biochim. Biophys. Acta*, 2003, **1557**, 51.

31. F. Schuler, T. Yano, S. Di Bernardo *et al.*, *Proc. Natl. Acad. Sci. U.S.A.*, 1999, **96**, 4149.

32. S. Magnitsky, L. Toulokhonova, T. Yano *et al.*, *J. Bioenerg. Biomembr.*, 2002, **34**, 193.

33. V. Yankovskaya, R. Horsefield, S. Törnroth *et al.*, *Science*, 2003, **299**, 700.

34. M. Jormakka, S. Törnroth, B. Byrne and S. Iwata, *Science*, 2002, **295**, 1863.

35. Z. Zhang, L. Huang, V. M. Shulmeister *et al.*, *Nature*, 1998, **392**, 677.

36. J. Deisenhofer and H. Michel, *Science*, 1989, **245**, 1463.

37. C.C. Page, C.C. Moser, X. Chen and P.L. Dutton, *Nature*, 1999, **402**, 47.

38. M.D. Esposti, *Biochim. Biophys. Acta*, 1998, **1364**, 222.
39. H. Miyoshi, *Biochim. Biophys. Acta*, 1998, **1364**, 236.
40. S. Dröse, K. Zwicker and U. Brandt, *Biochim. Biophys. Acta*, 2002, **1556**, 65.
41. M.S. Sharpley Ph.D. Thesis, Cambridge University, 2005.
42. X. Gong, T. Xie, L. Yu *et al.*, *J. Biol. Chem.*, 2003, **278**, 25731.
43. J.G. Okun, P. Lümmen and U. Brandt, *J. Biol. Chem.*, 1999, **274**, 2625.
44. F.G.P. Earley, S.D. Patel, C.I. Ragan.and G. Attardi, *FEBS Lett.*, 1987, **219**, 108.
45. T. Yagi and Y. Hatefi, *J. Biol. Chem.*, 1988, **263**, 16150.
46. E. Nakamura-Ogiso, K. Sakamoto, A. Matsuno-Yagi, H. Miyoshi and T. Yagi, *Biochemistry*, 2003, **42**, 746.
47. E. Darrouzet, J.P. Issartel, J. Lunardi and A. Dupuis, *FEBS Lett.*, 1998, **431**, 34.
48. V. Zickermann, M. Bostina, C. Hunte *et al.*, *J. Biol. Chem.*, 2003, **278**, 29072.
49. W.J. Ingledew and T. Ohnishi, *Biochem. J.*, 1980, **186**, 111.
50. T. Ohnishi in *Membrane Proteins in Energy Transduction*, R. Capaldi, ed. Marcel Dekker Inc. New York, 1979, 1–87.
51. T. Ohnishi, *Biochim. Biophys. Acta*, 1998, **1364**, 186.
52. T. Ohnishi, J.S. Leigh, C.I. Ragan and E. Racker, *Biochem. Biophys. Res. Commun.*, 1974, **56**, 775.
53. Y. Zu, R.J. Shannon and J. Hirst, *J. Am. Chem. Soc.*, 2003, **125**, 6020.
54. M.G. Bertero, R.A. Rothery, M. Palak, *et al.*, *Nat. Struct. Biol.*, 2003, **10**, 681.
55. A.J. Lambert and M.D. Brand, *Biochem. J.*, 2004, **382**, 511.
56. A. Osyczka, C.C. Moser, F. Daldal and P.L. Dutton, *Nature*, 2004, **427**, 607.
57. D. Stock, C. Gibbons, I. Arechaga, A.G.W. Leslie and J.E. Walker, *Curr. Opin. Struct. Biol.*, 2000, **10**, 672.
58. C. Toyoshima, H. Nomura and Y. Sugita, *FEBS Lett.*, 2003, **555**, 106.
59. P.L. Dutton, C.C. Moser, V.D. Sled, F. Daldal and T. Ohnishi, *Biochim. Biophys. Acta*, 1998, **1364**, 245.
60. A.C. Gemperli, P. Dimroth and J. Steuber, *Proc. Natl. Acad. Sci. U.S.A.*, 2003, **100**, 839.
61. U. Brandt, *Biochim. Biophys. Acta*, 1997, **1318**, 79.
62. K. Chen, J. Hirst, R. Camba *et al.*, *Nature*, 2000, **405**, 814.
63. U. Brandt, S. Kerscher, S. Dröse, K. Zwicker and V. Zickermann, *FEBS Lett.*, 2003, **545**, 9.
64. T. Ohnishi, *Eur. J. Biochem.*, 1976, **64**, 91.
65. P. Hellwig, D. Scheide, S. Bungert, W. Mäntele and T. Friedrich, *Biochemistry*, 2000, **39**, 10884.
66. D. Flemming, P. Hellwig and T. Friedrich, *J. Biol. Chem.*, 2003, **278**, 3055.
67. B. Böttcher, D. Scheide, M. Hesterberg, L. Nagel-Steger and T. Friedrich, *J. Biol. Chem.*, 2002, **277**, 17970.
68. L.A. Sazanov, J. Carroll, P. Holt, L. Toime, L.and I.M. Fearnley, *J. Biol. Chem.*, 2003, **278**, 19483.
69. A.A. Mamedova, P.J. Holt, J. Carroll and L.A. Sazanov, *J. Biol. Chem.*, 2004, **279**, 23830.
70. M. Yamaguchi, G.I. Belogrudov, and Y. Hatefi, *J. Biol. Chem.*, 1998, **273**, 8094.

CHAPTER 9

Structure of Photosystem II from Thermosynechococcus elongatus

KRISTINA N. FERREIRA AND SO IWATA

Department of Biological Sciences,
Imperial College
London SW7 2AZ, UK

1 Introduction

The integral membrane protein complex photosystem II (PSII) catalyzes the light-driven four-electron oxidation of water in plants, algae and cyanobacteria.[1,2] This oxygenic reaction occurs at a unique tetra-manganese oxygen-evolving cluster (OEC), which has not been accurately characterized by spectroscopic techniques. The reaction supplies the organism with electrons for the reduction of carbon dioxide during the Calvin cycle and contributes to the creation of a protonmotive force across the membrane. PSII from *Thermosynechococcus elongatus* (*T. elongatus*) can be purified as both a monomer and a dimer with the latter believed to be physiologically relevant. Each monomer comprises 19 or more subunits.

Initial insights to the structure have come from cryo-electron crystallography and single particle electron microscopy (EM) analyses. These techniques elucidated the shape of the dimeric PSII complex, the position of the transmembrane helices as well as the location of many of the chlorophyll and heme molecules, in plant, algae and cyanobacteria.[3-6] These were followed by two structures of cyanobacterial PSII determined by X-ray crystallography at 3.8 Å[7] and 3.7 Å[8] resolution, yielding a Cα and a poly-alanine trace, respectively, as well as revealing the position of the OEC. In the centre of the membrane-spanning domain of PSII is situated a reaction centre (RC) consisting of two subunits, D1 and D2, which form the core of each PSII monomer. The RC contains the electron-transfer cofactors and is surrounded by two light-harvesting transmembrane subunits, CP43 and CP47, as well as a number of membrane-spanning low molecular weight and extrinsic subunits. The composition of these subunits differs somewhat between prokaryotic and eukaryotic PSII, even though the overall structure of the PSII complex, as seen by EM, cryo-electron crystallography and X-ray crystallography, has been shown to be similar.

Recently, we have succeeded in crystallizing PSII from *T. elongatus* resulting in crystals that diffract to 3.3 Å resolution. This improved crystallization protocol includes optimization of the sample preparation, additive screening and crystallization procedure. Using the crystals, we have succeeded in obtaining a 3.5 Å resolution structure of PSII, which revealed most of the subunit structures including sidechains and the atomic arrangement of the oxygen-evolving centre[1,9] (PDB entry 1S5L). This chapter reviews: (1) the optimized crystallization protocol, (2) the structure of each subunit, and (3) details of the surrounding coordination sphere of the OEC and the implications for a possible oxygen-evolving mechanism.

2   Optimized Crystallization of Photosystem II from T. elongatus

2.1   Cell Culture

Cells of the thermophilic cyanobacterium *T. elongatus* were grown in a 30 L fermenter in medium D, pH 7.5, 56°C[10] with CO_2-enriched (5%) air continuously bubbled through the culture. After six days of growth under continuous intensity white light (250 μE m^{-2} s^{-1}), the cells reached an OD_{685} of 1.7–2.0 and were harvested, yielding ~65 g of wet cells. From here all work was performed in the dark or in dim light, to avoid photodamage to PSII.

2.2   Purification

Dimeric photosystem II (PSII) from *T. elongatus* was initially purified as previously described.[11] In brief, cells were disrupted using lysozyme (1 g L^{-1}, ~65 g cells wet weight, 40°C, 90 min) followed by two passages through a French pressure cell. The thylakoid membranes were washed and isolated by centrifugation and split into four aliquots prior to flash freezing in liquid nitrogen using 20% glycerol as a cryo-protectant. The membranes were thawed on liquid nitrogen-cooled ice overnight, washed with 0.05% *n*-dodecyl-β-D-maltoside (DDM) (w/v) (Biomol, Germany) and solubilized using 1.2% DDM (w/v) and 0.5% sodium cholate (w/v) (Sigma, England) at a chlorophyll concentration of 1 mg mL^{-1}. PSII was purified by hydrophilic interaction chromatography using POROS-ET perfusion chromatography media (Applied Biosystems, U.S.A.) followed by anionic exchange chromatography using a UnoQ-12 column (BioRad Laboratories, U.S.A.). The sample was washed in 20 mM MES pH 6.5, 10 mM $CaCl_2$ and 10 mM $MgCl_2$, concentrated to 1.75 (mg chlorophyll) mL^{-1} and stored on ice.

The scale of the preparation was crucial for sample and thus crystal quality. Decreasing the scale of the purification from ~60 mg chlorophyll to ~35 mg chlorophyll (corresponding to ~16 g of wet cells) was particularly important for the hydrophobic interaction chromatography. In general, crystals grew from protein sample that primarily contained dimeric PSII. However, the sample giving rise to the

3.3 Å resolution diffraction was estimated to contain ~30% monomers, as shown by analytical size-exclusion chromatography (data not shown), which seemed not to have affected the crystal quality at this stage. The crystallization itself could work as a final purification step, separating monomers from dimers.

2.3 Crystallization and Optimization

Crystals grew using a hanging drop method. The sample was mixed 1:1 with a crystallization solution with the addition of 20% glycerol (100 mM Hepes pH 7.5, 100 mM $(NH_4)_2SO_4$, 8.0–14.0% PEG-4000; twice as concentrated as the reservoir solution). The reduced precipitant concentration in the reservoir improved the shape of crystals significantly. Crystals appeared within two days at 17°C and grew to an average size of 0.2 mm × 0.2 mm × 0.1 mm over four days (Figure 1). Crystal diffraction was improved by using detergent and heavy metal additives in the crystallization reaction, with the best crystals grown from mother liquors containing 55 µM octyloxyethylene laurylether ($C_{12}E_8$), 0.5 mM methyl mercurial acetate or methyl mercurial chloride, or 0.5 mM trimethyl lead acetate. To remove protein aggregates, the sample was centrifuged in a bench top centrifuge at 1300 rpm immediately prior to drop set-up.

Centrifugation prior to crystallization was essential for the development of fewer but larger crystals, due to the decrease in protein aggregates causing too many nucleations. This is similar to earlier reports on various membrane proteins.[12,13] Centrifugation at 'bench top' speed was sufficient for limiting the crystal nucleation, whereas ultracentrifugation resulted in the loss of the entire protein sample in the pellet.

Optimization of the crystallization conditions involved the addition of $C_{12}E_8$ and heavy-metal compounds and was crucial for the development of large, diffraction-quality crystals.[13]

Figure 1 *Crystals of PSII from T. elongatus. Approximate size of the largest crystal in the drop is 0.2 mm × 0.2 mm × 0.05 mm*

2.4   Data Collection

Prior to cryo-cooling in liquid nitrogen, the crystals were cryo-protected in 25% glycerol added stepwise in artificial mother liquor. Data were collected to 3.5 Å at −180°C on beamline X06SA at the Swiss Light Source (SLS) using a MAR CCD detector. Data were processed with DENZO and scaled with SCALEPACK,[14] and the CCP4 suite of programs.[15]

The crystals displayed a maximal diffraction of 3.3 Å at beamline X06SA at the SLS. Data were collected, processed and scaled to 3.5 Å resolution. The crystals belonged to the orthorhombic space group $P2_12_12_1$ with $a = 135.0$ Å, $b = 228.8$ Å, and $c = 309.9$ Å. The overall R-merge of the data was 8% (43% for the last shell) with the completeness of 87.3% (80.9% for the last shell). PSII dimer in the asymmetric unit, yielded a Matthew's coefficient, V_M, of 3.9 Å,[3] and a solvent content of ~70%.

3   Structure of Photosystem II from T. elongatus

3.1   Overall Structure

The crystallographic asymmetric unit contains a dimer of PSII, where the two monomers in the dimer are almost identical. In the model, the monomer contains 19 protein subunits that have been clearly assigned to electron density except the putative assignment of PsbN, where the sidechains are not included in the model due to disorder. Each monomer contains 36 chlorophyll a (Chl) and 7 *all-trans* carotenoids assumed to be β-carotene molecules. However two of the Chl molecules are loosely attached to the complex and disordered. Additionally, there is unassigned electron density remaining that could be carotenoid molecules or the fatty-acid tails of lipid molecules. Each monomer model also includes one OEC, one heme b, one heme c, two plastoquinones, two pheophytins, one non-heme Fe and two bicarbonates (one is tentatively assigned as an unknown non-protein ligand at the OEC).

The PS II dimer (Figure 2a) has dimensions of 105 Å depth (45 Å in membrane), 205 Å length and 110 Å width. Compared to previous models,[7,8] the sequence assignment has been significantly improved. We have newly assigned or reassigned 3916 residues and built the side chains. The improvement is significant in the models for the extrinsic subunits, extrinsic domains of D1, D2, CP43 and CP47 and the small transmembrane subunits.

Within the dimer, the monomers are related by a non-crystallographic two-fold axis perpendicular to the membrane plane. Our model of PSII consists of 16 integral membrane subunits composed of 35 transmembrane helices and 3 lumenal subunits. The monomer is characterised by pseudo-two-fold symmetry, which relates the D1, CP47 and PsbI subunits to the D2, CP43 and PsbX subunits (Figure 2b).

3.2   D1 and D2 Proteins

The PSII reaction centre comprises the D1/D2 heterodimer, in which the two subunits are related by the pseudo-two-fold symmetry axis. Each subunit consists of five transmembrane helices (Figure 2), which together with the ones from the other sub-unit forms an ellipsoid that measures approximately 29 Å in width and 65 Å in length, as seen perpendicular to the membrane normal. The residues comprising the membrane-spanning helices in the structure are in reasonable agreement with those predicted by Trebst.[16] Five transmembrane helices of each subunit are organised in a manner almost identical to that of the L- and M-subunits of the reaction center of photosynthetic purple bacteria (bRC),[17,18] with an rms deviation between PSII and bRC from *Rhodopseudomonas viridis* of 1.9 Å for 395 C_α atoms. However, loops connecting transmembrane helices of the D1 and D2 proteins are considerably longer than those for bRC, especially on the lumenal side close to the OEC.

The reactions of PSII are powered by light-driven primary and secondary electron transfer processes across the reaction center (RC), composed of the D1 and D2 sub-units (Figure 3a). Upon illumination, an electron is ejected from the excited primary electron donor P680, a Chl located towards the lumenal surface (P_{D1} is likely to be P680 as discussed below). The electron is quickly transferred towards the stromal surface to the final electron acceptor, plastoquinone Q_B, *via* Chl_{D1}, pheophytin ($Pheo_{D1}$) and plastoquinone Q_A. After accepting two electrons and undergoing pro-tonation, Q_B is released from PSII into the membrane. The radical cation $P680^{\bullet+}$ is reduced by a redox active tyrosine, known as Tyr_Z (Tyr161 of D1 subunit), to gener-ate a neutral tyrosine radical $Tyr_Z^{\bullet}$ which acts as an oxidant for the water oxidation process at the OEC. The oxidation of two water molecules to produce a dioxygen proceeds in a step-wise manner as described in the S-state cycle.[19] With the excep-tion of Q_A, all the redox-active cofactors involved in the electron-transfer processes are located on the D1 side of the RC.

The radical cation $P680^{\bullet+}$ has a very high oxidizing potential, estimated to be 1.3–1.4 V,[20] which is required for water oxidation. This contrasts with 0.4 V pro-duced by the bacterial reaction centre consisting of a 'special pair' of bacteri-ochlorophylls (BChl), which is reduced by a cytochrome after photooxidation. The two chlorophyll molecules in PSII equivalent to the 'special pair' in the bRC are P_{D1} and P_{D2}.[7] The overall organisation of P_{D1} and P_{D2} is identical to the 'special pair', however, the relative orientation angle between their tetrapyrrole heads differs slightly from their bacterial equivalents. The head groups of P_{D1} and P_{D2} are in direct van der Waals contact with a Mg–Mg distance of 8.2 Å (Figure 3b). This is slightly longer than the 7.5 Å observed for the bRC 'special pair', which could explain why P680 shows more monomeric character and weaker electronic coupling than the bac-terial 'pair'.[21]

P_{D1} and P_{D2} are located in the vicinity of Chl_{D1} and Chl_{D2} (Figure 3), which are equivalent to the accessory BChls of the bRC. The P680 excited state is delocalised over the four chlorophylls [22,23] and Chl_{D1}, which is the chlorophyll closest to the active $Pheo_{D1}$. This is involved in the initial primary charge separation.[24,25] There is no obvious amino acid ligation for either of Chl_{D1} and Chl_{D2}; a factor that may influ-ence the electronic character of P680 in comparison to the bacterial 'pair', which have protein ligands in the form of L-His15 and M-His180 or M-His182 in *Rhodobacter spaeroides* and *Rhodopseudomonas viridis*, respectively. The nearest residues to Chl_{D1} and Chl_{D2} are D1-Tyr179 and D2-Ile178, respectively, both of which are situ-ated on the lumenal CD helices in each subunit. Water molecules might be involved

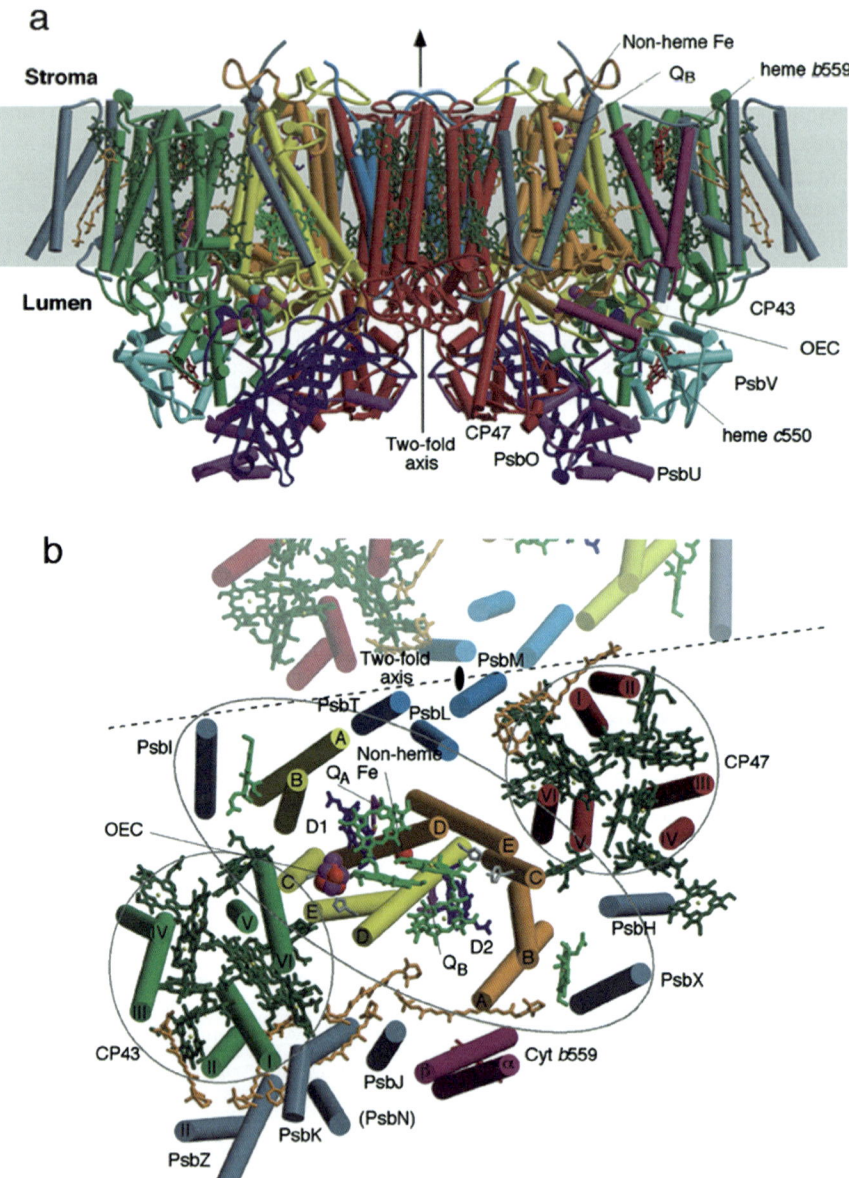

Figure 2 *Overall structure of PSII from T. elongatus. (a) View of the PSII dimer perpendicular to the membrane normal. Helices are represented as cylinders with D1 in yellow; D2 in orange; CP47 in red; Cyt b559 in wine red; PsbL, PsbM and PsbT in medium blue; and PsbH, PsbI, PsbJ, PsbK, PsbX, PsbZ, and the putative PsbN in grey. The extrinsic proteins are PsbO in blue; PsbU in magenta; and PsbV (Cyt c550 in cyan). Cofactors in the D1/D2 reaction centre are chlorophylls in light green; pheophytins in blue; β-carotenes in orange; non-heme iron (Fe) in red; plasto-quinones (Q_A, Q_B) in magenta. The components of the oxygen-evolving centre (OEC) are shown as red (oxygen atoms), magenta (Mn ions) and cyan (Ca^{2+}) balls. The*

in coordinating the Mg ions; however, it is not possible to confirm this due to the resolution of our structure.

The electrons are transferred from $Pheo_{D1}$ to Q_A, which is a firmly bound plastoquinone in a hydrophobic pocket in the D2 protein (Figure 4), containing residues D2-Ile213, D2-Thr217, D2-Met246, D2-Ala249, D2-Trp253, D2-Ala260 and D2-Leu267. Two residues are directly involved in hydrogen bonding to the ring head group of the molecule: D2-Phe261 by its main-chain amide group, and D2-His 214 (Figure 4b). Despite the differences in quinone electron acceptors between bacterial RC and PSII, the Q_A binding pockets are structurally similar with the ring head group approximately perpendicular to the membrane plane. Unlike previous PSII structures, our electron density map has allowed us to model the terminal electron acceptor in the PSII electron transport chain, Q_B. The hydrophobic cavity in which Q_B is bound includes D1-Met214, D1-Leu218, D1-Ala251, D1-Phe255 and D1-Leu271 (Figure 4c). Again two residues are directly involved in hydrogen bonding of the ring head group of the molecule: D1-Ser264 and D1-His215. Whereas the serine and histidine coordination is analogous to the equivalent quinone binding in bacterial RC (L-Ser223 and L-His190), the overall cavity is less conserved. This is because there is a conformational difference in the loops containing D1-Ser264 and the equivalent in bRC, which is caused by the insertion of one residue. As a consequence, the volume of the Q_B binding pocket, composed by residues D1-Met214, D1-Leu218, D1-Ala251, D1-Phe255 and D1-Leu271, is slightly larger in PSII than in the bRC. This could explain the difference in herbicide specificity between PSII and bRC.[26] In addition, the proton pathway connecting the Q_B site to the stromal space is significantly different in PSII compared with bRC. The PSII Q_B site is much closer to the surface than in bRC, because of the absence of an equivalent to the bRC H-subunit. Since D1-His252 is within hydrogen-bonding distance of D1-Ser264 this residue could aid Q_B protonation (Figure 4c).

The non-heme Fe that mediates electron transfer between Q_A and Q_B is located on the pseudo-two-fold axis of the D1/D2 heterodimer (Figure 4a). Four histidine residues (D1-His215, D1-His272, D2-His214 and D2-His268) coordinate the Fe ion in analogy to bacterial RC. However, contrary to bacterial RC, where a glutamate takes part in coordinating its non-heme Fe, there is no fifth amino acid ligand in PSII. It has been suggested that bicarbonate may function as the fifth ligand to the non-heme Fe in PSII[27] and that bicarbonate has a regulatory function controlling electron transfer from Q_A to Q_B hence facilitating the protonation of Q_B.[28] In the structure, the non-heme Fe is associated with an electron density that is sufficient to

---

chlorophyll molecules of the antenna complexes are in dark green, and the hemes are in red. The phytol tails of chlorophyll and pheophytin are omitted for the sake of clarity. (b) View of the PSII monomer along the membrane normal from the lumenal side. A part of the other monomer in the dimer is shown to emphasize the region of monomer-monomer interaction in the membrane plane (dotted line). The pseudo-twofold axis perpendicular to the membrane plane passing through the non-heme Fe relates the transmembrane helices of the D1/CP43-D2/CP47, and to a certain extent PsbI and PsbX, as emphasized by the black lines encircling these subunits. Coloring is the same as in (a). All figures in this review are created using Molscript[88], Bobscript[89] and Raster3D[90] programs

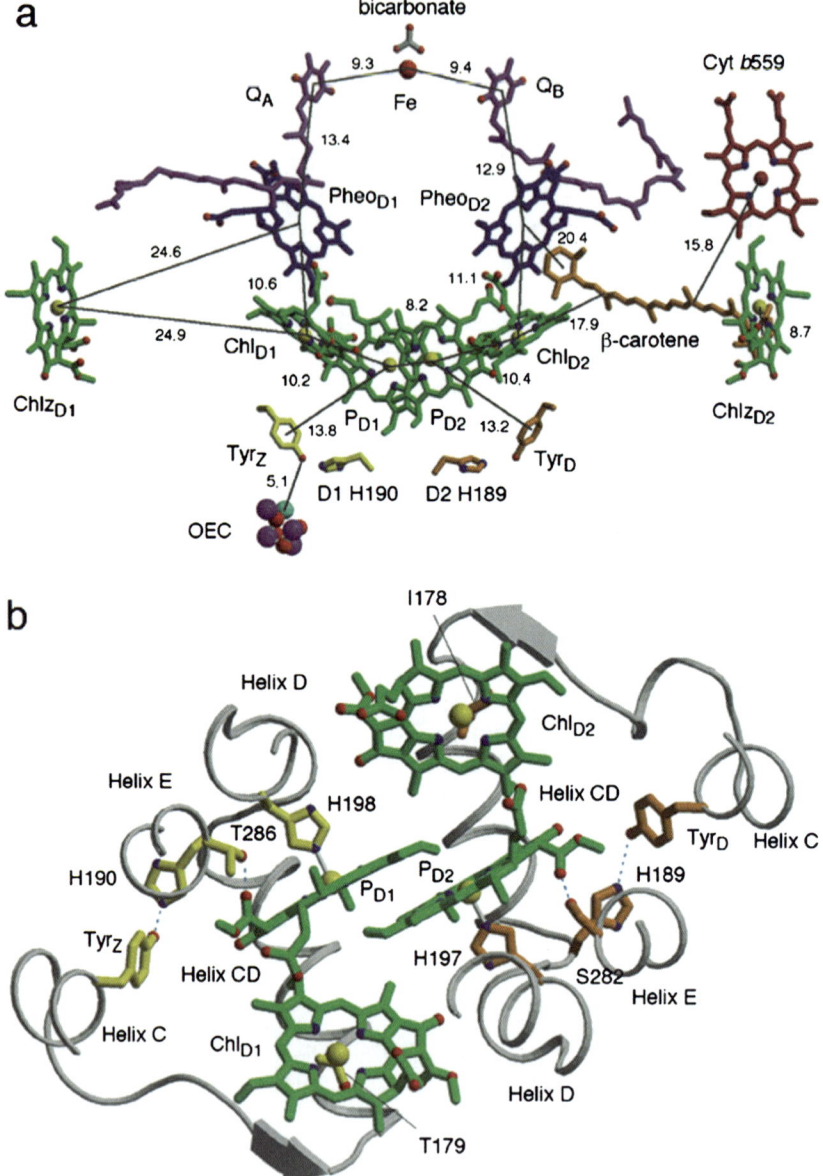

Figure 3 *Cofactors involved in electron transfer. (a) View perpendicular to the internal pseudo-two-fold symmetry axis. Coloring scheme is the same as in Figure 2. The phytol tails have been removed for clarity. The side chains of Tyr_Z (D1 Tyr161) and D1 His190 are shown in yellow, and Tyr_Y (D2 Tyr160) and D2-His189 are in orange. (b) Organization of the P680 chlorophyll molecules, including the P_{D1}/P_{D2} pair and the accessory Chl_{D1} and Chl_{D2}. The histidine ligands D1 His198 and D2 His197 are shown, as well as the redox-active Tyr_Z (D1 Tyr161)/D1 His190 and Tyr_D (D2-Tyr160)/D2 His189 pairs. The phytol-tails are omitted for the sake of clarity. The view is along the pseudo-two-fold axis from the stromal side*

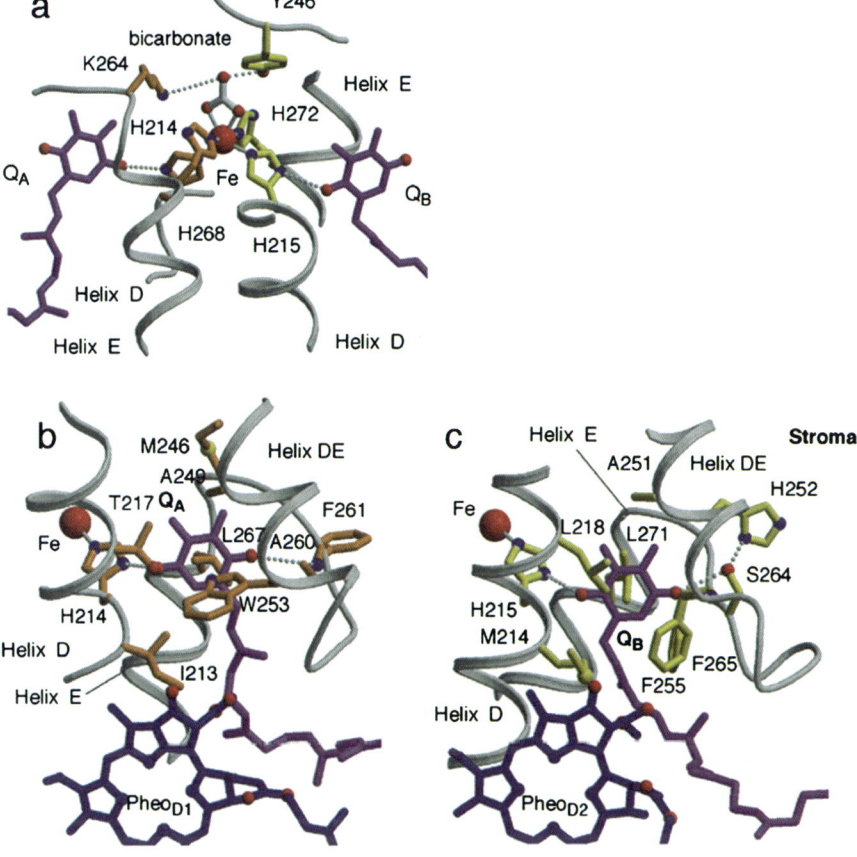

Figure 4 *Plastoquinone binding sites of PSII. (a) Overall view of the non-heme iron, Q_A and Q_B. Coloring scheme is as in Figure 2, with protein mainchains depicted in grey and with sidechain bonds and carbon atoms following the coloring of the protein subunit as used in Figure 2. The bicarbonate completing the coordination sphere of the non-heme Fe is shown with magenta bonds and is probably hydrogen-bonded to D2-Lys264 and D1-Tyr246. (b) The Q_A binding pocket. The hydrophobic residues forming this pocket are shown. The O_1 of the plastoquinone head group is likely to be hydrogen bonded to the non-heme Fe ligating D2-His214 via its δ– nitrogen, while the O_4 atom may hydrogen bond to the backbone amide nitrogen of D2-Phe 261. (c) The Q_B binding pocket. Q_B binds deep into a cavity lined with the hydrophobic residues. O_1 is likely to be hydrogen bonded to the δ-nitrogen of D1 His215, which also forms a ligand to the non-heme Fe, while O4 may form hydrogen bonds with the amide nitrogen of D1-Phe265 and the side chain γ-oxygen of D1-Ser264. D1-Ser264 appears to make further hydrogen-bonding contact with D1-His252. Probable hydrogen bonds are shown as dotted lines while solid lines represent ligands*

accommodate this anion. Close to this non-protein density are D1-Tyr246 and D2-Lys264, located and oriented so that they could stabilize the bicarbonate by hydrogen bonding.

3.3   CP43 and CP47 subunits

The two internal light-harvesting subunits, CP43 and CP47, are related by the same pseudo–two-fold axis, that relates D1 and D2 (Figure 2). This symmetry only applies for the membrane-spanning helices of CP43 and CP47, and not for the lumenal loop regions (Figure 5). Each subunit consists of six transmembrane helices,[29] which are arranged in three two-helix pairs around a pseudo-three-fold symmetry axis in analogy to the two N-terminal six-helix domain of the PSI subunits A and B.[30]

On the lumenal side of PSII, both internal light-harvesting subunits have long insertions. The V–VI lumenal loop in CP47 (186 residues) is considerably longer than the corresponding loop in CP43 (129 residues, Figure 5a). The loop structures follow neither the pseudo-two-fold symmetry of the whole complex nor the pseudo-three-fold symmetry of CP43 or CP47. Together with PsbE, PsbO and PsbU, the large CP47 loop shields the D2 protein from the solvent (Figure 2). PsbV and the main parts of PsbO and PsbU are located on the D1/CP43 side of the PSII monomer, hence the large CP43 loop participates in sheltering the OEC in the D1 protein from the solvent. The CP47 V–VI loop contains two long and four short α-helices, and three β-sheets. The long CP43 V–VI loop contains two long and two short α-helices, one 3_{10} helix and a two-stranded β-sheet (Figure 5b). The 3_{10} helix is composed of Gly-Gly-Glu-Thr-Met-Arg-Phe-Trp-Asp. This sequence is fully conserved in CP43 from all oxygenic photosynthetic organisms and contains CP43-Glu354 and CP43-Arg357 that are the sole non-D1 OEC ligands, as discussed later. This is in line with findings that this large extrinsic domain plays an important role in water oxidation.[31]

We have assigned 14 and 16 Chls bound to the membrane-spanning region of CP43 and CP47, respectively (Figure 6). Although most Chls are arranged in two layers on opposite sides of the membrane, two are positioned at an almost equal distance from both surfaces. As a result of this, they form a stack with adjacent Chls positioned on the stromal and lumenal sides (Figure 6). The function of these stacked Chls is not clear but they may facilitate the fast energy transfer known to occur within CP43 and CP47 and could be the origin of the long wavelength-absorbing Chls of these proteins.[32] Ten Chls of CP43 and thirteen Chls of CP47 are ligated by histidine and one in CP43 has an asparagine ligand. Both proteins have two Chls at equivalent positions associated with either methionine or serine sidechains. The Mg^{2+} ions of two loosely attached Chls are close to the mainchain carbonyl oxygen and cysteine sidechain respectively, but it is not clear whether these groups are ligands because of disorder. It is also possible that these molecules are not Chls but other types of ligand including lipids.

Seven β-carotenes are assigned to the density although it is possible that there are more as indicated by biochemical analyses.[33] According to our electron-density map, it is possible that another two or three β-carotene molecules may be located close to

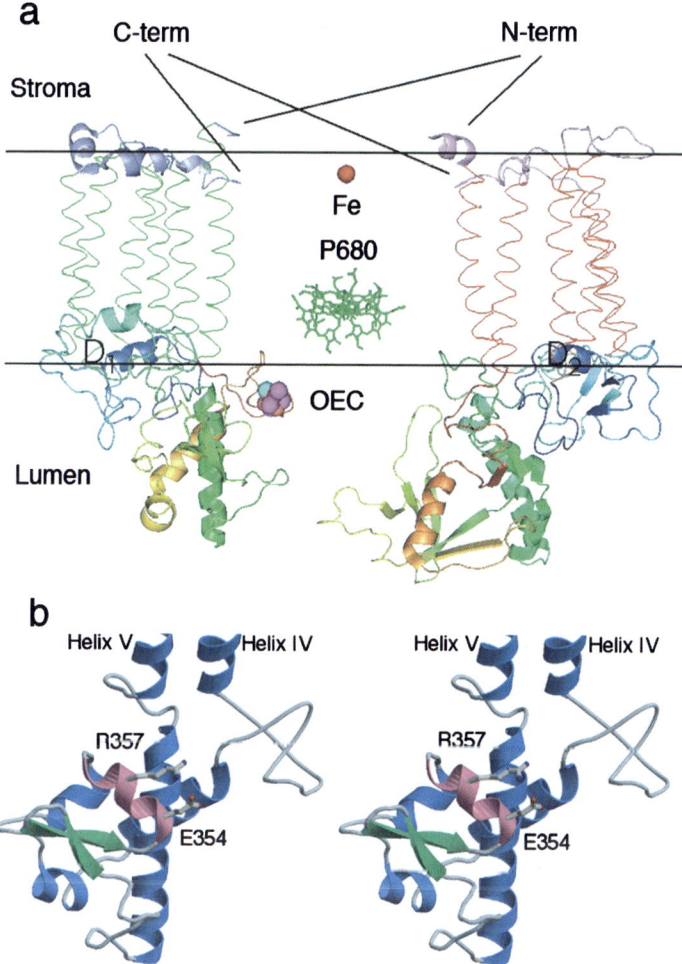

Figure 5 *The extrinsic loop regions of CP43 and CP47. (a) View perpendicular to the membrane normal from the stromal side. The N- and C-termini are located at the stromal side of the light-antenna subunits. Part of the III–IV loop in CP43 (turquoise helix) penetrates the membrane outside helix IV, below the helix in the IV–V interhelical loop. (b) A stereoview of the extrinsic V–VI loop domain in CP43. The domain contains two long and two short α-helices (in blue), one two-stranded β-sheet (in green), and one 3_{10} helix (in magenta). The last is composed of the conserved region Gly352-Gly-Glu-Thr-Met-Arg-Phe-Trp-Asp360, which is fully conserved in all CP43 sequences and contains residues involved in water oxidation at the OEC (Glu354 and Arg357)*

the dimer interface. All but one β-carotene are in contact with Chl head groups, as required for facilitating energy transfer from β-carotene to Chl and for quenching Chl triplets. β-carotene can be photooxidised when water splitting is inhibited with a ms time constant[34,35] suggesting that at least one carotenoid must be positioned about

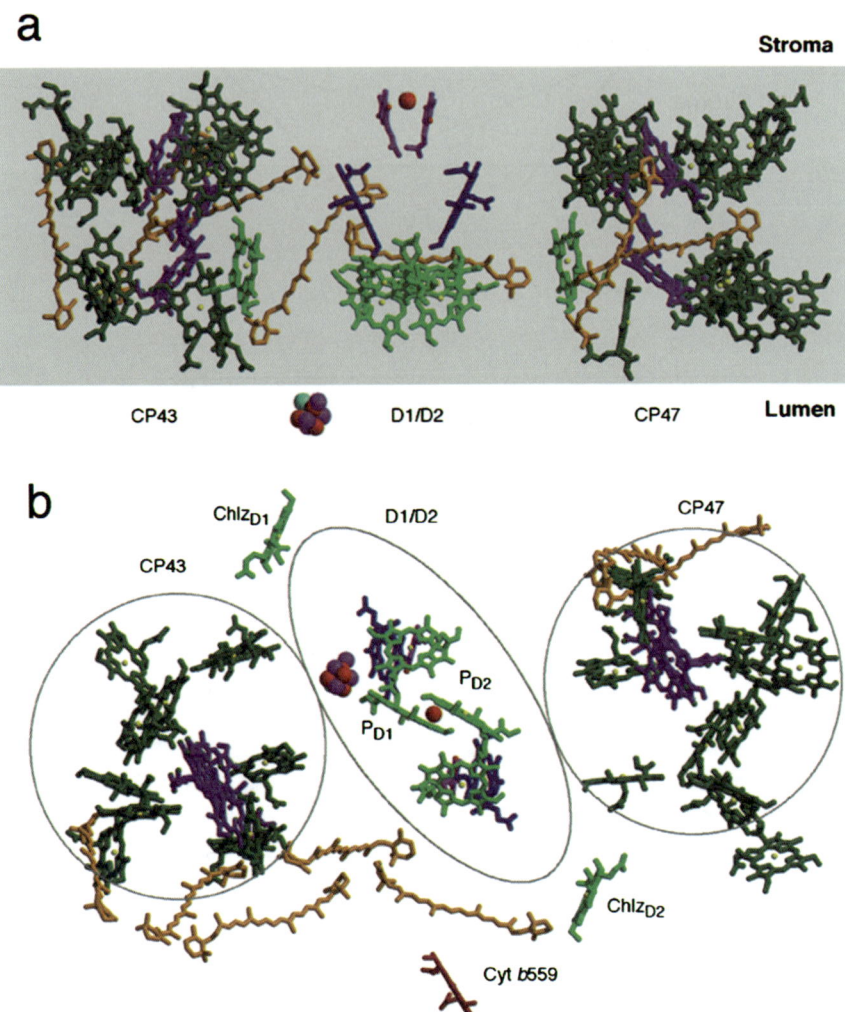

Figure 6 *Light-antenna cofactor organization in the PSII monomer complex. The Chl phytol tails are omitted for the sake of clarity. The area of the CP43 and CP47 transmembrane helices are indicated by green and red lines, respectively. The coloring scheme is as in Figure 2, apart from the stacked chlorophylls that are shown in violet. (a) View perpendicular to the membrane normal. (b) View along the pseudo-two-fold axis, perpendicular to the view in (a)*

20 Å from P680. Moreover there is spectroscopic evidence indicating that β-carotene facilitates long-distance electron flow from Cytb559 to P680$^{\bullet+36}$ and from Chl$_{ZD2}$ or Chl$_{ZD1}$.^{33,37} According to our assignment the head group of one *all-trans* β-carotene is in direct contact with Chl$_{ZD2}$ and is located between Cytb559 and the RC Chls (Figure 3). It is therefore likely that this β-carotene is involved in electron transfer

from Cytb559 and Chl_{ZD2} to P680. Some of the carotenoids may also serve to quench any singlet oxygen produced by P680 triplet.[38]

The overall distribution of Chls in PSII differs significantly from that in PSI. The distribution of the peripheral antenna Chls of PSI bound to the N-terminal domains of PsaA and PsaB is similar to that for CP43 and CP47. However, the C-terminal domains of PsaA and PsaB, together with other subunits, coordinate 43 Chls, which form a central antenna surrounding the electron-transfer system of the PSI RC, whereas PSII only coordinates two peripheral RC Chls (Chl_{ZD1} and Chl_{ZD2}). Figure 6a clearly shows that the locations of these Chls are not optimized to mediate energy transfer from CP43 and CP47 to the RC. This difference probably explains the well-known slow trapping of excitation energy in PSII compared with PSI.[39]

3.4 Low Molecular Weight Subunits

3.4.1 PsbE and PsbF (Cyt b559)

The two subunits of the heterodimer Cyt b559, PsbE and PsbF, are expressed from a widespread gene cluster, the *psbEFJL* operon.[40] They correspond to the α and β subunits of Cyt b559, respectively[41] (Figure 7), where the heme b is ligated by PsbE-His22 and PsbF-His23.[42] These residues are located on the stromal side, where both helices have their N-termini. On the stromal side of its transmembrane α-helix (residues 17 to 40), PsbE interacts with PsbJ. The lumenal part of PsbE includes two helices; one short (residues 42 to 47) immediately after the transmembrane helix, and one longer (residues 72 to 84) located at the C-terminal end (Figure 7). This C-terminal extension has a contact with PsbH, D1, D2 and PsbV subunits. The PsbF α-helix (residues 17 to 45) lies parallel to transmembrane helix A of the D1 protein, enclosing a β-carotene molecule in between itself and the helix A. The C-terminus of PsbF is sandwiched between PsbE and the C-terminal of PsbJ, reaching towards the N-terminal part of PsbV.

3.4.2 PsbL, PsbM and PsbT

We have assigned three α-helices at the dimer interface to PsbL, PsbM and PsbT (Figure 2). The N-terminal end of PsbL is located in the stroma, whereas those of PsbM and PsbT are in the lumen. PsbL is expressed from the *psbEFJL* operon,[40] and has a transmembrane helix (residues 14 to 32) following the stromal loop (residues 1 to 13) interacting with CP47. It has been suggested that PsbL is involved in dimer formation, as detected in a dimeric form of an algal CP47-RC subcomplex only.[43] PsbT in cyanobacteria is equivalent to $PsbT_C$, a chloroplast-encoded protein that is co-purified with a monomeric form of the CP47-RC subcomplex,[43] which indicates a close association of PsbT with CP47, D1, D2 and Cyt b559. This is consistent with our findings, that the PsbT transmembrane helix (residues 1 to 22) is located next to the A helix of D1 (Figure 2). PsbM, like PsbT, has not been assigned in any previous X-ray structures. Besides the association with PsbL and PsbT, PsbM interacts with its equivalent from the other PSII monomer (Figure 2).

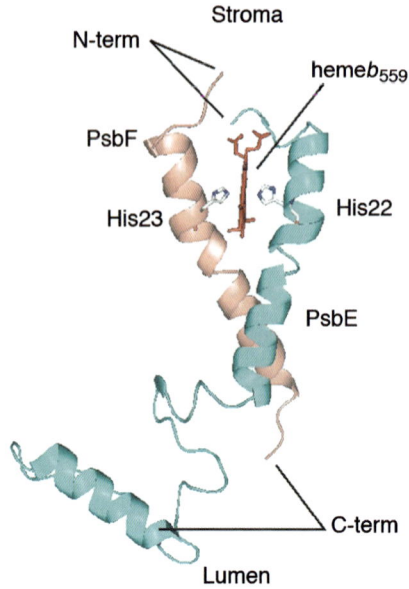

Figure 7 *The α- and β-subunits of Cyt b559, PsbE and PsbF. PsbE and PsbF are shown in light blue and pink, respectively. View perpendicular to the membrane normal, from outside the PSII complex. The heme b group between PsbE-His22 and PsbF-His23 is shown*

3.4.3   PsbI

PsbI is associated with helices A and B of the D1 protein (Figure 2). The subunit has a long C-terminal extension with 14 residues on the stromal side (residues 25 to 38) interacting with the D1 protein. PsbI is involved in the binding of Chlz$_{D1}$, whose Mg ion is coordinated by D1-His118, with Val8 and Val12. This subunit is co-purified with the reaction centre of PSII in pea as the D1/D2-Cyt b559-PsbI subcomplex,[44] and in spinach as the CP47-RC monomer,[43] which supports our assignment. However, our placement of PsbI cannot explain the PsbI cross-linking with both D2 and α-Cyt b559 in spinach PSII.[45]

3.4.4   PsbH and PsbX

The PsbH and PsbX subunits are associated with the D2 protein and CP47 (Figure 2). The transmembrane helix of PsbX (residues 14 to 44) is located between Cyt b559 and CP47. The N-terminal residues (residues 1 to 10) are not assigned in our structure, whereas the C-terminal extension (residues 45 to 50) interacts with the D2 protein. PsbX is involved in the binding of Chlz$_{D2}$, whose Mg ion is coordinated by D2-His117. Single particle EM experiments on the gold-labeled algal PSII locate PsbH in the vicinity of the Cyt b559 heterodimer.[46] This observation, together with the report that PsbX is cross-linked to α-Cyt b559,[47] suggests a possible close association of PsbH and PsbX, which is consistent with our assignment.

3.4.5 PsbJ

The transmembrane helix of PsbJ (residues 10 to 34) is surrounded by Cyt *b*559 (PsbE and PsbF), PsbK, and the putatively assigned PsBN. The N-terminal extension of PsbJ (residues 1 to 9) is located on the stromal side and the C-terminal extension (residues 35 to 40) on the lumenal side interacts with CP43 and the D1 protein. Also, the C-terminus interacts with the surface of Cyt *c*550.

3.4.6 PsbK

The transmembrane helix of PsbK (residues 10 to 32) has kinks caused by two proline residues (PsbK-Pro17 and PsbK-Pro20) (Figure 2). Zheleva *et al.*,[43] have proposed that PsbK, together with PsbL, is involved in dimer formation, however, it is unlikely based on our structure. According to our assignment, the subunit is surrounded by helix I of CP43, PsbJ, putative PsbN and PsbZ; these subunits seem to facilitate β-carotene and Chl binding (see Figure 2).

3.4.7 PsbZ

Unlike other low molecular weight subunits, PsbZ has two transmembrane helices (residues 1 to 28; 33 to 58) as shown in Figure 2.[48] PsbZ is associated with helix II of CP43, PsbK and putative PsbN. The subunit interacts with three β-carotene molecules.

3.4.8 Putative PsbN

We have putatively assigned one transmembrane helix to PsbN. The helix is surrounded by PsbJ, PsbK and the first transmembrane helix of PsbZ (Figure 2). This helix is also observed in other X-ray structures but it is absent in the plant EM model. In our model, this helix seems to be involved in positioning one β-carotene molecule. Because of disordered density, the protein is modeled only by its Cα backbone; it is possible that this subunit could be PsbY.

3.5 Extrinsic Subunits

The three extrinsic subunits in *T. elongatus* form a protein cover shielding the OEC from the lumen, although none are directly involved in the OEC ligation. The extrinsic subunits are mainly located on the D1/CP43 side of the PSII heterodimer. PsbO and PsbV (Cyt *c*550) are not in direct contact with each other but 18 Å apart, with Psb U as a bridge in between.

3.5.1 PsbO

PsbO, which is also known as the manganese-stabilizing protein, is highly conserved between oxygenic photosynthetic organisms. The X-ray structure clearly shows that the PSII monomer contains only one copy of PsbO in spite of there being some controversy over whether there are one or two copies of PsbO per PSII monomer (as reviewed in Bricker and Frankel[49]) (Figure 8a). The removal of PsbO from PSII is associated with decreased binding of manganese,[50] loss of oxygen evolution activity and increased sensitivity to photodamage;[51] these phenomena are reverted by adding PsbO back to the complex.[52] However, deletion of the *psbO* gene in a *Synechocystis* species does not inhibit oxygen evolution or photoautotrophic growth.[53–55] Thus, the real function of PsbO is yet to be confirmed.

The inside of the eight-stranded β-barrel, which makes up the bulk of PsbO, is filled with large, hydrophobic amino-acid residues. The two cysteine residues in the PsbO sequence, Cys19 and Cys44, form a disulphide bridge that holds the N-terminal extension to the β-barrel (Figure 8a). On reduction of this disulphide bond, the protein unfolds and loses activity.[56] Replacement of one cysteine with a serine in a cyanobacterium induces a phenotype with no detectable accumulation of PsbO and impaired O_2 evolution activity.[57] However, the removal of this disulphide bridge in recombinant spinach PsbO did not disrupt its binding and its functional properties.[58]

Spinach PsbO contains two amino-acid sequences (Lys4-Arg-Leu-Thr-Tyr-Asp-Glu10) and (Thr15-Tyr-Leu-Glu18) at the N-terminal, which are required for functional

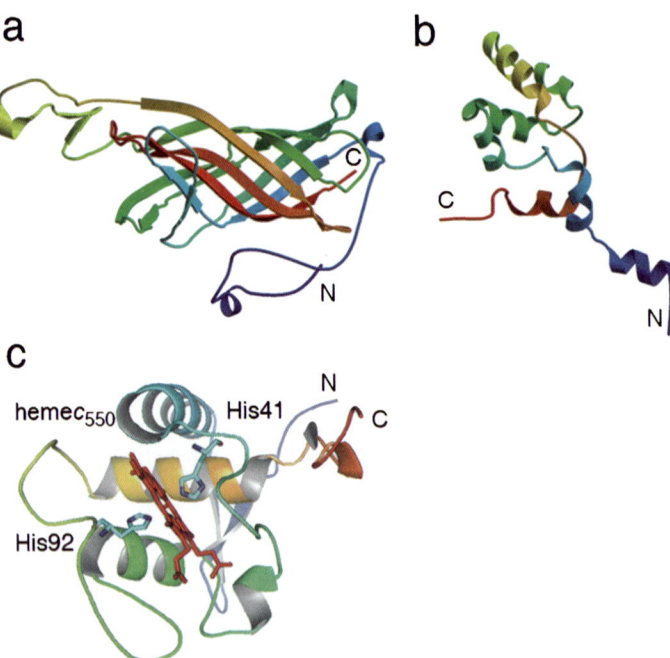

Figure 8 *The PsbO, PsbU and PsbV subunits. (a) In the β-barrel shaped PsbO (shown in rainbow coloring), Cys19 and Cys44 form a disulphide bridge near the N-terminal end. In our structure, the N-terminus is exposed to the crystal solvent/stromal phase. The areas of interaction with D1, D2, CP43 and CP47 are indicated. (b) The PsbU protein (shown in rainbow coloring). (c) The PsbV protein (shown in rainbow coloring) as seen from the membrane side. The heme ligands, His41 and His92, are shown*

binding of PsbO to PSII.[59,60] In the *T. elongatus psbO* gene, the first sequence (Asn3-Thr-Leu-Thr-Tyr-Asp-Asp9) contains mutations and the second sequence is completely missing. We found that only PsbO-Tyr7 and PsbO-Asp8 of the sequence are involved in the subunit binding to CP43 in the structure of *T. elongatus* PsbO.

In spinach PSII, the PsbO C-terminal region also seems to be important for the tertiary structure and for O_2 evolving activity, as studied using deletion mutants of PsbO.[61] However, the C-terminal region in the structure is situated on the opposite side from the PSII protein bulk and is facing towards the solvent. The deletion most likely impairs the folding of the PsbO.

3.5.2   PsbU

PsbU is located above the OEC (Figure 8b). The protein has a N-terminal extension (residues 30 to 43) protruding from the globular C-terminal domain (residues 44 to 134) mainly containing helices. The N-terminal extension interacts with PsbO and CP47, whereas the C-terminal domain is associated with CP43 and PsbV, as well as the D1/D2 proteins. The C-terminal domain may stabilize the D1 terminal loop that provides several residues for coordination of the OEC.

3.5.3   PsbV (Cyt c550)

The PsbV subunit was modeled to the electron density map guided by a cyanobacterial Cyt *c*550,[62] which has a typical *c*-type cytochrome fold like other Cyt *c*550s [63,64] as seen in Figure 8c. The cytochrome has bis-histidine heme ligands (His44 and His92). Besides interacting with the C-terminus of PsbU, PsbV participates in contacts with six other subunits: PsbE, PsbF, PsbJ, D1, D2 and CP43. The C-terminal sequence of PsbV, which has a different conformation in the structure of the separate protein, seems to play the role of anchor to other subunits.

3.6   The Oxygen-evolving Centre (OEC)

As reported previously,[7,8] the 'pear-shaped' density for the OEC is located close to the lumenal surface of the D1 protein (Figure 9). Previous X-ray studies were interpreted to have three Mn ions in the larger domain and one in the connected small domain. However, we could clearly assign four metal ions of the size of Mn at the corners of a tetrahedron in the large globular density, but one metal in the centre of the extended region. The simulated-annealing omit maps confirmed this arrangement, which shows the peak for the respective metal ion when each metal ion of the cluster is omitted from the model (Figure 10). One of the five metal ions is likely to be Ca^{2+}, which is associated with the OEC.[65] We have identified the metals using X-ray anomalous difference maps at the Mn absorption edge (1.89 Å) and at 2.25 Å wavelength, where the Ca^{2+} ion has 3.9 times stronger anomalous difference (f") than the Mn ion (Figure 10). The Mn anomalous difference map only covers one metal in the small domain and three in the large globular domain, whereas the 2.25 Å wavelength map correlates with the remaining one metal ion in the large domain. Based on the approximate positions of the metal ions and their coordination properties,

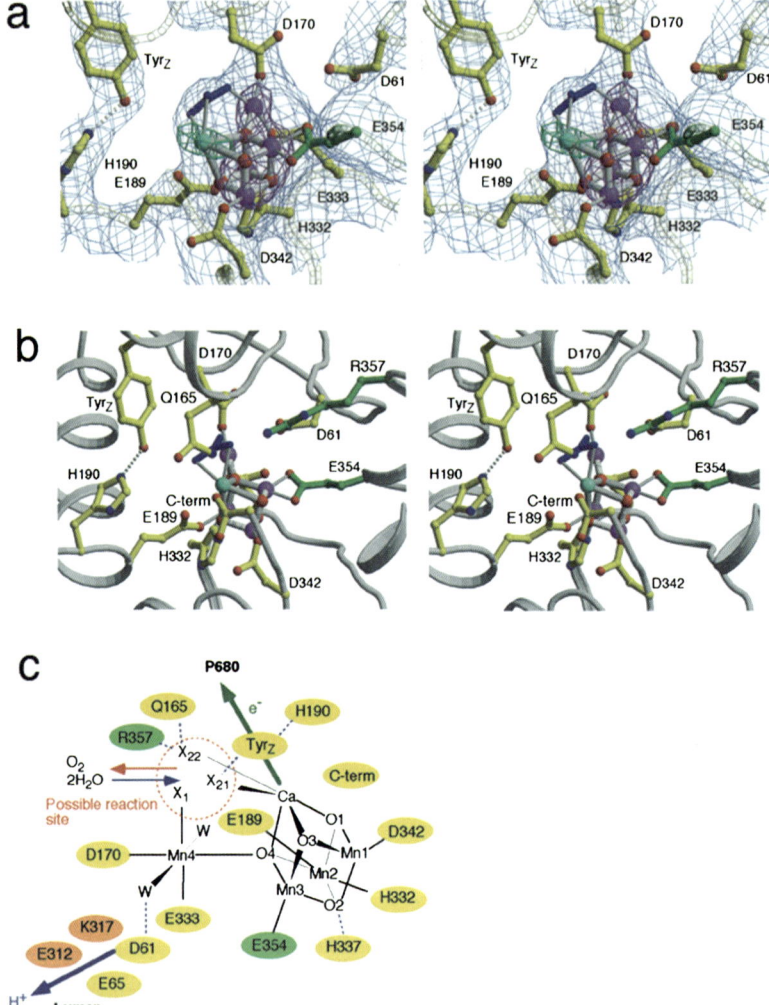

Figure 9 *The oxygen-evolving center (OEC). (a) Stereo view of the OEC with sidechain ligands and possible catalytically important sidechain residues. Mn ions, Ca^{2+} and oxygen atoms are shown in magenta, cyan and red, respectively. One unidentified non-protein ligand to the OEC is coloured in green. Protein mainchain is depicted in light grey while the sidechain bonds and carbon atoms follow the coloring of the protein subunits (D1: yellow, CP43: green). σ_A- weighted $2|F_o| - |F_c|$ density is shown as a light blue wire mesh contoured at 1.5 σ. Anomalous difference Fourier maps at 1.89340 Å (Mn edge, contoured at 10 σ.) and 2.25430 Å (highlights Ca^{2+}, contoured at 7 σ) wavelengths are shown in magenta and blue-green, respectively. (b) The same as (a) but with a rotation around the y-axis of 40° and without electron density and anomalous difference maps. (c) Schematic view of the OEC. Residues in D1, D2 and CP43 subunits are shown in yellow, orange and green, respectively. X_{11}, X_{21} and X_{22} are possible substrate water binding positions to Mn4 (X_{11}) and to Ca^{2+} (X_{21} and X_{22}) identified from the position of the non-protein ligand and coordination pattern of Mn and Ca^{2+} ions. Possible water molecules, which are not visible at the current resolution, are indicated as W. Possible hydrogen bonds are shown as light-blue dotted lines*

we suggest that the OEC is a cubane-like Mn_3CaO_4 cluster with each metal ion in this cluster having tri-μ-oxo bridges (the large domain) connected to another Mn ion by a mono-μ-oxo bridge in the extended region (Figure 9). This 3 + 1 or 'dangler model' organisation for the four Mn ions was suggested by both extended X-ray absorption fine structure (EXAFS) and pulsed electron paramagnetic resonance (EPR) studies.[66,67] Metal distances in the current model, 2.7 Å for Mn-di-μ-oxo-Mn, 3.3 Å for Mn-mono-μ-oxo-Mn and 3.4 Å for Mn-di-μ-oxo-Ca are consistent with those measured by EXAFS.[67–70]

Six sidechains in close proximity to the OEC are assigned as ligands; D1-Asp342 for Mn1, D1-Glu189 and D1-His332 for Mn2, CP43-Glu354 for Mn3, D1-Asp170 and D1-Glu333 for Mn4 (Figures 9 and 10). The assignment of these protein ligands is mostly consistent with mutational studies recently reviewed by Debus[71] and Diner.[72] Additionally the carboxyl group of the C-terminal D1-Ala344 is located close to Ca^{2+}. Various studies have indicated the C-terminal carboxy group is a Mn ligand.[71,73,74] In our model, Ala344 is not a Mn ligand, but possibly a ligand of Ca^{2+} at some stage during the S-state cycle. Mn2 has two protein ligands D1-Glu189 and D1-His332. The identification of CP43-Glu354 as a ligand was unexpected although a mutational study of the residue did hint that it might be involved in PSII activity.[75] It has been reported that mutants of D1-Glu189 do not affect the electron transfer to the OEC.[76] The result suggests that the coordination of D1-Glu189 to the OEC may not be essential.

Between Mn4 and Ca^{2+}, a non-protein density indicative of an anionic ligand is observed (Figure 10). According to the coordination geometry of Ca^{2+} and Mn and a simulated-annealing omit map, we tentatively assigned a bicarbonate (or carbonate) molecule to this density, where it seems to be acting as a tridentate bridge between Mn4 and Ca^{2+}; bicarbonate has been suggested to play a role in the assembly and functioning of the OEC.[77,78] It is very likely that this is the water oxidation site; in the active state, this non-protein ligand is likely replaced by water molecules, X_1 for Mn_4 and X_{21} and X_{22} for Ca^{2+} (Figure 9) although one of the Ca^{2+} ligands could be a Cl^-, which is a cofactor for the OEC.[79]

4 Mechanism of Water Oxidation

Various schemes for water oxidation chemistry have been proposed over the years.[80,81] Among them, the new OEC structure is compatible with those, which suggest that one of the substrate water molecules is bound to one Mn ion, while the other is bound within the coordination sphere of Ca.[2+ 82–85] The organisation of the Mn ions in our OEC model shows that Mn4 is possibly the site for one of the substrate water molecules. It has been suggested that this Mn ion is oxidized to Mn(V) such that the ligated oxo group, remaining after deprotonation of the water molecule becomes highly electrophilic.[83,85] This oxo group is a candidate for nucleophilic attack by the oxygen of the second substrate water molecule bound to Ca^{2+}. The deprotonation of the substrate water molecules and the stabilization of intermediates could be aided by other ions including Cl^- and the hydrogen-bonding networks provided by protein side chains.[83] Tyr_Z (D1-Tyr161), D1-Glu189 (both on the X_{21} side), CP43-Arg357 and D1-Gln165 (both on the X_{22} side) are closely associated with the

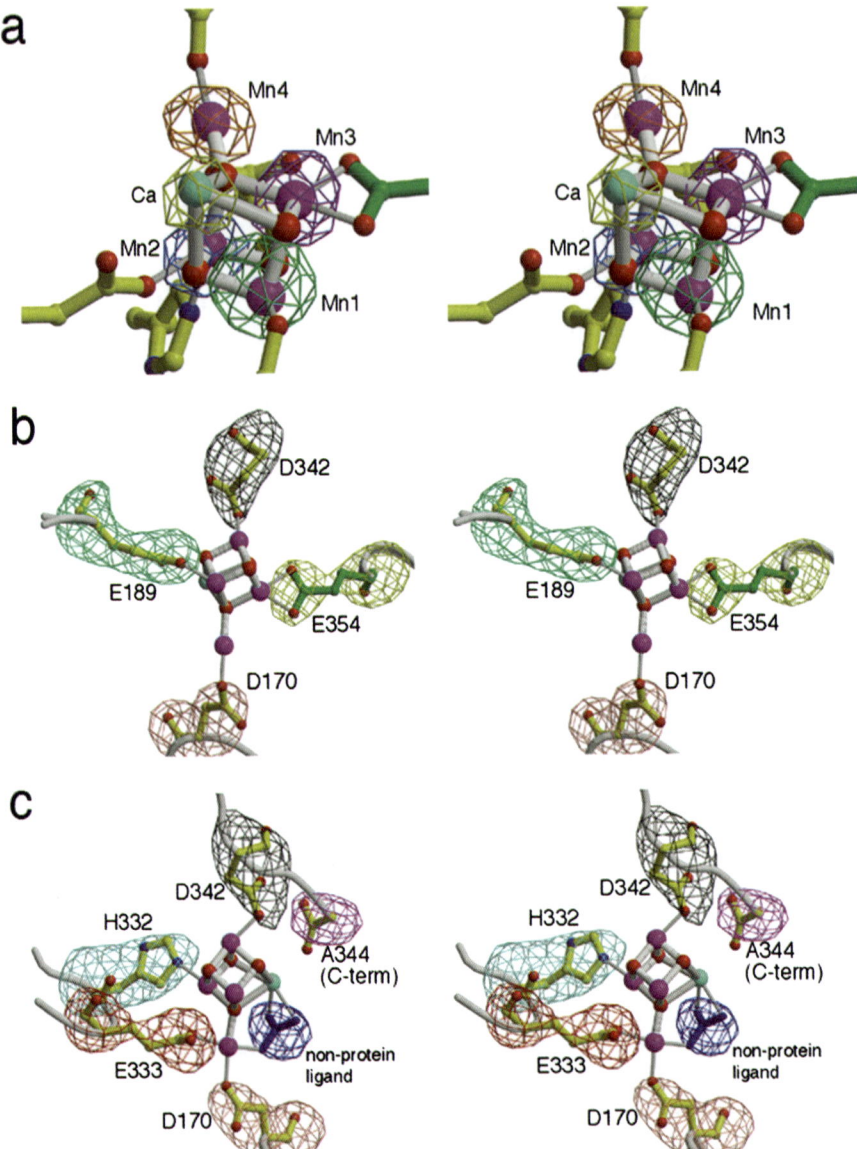

Figure 10 *Stereoviews of electron-density maps for the OEC metals and ligands. For the omit maps, simulated annealing using the 'slow cooling' procedure (from 1000 K) was performed prior to map calculation using the program CNS.[91] (a) Maps calculated after the serial omission of each metal ion of the OEC. The map calculated after the omission of Mn1 is shown in green, after the omission of Mn2 is shown in blue, after the omission of Mn3 is shown in purple, after the omission of Mn4 is shown in orange and after the omission of Ca²⁺ is shown in yellow. Maps are contoured at 8 σ for Mn1, Mn2, Mn3, and 7 σ for Mn4 and Ca²⁺. (b) Maps calculated after the omission of each OEC ligand. The maps calculated after the omission of: Asp170 from D1 are shown in pink; Glu189 from D1 are shown in green; Asp342*

non-protein ligand-binding site thus possibly forming a hydrogen bond to substrate water molecules during the reaction cycle.

The nucleophilic attack of the oxygen of the second water substrate molecule on the highly electrophilic oxo to facilitate above O=O bond formation is a relatively simple concept and is essentially the reverse of the first step in the reduction of oxygen to water in the binuclear centre of cytochrome *c* oxidase (for example,[86]). The first step of the oxygen reduction reaction in cytochrome oxidase involves the breaking of the O=O bond to form an oxo ligand to the Fe and an OH ligand to the Cu of the binuclear centre. For this reaction to occur both Fe and Cu must be in their most reduced states; Fe^{2+} and Cu^+ and the O=O bond breakage is driven by their oxidation. This, therefore, creates a comparable species to that of the final step proposed for O=O bond formation in PSII. Further comparative study seems to be extremely important to understand the oxygen chemistry of both enzymes.

5   Conclusion

The recent X-ray structures of PSII have provided us with a substantial knowledge base to elucidate the many facets of PSII structure and function including detailed subunit structures and the water-splitting reaction. As we have shown, fine-tuning of the crystallization process was essential for this achievement. However, it is very important to provide even higher resolution structures for further kinetic crystallography to show different intermediate states. Although the OEC cluster has a very high redox potential and thus can not be extremely radiation sensitive compared to other redox centres, the oxygen intermediates can be easily reduced by X-rays, thus addressing this problem will be the key to successful experiments (for example, see [87]). It might take a long time to see any intermediate structures by X-ray crystallography but, certainly, we are now at the last stage of the challenge to reveal the molecular events of a reaction on which virtually all life depends and which makes our planet very special.

Abbreviations

bRC: Bacterial photosynthetic reaction centre
BChl: Bacteriochlorophyll
CP43: 43kDa chlorophyll protein

---

from D1 are in dark grey; Glu354 from CP43 are shown in yellow. His332 from D1, Glu333 from D1, Ala344 from D1 and a putative non-protein ligand are in the front and back of this view and are thus omitted from the panel for clarity. All maps are contoured at 5 σ. (c) Maps calculated after the omission of cluster ligands. The view is 90° away from (b) such that the remainder of the ligands can be easily observed. Glu189 from D1 and Glu354 from CP43 are in the front and back of this view, so are omitted from the panel for clarity. The maps calculated after the omission of: Asp170 from D1 are shown in pink; His332 from D1 are shown in cyan; Glu333 from D1 are shown in red; Asp342 from D1 are shown in dark grey; Ala344 (C-terminus) from D1 are shown in magenta; and the putative non-protein ligand (modelled as bicarbonate) is shown in blue

CP47: 47kDa chlorophyll protein
CCP4: Collaborative Computational Project, Number 4
CCD: Charge-coupled device
$C_{12}E_8$: Octaethyleneglycol dodecylether
Chl: Chlorophyll
DDM: *n*-Dodecyl-β-D-maltoside
EM: Electron microscopy
OEC: Oxygen-evolving centre
P680: Primary-electron-donor chlorophyll
Pheo: Pheophytin
PSI: Photosystem one
PSII: Photosystem two
Q_A: Quinone A
Q_B: Quinone B
RC: Reaction centre
T. elongatus: Thermosynechococcus elongatus
Tyr_Y: D2-Tyr160
Tyr_Z: D1-Tyr161

Acknowledgements

This manuscript is based on collaborations with Prof James Barber, Imperial College London. Dr Tina M. Iverson contributed to the structural determination and refinement of the PSII structure. SI acknowledges support from the Biotechnology and Biological Research Science Council and Japan Science and Technology Corporation. We thank the Centre for Structural Biology and Bioinformatics Facility at Imperial College, London for technical support and Drs Clemens Schulze-Briese and Takashi Tomizaki at PX06SA/SLS, Paul Scherrer Institute, Villigen, Switzerland, Dr Bill Shepard at ID29 and Momi Iwata of the ERATO ATPase project for help with data collection for solving the structure of PSII, which is the basis of this review.

References

1. S. Iwata and J. Barber, *Curr. Opin. Struct. Biol.*, 2004, **14**, 447.
2. J. Barber, *Curr. Opin. Struct. Biol.*, 2002, **12**, 523.
3. K.H. Rhee, E.P. Morris, J. Barber and W. Kuhlbrandt, *Nature*, 1998, **396**, 283.
4. B. Hankamer, E.P. Morris and J. Barber, *Nature Struct. Biol.*, 1999, **6**, 560.
5. H. Kuhl, M. Rogner, J.F. van Breemen and E.J. Boekema, *Eur. J. Biochem.*, 1999, **266**, 453.
6. J. Nield, O. Kruse, J. Ruprecht, P. da Fonseca, C. Buchel and J. Barber, *J. Biol. Chem.*, 2000, **275**, 27940.
7. A. Zouni, H.T. Witt, J. Kern, P. Fromme, N. Krauss, W. Saenger and P. Orth, *Nature*, 2001, **409**, 739.

8. N. Kamiya and J.R. Shen, *Proc. Natl. Acad. Sci., U.S.A.*, 2002, **100**, 98.
9. K.N. Ferreira, T.N. Iverson, K. Maghlaoui, J. Barber and S. Iwata, *Science*, 2004, **303**, 1831.
10. R.W. Castenholz, *Bacteriol. Rev.*, 1969, **33**, 416.
11. H. Kuhl, J. Kruip, A. Seidler, A. Krieger-Liszkay, M. Bunker, D. Bald, A.J. Scheidig and M. Rogner, *J. Biol. Chem.*, 2000, **275**, 20652.
12. *"Methods and Results in Membrane Protein Crystallization"*, S. Iwata (ed.), International University Line, La Jolla, California, 2003.
13. R. Horsefield, V. Yankovskaya, S. Törnroth, C. Luna-Chaves, E. Stambouli, J. Baber, B. Byrne, G. Cecchini and S. Iwata, *Acta Cryst.*, 2003, **59**, 600.
14. Z. Otwinowski and W. Minor, W. *Methods Enzymol.*, 1997, **276**, 307.
15. Collaborative Computational Project, Number 4, *Acta Cryst.*, 1994, **D50**, 760.
16. A. Trebst, *Zeitschrift Fur Naturforschung*, 1986, **41,** 240.
17. J. Deisenhofer, O. Epp, K. Miki, R. Huber and H. Michel, *Nature*, 1985, **318**, 618.
18. J.P. Allen, G. Feher, T.O. Yeates, D.C. Rees, J. Deisenhofer, H. Michel and R. Huber, *Proc. Natl. Acad. Sci. U.S.A.*, 1988, **85**, 8487.
19. B. Kok, B. Forbush and M. McGloin, *Photochem. Photobiol.*, 1970, **11**, 457.
20. F. Rappaport, M. Guergova-Kuras, P.J. Nixon, B.A. Diner and J. Lavergne, *Biochemistry*, 2002, **41**, 8518.
21. L.M.C. Barter, J.R. Durrant and D.R. Klug, *Proc. Natl. Acad. Sci. U.S.A.*, 2003, **100**, 946.
22. J.R. Durrant, D.R. Klug, S.L. Kwa, R. van Grondelle, G. Porter and J.P. Dekker, *Proc. Natl. Acad. Sci. U.S.A.*, 1995, **92**, 4798.
23. J.P. Dekker and R. van Grondelle, *Photosyn. Res.*, 2000, **63**, 195.
24. J. Barber and M.D. Archer, *Photochem. Photobiol. A-Chem.*, 2001, **142**, 97.
25. B.A. Diner, E. Schlodder, P.J. Nixon, W.J. Coleman, F. Rappaport, J. Lavergne, W.F. Vermaas and D.A. Chisholm, *Biochemistry*, 2001, **40**, 9265.
26. W. Oettmeier, In *The Photosystems: Structure, Function and Molecular Biology*. J. Barber, (ed.) Elsevier, Amsterdam, The Netherlands, 1992, **11**, 349.
27. R. Hienerwadel, C. Berthomieu, *Biochemistry*, 1995, **34**, 16288.
28. Govindjee, J.J.S. van Rensen, In *The Photosynthetic Reaction Center*, Vol 1, Academic Press, San Diego, CA, 1993, 357.
29. T.M. Bricker, *Photosynthesis Research*, 1990, **24**, 1.
30. P. Jordan, P. Fromme, H.T. Witt, O. Klukas, W. Saenger and N. Krauss, *Nature*, 2001, **411**, 909.
31. C. Rosenberg, J. Christian, T.M. Bricker and C. Putnam-Evans, *Biochemistry*, 1999, **38**, 15994.
32. F.L. de Weerd, I.H. van Stokkum, H. van Amerongen, J.P. Dekker and R. van Grondelle, *Biophys. J.*, 2002, **82**, 1586.
33. C.A. Tracewell, J.S. Vrettos, J.A. Bautista, H.A. Frank and G.W. Brudvig, *Arch. Biochem. Biophys.*, 2001, **385**, 61.
34. A. Telfer, J. De Las Rivas and J. Barber, *Biochim. Biophys. Acta*, 1991, **1060**, 106.
35. J. Hanley, Y. Deligiannakis, A. Pascal, P. Faller and A.W. Rutherford, *Biochemistry*, 1999, **38**, 8189.
36. P. Faller, A. Pascal and A.W. Rutherford, *Biochemistry*, 2001, **40**, 6431.

37. C.A. Tracewell, A. Cua, D.H. Stewart, D.F. Bocian and G.W. Brudvig, *Biochemistry*, 2001, **40**, 193.
38. A. Telfer, S. Dhami, S.M. Bishop, D. Phillips and J. Barber, *Biochemistry*, 1994, **33**, 14469.
39. F.L. de Weerd, I.H. van Stokkum, H. van Amerongen, J.P. Dekker and R. van Grondelle, *Biophys. J.*, 2002, **82**, 1586.
40. T.S. Mor, I. Ohad, J. Hirschberg and H.B. Pakrasi, *Mol. Gen. Genet.*, 1995, **246**, 600.
41. W.R. Widger, W.A. Cramer, M. Hermodson and R. G. Herrmann, *FEBS Lett.*, 1985, **191**, 186.
42. G.T. Babcock, W.R. Widger, W.A. Cramer, W.A. Oertling and J.G. Metz, *Biochemistry*, 1985, **24**, 3638.
43. D. Zheleva, J. Sharma, M. Panico, H.R. Morris and J. Barber, *J. Biol. Chem.*, 1998, **273**, 16122.
44. A.N. Webber, L.C. Packman, D.J. Chapman, J. Barber and J.C. Gray, *FEBS Lett.*, 1989, **242**, 259.
45. T. Tomo, I. Enami and K. Satoh, *FEBS Lett.*, 1993, **323**, 15.
46. C. Buchel, E. Morris, E. Orlova and J. Barber, *J. Mol. Biol.*, 2001, **312**, 371.
47. L.X. Shi, S.J. Kim, A. Marchant, C. Robinson and W.P. Schroder, *Plant Mol. Biol.*, 1999, **40**, 737.
48. M. Swiatek, R. Kuras, A. Sokolenko, D. Higgs, J. Olive, G. Cinque, B. Muller, L.A. Eichacker, D.B. Stern, R. Bassi, R.G. Herrmann and F.A. Wollman, *Plant Cell*, 2001, **13**, 1347.
49. T.M. Bricker and L.K. Frankel, *Photosynthesis Research*, 1998, **56**, 157.
50. H.A. Chu, A.P. Nguyen and R.J. Debus, *Biochemistry*, 1999, **33**, 6150.
51. J. Komenda, and J. Barber, *Biochemistry*, 1995, **34**, 9625.
52. T.A. Ono and Y. Inoue, *FEBS Lett.*, 1984, **166**, 381.
53. S.R. Mayers, K.M. Cook, S.J. Self, Z.H. Zhang and J. Barber, *Biochim. Biophys. Acta*, 1991, **1060**, 1.
54. R.L. Burnap and L.A. Sherman, *Biochemistry*, 1991, **30**, 440.
55. J.B. Philbrick, B.A. Diner and B.A. Zilinskas, *J. Biol.Chem.*, 1991, **266**, 13370.
56. S. Tanaka, Y. Kawata, K. Wada and K. Hamaguchi, *Biochemistry*, 1989, **28**, 7188.
57. R.L. Burnap, M. Qian, J.R. Shen, Y. Inoue and L.A. Sherman, *Biochemistry*, 1994, **33**, 13712.
58. S.D. Betts, J.R. Ross, K.U. Hall, E. Pichersky and C.F. Yocum, *Biochim. Biophys. Acta*, 1996, **1274**, 135.
59. H. Popelkova, M.M. Im, J. D'Auria, S.D. Betts, N. Lydakis-Simantiris and C.F. Yocum, *Biochemistry*, 2002, **41**, 2702.
60. H. Popelkova, M.M. Im and C.F. Yocum, *Biochemistry*, 2003, **42**, 6193.
61. S.D. Betts, N. Lydakis-Simantiris, J.R. Ross and C.F. Yocum, *Biochemistry*, 1998, **37**, 14230.
62. C. Frazao, F.J. Enguita, R. Coelho, G.M. Sheldrick, J.A. Navarro, M. Hervas, M.A. De la Rosa and M.A. Carrondo, *J. Biol. Inorg. Chem.*, 2001, **6**, 324.

63. M.R. Sawaya, D.W. Krogmann, A. Serag, K.K. Ho, T.O. Yeates and C.A. Kerfeld, *Biochemistry*, 2001, **40**, 9215.

64. C.A. Kerfeld, M.R. Sawaya, H. Bottin, K.T. Tran, M. Sugiura, D. Cascio, A. Desbois, T.O. Yeates, D. Kirilovsky and A. Boussac, *Plant and Cell Physiology*, 2003, **44**, 697.

65. B.A. Diner, G.T. Babcock, In *Advances in Photosynthesis: The Light Reactions*. D.R. Ort and C.F. Yocum, (eds.) Kluwer Academic Publ., Dordrecht, The Netherlands, 1996, **4**, 213.

66. J.M. Pelloquin and R.D. Britt, *Biochim. Biophys. Acta*, 2001, **1503**, 96.

67. J.H. Robblee, J. Messinger, R.M. Cinco, K.L. McFarlane, C. Fernandez, S.A. Pizarro, K. Sauer and V.K. Yachandra, *J. Am. Chem. Soc.*, 2002, **124**, 7459.

68. R.M. Cinco, K.L. McFarlane-Holman, J.H. Robblee, J. Yano, S.A. Pizarro, E. Bellacchio, K. Sauer and V.K. Yachandra, *Biochemistry*, 2002, **43**, 12928.

69. V.J. DeRose, I. Mukerji, M.J. Latimer, V.K. Yachandra, K. Sauer and M.P. Klein, *J. Am. Chem. Soc.*, 1994, **116**, 5239.

70. V.K. Yachandra, *Phil. Trans. R. Soc. Lond.*, 2002, **B357**, 1347.

71. R.J. Debus, *Biochim. Biophys. Acta*, 2001, **1503**, 164.

72. B.A. Diner, *Biochim. Biophys. Acta*, 2001, **1503**, 147.

73. B.A. Diner, P.J. Nixon and J.W. Farchaus, *Current Opinion Struct. Biol.*, 1991, **1**, 546.

74. H.A. Chu, W. Hiller and R.J. Debus, *Biochemistry*, 2004, **43**, 3152.

75. C. Rosenberg, J. Christian, T.M. Bricker and C. Putnam-Evans, *Biochemistry*, 1999, **38**, 15994.

76. J. Clausen, S. Winkler, A.M. Hays, M. Hundelt, R.J. Debus and W. Junge, *Biochim. Biophys. Acta*, 2001, **1506**, 224.

77. S.V. Buranov, G.M. Ananyev, V.V. Klimov and G.C. Dismukes, *Biochemistry*, 2000, **39**, 6060.

78. V.V. Klimov and S.V. Baranov, *Biochim. Biophys. Acta*, 2001, **1503**, 187.

79. R.J. Debus, *Biochim. Biophys. Acta*, 1992, **1102**, 269.

80. C. Goussias, A. Boussac and A.W. Rutherford, *Phil. Trans. Roy. Soc. Lond.*, 2002, **B357**, 1369.

81. R.D. Britt, K.A. Campbell, J.M. Peloquin, M.L. Gilchrist, C.P. Aznar, M.M. Dicus, J. Robblee and J. Messinger, *Biochim. Biophys. Acta*, 2004, **1655**, 158.

82. J. Messinger, M. Badger and T. Wydrzynski, *Proc. Natl. Acad. Sci. U.S.A.*, 1995, **92**, 3209.

83. J.S. Vrettos, J. Limburg and G.W. Brudvig, *Biochim. Biophys. Acta*, 2001, **1503**, 246.

84. P.E.M. Siegbahn, *Current Opinions Chem. Biol.*, 2002, **6**, 227.

85. V.L. Pecoraro, M.J. Baldwin, M.T. Caudle, W.Y. Hsieh and N.A. Law, *Pure and Appl. Chem.*, 998, **70**, 925.

86. M. Svensson-Ek, J. Abramson, G. Larsson, S. Tornroth, P. Brzezinski and S. Iwata, *J. Mol. Biol.,* 2002, **321**, 329.

87. G.I. Berglund, G.H. Carlsson, A.T. Smith, H. Szoke, A. Henriksen and J. Hajdu, *Nature*, 2002, **417**, 463.

88. P. J. Kraulis, *J. Appl. Cryst.*, 1991, **24**, 946.
89. R. Esnouf, *J. Mol. Graph.*, 1997, **15**, 133.
90. E.A. Merritt and M.E.P. Murphy, *Acta Crystallogr.*, 1994, **D50**, 869.
91. A.T. Brunger, P.D. Adams, G.M. Clore, W.L. DeLano, P. Gros, R.W. Grosse-Kunstleve, J.S. Jiang, J. Kuszewski, M. Nilges, N.S. Pannu, R.J. Read, L.M. Rice, T. Simonson and G.L. Warren, *Acta Cryst.*, 1998, **D54**, 905.

CHAPTER 10

A Structural View of Proton Transport by Bacteriorhodopsin

JANOS K. LANYI

Department of Physiology & Biophysics,
University of California,
Irvine, CA 92697, USA

1 Introduction

Bacteriorhodopsin,[1–4] and the more recently discovered growing number of other retinal-based light-driven proton pumps,[5–7] are transporters structurally closely related to G-protein coupled receptors. They are small (*ca.* 24 kDa) heptahelical transmembrane proteins that contain a buried retinal much like in visual rhodopsin but in the all-*trans* isomeric state. Photoisomerization to 13-*cis*,15-*anti* sets off a sequence of spectroscopically, crystallographically, and electrically detectable reactions[8] with a turn-over of about 100 s^{-1}, in which protons are transferred between the retinal Schiff base and protein residues and between the protein and the bulk, protein conformational changes of increasing magnitude occur, and in the end the retinal regains its original isomeric state. The net result of this reaction cycle is the active translocation of a proton across the membrane, against an electrochemical potential well over what is required for physiological purposes like ATP synthesis. The events in this 'photocycle' and the structural changes that accompany them have been studied by many groups and in great detail, with the hope that they will reveal how the proton transport is accomplished. Recent progress in this effort is impressive. FTIR spectra are providing more and more specific mechanistic insights.[9–13] As described below, X-ray diffraction is yielding extraordinarily high-resolution maps for the protein and many of the intermediate states of the photocycle, showing the location of water molecules and the movements of the retinal and residue side-chains.[3,14,15] Solid-state NMR is producing accurate inter-atomic distances and bond angles at locations of functional importance, and their changes.[16,17] The transport mechanism that is emerging from these studies describes not only the donors and acceptors along the pathway of the proton trajectory inside the protein and at the two membrane surfaces, but also the means by which the proton transfers are coupled to the transformations of the retinal. Today, proton transport in bacteriorhodopsin is understood at a level unimaginable only a few years ago. It has implications not only for ion translocation in other pumps but also for the way protein conformational changes arise when receptors bind their agonists.[18]

The author's version of the transport mechanism will be told below, but this is a contentious field. Other, differing, views will not be discussed here, but the reader may wish to consult several recent publications where the controversies have been aired.[19–23]

2   Structure of Bacteriorhodopsin

Determined first with cryo-electron microscopy,[24–26] the heptahelical structure of bacteriorhodopsin is a well-known motif because it has since become the prototype for the very large family of G-protein coupled receptors. The seven transmembrane helices A through G, inclined at small angles to the membrane normal, form a bundle that surrounds the transversely lying all-*trans* retinal in a protonated Schiff base linkage to lys-216 at about the middle of helix G. The active site is the Schiff base, with a hydrogen bond to water 402 that donates hydrogenbonds to the anionic asp-85 and asp-212.[27–29] It divides the structure into two halves. The extracellular half contains the two aspartate residues and a 3-dimensional hydrogen-bonded network of numerous polar and charged protein groups and seven water molecules. This network connects the active site to the membrane surface, and its rearrangements during the photocycle cause the release of a proton to the bulk when asp-85 becomes protonated by the Schiff base. The cytoplasmic half, in contrast, contains few polar groups, only three water molecules, and no network. It contains asp-96 that will be the proton donor to the Schiff base as the initial state begins to recover in the second part of the photocycle.

Thus, the structure contains strong clues to how the pump operates.[30] The extracellular half contains a largely pre-existing proton-transfer pathway between the retinal Schiff base and the membrane surface. The cytoplasmic half is, in contrast, a hydrophobic barrier that prevents the leak of protons across the resting pump. Proton transfer through this region, between the Schiff base and asp-96 and between the membrane surface and asp-96, must be at the expense of conformational changes that create a hydrogen-bonded chain. The only way this could be visualized is by entry of water to form a transient network when it is needed.

3   Photochemical Reaction Cycle

The transformations of the retinal and the protein after absorption of a photon are described by at least nine spectroscopically detectable states. Although the nomenclature varies somewhat, many agree that it is a linear sequence of intermediates, *e.g.* $BR - h\nu \rightarrow K \leftrightarrow L \leftrightarrow M_1 \leftrightarrow M_2 \rightarrow M_2' \leftrightarrow N \leftrightarrow N' \leftrightarrow O \rightarrow BR$.[31–33] The rise and decay of these states have more complex time-courses than a sequential scheme would predict, and a fully satisfactory kinetic model is yet to be found. Some of the complications are accounted for by the reversible reactions, as well as the pH dependent protonation states of crucial groups.[34, 35] In spite of the unresolved issues, the vast efforts made by many groups to understand the details of the photocycle have been essential and rewarding. Static (low-temperature) and time-resolved spectroscopy including FTIR and Raman, solid-state NMR, EPR, together with site-specific mutagenesis, have identified the main molecular properties of the different states. Even in the absence of high-resolution crystal structures of the intermediates, these studies have established the underlying events of the photochemical cycle and the path of the transported proton, as follows.

After photoisomerization of the retinal from all-*trans* to 13-*cis*,15-*anti* in K and relaxation of its twist in L, the Schiff base proton is transferred to asp-85 on the extracellular side in the L → M_1 reaction. This is followed by the M_1 → M_2 transition where access to the Schiff base is thought to change from the extracellular to the cytoplasmic side. A proton is then released to the bulk in M_2', from a site that appears to be an aqueous network stabilized by arg-82, glu-194, glu-204, and a few other residues. Reprotonation of the Schiff base in N is from asp-96 located on the cytoplasmic side, aided by tilts of the cytoplasmic ends of helices F and G and, as thought by many, the resulting entry of water into this region. Reprotonation of the Schiff base through the proposed chain of water molecules is followed by reprotonation of asp-96 from the cytoplasmic surface in N′ and reisomerization of the retinal to all-*trans* in the O state. Finally, transfer of a proton from asp-85 to the vacant proton release site completes the cycle.

Although this is a more detailed model than available for other ion pumps, until very recently many urgent questions remained unanswered. How is energy stored in the retinal and how is it transferred to the protein? What is the molecular basis for the 'switch' (or 'gate') that gives the pump its directionality? What is the barrier to retinal reisomerization directly from the L state, *i.e.* to a shunt that would by-pass the transport events in the cycle? What is the cause of the changes in the pK_as of donor and acceptor groups? If a hydrogen-bonded chain of water develops to facilitate reprotonation of the Schiff base, as postulated, where are these water molecules and how does the chain form and then dissipate? As discussed below, the answers to many of these questions are now at hand.

4   Evaluating the Crystallography

It is well known that the crystallographic description of a membrane protein and the sequence of the various states during its functional cycle must overcome formidable problems. Growing high-quality crystals of integral membrane proteins has proved to be very difficult, and for bacteriorhodopsin it became possible only with the cubic lipid phase method.[36,37] The best of the thin and fragile hexagonal plates produced diffract to 1.4 Å,[38] and even this was barely sufficient to detect some of the subtle structural changes of interest. Because in the first few intermediates the main questions concern the twists and bond rotations of the retinal chain that do not greatly alter the electron density map,[39] and the intermediate states can be trapped (by illumination at an appropriate temperature) with no more than partial occupancy,[22] extraordinary care must be exercised to ascertain that the observed changes are real and reproducible.

Indeed, there are serious disagreements about the crystal structures of the K, L and M states as determined by various groups.[21,33,38,40–45] Some of the structures for these states predicted from theoretical considerations are different still.[46] How to decide whether a small structural change in the map is reproducible and meaningful? A conservative approach would be to determine the most important inter-atomic distances, bond angles, and torsion angles, by an entirely different method, such as solid-state NMR, and compare the results. To an increasing degree, such comparisons are becoming possible (see below). Until all the relevant details are confirmed by at least two methods, however, we have to rely on the crystallography. There exist tough criteria for passing judgment over the crystallographic structures, as follows.

The most stringent demand on the crystallography is to insist that the electron-density maps show a recognizable isomerized retinal in those states where vibrational spectroscopy indicates that the retinal is 13-*cis*. If this criterion is satisfied, changes detected elsewhere along the retinal chain and in the protein (or lack of changes) should gain credibility. For example, Figure 1 shows a comparison of a bias-free extrapolated map for the L intermediate (see legend) with the map for the BR state, from diffraction data with matching resolutions (1.6 Å in this case). Both maps were calculated directly from the observed diffraction intensities before refinement of a model (F_{obs} maps), and the phases were from the BR coordinates that contain all-*trans* retinal. Figure 1, panels a and c compare the region of the retinal and its immediate surroundings with asp-85 and asp-212 in BR and the L state obtained in this way. The small differences in the positions of side-chains and bound water between the two maps make the differences between the retinal at the $C_{13}=C_{14}$ bond stand out (arrows). This region is shown enlarged in panels *b* and *d*, with the refined models digitally removed so as not to lead the eye. The changes in the map that reveal the expected double-bond rotation are clearly evident, in spite of the lack of any manipulation that would introduce a bias for the 13-*cis* isomeric state. Another test with the same intent was performed for the K state elsewhere,[38] for maps calculated before refinement and without any phase contribution from the retinal (omit F_{obs} maps) at 1.4 Å resolution. Correlation coefficients for grid points between these maps for an illuminated and a non-illuminated crystal revealed differences caused by the illumination, in the form of decreased correlation coefficients specifically at the $C_{13}=C_{14}$ bond, as expected.[38] The same comparison between two illuminated and two non-illuminated crystals, as controls, showed virtually no differences.

If the crystallographic data do contain the expected changes at the retinal, as it appears, refinement of a model for the intermediate state present is justified. The refinement should be done with extraordinary care to avoid any input that would influence the resulting configuration of the retinal. If the photoisomerized retinal is twisted and distorted in some of the intermediates, as indicated by the large-amplitude hydrogen-out-of-plane (HOOP) bands in the K and L states (see below), the model will be free of bias only if the usual restraints on bond angles and out-of-plane atom displacements are loosened or removed.[23] If rotation around the $C_{13}=C_{14}$ double bond and the flatness of the retinal chain are both enforced, the retinal geometry will be decided more by the restraints than by the data. On the other hand, in the absence of restraints on retinal atoms a greater than usual degree of noise will be present in the model.[33,38] This must be accepted as the price of avoiding bias, and the

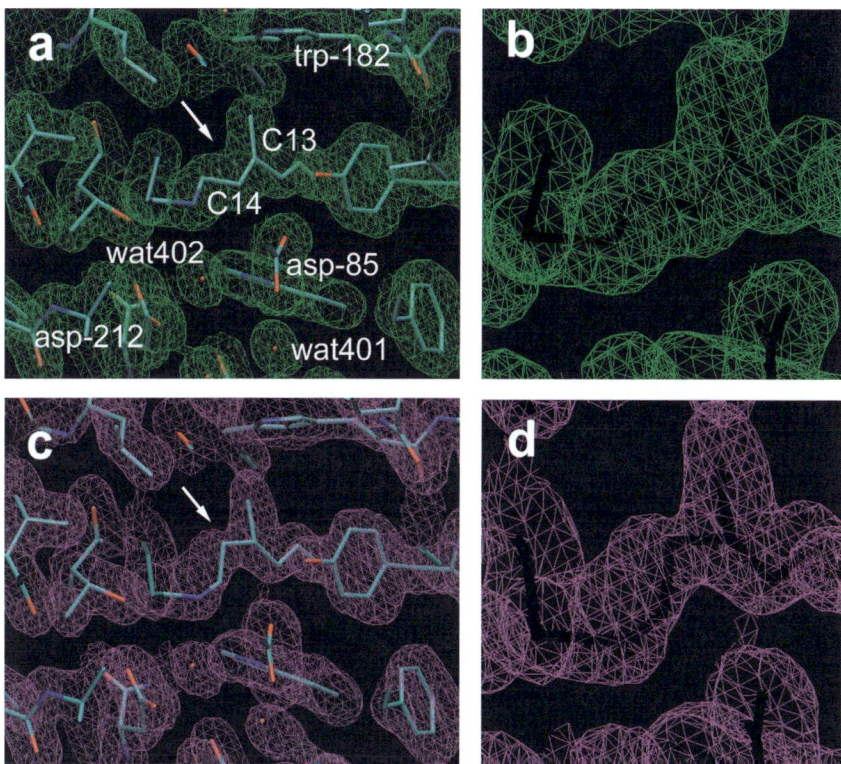

Figure 1 *Comparison of the electron density map for the BR state with a bias-free extrapolated map for the L state at 1.6 Å resolution. The intent is to test if the diffraction data are sufficient to show, without any refinement, that the retinal is isomerized from all-*trans *to 13-*cis *in the L intermediate, as explained in the text. Shown are* F_{obs} *maps calculated directly from the measured reflection intensities and the coordinates of the BR state (with all-*trans *retinal). The extrapolated map for L, accumulated in the crystal to 60% occupancy, was calculated by subtracting from the map coordinates for the illuminated crystal 0.4 × the map coordinates of the non-illuminated crystal, and dividing the result by 0.6. Panels a and c, region of the retinal in BR and L, respectively, with some residues and water labeled for orientation. The models in the densities are from the refined coordinates of the two states. Panels b and d, enlargement of the region of the* $C_{13} = C_{14}$ *bond (pointed out by arrows in a and c), with the models digitally removed*

resulting noise means that multiple determinations of crystal structure with the statistics for the reproducibility they yield assume greater significance in this kind of work than usual in protein crystallography.

The quality of the crystals is important because the information content of diffraction data is inversely proportional to the cube of the crystallographic resolution. The structure of a protein the size of bacteriorhodopsin, together with the lipid annulus, is defined, mathematically, by approximately 8,000 structure factors. In the P6$_3$ unit cell this number of reflections will be produced at about 2.4 Å resolution. Bacteriorhodopsin crystals grown in cubic phase tend to be twinned,[27,47] which

lowers the information content of the diffraction data somewhat but does not preclude bias-free refinement if a twinned model is refined against the twinned data
without attempts at prior de-twinning.[48,49] A high degree of twinning means that a
somewhat better resolution is required, but in either case about 2 Å may be regarded
as the minimum acceptable resolution because over-determination of the structure,
which increases steeply at better resolutions, is essential for detecting structural
details. For finding changes smaller than the large swings of side-chains often seen
in other proteins, a much better resolution is needed. In a test, when the data for the
K state was truncated from the collected 1.4 Å resolution to 2 Å, it no longer contained detectable changes in the retinal.[38]

If the expected structural changes are subtle, the possibility of artifacts from X-ray
radiation damage to the protein gains significance. Such damage to bacteriorhodopsin
crystals at the third-generation synchrotron source SPring-8 was described in a series
of three papers.[42,43,45] Degradation of the electron density map and spectral shifts of
the chromophore were reported when the total exposure during the diffraction measurement was greater than a 'safe' level of 3×10^{14} X-ray photons/mm^{-2}. This is
equivalent to about 1.2×10^{13} photons received by the crystal tested in that report.[43]
At other synchrotrons, however, the crystals are exposed to much less X-ray flux. For
example, at beamline 9.1 of the Stanford synchrotron (SSRL), used by the author, a
typical experiment records 150 oscillation frames with 20–30 second duration each,
which add up to a total exposure of only $4-6 \times 10^{12}$ photons.

5   Trapping the Right Structure and the Right Intermediate State

Being inserted into the lipid bilayer in their native state, integral membrane proteins
experience an environment usually not duplicated in protein crystals. In most cases,
the lipid bilayer is replaced by single detergent molecules that cover the hydrophobic
surfaces that would contact the hydrophobic core of the lipid chains. It is easy to
imagine how the tertiary structure (*e.g.* helix-helix interactions, helical tilts or bends,
and their consequences for the active sites), may not be the quite the same in the crystals as in the membranes. For an overall assessment of the structure such discrepan-
cies might not matter much, but they would be seriously detrimental to conclusions
drawn from structural changes about reaction mechanisms. It is fortunate, therefore,
that the *a–b* plane of the P6$_3$ unit cell of bacteriorhodopsin is essentially identical to
the two-dimensional P$_3$ lattice of the *in vivo*-produced purple membrane (lattice constant 60–61 Å). Further, maps reveal the original branched-chain lipid chains at the
periphery of the monomers,[28,50] and at the three-fold axis of the trimers in another
kind of crystal,[51] that were not lost during detergent solubilization and crystallization.
In the *a–b* plane, therefore, there will be few forces that would alter the protein in the
crystals. On the other hand, the crystal contacts along the *c*-axis do not exist *in vivo*,
and the conformations of interhelical loops should be evaluated with this in mind.

In order to minimize radiation damage, X-ray and electron diffraction are measured at a temperature no higher than 100 K. The assumption is that the frozen samples will contain the same ensemble of slightly different conformational states that

exists at ambient temperature, even if their interconversions are blocked. Although cooling of the very small crystals by a cold stream of nitrogen is rapid, questions have been raised about the possibility of conformational re-equilibration and dehydration as the sample passes the critical temperature region of about 200 K.[52] This is a potential problem, but if large-scale protein conformational changes can be ruled out because of resistance from the crystal lattice (see below), particularly at low temperature, water molecules are unlikely to move rapidly in or out of the protein.

A different question is whether the photostationary mixtures created at cryogenic temperatures, *i.e.* for K and L, contain the same intermediates as produced transiently at ambient temperature.[39] Direct comparisons between infrared[53] spectra measured under the two conditions reveal striking similarities. The same spectra show differences in the hydrogen-out-of-plane (HOOP) bands, however, with frequency shifts and lower amplitudes when measured at low temperature. Is the twist of the retinal that gives rise to these bands less or different under physiological conditions? Warming the mixture containing L from 170 K to 230 K does give rise to M, as in the ambient temperature photocycle, although with a yield less than expected because at low temperatures an additional decay pathway exists from L to the BR state that by-passes the rest of the cycle.[54,55]

The lattice forces in a three-dimensional crystal will oppose large-scale conformational changes in the photocycle if they change the overall dimensions or shape of the protein. There are indications that this is the case. Attempts to trap the N (but not N') and O states in crystals, under the conditions where they accumulate in large quantities in membrane suspensions (mutations, pH, wavelength of illumination, *etc.*), have been unsuccessful (unpublished results of the author). One would expect that if these states can be populated in lesser amounts, the photocycle kinetics would be affected. Several-fold slowing of steps in the second half of the photocycle in the crystals have been reported,[32] consistent with this. The two-dimensional crystals will be subject to less steric constraints than three-dimensional crystals, and global conformational changes not seen in X-ray structures are indeed measurable in purple membrane sheets by cryo-electron microscopy[56] and with non-crystallographic methods like spin-labeling.[57–61] However, perhaps for the same kinetic reasons as in the crystals, under most conditions N and O do not accumulate in large amounts in the wild-type photocycle in membranes either.

6   Retinal Motions: The Pump

It appears that the events essential for the pump occur at the retinal. Structural models from high-resolution (1.4 to 2 Å) diffraction data are available for the trapped K[38] and L[33] states, as well as three trapped M states that correspond, arguably, to M_1[32, 62], M_2[31] and M_2'.[63] They show that the retinal undergoes progressive motions as it passes through these intermediates after its photoisomerization to 13-*cis*,15-*anti*. The images constitute a revealing atomic level 'movie' of the way this pump works (Figure 2).

The changes in the geometry of the retinal are affected by the protein matrix around it. When in solution, isomerization from all-*trans* to 13-*cis*,15-*anti* results of a distinctive change in the shape of the retinal chain, from linearly extended to bent at C_{13}. The polyene remains flat but the direction of the Schiff base N–H bond is re-oriented

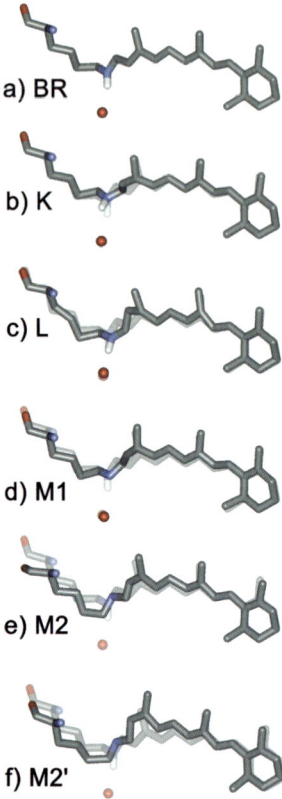

Figure 2 *Sequence of structural changes of the retinal plus the attached lys-216 in the first half of the photocycle, a-f, from BR to M2' as labeled. Water 402 is also shown. The retinal in the BR state is laid under, at a lower opacity, all the other images, to illustrate the changes. Coordinate sets used 1C3W, 1M0K, 1O0A, 1M0M, 1F4Z, and 1C8S, for a through f, respectively*
(Reproduced with permission from reference 33, Elsevier, Inc.)

by 180°. There being no constraints by the solvent, the polyene chain rapidly assumes its lowest energy state. In the K state of bacteriorhodopsin, however, these changes of the retinal are hindered by the binding site and the movements are restricted to the immediate vicinity of the $C_{13}=C_{14}$ bond. Rotation of this double bond is complete, but the retinal is not bent. Retaining the longer end-to-end distance of the all-*trans* state is achieved at the expense of considerable strains of the retinal skeleton. According to the crystal structure, the strain is localized most obviously in an increased bond angle at C_{13}. Because the N–H bond continues to point roughly in the original direction, there are also out-of-plane twists, *i.e.* torsion angles that deviate from 0 or 180°. The distorted bond angle at C_{13} and any twist of the $C_{13}=C_{14}$ and $C_{15}=NZ$ double bonds will store energy. Figure 3 shows that both the C_{13} bond angle and the sum of the two double-bond twists are increased in K over the BR state. These distortions of the retinal are maximal in the K and decrease, step by step, as the photocycle passes through the K, L, and the three M states.

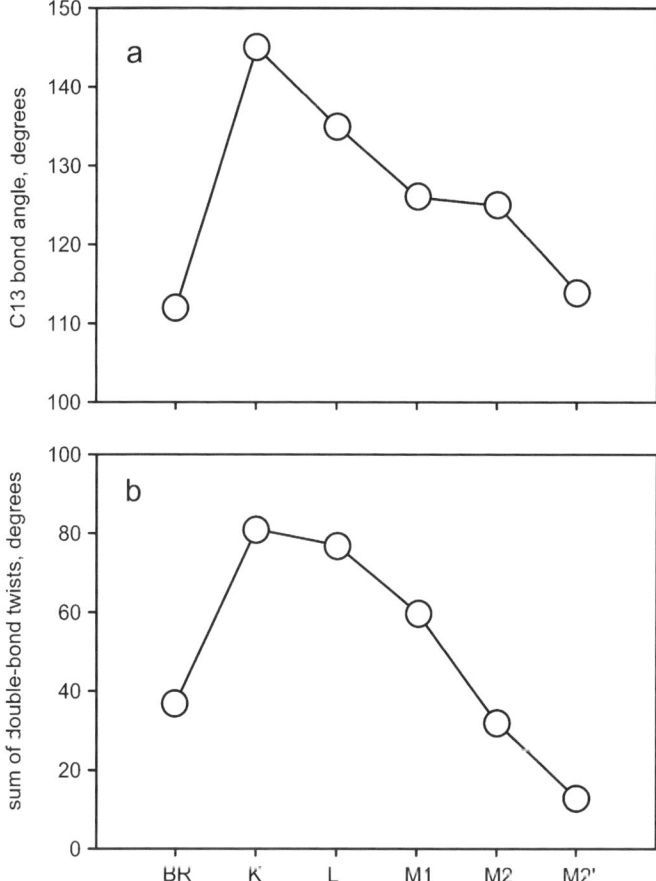

Figure 3 *Relaxation of the increased bond angle at C_{13} of the retinal (Panel a) and the sum of the out-of-plane twists of the C_{13}=C_{14} and C_{15}=NZ double bonds (Panel b), in the intermediates of the first half of the photocycle. Coordinate sets used are the same as in Figure 2*

What are the changes in the binding site that permit the relaxation of the strained configuration of the retinal, and how do they result in the transport of a proton? There are two main barriers to overcome if the isomerized retinal is to approach a solution-like low-energy configuration. First, the positively charged protonated Schiff base is fixed in its extracellular orientation by its hydrogen bond to water 402 and electrostatic interaction with the anionic asp-85 and asp-212. These interactions must be broken for rotation of the C_{15}=NZ-CE segment to reorient the N–H bond. It could be accomplished by transfer of the Schiff base proton to asp-85 in the L $\rightarrow$ (M_1, M_2, M_2') transition, converting both to uncharged species. Second, the β-ionone ring and the lys-216 peptide being relatively fixed in the protein, the acute angle of the 13-*cis* retinal at C_{13} can develop only by a thrust of the chain at C_{13}, along with the 13-methyl group, in the cytoplasmic direction. This would be possible if the indole ring of trp-182 and other neighboring side-chains between helices

F and G moved out of the way. The structures in Figure 2 reveal that both of these changes of the retinal occur.

In the K state the retinal is wound up, in a manner of speaking, by the $C_{13}=C_{14}$ bond rotation, but its unwinding through transfer of the Schiff base proton to asp-85 is blocked because the hydrogen bond of the Schiff base to water 402, that would connect it to asp-85, develops an unfavorably low angle. As the distortion is redistributed in the L state along the retinal chain to bonds on both sides of the initial bond rotation, however, the hydrogen bond recovers sufficiently to allow the proton transfer. In the M_1 state that results, the orientation of the now uncharged deprotonated Schiff base is still toward the protonated asp-85, but the electrostatic interaction that stabilized this geometry is removed. Water 402 has moved by 1 Å toward the two aspartic acids, further disconnecting the Schiff base from the extracellular region. In the $M_1 \rightarrow M_2$ transition that follows, the changes in M_1 allow rotation of the Schiff base nitrogen to face in the cytoplasmic direction. In the same intermediate states and continuing in M_2', the overall shape of the retinal gradually shifts to assume the expected bent contour of the 13-*cis*,15-*anti* configuration. The reactions that follow M_1 shift the protonation equilibrium fully toward deprotonation of the Schiff base. In the process, C_{13} and the 13-methyl group are thrust in the cytoplasmic direction, accompanied by distortion of the attached side-chain of lys-216 and its peptide segment (Figure 2). The latter results in the partial recovery of an α-helical hydrogen-bond pattern at the π-bulge of helix G.[31]

The 'local access' hypothesis was the outcome of an attempt to find a unifying interpretation for a large array of kinetic and spectroscopic experiments designed to determine the cause of directionality in the proton transport.[64,65] It attributes the protonation switch to bidirectional access of the retinal Schiff base that arises after its deprotonation. Kinetically, it corresponds to a rapid equilibrium that was found to develop between M_1 and M_2.[34] Combined with proton affinity changes in the extracellular region (where asp-85 is located), and the cytoplasmic region (containing asp-96), that may be regarded, conceptually, as modulated proton conductivities in the two half-channels, and with a strongly unidirectional last step as the initial state is re-set, this kind of switch will generate the pump activity observed. The changing geometry of the retinal in Figure 2 gives a structural basis to the bidirectional access: the M_1/M_2 equilibrium is now seen as the flickering of the Schiff base nitrogen between its extracellular and cytoplasmic orientations.

7 Conformational Cascades in Response to Relaxation of the Retinal

The above-described events at the retinal ensure that transport will occur, because they move a proton to an inherently low-affinity site (asp-85) on the extracellular side of the active center, and create an unoccupied inherently high-affinity site on the other side (the unprotonated Schiff base facing the cytoplasmic region). All that remains to complete the full translocation is for a proton to be released to the extracellular surface and another taken up from the cytoplasmic surface to reprotonate the Schiff base. The requirements for passage of the protons to and from the two membrane surfaces do not seem to be very stringent. These proton transfers can be slowed by many residue replacements, but few of the over 1,000 site-specific mutations in the bacteriorhodopsin literature completely block them.

In fact, the same regions will conduct a chloride ion in the related retinal protein, halorhodopsin,[66] and in the opposite direction from the movement of protons in bacteriorhodopsin. In the active center of halorhodopsin the homologue of asp-85 is a threonine. The Schiff base is not deprotonated in the photocycle, and it seems likely that the lysine NZ transfers the bound chloride ion from the extracellular to the cytoplasmic side in a rotational motion similar to what occurs in bacteriorhodopsin. Remarkably, the D85T and D85S mutants of bacteriorhodopsin exhibit chloride-dependent shifts of the chromophore absorption maximum, a halorhodopsin-like photocycle, and transport of chloride (and bromide) in the extracellular to cytoplasmic direction.[67,68] The crystal structure of the D85S mutant with and without bound bromide[69,70] indicates that the halide replaces water 402 of bacteriorhodopsin, and there are large-scale changes of the tilt of the extracellular ends of helices A, D, and E to accommodate the structural rearrangement of the extracellular region when the halide is absent. The turn-over of the pump for chloride at physiological salt concentrations is about the same as in bacteriorhodopsin for protons.[67,68] It is not yet clear how the traffic of halide ions to and from the active center is accomplished and whether it relates to how protons are transferred through the same regions. However, in bacteriorhodopsin at least, there are efficient mechanisms that move the proton in a specific manner, as follows.

As the photocycle progresses through the three successive M states, the Schiff base-asp-85 protonation equilibrium shifts toward complete deprotonation of the Schiff base. Protonation of asp-85 causes collapse of the extended extracellular hydrogen-bonded network. After the $M_1 \rightarrow M_2$ reaction water molecules 402 and 406 that link the Schiff base to asp-85 and asp-85 to arg-82, are no longer evident in the structure, and the arg-82 side-chain flips away from the active site.[44,45,63] The movement of its positive charge toward the extracellular surface and the proton release site (Figure 4) appears to be the means of the earlier proposed[71,72] coupling between the protonation states of asp-85 and the proton release site.

The identity of the proton release group has long been controversial. In the wild-type protein neither glu-194 nor glu-204[73,74] is the origin of the proton as a depletion COOH band is not detectable in the M state,[75] and recent work indicates in any case that they are both anionic in the BR state.[76] Arg-82 had been considered as the release site, and its deprotonation is supported by a recent analysis of the FTIR spectrum of the M state.[77] However, a very considerable amount of other experimental evidence[75,76] and theoretical calculation from the crystal structure[78,79] favor the idea that the released proton is the delocalized (Zundel) proton in $H_5O_2^+$ within a network of five water molecules stabilized by the side-chains of tyr-57, arg-82, tyr-83, ser-193, glu-194, glu-204, and the main-chain carbonyls of pro-77, tyr-79, and thr-205. Thus, no single group dissociates upon proton release because the proton comes from a hydrogen-bonded aqueous network. Loss of the delocalized proton is detected by decrease of the amplitude of an infrared continuum band, correlated in its kinetics with the appearance of the proton at the extracellular surface.[76] Interestingly, this mechanism is perturbed by mutations in the extracellular region in

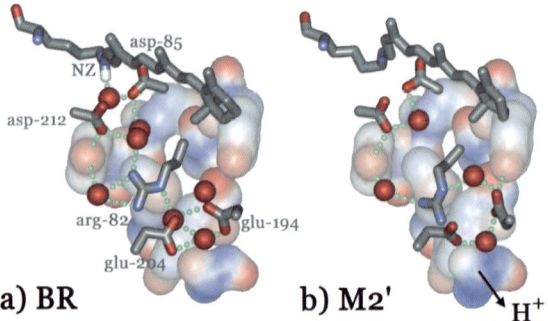

Figure 4 *Structural changes in the extracellular region between the BR (a) and the M₂' (b) states. Note movement of the side-chain of arg-82 toward the extracellular surface (downward). Coordinate sets used 1C3W and 1C8S (Reproduced with permission from reference 18, American Chemical Society)*

various ways. In the E194D mutant, for example, the continuum change associated with proton release is absent,[75] and the proton originates from glu-204 instead, with asp-194 first protonating and then deprotonating coincidentally with the proton being detected at the surface.[80]

The events on the cytoplasmic side have been described also. The Schiff base is reprotonated in the $M_2' \rightarrow N$ reaction by asp-96 near the cytoplasmic surface, through a hydrogen-bonded chain of water molecules (Figure 5). This chain is formed transiently during the photocycle, and begins as a cluster of water molecules at asp-96 in the M_1 and M_2 states. There is much evidence for the outward and inward tilts of the cytoplasmic ends of helices F and G, respectively, in both the late M and N states. These tilts are evident in projection maps from cryo-electron microscopy,[56,81] X-ray[82–84] and neutron diffraction,[85] movements of heavy-atom labels,[86] environmental and distance changes of spin-labels,[57–61] and changes of cysteine reactivity along the E-F interhelical loop.[87] They appear to be the result of repacking of side-chains between helices F and G in response to the displacements of the indole ring of trp-182 and the side-chains of leu-93 and leu-181 in contact with the retinal polyene chain.[31] The conformational changes gradually create cavities of increasing size,[32,88] and it was suggested that the single file chain of four hydrogen-bonded water molecules, that connects asp-96 with the Schiff base in the X-ray structure of the N' state trapped in the V49A mutant,[32] forms because the cavities become large enough to be filled with water molecules. Indeed, when a cavity is created by replacing the bulky side-chain of phe-219 with a leucine, two additional water molecules are recruited by water 501 to fill the cavity, and the three water molecules form a hydrogen-bonded cluster.[32]

A low-resolution three-dimensional cryo-electron microscopy structure for the N state accumulated by illuminating the F219L mutant with slow N decay[89] confirmed the tilts of helices F and G in the projection maps. A somewhat higher resolution structure, also from cryo-electron microscopy, is available for the non-illuminated state of the D96G/F171C/F219L triple mutant,[90] in which, as in D85N at high pH and D85N/D96N already at neutral pH,[82] the tilts arise, apparently, from destabilization

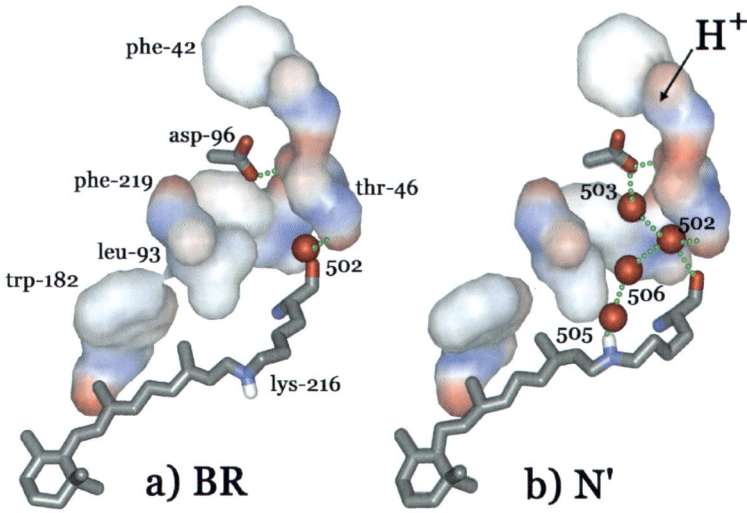

Figure 5 *Structural changes in the cytoplasmic region between the BR (a) and the N' (b) states. Note assembly of a hydrogen-bonded chain of water molecules between the retinal Schiff base and asp-96. Coordinate sets used 1C3W and 1P8U (Reproduced with permission from reference 2, Annual Review of Physiology)*

of the normal BR structure. However, projection maps from time-resolved X-ray diffraction indicate that the structural changes in the late M and early N states are not equivalent.[91]

The large-scale helical tilts in N are not present in the N' state,[32] probably because they reverse once asp-96 becomes protonated. Asp-96 is about 6 Å from the cytoplasmic surface, and its reprotonation from the bulk must be also through a transient chain of water molecules. The two water molecules detected in this region in the crystal structure of N', one at asp-96 and another near the surface,[32] may be the remnants of this chain. Capture of the proton from the bulk was proposed to be by asp-38,[92] or a 'proton antenna' formed by asp-36, asp-38, asp-102, and asp-104[93,94,94,95] on the cytoplasmic water-protein interface, but replacement of these residues, individually or in groups, by asparagine caused only minor changes in the photocycle.[96] However, electrostatic effects at the surface are likely to be important. The great slowing of the proton uptake in the photocycle of the D96N mutant was suggested to have its origin in the absence of the negative charge of the anionic asp-96 beneath the cytoplasmic surface.[97]

Reprotonation of asp-96 allows the rapid reisomerization of the retinal to all-*trans* to form the O state. The existence of this coupling becomes evident when it is disrupted by mutations at either near asp-96 or at the retinal, and the reisomerization becomes both independent of the protonation of asp-96 and very slow.[98] The coupling is demonstrated also by an experiment without illumination. When an N-like state containing 13-*cis*,15-*anti* retinal was produced by replacing asp-85 with asparagine and lowering the pK_a of asp-96 by a nearby mutation so as to make it anionic at pH 9 (see also below), a pH jump from 9 to 6 to reprotonate asp-96 caused isomerization to all-*trans* within 30 ms,[99] consistent with the reisomerization kinetics in the photocycle of the wild-type protein.

During the final step of the photocycle, asp-85 reprotonates the vacant proton release site. Presumably, this occurs because asp-85 regains its very low pK_a in the BR state (about 2.5), but there is little information on how it happens. Arguably, the crystallographic structure of the D85S mutant in the absence of halide resembles the O state.[69] If so, the proton transfer is the result of the new set of helical motions detected in the mutant, where the extracellular ends of helices A, D, and E tilt outward. On the other hand, there is a direct correlation between the rates of the O decay and of the deprotonation of asp-85 upon rapidly raising the pH from 2 to 6, in the dark, in various mutants in the extracellular region.[100] This would argue both for the deprotonation of asp-85 being the rate limiting step and for a BR-like structure during the final step of the photocycle.

8   Crystallographic *vs.* Non-crystallographic Evidence

The constraint of the crystal lattice, which will oppose the large-scale conformational changes detectable in membranes, is one of the limitations of the crystallographic method. Another, at the other end of the scale, is an uncertainty when describing very small changes in geometry at all but the highest resolutions (see above). Thus, when the atomic displacements are small, the crystallographic evidence must be evaluated against information from non-crystallographic measurements. Vibrational spectra are in some cases difficult to interpret, and assignment of O–H stretch bands to specific bound water molecules in bacteriorhodopsin is not straightforward, but this valuable method can focus on changes too small to be detected at the crystallographic resolutions available. Solid-state NMR requires expensive isotope labeling of large quantities of protein specific for each measurement, but produces very accurate information on structural changes. Photoelectric measurements of bacteriorhodopsin do not lend themselves directly to a molecular interpretation, but provide unique information that any transport model must account for. Local information that can add up to a global description of conformational shifts otherwise difficult to obtain can be gained from spin-labels or cysteine accessibility scanning. In the following, the conclusions from the X-ray diffraction are compared to some of these results.

The C=N stretch frequency of the protonated Schiff base is mixed with the N–H rocking mode, and thus contains information about the strength of its hydrogen bond. The latter can be assessed by comparing the C=N stretch frequency in H_2O (where the modes mix) and D_2O (where they do not). In the BR state the N–H rock contributes a 20 cm^{-1} upshift, suggesting strong hydrogen bonding. In K this shift nearly completely disappears, but recovers in L,[101,102] consistent with the breaking and reformation of the hydrogen bond of the Schiff base to water 402 in the crystal structures (Figure 2). The same conclusion about the breaking of the hydrogen bond of the Schiff base in K may be drawn from the change of the angle of the N–D bond (in D_2O), as measured with polarized FTIR.[103] FTIR spectra provide support also for the changes in the extracellular region later in the photocycle. Polarized FTIR measurements of the N–D stretching bands assigned to the guanidinium group of arg-82 hydrogen bonded to water 406 indicate that their orientation changes in M (produced at 230K) but not L,[104] consistent with the crystal structures[33] that indicate movement of the arg-82 side-chain and water 406 in M_2 but not in L and M_1.

On the other hand, there is disagreement between the structural models in Figures 1 and 2 and other FTIR data that address the location of water 402. In the crystal structures,[33] water 402 remains a bridge between the Schiff base and asp-85 until M_1. FTIR studies aimed at defining the location of this water in L suggested,[105–107] however, that it moves to the cytoplasmic side of the Schiff base before protonation of asp-85. Although this movement of water 402, to near leu-93 and trp-182, could be deduced only from the changed influence of site-specific mutation of these residues, and others nearby, on its O–H band relative to the BR state, this is a clear contradiction that has not been resolved. There are various alternatives. The principle that shifts of O–H stretch frequencies, or even the absence of some O–H bands, in response to the local perturbations in mutants will reveal the location of water molecules seems reasonable, but long-range effects cannot be excluded. Several examples for the appearance of new and unsuspected water molecules in the crystal structure of mutants are available,[31,32,63] and argue for caution in assigning O–H stretch bands. Finally, a disordered water molecule, or water with partial occupancy, is difficult to locate in the crystal structures and might have been missed.

Solid-state NMR provides accurate local information about torsion angles and inter-atomic distances, and describes the changes in geometry in the protein. Where NMR information is available, it agrees reasonably well with the crystal structures. Thus, from NMR[108] the torsion angles for the C_{14}–C_{15} bond in the BR, early M (M_1), and late M (M_2') states are $164 \pm 4°$, $147 \pm 10°$, and $150 \pm 4°$, and for the C_{15}=NZ bond in BR it is $161 \pm 5°$ (absolute values, as the direction of the torsion angle is not defined here). From the crystallography (Figure 2), these values are 178°, $110 \pm 42°$, $-166°$, and $-168°$, respectively. Likewise, the NMR results and the crystal structures agree that in M the 13-methyl group of the retinal approaches the indole NE1 of trp-182. From NMR[109] this distance decreases from 3.36 ± 0.2 Å in BR to 3.16 0.4 Å in early M (M_1), while from crystallography these values are 3.33 Å and 3.03 ± 0.02 Å, respectively.[33]

There are many other results that offer some degree of support to the crystal structures and their functional interpretation. The results of spin-labeling the cytoplasmic ends of helices to detect distance changes were discussed above. A final example is on the role of water in the photocycle. The involvement of water in conducting the proton from asp-96 to the retinal Schiff base was suspected long before the crystal structure revealed the predicted hydrogen-bonded chain, but the experimental evidence was indirect. Suggestively, however, osmotic agents like glucose and elevated hydrostatic pressure were found to decrease the rate of specifically the M $\leftrightarrow$ N reaction in the photocycle.[97,110] The inhibition was attributed to removing water molecules bound in the protein interior that facilitate this reaction.

9 Pump Energetics

Ion pumps are devices that convert energy supplied by the driving reaction into the work of moving a charged species, here protons, against an electrochemical gradient across a membrane. Without elucidation of the flow of energy through the system, and its transformations at critical reactions in the cycle, we cannot claim full understanding of how the pump works. The bacteriorhodopsin transport mechanism described here contains many clues to its energetics.

The energy gain from the absorbed photon in K must be in the distortions of the retinal[38,42] that conserve part of the energy of the excited state, and are achieved at the expense of breaking the Schiff base-asp-85 hydrogen bond.[111] A Gaussian calculation[38] for the increased C_{13} bond angle that assumes, for the sake of simplicity, no interaction of the retinal with its surroundings, yields a configurational energy roughly equivalent to the measured enthalpy rise of about 50 kJ mol^{-1} in K. Energy will be conserved in this strain plus the two double bond distortions only if the retinal binding site is not compliant, *i.e.* it does not accommodate the changed geometry that an unhindered 13-*cis*,15-*anti* polyene chain would assume. By the same token, the conserved energy can be used for uphill proton transport only if this energy is transferred to the protein. In this process, the first step is the redistribution of strain along the retinal chain that allows the re-formation of the hydrogen bond between the Schiff base and water 402 (Figure 2). The kinetics indicate that the free energy dissipation associated with this step is about 6 kJ mol^{-1}.[112] As discussed above, further relaxation of the retinal depends on the proton transfer to asp-85 that can occur *via* this hydrogen bond, and displacements of side-chains that allow the 13-methyl group to move upward as the 13-*cis*,15-*anti* retinal assumes its bent shape.[33] Passing through the M_1 and M_2 states is with a nearly zero net ΔG. The pK_a for proton release being about 5–6,[34] at a physiological pH (7.5) the M_2' state is produced from M_2 with a 10 kJ mol^{-1} free energy loss (earlier this loss was ascribed to the $M_1 \rightarrow M_2$ reaction[112]). In intact cells this, together with part of the rest of the energy dissipated in the open membrane sheets, will be conserved in the protonmotive force created.

It appears that once the N state is reached, however, the photoisomerized 13-*cis*,15-*anti* retinal is fully accommodated by the changed binding site and the changed electrostatics of the active site and the protein. In the D85N/F42C mutant at pH 9 asp-96 is anionic (because its pK_a is lowered from > 11 to about 8 by the mutation of phe-42) and residue 85 is neutral (because it is changed to an asparagine) as in N, and the retinal was found to be in the 13-*cis*,15-*anti* configuration without illumination.[99] Light-dependent FTIR difference spectra indicate that in this protein it is the BR-like state that is transient, and it returns to the N-like state once illumination is terminated. The implication of these findings is that once the N state is reached the excess free energy that allows recovery of the BR state is no longer in the retinal but in the protein. Thus, while the transport is driven by the relaxation of the retinal that causes the required proton transfer reactions, the recovery of the initial state depends on relaxation of the protein. According to the kinetics the steps that lead to O, including reprotonation of asp-96 at a physiological pH near its pK_a at this stage, occur with nearly zero ΔG. The lack of a measurable backreaction in the last step of the photocycle, that recovers the BR state, suggests on the other hand that it is accompanied by dissipation of at least 15 kJ mol^{-1}. Presumably, this reflects, at least partly, the energetically favorable reprotonation of the proton release group by asp-85.[100,113] If so, the last step will account for most of the free energy dissipation in the cycle, because the ΔpK_a between asp-85 and the proton release group in the BR state is about 6,[71] amounting to about 35 kJ mol⁻¹.

If the kinetic model is correctly chosen, the free energy diagram of the photocycle can be deduced from the calculated rate constants and their temperature dependencies. An early attempt[112,114] yielded a scheme that still applies, although with modifications to account for updates to the model. The main conclusions are that: a) the first half of the photocycle is driven mainly by enthalpy differences of the intermediate states, while the second half by entropic forces; and b) when the protein is in the two-dimensional lattice of the purple membrane the barriers to the interconversions of the intermediates are mainly enthalpic but when it is in detergent micelles the barriers are mainly entropic. While these statements are easily rationalized in terms of local interactions that dominate in the initial steps and global conformational changes in the recovery of the BR state, as well as increased conformational heterogeneity when the protein is removed from immobile lattice, a complete physical interpretation of the free energy levels and barriers is still lacking and will need information from both kinetics and crystallographic structures.

An example is the proton transfer from asp-96 to the unprotonated Schiff base in the $M_2' \rightarrow N$ reaction. The pK_as of the acceptor and donor at this stage in the cycle were reported to be 8.3[115] and 7–7.5,[98,116] respectively. The kinetics indicate that the ΔG of the M_2'/N equilibrium is about 4 kJ mol⁻¹,[112] *i.e.* the ΔpK_a is about 0.7, in surprisingly good agreement with the independently determined pK_a values. The nature of the barrier is more problematic. The proton transfer, over a linear chain of four hydrogen-bonded water molecules,[32] occurs in a few ms, which is 8–9 orders of magnitude slower than predicted from considerations of how protons move along such chains.[117] The rate-limiting step might be the protein conformational change that allows formation of the hydrogen-bonded chain. However, the tilts of helices F and G occur, or begin to occur, in the last M state (M_2' in the notation used here),[63] *i.e.* before the rise of N. There is a more likely explanation. Replacement of asp 96 with an asparagine was found to lower the enthalpy of activation for the reprotonation of the Schiff base (presumably through the same pathway but from the cytoplasmic surface without aid of asp-96) by 42 kJ mol⁻¹. The reaction is slower and pH dependent,[97,118,119] probably because without the anionic asp-96 near the surface the bulk proton is more difficult to capture. If there is an additional barrier of 42 kJ mol⁻¹ in the wild-type protein, it should correspond to the enthalpy cost of separating the proton from asp-96.[97] This electrostatic barrier will slow the reaction by 7–8 orders of magnitude from what would be expected for proton transfer *via* a hydrogen-bonded chain, which would account for most, and perhaps all, of the observed rate.

References

1. U. Haupts, J. Tittor and D. Oesterhelt, *Annu. Rev. Biophys. Biomol. Struct.*, 1999, **28**, 367.
2. J.K. Lanyi, *Annu. Rev. Physiol.*, 2004, **66**, 11.
3. H. Luecke and J.K. Lanyi, *Adv. Protein Chem.*, 2003, **63**, 111.
4. R. Neutze, E. Pebay-Peyroula, K. Edman, A. Royant, J. Navarro and E.M. Landau, *Biochim. Biophys. Acta*, 2002, **1565**, 144.

5. O. Beja, L. Aravind, E.V. Koonin, M.T. Suzuki, A. Hadd, L.P. Nguyen, S.B. Jovanovich, C.M. Gates, R.A. Feldman, J.L. Spudich, E.N. Spudich and E.F. DeLong, *Science*, 2000, **289**, 1902.
6. G. Sabehi, R. Massana, J.P. Bielawski, M. Rosenberg, E.F. DeLong and O. Beja, *Environ. Microbiol.*, 2003, **5**, 842.
7. J.C. Venter, K. Remington, J.F. Heidelberg, A.L. Halpern, D. Rusch, J.A. Eisen, D. Wu, I. Paulsen, K.E. Nelson, W. Nelson, D.E. Fouts, S. Levy, A.H. Knap, M.W. Lomas, K. Nealson, O. White, J. Peterson, J. Hoffman, R. Parsons, H. Baden-Tillson, C. Pfannkoch, Y.H. Rogers and H.O. Smith, *Science*, 2004, **304**, 66.
8. J.K. Lanyi and G. Varo, *Isr. J. Chem.*, 1995, **35**, 365.
9. A. Maeda, H. Kandori, Y. Yamazaki, S. Nishimura, M. Hatanaka, Y.S. Chon, J. Sasaki, R. Needleman and J.K. Lanyi, *J. Biochem. (Tokyo)*, 1997, **121**, 399.
10. H. Kandori, *Biochim. Biophys. Acta*, 2000, **1460**, 177.
11. H. Kandori, *Biochim. Biophys. Acta*, 2004, **1658**, 72.
12. A.K. Dioumaev, *Biochemistry (Mosc.)*, 2001, **66**, 1269.
13. A. Maeda, *Biochemistry (Mosc.)*, 2001, **66**, 1256.
14. J.K. Lanyi, *J. Phys. Chem. B*, 2003, **104**, 11441.
15. J.K. Lanyi and B. Schobert, *Biochemistry*, 2004, **43**, 3.
16. J. Herzfeld and B. Tounge, B*iochim. Biophys. Acta*, 2000, **1460**, 95.
17. J. Herzfeld and J.C. Lansing, *Annu. Rev. Biophys. Biomol. Struct.*, 2002, **31**, 73.
18. J.K. Lanyi and B. Schobert, *Biochemistry*, 2004, **43**, 3.
19. J.K. Lanyi and H. Luecke, *Curr. Opin. Struct. Biol.*, 2001, **11**, 415.
20. A. Royant, K. Edman, T. Ursby, E. Pebay-Peyroula, E.M. Landau and R. Neutze, *Photochem. Photobiol.*, 2001, **74**, 794.
21. K. Edman, A. Royant, G. Larsson, F. Jacobson, T. Taylor, S.D. Van Der, E.M. Landau, E. Pebay-Peyroula and R. Neutze, *J. Biol. Chem.*, 2004, **279**, 2147.
22. S.P. Balashov and T.G. Ebrey, *Photochem. Photobiol.*, 2001, **73**, 453 .
23. J.K. Lanyi, *Biochim. Biophys. Acta*, 2004, **1658**, 14.
24. R. Henderson and P.N. Unwin, *Nature*, 1975, **257**, 28.
25. N. Grigorieff, T.A. Ceska, K.H. Downing, J.M. Baldwin and R. Henderson, *J. Mol. Biol.*, 1996, **259**, 393.
26. T.A. Ceska, R. Henderson, J.M. Baldwin, F. Zemlin, E. Beckmann and K. Downing, *Acta Physiol. Scand.* , *Suppl.*, 1992, **146**, 31.
27. H. Luecke, H.T. Richter and J.K. Lanyi, *Science*, 1998, **280**, 1934.
28. H. Luecke, B. Schobert, H.T. Richter, J.P. Cartailler and J.K. Lanyi, *J. Mol. Biol.*, 1999, **291**, 899.
29. H. Belrhali, P. Nollert, A. Royant, C. Menzel, J.P. Rosenbusch, E.M. Landau and E. Pebay-Peyroula, *Structure. Fold. Des*, 1999, **7**, 909.
30. M. Wikstrom, *Curr. Opin. Struct. Biol.*, 1998, **8**, 480.
31. H. Luecke, B. Schobert, H.T. Richter, J.-P. Cartailler, A. Rosengarth, R. Needleman, and J.K. Lanyi, *J. Mol. Biol.*, 2000, **300**, 1237.
32. B. Schobert, L.S. Brown and J. K. Lanyi. *J. Mol. Biol.*, 2003, **330**, 553.
33. J.K. Lanyi and B. Schobert, *J. Mol. Biol.*, 2003, **328**, 439.
34. L. Zimanyi, G. Varo, M. Chang, B. Ni, R. Needleman and J.K. Lanyi, *Biochemistry*, 1992, **31**, 8535.

35. S.P. Balashov, *Biochim.*, *Biophys. Acta*, 2000, **1460**, 75.
36. G. Rummel, A. Hardmeyer, C. Widmer, M.L. Chiu, P. Nollert, K.P. Locher, I. Pedruzzi, E.M. Landau and J.P. Rosenbusch, *J. Struct. Biol.*, 1998, **121**, 82.
37. M.L. Chiu, P. Nollert, M.C. Loewen, H. Belrhali, E. Pebay-Peyroula, J.P. Rosenbusch and E.M. Landau, *Acta Crystallogr. D. Biol. Crystallogr.*, 2000, **56**, 781.
38. B. Schobert, J. Cupp-Vickery, V. Hornak, S. Smith and J. Lanyi, *J. Mol. Biol.*, 2002, **321**, 715.
39. B.W. Edmonds and H. Luecke, *Front Biosci.*, 2004, **9**, 1556.
40. K. Edman, P. Nollert, A. Royant, H. Belrhali, E. Pebay-Peyroula, J. Hajdu, R. Neutze and E.M. Landau, *Nature*, 1999, **401**, 822.
41. A. Royant, K. Edman, T. Ursby, E. Pebay-Peyroula, E.M. Landau and R. Neutze, *Nature*, 2000, **406**, 645.
42. Y. Matsui, K. Sakai, M. Murakami, Y. Shiro, S. Adachi, H. Okumura and T. Kouyama, *J. Mol. Biol.*, 2002, **324**, 469.
43. T. Kouyama, T. Nishikawa, T. Tokuhisa and H. Okumura, *J. Mol. Biol.*, 2004, **335**, 531.
44. H.J. Sass, G. Buldt, R. Gessenich, D. Hehn, D. Neff, R. Schlesinger, J. Berendzen and P. Ormos, *Nature*, 2000, **406**, 649.
45. K. Takeda, Y. Matsui, N. Kamiya, S. Adachi, H. Okumura and T. Kouyama, *J. Mol. Biol.*, 2004, **341**, 1023.
46. A.N. Bondar, M. Elstner, S. Suhai, J.C. Smith and S. Fischer, *Structure*, 2004, **12**, 1281.
47. A. Royant, S. Grizot, R. Kahn, H. Belrhali, F. Fieschi, E.M Landau and E. Pebay-Peyroula, *Acta Crystallogr. D Biol. Crystallogr.*, 2002, **58**, 784.
48. G.M.S.T. Sheldrick, *Methods Enzymol.*, 1997, **277**, 319.
49. G.B. Jameson, *Acta Crystallographica Section A*, 1982, **38**, 817.
50. J.P. Cartailler and H. Luecke, *Annu. Rev. Biophys. Biomol. Struc.*, 2003, **32**, 285.
51. L. Essen, R. Siegert, W.D. Lehmann and D. Oesterhelt, *Proc. Natl. Acad. Sci. U.S.A.*, 1998, **95**, 11673.
52. B. Halle, *Proc. Natl. Acad. Sci. U.S.A.*, 2004, **101**, 4793.
53. O. Weidlich and F. Siebert, *Appl. Spectroscopy*, 1993, **47**, 1394.
54. F.F. Litvin, S.P. Balashov and V.A. Sineshchekov, *Bioorganicheskaya Khimiya*, 1975, **1**, 1767.
55. T. Iwasa, F. Tokunaga and T. Yoshizawa, *Biophys. Struct. Mech.*, 1980, **6**, 253.
56. S. Subramaniam, M. Lindahl, P. Bullough, A.R. Faruqi, J. Tittor, D. Oesterhelt, L. Brown, J. Lanyi and R. Henderson, *J. Mol. Biol.*, 1999, **287**, 145.
57. T. Rink, M. Pfeiffer, D. Oesterhelt, K. Gerwert and H.J. Steinhoff, *Biophys. J.*, 2000, **78**, 1519.
58. R. Mollaaghababa, H.J. Steinhoff, W.L. Hubbell and H.G. Khorana, *Biochemistry*, 2000, **39**, 1120.
59. N. Radzwill, K. Gerwert and H.J. Steinhoff, *Biophys. J.*, 2001, **80**, 2856.
60. T.E. Thorgeirsson, W. Xiao, L.S. Brown, R. Needleman, J.K. Lanyi and Y.K. Shin, *J. Mol. Biol.*, 1997, **273**, 951.
61. W. Xiao, L.S. Brown, R. Needleman, J.K. Lanyi and Y.K. Shin, *J. Mol. Biol.*, 2000, **304**, 715.

62. J.K. Lanyi and B. Schobert, *J. Mol. Biol.*, 2002, **321**, 727.
63. H. Luecke, B. Schobert, H.T. Richter, J.P. Cartailler and J.K. Lanyi, *Science*, 1999, **286**, 255.
64. L.S. Brown, A.K. Dioumaev, R. Needleman and J.K. Lanyi, *Biophys. J.*, 1998, **75**, 1455.
65. L.S. Brown, A.K. Dioumaev, R. Needleman and J.K. Lanyi, *Biochemistry*, 1998, **37**, 3982.
66. D. Oesterhelt, *Israel J. Chem.*, 1995, **35**, 475.
67. J. Sasaki, L.S. Brown, Y.S. Chon, H. Kandori, A. Maeda, R. Needleman and J.K. Lanyi, *Science*, 1995, **269**, 73.
68. L.S. Brown, R. Needleman and J.K. Lanyi, *Biochemistry*, 1996, **35**, 16048.
69. S. Rouhani, J.P. Cartailler, M.T. Facciotti, P. Walian, R. Needleman, J.K. Lanyi, R.M. Glaeser and H. Luecke, *J. Mol. Biol.*, 2001, **313**, 615.
70. M.T. Facciotti, V.S. Cheung, D. Nguyen, S. Rouhani and R.M. Glaeser, *Biophys. J.*, 2003, **85**, 451.
71. R. Govindjee, S. Misra, S.P. Balashov, T.G. Ebrey, R.K. Crouch and D.R. Menick, *Biophys. J.*, 1996, **71**, 1011.
72. H.T. Richter, L.S. Brown, R. Needleman and J.K. Lanyi, *Biochemistry*, 1996, **35**, 4054.
73. S.P. Balashov, E.S. Imasheva, T.G. Ebrey, N. Chen, D.R. Menick and R.K. Crouch, *Biochemistry*, 1997, **36**, 8671.
74. L.S. Brown, J. Sasaki, H. Kandori, A. Maeda, R. Needleman and J.K. Lanyi, *J. Biol. Chem.*, 1995, **270**, 27122.
75. R. Rammelsberg, G. Huhn, M. Lubben and K. Gerwert, *Biochemistry*, 1998, **37**, 5001.
76. F. Garczarek, L. Brown, J.K. Lanyi and K. Gerwert, *Proc. Natl. Acad. Sci. U.S.A.*, 2005, **102**, 3633.
77. M.S. Hutson, U. Alexiev, S.V. Shilov, K.J. Wise and M.S. Braiman, *Biochemistry*, 2000, **39**, 1318.
78. V.Z. Spassov, H. Luecke, K. Gerwert and D. Bashford, *J. Mol. Biol.*, 2001, **312**, 203.
79. C. Kandt, J. Schlitter and K. Gerwert *Biophys. J.*, 2004, **86**, 705.
80. A.K. Dioumaev, H.T. Richter, L.S. Brown, M. Tanio, S. Tuzi, H. Saito, Y. Kimura, R. Needleman and J.K. Lanyi, *Biochemistry*, 1998, **37**, 2496.
81. S. Subramaniam, M. Gerstein, D. Oesterhelt and R. Henderson, *EMBO J.*, 1993, **12**, 1.
82. M. Kataoka, H. Kamikubo, F. Tokunaga, L.S. Brown, Y. Yamazaki, A. Maeda, M. Sheves, R. Needleman and J.K. Lanyi, *J. Mol. Biol.*, 1994, **243**, 621.
83. H. Kamikubo, M. Kataoka, G. Varo, T. Oka, F. Tokunaga, R. Needleman and J.K. Lanyi, *Proc. Natl. Acad. Sci. U.S.A.*, 1996, **93**, 1386.
84. H. Kamikubo, T. Oka, Y. Imamoto, F. Tokunaga, J.K. Lanyi and M. Kataoka, *Biochemistry*, 1997, **36**, 12282.
85. N.A. Dencher, D. Dresselhaus, G. Zaccai and G. Bueldt, *Proc. Natl. Acad. Sci. USA*, 1989, **86**, 7876.
86. T. Oka, H. Kamikubo, F. Tokunaga, J.K. Lanyi, R. Needleman and M. Kataoka, *Biophys. J.*, 1999, **76**, 1018.

87. L.S. Brown, R. Needleman and J.K. Lanyi, *J. Mol. Biol.*, 2002, **317**, 471.
88. R. Friedman, E. Nachliel and M. Gutman, *Biophys. J.*, 2003, **85**, 886.
89. J. Vonck, *EMBO J.*, 2000, **19**, 2152.
90. S. Subramaniam and R. Henderson, *Nature*, 2000, **406**, 653.
91. T. Oka, N. Yagi, T. Fujisawa, H. Kamikubo, F. Tokunaga and M. Kataoka, *Proc. Natl. Acad. Sci. U.S.A.*, 2000, **97**, 14278.
92. J. Riesle, D. Oesterhelt, N.A. Dencher and J. Heberle, *Biochemistry*, 1996, **35**, 6635.
93. Y. Kimura, D.G. Vassylyev, A. Miyazawa, A. Kidera, M. Matsushima, K. Mitsuoka, K. Murata, T. Hirai and Y. Fujiyoshi, *Nature*, 1997, **389**, 206.
94. S. Checover, Y. Marantz, E. Nachliel, M. Gutman, M. Pfeiffer, J. Tittor, D. Oesterhelt and N.A. Dencher, *Biochemistry*, 2001, **40**, 4281.
95. E. Nachliel, M. Gutman, J. Tittor and D. Oesterhelt, *Biophys. J.*, 2002, **83**, 416.
96. L.S. Brown, R. Needleman and J.K. Lanyi, *Biochemistry*, 1999, **38**, 6855.
97. Y. Cao, G.Varo, M. Chang, B. Ni, R. Needleman and J.K. Lanyi, *Biochemistry*, 1999, **30**, 10972.
98. A.K. Dioumaev, L.S. Brown, R. Needleman and J.K. Lanyi, *Biochemistry*, 2001, **40**, 11308.
99. A.K. Dioumaev, L.S. Brown, R. Needleman and J.K. Lanyi, *Biochemistry*, 1998, **37**, 9889.
100. H.T. Richter, R. Needleman, H. Kandori, A. Maeda and J.K. Lanyi, *Biochemistry*, 1996, **35**, 15461.
101. A. Maeda, J. Sasaki, J.M. Pfefferle, Y. Shichida and T. Yoshizawa, *Photochem. Photobiol.*, 1991, **54**, 911.
102. S.P.A. Fodor, W.T. Pollard, R. Gebhard, E.M.M. Van den Berg, J. Lugtenburg and R.A. Mathies, *Proc. Natl. Acad. Sci. U.S.A.*, 1988, **85**, 2156.
103. H. Kandori, M. Belenky and J. Herzfeld, *Biochemistry*, 2002, **41**, 6026.
104. T. Tanimoto, M. Shibata, M. Belenky, J. Herzfeld and H. Kandori, *Biochemistry*, 2004, **43**, 9439.
105. A. Maeda, S.P. Balashov, J. Lugtenburg, M.A. Verhoeven, J. Herzfeld, M. Belenky, R.B. Gennis, F.L. Tomson and T.G. Ebrey, *Biochemistry*, 2002, **41**, 3803.
106. A. Maeda, F.L. Tomson, R.B. Gennis, S.P. Balashov and T.G. Ebrey, *Biochemistry*, 2003, **42**, 2535.
107. A. Maeda, J. Herzfeld, M. Belenky, R. Needleman, R.B. Gennis, S.P. Balashov and T.G. Ebrey, *Biochemistry*, 2003, **42**, 14122.
108. J.C. Lansing, M. Hohwy, C.P. Jaroniec, A.F. Creemers, J. Lugtenburg, J. Herzfeld and R.G. Griffin, *Biochemistry*, 2002, **41**, 431.
109. A.T. Petkova, M. Hatanaka, C.P. Jaroniec, J.G. Hu, M. Belenky, M. Verhoeven, J. Lugtenburg, R.G. Griffin and J. Herzfeld, *Biochemistry*, 2002, **41**, 2429.
110. G. Varo and J.K. Lanyi, *Biochemistry*, 1995, **34**, 12161.
111. S. Hayashi, E. Tajkhorshid, H. Kandori and K. Schulten, *J. Am. Chem. Soc.*, 2004, **126**, 10516.
112. G. Varo and J.K. Lanyi, *Biochemistry*, 1991, **30**, 5016.
113. H. Kandori, Y. Yamazaki, M. Hatanaka, R. Needleman, L.S. Brown, H.T. Richter, J.K. Lanyi and A. Maeda, *Biochemistry*, 1997, **36**, 5134.

114. G. Varo and J.K. Lanyi, *Biochemistry*, 1991, **30**, 7165.

115. L.S. Brown and J.K. Lanyi, *Proc. Natl. Acad. Sci. U.S.A.*, 1996, **93**, 1731.

116. C. Zscherp, R. Schlesinger, J. Tittor, D. Oesterhelt and J. Heberle, *Proc. Natl. Acad. Sci. U.S.A.*, 1999, **96**, 5498.

117. C. Dellago, M.M. Naor and G. Hummer, *Phys. Rev. Lett.*, 2003, **90**, 105902.

118. H. Otto, T. Marti, M. Holz, T. Mogi, M. Lindau, H.G. Khorana and M.P. Heyn *Proc. Natl. Acad. Sci. U.S.A.*, 1989, **86**, 9228.

119. A. Miller and D. Oesterhelt, *Biochim. Biophys. Acta*, 1990, **1020**, 57.

CHAPTER 11

The Dynamics of Proton Transfer Across Bacteriorhodopsin Explored by FT-IR Spectroscopy

JOACHIM HEBERLE

Forschungszentrum Jülich
IBI-2: Structural Biology 52425 Jülich,
Germany

1 Introduction

Understanding of the physiological mechanism of a membrane protein emerges from structural details about key intermediate steps of enzymatic action. Yet, many of the important structural changes associated with the catalytic mechanism, such as protonation-state changes of individual amino-acid side chains, often remain unresolved in the crystallographic models. Therefore, an alternative technique must be used to gain information about the coordinated involvement of particular residues and the chronological sequence of steps spanning the reaction mechanism. Vibrational spectroscopy (infrared absorption and Raman scattering) complements the understanding of the catalytic mechanism of a membrane protein because it provides high structural sensitivity along with the possibility to record the associated dynamics.

Infrared spectroscopy senses bond strengths and masses by exciting vibrational levels, and thus has long been an established tool to identify small molecules by their characteristic vibrational 'fingerprint'. Moreover, it senses non-covalent forces (from hydrogen bonding, van-der-Waals or electrostatic interactions) which fine-tune vibrational frequencies, and thus can probe structural properties of proteins. Due to the complexity of biological samples, resolution of FT-IR spectroscopy is hampered by the appearance of innumerable vibrational bands. This can be simply illustrated by calculating the number of vibrational degrees of freedom $F = 3 \times N - 6$ where N denotes the number of atoms vibrating. N can be of the order of $> 3,000$, even for a mid-sized protein, like bacteriorhodopsin! Moreover, in condensed phase, bands are broadened and therefore strongly overlap. Difference spectroscopy has proved to be an optimal tool to study transitions from one state to another since only those vibrations change during the action of the protein (*vide infra*).[1-3]

The first IR study covering the spectral region of protein modes [1000–1800 cm^{-1}] has been performed on rhodopsin using a conventional dispersive IR spectrophotometer.[4] Consequent application of this technique for the investigation of bacteriorhodopsin photoreactions led to the identification of the C=O modes from transiently protonated carboxylic-acid side chains involved in light-driven proton transfer.[5,6] In parallel, the possibility of using Fourier transform infrared (FT-IR) spectroscopy to obtain steady-state difference spectra of intermediates in the bacteriorhodopsin photocycle was recognized in the early 80s.[7-9]

Starting from these first pilot investigations, a whole body of studies evolved on isotope-modified or structurally varying retinals, isotope-modified protein, and particularly on site-directed mutants. The pioneering studies on rhodopsins were soon expanded and other proteins were investigated. Particularly, membrane proteins involved in energy conversion have been a major target. Photosynthetic[10-22] and respiratory complexes[23-43] have been extensively studied by FT-IR spectroscopy which yielded valuable information on the reaction mechanism of these intricate molecular machineries.

2 FT-IR Spectroscopy

2.1 Selecting a Single Vibration from a Protein: Difference Spectroscopy

The size and the heterogeneity of a protein leads to an intractable number of overlapping vibrational bands and the extraction of molecular events from the IR absorption spectrum is hardly possible. This problem is simplified by the use of *reaction-induced difference spectroscopy* (reviewed in[38,44-47]), which examines the differences in absorbance before and after inducing a perturbation that changes the state of the protein. This biochemical reaction alters a subset of the total vibrations; only those vibrations altered in the transition appear in the difference spectrum. Difference bands of the order of 10^{-6} absorbance units are observable demonstrating the superb sensitivity of this technique. Thus, bands of single vibrations are discernible in front of the strong background absorbance of the sample which comprises the entire protein, the lipids of the membrane, buffer, detergent, and, finally, water as the solvent.

Light is the ideal trigger in reaction-induced difference spectroscopy as the sample must not be touched in-between the recording of the two protein states. Moreover, the reaction can be started remotely without any interference with the experimental setup. However, only a few proteins respond to light but are rather stimulated by electrons, membrane potential, binding of external ligands or other proteins, *etc*. Triggering by the former has been made possible by the development of a spectroelectrochemical thin layer cell[48] where the potential variation leads to redox-induced difference spectra of electron-transferring proteins. The possibility to initiate electron transfer at an electrode allows for pulsed experiments to be performed. Ultimately, such experiments can be combined with time-resolved approaches. However, only the layer of protein molecules which is in immediate contact with the electrode will accept electrons. The next layer is too distant from the electrode and direct electron transfer becomes virtually impossible. Therefore, the sensitivity of FT-IR difference spectroscopy must be pushed to detect the vibrational changes of a protein monolayer. This has been recently achieved by the development of surface-enhanced infrared difference spectroscopy (SEIDA) which is very useful for studies on proteins that respond to electrons[49-51] or to transmembrane-potential differences.

2.2 Finding a Suitable Environment: Sampling Techniques

2.2.1 Transmission Spectroscopy

Most commonly, IR spectrosocopy is performed in the transmission configuration, in which the IR radiation passes directly through the sample. Biological samples require water to be present, and the strong water absorbance obscures too much of the IR spectrum. For this purpose, concentrated samples of protein are deposited on an IR-transmissible window (usually CaF_2 or BaF_2) and slowly dried to reduce the water content. A homogeneous protein film is formed which is placed in a humidity chamber and rehydrated. A second window seals the hydrated protein sample from the dry atmosphere of the IR spectrometer. Alternatively, highly concentrated protein samples ($\sim 30\ \mu M$) can be obtained by ultracentrifugation. The resulting pellet is squeezed in-between two IR-transmissible windows.[52]

2.2.2 Attenuated Total Reflection (ATR)

Another way of circumventing the strong absorption of water is the application of the attenuated total reflection (ATR) technique. This technique has proven to be extremely useful for studies on membrane proteins. When the IR radiation is passed through an IR transmissive crystal material called an *internal reflection element* (IRE) at greater than the critical angle (Figure 1, left panel), it is totally internally reflected. A sample applied to the surface of this IRE receives IR radiation from evanescent waves generated at the interface between the (optically denser) IRE and the (optically rarer) sample.[53] These evanescent waves provide the IR energy for absorption by the sample.

The decay of the intensity of the evanescent wave is exponential. The depth to which the evanescent waves penetrate the sample is characterized by the distance at which the intensity has decayed to $1/e$:

$$d_p = \frac{\lambda/n_1}{2\pi[\sin^2\Theta - (n_2/n_1)^2]^{1/2}} \tag{1}$$

(with d_p = penetration depth, λ = wavelength, n_1 = index of refraction of the IRE, n_2 = index of refraction of the sample, and Θ = angle of incidence)

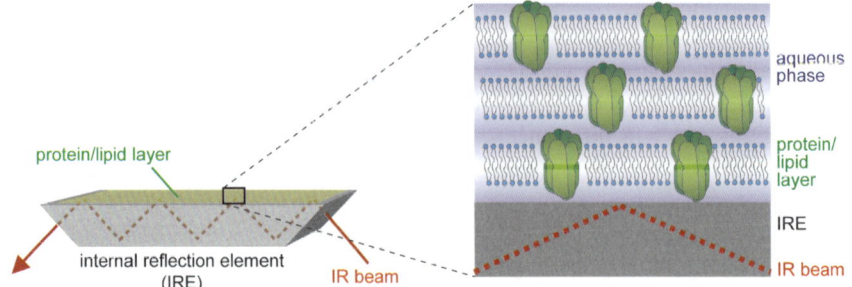

Figure 1 *The geometric configuration of the path of the infrared radiation through the internal reflection element (IRE). At each of the reflection points, an evanescent wave is generated on the top surface of the IRE penetrating into the adsorbed protein/lipid layer. This evanescent wave IR radiation is absorbed by the membrane protein samples. The schematic enlargement illustrates the interface between the IRE and the adsorbed stack of layers of membrane protein reconstituted with artificial lipids. A buffer solution bathing the sample can be exchanged for one of different pH, ionic strength, or chemical composition*

If a sample is enriched at the surface of the IRE, only the sample material is probed by the evanescent wave, provided sample thickness is larger than the penetration depth of radiation. Sample enrichment is usually achieved by gently drying a concentrated membrane suspension on the surface of the IRE. Under appropriate conditions (ionic strength and pH) the resulting film does not dissolve when rewetted with an aqueous solution (Figure 1, right panel). Water penetrates in between the membrane layers, swelling the stack.[54]

The great advantage of using evanescent waves lies in the reservoir of aqueous buffer. This can be exchanged for another solution of differing pH or chemical composition. This possibility of perfusion exchange permits spectra of the same sample to be recorded under a wide variety of solution conditions. The sample must be prepared in such a way that it is adhesive to the IRE and can withstand exchange of the solution. Membrane proteins are naturally well suited for adhesion to the IRE; some can be purified with their natural lipids, facilitating their adhesion to the IRE (*e.g.* bacteriorhodopsin residing in the native purple membrane). Yet, most membrane proteins are purified in detergent, which does not adhere to the IRE. In such cases, detergent can be removed or exchanged for lipid in order to make the protein sample adhesive to the IRE[36] (Figure 1, right panel). Despite the inevitable loss of photons in a reflection experiment, ATR spectroscopy has been successfully combined with time-resolved step-scan spectroscopy to record the reaction dynamics of bacteriorhodopsin under a wide variety of external parameters.[54–58]

2.2.3   Surface-enhanced Infrared Absorption (SEIRA)

Surface-enhanced spectroscopy, as an advance in ATR methodology, can be used to probe proteins on the level of a single monolayer. There, the noble metal surface exhibits a strong electromagnetic field that leads to the enhancement of vibrational transitions of bound molecules.[59,60] The enhancement decays rapidly with the distance from the surface ($<$ 8 nm) which eliminates the bulk contribution to the IR spectrum and selectively detects signals from the adsorbed monolayer even when the surface is immersed in water.[49,61–65] On this basis, we have recently developed SEIDA (surface-enhanced infrared difference absorption) spectroscopy and detected electrochemically induced changes in a cytochrome *c* monolayer adsorbed on to a modified gold surface.[49,51] Even time-resolved experiments on electron transferring proteins are now feasible as the protein action can be triggered by pulsed electron injection from the gold electrode into the protein monolayer.

The short range effect of SEIRA requires that the molecule resides close to the metal surface. Gold can be functionalized to attach virtually any protein to the metal surface *via* Ni-NTA[a] modification.[50,66–71] Once tethered to the surface, the detergent-solubilized membrane protein is reconstituted directly at the surface into a lipid layer. Such a biomimetic membrane is fully functional in catalysis.[50,72,73] Protein–protein interactions can be studied on the monolayer level with molecular resolution provided by vibrational spectroscopy.

2.3 Resolving the Reaction Dynamics: Time-resolved Step-scan Spectroscopy

The time taken to record an IR spectrum with an interferometric spectrometer is limited by the speed of the movable mirror. With fast-scanning spectrometers, the limit is at about 10 ms with a spectral resolution of 4 cm^{-1}. Beyond that, the co-planarity between the two mirrors of the interferometer is lost which prevents faster scanning velocities. For significantly higher time-resolution, time-resolved step-scan spectroscopy was developed about two decades ago.[74,75] In contrast to continuously scanning spectrometers, the movable mirror is set fixed at a certain pathlength difference and the reaction is initiated. Alike to flash photolysis, the time-course of the intensity changes at this mirror position is recorded by a transient digitizer. Then, the mirror is moved to the next pathlength difference and the procedure is repeated. By stepping to all pathlength differences required to record a complete interferogram, a three-dimensional data matrix is generated with the time traces of each retardation point. The matrix is rearranged to yield interferograms at certain time points after initiation of the reaction. These interferograms are Fourier-transformed to give spectra at the respective time point. With this principle, the speed of the detector electronics is the only limiting factor for the time-resolution. State-of-the-art step-scan spectroscopy allows the recording of data with ns time-resolution.[76–82]

It is obvious that the step-scan technique requires strict reversibility of the reaction under study. This is true for proteins that undergo a cyclic reaction; *i.e.* they eventually return to their initial state after the reaction has been started (for example bacteriorhodopsin, see Section 3). However, it should be considered that the number of reaction cycles to be initiated for a time-resolved step-scan experiment is enormous. An interferogram recorded at a typical frequency resolution of 4 cm^{-1},

---

[a] Nickel nitrilotriacetic acid.

requires about 1000 pathlength differences[b]. To achieve a noise level of lower then $\Delta A = 10^{-3}$, at least 20 averages at each pathlength difference have to be recorded. This results in 20,000 repetitions of the reaction for a single step-scan experiment! If the reaction to be studied is completed within one second, then it takes 20,000 seconds (or 5.5 hours) to perform such a step-scan experiment.

If one has overcome the experimental difficulties and succeeded in performing a step-scan experiment, the huge amount of information is both boon and bane. As a representative example, Figure 2 depicts the time-resolved absorbance changes of bacterior-hodopsin starting from 10 μs after the pulsed laser initiates the cyclic photoreaction. The measured data matrix comprises about two million data points. It is clear that massive data reduction has to be applied without losing significant information.[54,55] Details of the vibrational assignment of the difference bands and the consequences for the catalytic mechanism of this proton pump will be given in the next section.

3   Bacteriorhodopsin

3.1   The Molecule and its Photoreaction

Bacteriorhodopsin (bR) is the simplest biological machinery that converts light energy into a proton gradient (see[83–85] for reviews). Bacteriorhodopsin resides in the

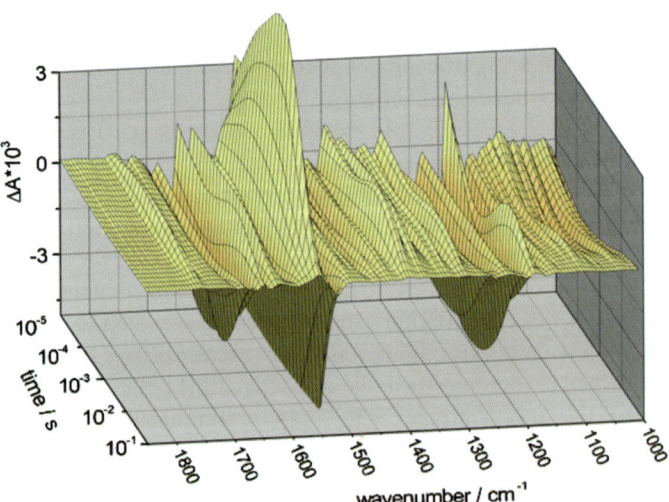

Figure 2 *IR absorbance changes of bacteriorhodopsin after excitation with a nanosecond laser pulse. Data have been recorded by time-resolved FT-IR step-scan spectroscopy in combination with the ATR technique and fitted to a sum of exponentials. The large negative difference band at 1525 cm⁻¹ is cut to facilitate the observation of smaller band features. The protein film is immersed in a phosphate-buffered solution of 1M KCl, pH 7.4 at 20°C*

---

[b] If the free spectral range is optically limited to frequencies below 1950 cm⁻¹ by the use of a broadband interference filter.

cell membrane of the ancient *Halobacterium salinarum*. The single polypeptide chain (248 amino acids) traverses the membrane by forming a seven α-helical bundle within the membrane layer (Figure 3A). The cofactor all-*trans* retinal is covalently bound to K216 via a protonated Schiff-base linkage. The three-dimensional structure of bR has been resolved down to a resolution of 1.55 Å[86] which is the highest resolution obtained so far with a membrane protein.

Absorption of light is accomplished by the chromophore retinal. Photon excitation of all-*trans* retinal leads to a series of isomerization reactions (Figure 3B) which put stress on the surrounding protein moiety. To compensate for the energy input the protein responds by structural alterations. Thereby, the acidity of protonable residues is altered and intraproteinous proton transfer takes place. Figure 3B displays the photocycle with the nomenclature of the intermediates.[87–89] They are characterised by their absorption maximum and their lifetimes span the enormous range of 11 decades.

Excitation of the bR ground state leads to the formation of a sub-picosecond product, termed the J intermediate. The transition of the vibrationally hot J intermediate to the subsequent K state proceeds with a time constant of 3 ps. Relaxation of the configurational strain of the retinal molecule leads to the L intermediate. Thus, the initial photoinduced *trans/cis* isomerization of retinal is finally settled in the L intermediate. The subsequent M intermediate represents the key intermediate of the photocycle as

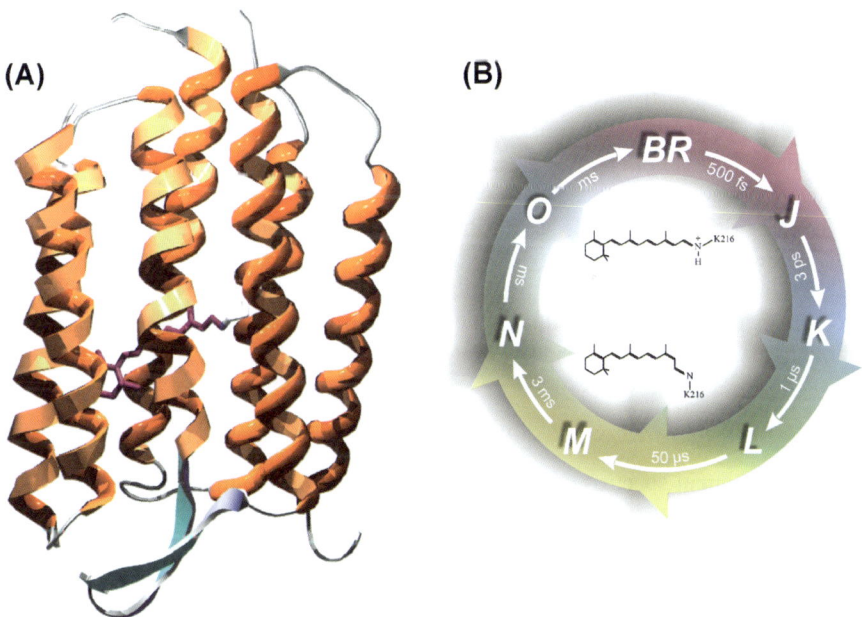

Figure 3 (A) *Crystallographic model of the three-dimensional structure of bacteriorhodopsin (PDB entry: 1C3W). (B) Photocycle of bR showing the light-adapted ground state BR, and the intermediates states J, K, L, M, N, and O along with their lifetimes. The background represents the color of the protein in the respective state. The configurations of all-*trans and 13-*cis retinal are depicted in the center of the photocycle*

the retinal Schiff base (RSB) has lost its proton. Deprotonation of the Schiff base initiates a series of proton transfer reactions within the protein that finally leads to proton translocation across the whole membrane. The late intermediates, N and O, are characterised by a reprotonated RSB. The color difference between these intermediates is primarily due to the different retinal configuration which is *13-cis* in N and all-*trans* in the O state. Finally, the parent state is recovered to complete the photocycle.

The photoactivity of bR leads to the transfer of one proton per photocycle from the cytoplasm to the extracellular medium.[90] Proton translocation across the whole membrane (membrane thickness: ~ 5 nm) can be broken down into distinct proton-transfer steps from donor to acceptor residues within bR. It is one of the major accomplishments of FT-IR spectroscopy to resolve these single steps spatially as well as temporally.

3.2 Vibrational Differences and Their Assignments

The IR difference spectrum contains information about specific structural changes to individual residues of the protein. In order to extract this information, the vibrational bands must be identified. Although the IR difference spectrum is greatly simplified as compared to the absorbance spectrum, the assignment of each feature in the spectrum is still a formidable task.

For more than 25 years, vibrational spectroscopy has added molecular information to the understanding of the proton pump bacteriorhodopsin.[1,91,92] With static and time-resolved FT-IR spectroscopy highly resolved difference spectra of the intermediate states are accessible. As an example, difference spectra between ground-state bR (negative bands) and the intermediates K, L, M, N, and O (positive bands) are displayed in Figure 4. It was a major contribution of time-resolved ATR/FT-IR spectroscopy to obtain pure difference spectra of the L, M, N, and O intermediate. This was achieved by variation of temperature and pH to extract the respective difference spectrum without admixtures from other intermediates.[55]

The large difference bands are attributed to the C=C and C−C stretching vibrations of the cofactor retinal. They arise from the change in dipole moment associated with the light-induced all-*trans* to 13-*cis* isomerization of retinal. The largest difference bands are in the M-BR difference spectrum because of the drastic change in dipole moment upon deprotonation of the retinal Schiff base.

3.2.1 Retinal Modes

Resonance Raman spectroscopy was the forerunner for assigning the retinal vibrations[92,93] and most of these assignments have been confirmed by infrared spectroscopy. The retinal modes of bR have been primarily assigned with the help of isotopically labeled retinal analogs. This experimental approach took advantage of the fact that the cofactor retinal is conveniently removed by hydroxylamine treatment of bacteriorhodopsin under intense illumination.[94] Bacteriorhodopsin readily incorporates chemically synthesized retinal analog by incubation under physiological conditions.[93,95]

The most intense bands in the IR difference spectrum are observed in the 1510–1560 cm^{-1} region and correspond to the (degenerate) C=C double bond

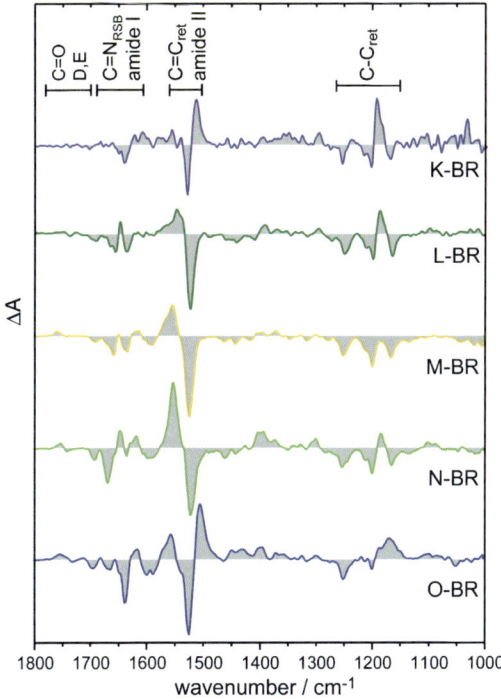

Figure 4 *Infrared difference spectra of the photocycle intermediates of bR. All but the K-BR spectrum have been extracted from time-resolved ATR/FT-IR experiments under the following conditions: L-BR (10 μs, pH 6.6, 20 C), M-BR (300–400 μs, pH 8.4, 20° C), N-BR (80–100 μs, pH 8.4, 20°C), and O-BR (5–10 μs, pH 4.0, 40°C). The K-BR difference spectrum was obtained under steady-state conditions (continuous illumination of a bR microcrystal with λ = 514 nm from an argon ion laser, sample kept at 100 K). Spectra are scaled to yield identical negative absorbance at 1252 cm⁻¹. Regions of marker bands are indicated by horizontal bars (D = aspartic acid; E = glutamic acid; RSB = retinal Schiff base; ret = Retinal)*

stretching vibration of the retinylidene moiety (Figure 4). The vibrational frequency correlates to the absorption maximum of the respective state in the visible wavelength region.[96] The bands of the C–C single bond stretching vibrations (1160–1260 cm⁻¹) are diagnostic for the retinal conformation. In the 13-*cis* intermediates, a positive band appears at 1192 cm⁻¹ for K, at 1189 cm⁻¹ for L and at 1185 cm⁻¹ for the N state. In the O-state with all-trans retinal, a positive band at 1168 cm⁻¹ is observed.[55] The light-induced conformational alterations also lead to changes in the C–H in-plane bending of the retinal. They appear in the 1300–1400 cm⁻¹ region albeit coupled to the N–H bend of the retinal Schiff base. The corresponding hydrogen-out-of-plane (HOOP) bending modes are marker bands for a distortion in the planarity of the retinal. Bands at 985 and 957 cm⁻¹ have been assigned to the C_{15}-HOOP mode and indicate a twisted 13-*cis* geometry in the K state as compared to the subsequent L state.[76] Finally, the C=N stretching vibration of the retinal Schiff base in ground-state bR has been assigned[97] to a band at 1640 cm⁻¹ which characteristically shifts in the various photocycle intermediates due to its H-bonding sensitivity.[98] Despite the strong polarity of the C=N bond, the assignment of the band has been performed by resonance Raman spectroscopy because the strong overlap with protein modes complicates the detection by IR difference spectroscopy (*vide infra*).

3.2.2   *Vibrational Bands of Amino-acid Side Chains*

With the advent of molecular genetics the assignment of vibrations on the single amino-acid level became feasible. However, due to the peculiar difficulties of expressing, purifying and often reconstituting membrane proteins it took until 1988 when sufficient quantities of mutated bacteriorhodopsin were available for FT-IR spectroscopy.[99,100] Since that time rapid co-evolution has made the combination of FT-IR spectroscopy and molecular genetics a powerful tool in investigating protein function in detail. Actually, bR is the best example in this assignment strategy.

Ideally, if a point mutant is used, the vibration of the exchanged residue will disappear from the spectrum. However, special care has to be taken when selecting point mutations. A conservative exchange should be preferred over a more drastic change in size and/or charge. Otherwise, indirect effects might lead to changes in interactions between other residues which, in turn, result in differences in other spectral regions. Moreover, a point mutation that is lethal to the enzyme would render it impossible to record a difference spectrum.

The pioneering work of Engelhard *et al.*[6] revealed that besides the retinal Schiff base, aspartic acids play a dominant role in proton transfer within bR. IR spectroscopy is particularly useful when studying the role of acidic amino acids because the C=O stretching frequency is well isolated from other vibrations of the protein. Figure 5 shows a three-dimensional representation of the time-course of absorbance changes in the carbonyl region where protonation changes of aspartates (or aspartic

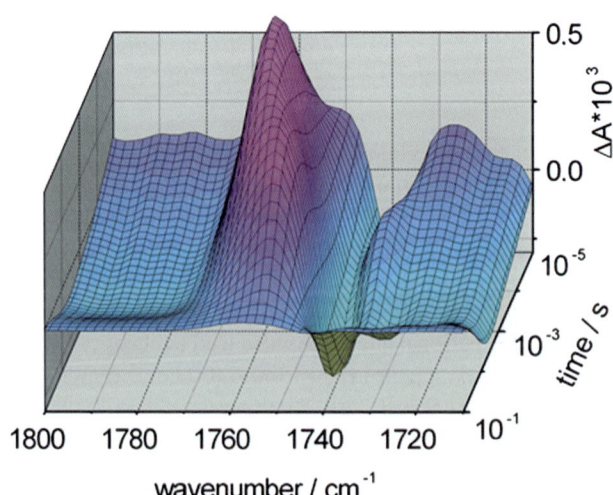

Figure 5 *Zoom-out of the absorbance changes of bacteriorhodopsin in the carbonyl region of the mid-IR range (see Figure 2 for the experimental conditions)*

acids, respectively) are monitored. Difference spectra in the carbonyl region of the photocycle intermediates L, M, N, and O are displayed in Figure 6.

The first application of site-directed mutagenesis to band assignment of infrared difference spectra of bR was published by Braiman *et al.*[99] In the M-BR difference spectrum of the D85E mutant, a strong band appears at 1725 cm^{-1} and the band of wild-type bR at 1761 cm^{-1} is absent (Figure 7). By these means, D85 was clearly identified to be the primary acceptor of the proton released by the retinal Schiff base in M. The large downshift of the carbonyl stretch indicates that the glutamic acid is more strongly hydrogen bonded in the D85E mutant than the aspartic acid of the wild-type which resides in a rather hydrophobic environment during the M state.[101] In other terms, the frequency of the C=O stretch correlates to the pK_a of the aspartic (or glutamic) acid side chain.[102–104]

Besides the assignment of the C=O stretching mode of D85, an enormous amount of effort has been spent on assigning difference bands of other acidic residues.[99,105–112] One of the most impressive works on assignment of vibrational bands by amino-acid substitution was performed by Sasaki *et al.*[113] They carried out a thorough investigation of the smaller difference bands in the carbonyl stretching region. The analysis

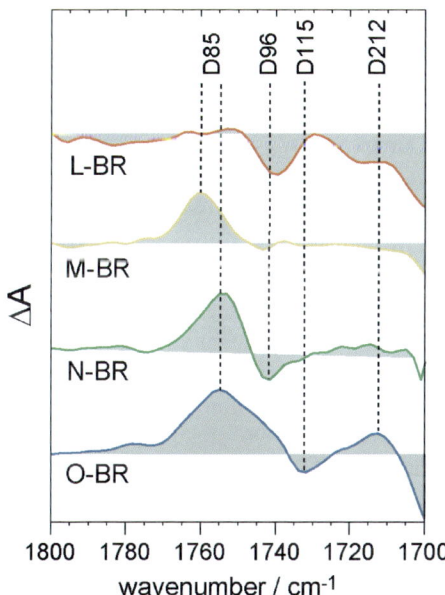

Figure 6 *Time-resolved ATR/FT-IR difference spectra in the carbonyl region of the photore-action of bacteriorhodopsin. The difference spectra of the respective intermediate states (L, M, N, and O versus bR) have been recorded under different environmental conditions to obtain clean difference spectra without contributions from other intermediate states (see Figure 4 for details). The band assignment (dashed vertical lines) illustrates the transient acid/base reactions of particular amino acids along with the environmental changes in the vicinity of the respective carboxylic acid. Negative bands are due to the deprotonation of an aspartic acid whereas the protonation is indicated by a positive band*

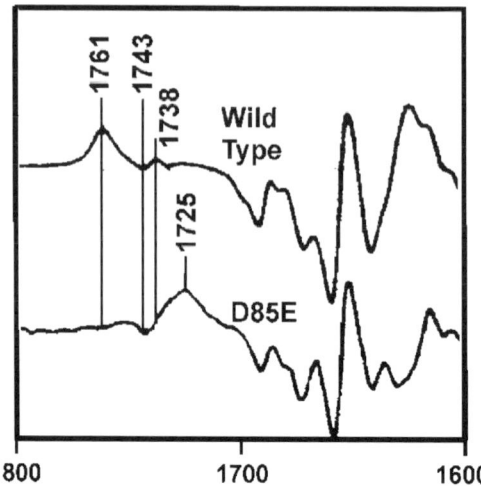

Figure 7 *Effect of the D85E mutation (lower spectrum) on the C5O vibration of wild-type bR*
(upper spectrum)
(Data have been taken from Braiman et al.[99] with permission of the American
Chemical Society)

was hampered by the fact that the absorbance of these difference bands is much
lower than the aforementioned C=O mode of D85.

Figure 8 depicts the M-BR difference spectrum. Mutations at positions 96 and 115
have been used to assign the band structure between 1742 cm^{-1} and 1700 cm^{-1}. In
the M-BR difference spectrum of the wild-type bR band maxima at 1740 cm^{-1} (positive) and at 1732 cm^{-1} (negative), respectively, appear (spectrum (a)). In the D96N
mutant (spectrum (b)) the band maximum at 1740 cm^{-1} is lowered in amplitude and
shifted in frequency to 1742 cm^{-1}. The negative band is also up-shifted by 2 cm^{-1}
and higher in amplitude. The D115N mutant exhibits a minimum at 1742 cm^{-1} and
a maximum at 1736 cm^{-1} (spectrum (c)). Therefore, the band structure of the wild-
type bR is well explained by a contribution of a positive band at 1736 cm^{-1} and a
negative band at 1742 cm^{-1} due to D96, superimposed by a positive band at 1742
cm^{-1} and a negative band at 1734 cm^{-1} of D115 (Table 1). Spectral contributions
from other acidic residues are thus excluded. In order to confirm the above interpretation the double mutant D96N/D115N was used. Besides the C=O stretch of D85,
no bands above noise level were detected (spectrum (d)). Moreover, the summation
(spectrum (e)) of the spectra of the single-site mutation (b) and (c) shows the same
band pattern as the wild-type. By this means, it is clearly demonstrated that the band
pattern around 1740 cm^{-1} of the wild-type difference spectrum (a) is a linear super-
position of changes in the C=O modes of D96 and D115.

As described above, ATR spectroscopy allows for manipulations in the aqueous
subphase during FT-IR experiments. Szaráz et al.[111] exploited this fact to perform a
pH titration of bR. They could determine the apparent pK_a of D96 to be as high as
11.4 in ground-state bR. With time-resolved ATR/FT-IR spectroscopy at various pH
values, Zscherp et al.[56] even determined the *change* in pK_a of D96 during the

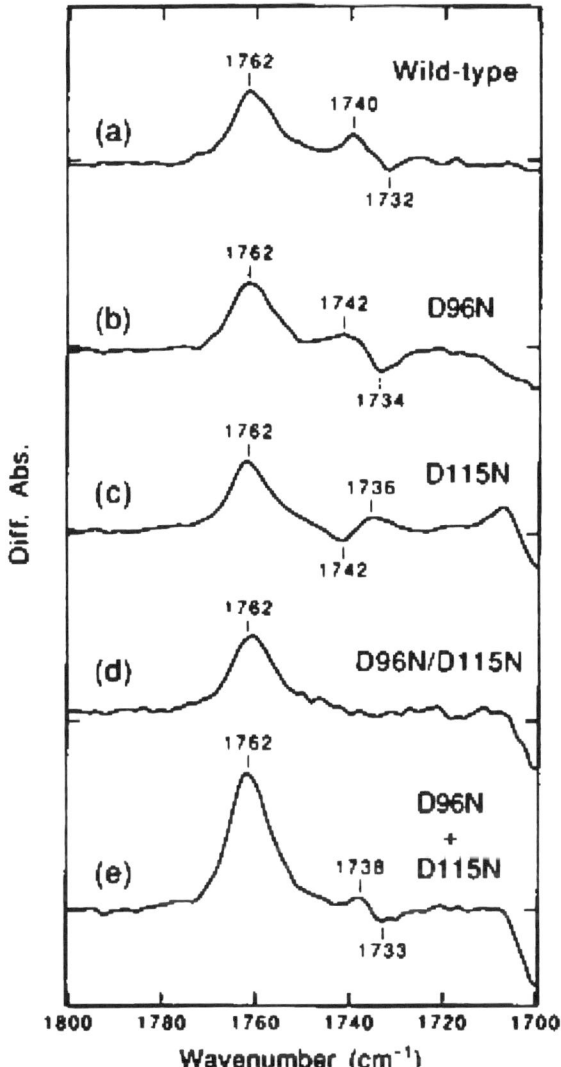

Figure 8 *Difference spectra of the photocycle intermediate M (left panel) of wild-type (a), D96N (b), D115N (c), and the double mutant D96N/D115N (d). (e) is a spectrum calculated by adding spectra (b) and (c).*
(Reprinted from Sasaki *et al. 113 with permission of the American Chemical Society)*

enzymatic activity of bR (*in situ*) which is a prerequisite for the chemical under-standing of proton transfer. They could demonstrate that the pK_a is decreased from 11.2 to 7.1 during the lifetime of the N intermediate.

It is obvious that a difference band only appears if an intermediate state has a suf-ficient lifetime, *i.e.* for the study of proton transfer reactions the residence time of a proton at a specific residue is the crucial factor. A member of the proton transfer

Table 1 *Frequencies (in cm^{-1}) of the carbonyl (C=O) and the corresponding symmetric carboxylate (COO$^-_{sym}$) mode of functionally relevant aspartic acids and aspartates, respectively, in ground-state bR and in the photocycle intermediates K, L, M, N, and O. Data have been compiled from*[56,58,99,105,108,113,115,117,143]

	D85		D96		D115		D212	
	C=O	COO$^-_{sym}$	C=O	COO$^-_{sym}$	C=O	COO$^-_{sym}$	C=O	COO$^-_{sym}$
BR	—	1385	1742	—	1734	—	—	1386
K	—	n.d.	1742	—	1731	—	—	n.d.
L	—	n.d.	1748	—	1729	—	—	n.d.
M	1761	—	1736	—	1742	—	—	n.d.
N	1755	—	—	1402	1740	—	—	n.d.
O	1753	—	1742	—	1740	—	1712	—

chain will release a proton as soon as it is protonated. Hence, the residence time for a proton is very short and it is impossible to detect any IR difference band. This happened to be the case for D38 where vibrational difference bands of this residue were absent in the time-resolved experiment.[114]

In the final stage of the proton transfer sequence across bR, D212 receives the proton from D85. This state is hardly accumulated in wild-type bR because deprotonation of D212 is fast compared to the protonation reaction. Consequently, the C=O stretch of the protonated species at 1712 cm^{-1} is small in amplitude (see O-BR in Figure 6[55]). In the E194Q mutant, however, deprotonation of D212 is slowed down and the C=O stretch of D212 gets much larger in amplitude. The band assignment was done by feeding the bacteria with ^{13}C-Asp. In the isolated E194Q mutant of bR, the band at 1712 cm^{-1} was characteristically shifted to lower frequency.[115] Because D212 is the only aspartate in the extracellular half-channel of bR, the band at 1712 cm^{-1} must have been due to the C=O stretch of this residue. Another assignment strategy used a quadruple mutant where all of the extracellular glutamates were replaced by glutamines.[58] Since the band at 1712 cm^{-1} persisted in this mutant, it was straightforward to conclude that it was from D212, the only carboxylic residue left in the extracellular proton transfer chain.

The carbonyl stretching vibration of aspartic and glutamic acids occurs in the region above 1700 cm^{-1} where hardly any other vibration absorbs. However, frequency changes of glutamines or asparagines due to electrostatic alterations in the vicinity of such residues should always be considered. Carboxylate vibrations of aspartates and glutamates are much harder to assign than the carbonyl vibration of the corresponding acids. The symmetric COO$^-$ stretch absorbs in the region around 1300–1400 cm^{-1}.[116] This frequency range is very crowded, predominantly from in-plane bending vibrations of other residues, and strong overlap is very likely to occur. Despite these obstacles, it was possible to assign the symmetric carboxylate vibration of some of the important aspartic residues of bR. The symmetric carboxylate vibration of D96 has been assigned to a band at 1402 cm^{-1} which was confirmed by ^{13}C-labeling experiments[108] and by the pH-dependence of the transient at this frequency.[56]

The symmetric carboxylate vibration of D85 absorbs at 1385 cm^{-1}.[56,117] This frequency is almost identical to that of the carboxylate vibration of D212 (1386 cm^{-1}).[58,115] Identical frequencies suggest an analogous environment of D85 and D212 in ground-state bR which agrees well with the crystallographic structure (see Figure 9) where a water molecule (water 402) symmetrically donates hydrogen bonds to the carboxylates of the respective aspartates. The other oxygens of the two carboxylates accept hydrogen bonds from a phenolic OH (from T89 in the case of D85, and from Y185 in the case of D212; Figure 9).

The complementary asymmetric vibration of the carboxylate absorbs around 1560 cm^{-1}.[116] This is the range of the very strong C=C stretching vibrations of retinal and the amide II modes. The overlap of these large bands with the asymmetric carboxylate vibration makes the latter very difficult to observe experimentally.[57,58]

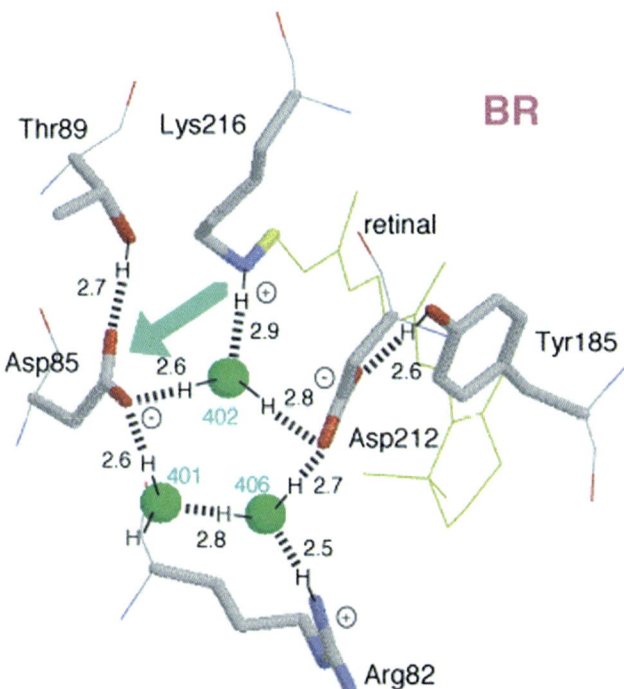

Figure 9 *X-ray crystallographic structures of the Schiff base region in BR. The membrane normal is approximately in the vertical direction of this figure. Upper and lower regions correspond to the cytoplasmic and extracellular sides, respectively. Green spheres (401, 402, and 406) represent water molecules which form a roughly pentagonal cluster with an oxygen from Asp85 and an oxygen from Asp212. The water-containing pentagonal cluster structure stabilizes the electric quadrupole in this region. Hydrogen atoms and hydrogen bonds (dotted lines) are supposed from the structure, while the numbers correspond to the hydrogen-bonding distances in Å. After photoisomerization of the retinal, the Schiff-base proton is transferred to Asp85 in the L-to-M transition (arrow), although the two aspartates (D85 and D212) are symmetrically arranged. (Taken from* Kandori[127] *with permission of Elsevier)*

Arginine vibrations give rise to strong bands in the mid-IR. In fact, the extinction coefficient of the symmetric and the asymmetric C=N stretching vibration of the terminal guanidine group are even higher then those of the C=O stretch of aspartic or glutamic acid.[116] However, their frequencies overlap with the amide I mode of the peptide bond and are, therefore, very difficult to disentangle. Braiman's group has studied the vibrational bands of arginine by an arsenal of (bio-) chemical approaches.[118–123] They have recently demonstrated that R82 is (at least partially) deprotonated in the M state of bR.[123] In ground-state bR, the asymmetric C=N stretch of the protonated guanidine side chain absorbs at 1660 cm^{-1}.[c] In the M-BR difference spectrum, this band is lost and a new band at 1556 cm^{-1} appears which indicates the presence of the neutral arginine residue. It is obvious that the frequency of the C=N vibration of the cationic arginine is very sensitive to the electrostatics in the immediate environment.[122] The frequency has been found to be as high as 1695 cm^{-1} for R108.[118,124]

3.2.3 Internal Water Molecules

The detection of vibrational changes of single water molecules within a protein has been pioneered by Maeda *et al.*[125] Usually, the strong absorption of solvent water renders it impossible to record a difference spectrum in the 3700–3000 cm^{-1} region where the O–H stretching vibration of water absorbs. However, under proper hydration conditions the functionality is not strongly impaired and the lowered background absorption allows for the analysis of the difference spectrum.

The selective identification of the O–H vibrations of water molecules (symmetric and asymmetric stretching vibration) is facilitated by the fact that the respective bands are down-shifted by ~ 10 cm^{-1} upon exchange by $H_2^{18}O$. By this procedure, vibrational contributions from amino-acid side chains with OH-groups (Tyr, Ser, Thr and protonated Asp and Glu) are elegantly excluded. The assignment to a particular water molecule and its position is carried out by point mutations where a putative hydrogen-bonding amino-acid side chain is exchanged. As a consequence, the H-bond is altered giving rise to a frequency shift in the O–H stretching vibration of the adjacent water molecule. Moreover, the frequency of the O–H stretching vibration is strongly dependent on the hydrogen-bonding environment making it an exquisite tool to probe the local environment around the water molecule. This strategy has been applied to identify the structure and the functional role of several water molecules in the proton transfer pathway of bR (see Figure 9 and reviewed in[126,127]).

A cavity close to the extracellular membrane surface accommodates a local area network (LAN) of hydrogen-bonded water molecules and amino-acid side chains (R82, E204, and E194). This LAN houses an excess proton which is released after photoexcitation of bacteriorhodopsin. Such protonated networks give rise to broad continuum absorbancies across the whole mid-IR region.[128] Gerwert's group has demonstrated that the absorbance change in the region of 1800–1900 cm^{-1} is indicative of the proton release from a protonated water network stabilized by polar amino acids.[129,130]

---

[c] Though not explicitly noted in the publication,[123] the symmetric C--N stretch is discernible at around 1640 cm^{-1} from their Fig. 4.

3.3 The Sequence of Proton Transfer Steps

Proton translocation by bR is initiated by the light-induced isomerization of the chromophore retinal. Isomerization causes an electronic redistribution of retinal which leads to an increase in acidity of the retinal Schiff base (RSB).[131] In addition, the Schiff-base nitrogen is transferred into a less hydrogen-bonded environment[132] which destabilizes the positive charge on the RSB and the high pK_a of the RSB (pK_a(RSB) = 13[133]) is decreased.

The electronic as well as the conformational changes of the retinal moiety trigger the following proton transfer reactions. These reactions can be subdivided into the proton release to the extracellular surface which proceeds with a time constant of 70 μs,[134] and proton uptake from the cytoplasm which takes 11 ms.[90] Finally, the initial proton release group is reprotonated which represents the last step in the sequence of proton transfer events. As demonstrated in the previous section, the assignment of the vibrational bands in the IR difference spectrum leads to an atomic description of the singular proton transfer steps (Figure 10) that comprise proton translocation across bR.

3.3.1 Proton Release

As a consequence of the decreased pK_a, the proton of the RSB is transferred to D85 (step 1 in Figure 10) and the M intermediate is established. This proton-transfer

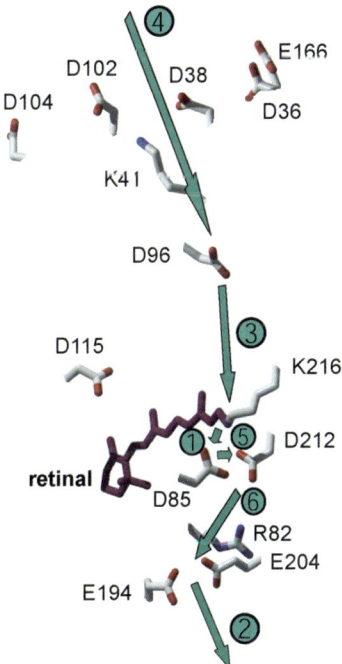

Figure 10 *Sequence of proton transfer events in bacteriorhodopsin (steps 1–6). The dark green arrows denote proton transfer reactions taking place among particular amino acids and the retinal. The residues and their role are discussed in the text*

reaction is monitored on the single vibrational level by FT-IR spectroscopy. The band at 1761 cm^{-1} in M (Figure 6) has been assigned to the primary acceptor of the Schiff-base proton, D85 (see above). The time constant of the proton transfer to D85 is 50 μs. Although D85 and D212 are symmetrically arranged in the ground-state structure of bR (see Figure 9), the movement of a water molecule towards D85 in the L state favours it as the acceptor of the Schiff-base proton in the following M state (*hydration switch model*[127]).

Neutralisation of D85 after proton transfer removes the initial electrostatic inter-action with R82. Thus, the strained conformation of the R82 is relieved and the side chain swings to interact with the proton-release group.[135,136] The proton-release group is not thought to be a single site but is rather represented by a hydrogen-bonded network comprising at least E194, E204, and several water molecules among which a delocalized proton is shared. The approach of the positively charged R82 'kicks' the proton out to the extracellular membrane surface (step 2 in Figure 10). Experimental support for this model has been achieved from the change in the con-tinuum band which occurs in the timescale of proton release to the membrane sur-face.[130] Moreover, no other carboxylic group besides D85 changes its protonation state during proton release (E9, E74, E194,[56] and E204[56,130]). Yet, R82 may at least partially deprotonate upon interaction with the proton-release group.[123]

3.3.2   Proton Uptake

Deprotonation of the RSB leaves a proton 'defect' which is refilled by proton transfer from D96[105] (step 3 in Figure 10). The change in protonation state of D96 is detected in the N-BR difference spectrum by a negative band at 1741 cm^{-1} (Figure 6) which occurs with a time constant of 3 ms. The unusually high pK_a of 11.4[111] drops by more than 4 pH-units to 7.1[56] enabling D96 to reprotonate the RSB. This reaction defines the M-to-N transition. The increase in acidity of D96 is caused by structural rearrange-ments in the protein backbone observed by the appearance of difference bands in the amide I region (see the N-BR difference spectrum in Figure 4) which leads to an influx of water molecules that hydrate D96.[137]

D96 itself is reprotonated from the cytoplasm (step 4 in Figure 10). K41 and D38 are candidates for important residues lining the putative proton pathway from the cytoplasmic surface to D96. In fact, mutations of D38 exert drastic effects on the proton-uptake reaction.[114] Unexpectedly, however, it is the reprotonation reaction of the RSB that is affected and not the reprotonation of D96. This puzzling result sug-gests an involvement of D38 in the conformational changes of the protein. Indeed, changes in tertiary structure during the photocycle are inhibited in the D38R mutant at neutral pH.[114,138–140] Mutations of K41, which is just one helix-turn further into the membrane (Figure 10), affected neither the kinetics of the photo- nor of the proton cycle.[85] It is concluded that though spatial position and presumed charge favours the interaction with D38, K41 is dispensable for proton pumping.

The cytoplasmic surface of bR comprises many charged residues. Point mutations of several acidic amino acids (D36, D102, D104, E161) exert a slight deceleration of proton uptake by bR.[114] The role of these amino acids was ascribed to efficiently collecting protons from the aqueous bulk phase and funneling them to the entrance of the cytoplasmic proton pathway where D38 is located. Such a proton-collecting antenna was also found in cytochrome *c* oxidase[141] and photosynthetic reaction centers[142] which points to a general property of proton pumps.

3.3.3 Reprotonation of the Proton-release Group

In the final steps of proton translocation by bR, D85 is deprotonated. However, the proton is not directly delivered to the proton-release group but nearby D212 is protonated as an intermediate step (see step 5 in Figure 10, and Figure 6 for the data). With proton transfer from D212 to the proton release group (step 6 in Figure 10), all proton donors and acceptors are reset to their initial states and the next sequence of proton transfer steps can be initiated by light.

References

1. A. Maeda, *Isr. J. Chem.*, 1995, **35**, 387.
2. K.J. Rothschild, *J. Bioenerg. Biomembr.*, 1992, **24**, 147.
3. F. Siebert, *Methods Enzymol.*, 1995, **246**, 501.
4. F. Siebert, W. Mäntele and W. Kreutz, *Biophys. Struct. Mech.*, 1980, **6**, 139.
5. F. Siebert, W. Mäntele and W. Kreutz, *FEBS Lett.*, 1982, **141**, 82.
6. M. Engelhard, K. Gerwert, B. Hess, W. Kreutz and F. Siebert, *Biochemistry*, 1985, **24**, 400.
7. K.J. Rothschild, M. Zagaeski and W.A. Cantore, *Biochem. Biophys. Res. Commun.*, 1981, **103**, 483.
8. K. Bagley, G. Dollinger, L. Eisenstein, A.K. Singh and L. Zimanyi, *Proc. Natl. Acad. Sci. U.S.A.*, 1982, **79**, 4972.
9. F. Siebert and W. Mäntele, *Eur. J. Biochem.*, 1983, **130**, 565.
10. J. Breton, *Biochim. Biophys. Acta*, 2001, **1507**, 180.
11. W.G. Mäntele, A.M. Wollenweber, E. Nabedryk and J. Breton, *Proc. Natl. Acad. Sci. U.S.A.*, 1988, **85**, 8468.
12. E. Nabedryk, W. Leibl and J. Breton, *Photosynth. Res.*, 1996, **48**, 301.
13. S. Kim and B.A. Barry, *J.Phys.Chem.B*, 2001, **105**, 4072.
14. R. Brudler, H.J. de Groot, W.B. van Liemt, W.F. Steggerda, R. Esmeijer, P. Gast, A.J. Hoff, J. Lugtenburg and K. Gerwert, *EMBO J.*, 1994, **13**, 5523.
15. A. Remy and K. Gerwert, *Nat. Struct. Biol.*, 2003, **10**, 637.
16. C. Berthomieu, R. Hienerwadel, A. Boussac, J. Breton and B.A. Diner, *Biochemistry*, 1998, **37**, 10547.
17. R. Hienerwadel, S. Grzybek, C. Fogel, W. Kreutz, M.Y. Okamura, M.L. Paddock, J. Breton, E. Nabedryk and W. Mäntele, *Biochemistry*, 1995, **34**, 2832.
18. Y. Kimura, N. Mizusawa, A. Ishii, T. Yamanari and T.A. Ono, *Biochemistry*, 2003, **42**, 13170.
19. G. Hastings and V. Sivakumar, *Biochemistry*, 2001, **40**, 3681.
20. H.A. Chu, H. Sackett and G.T. Babcock, *Biochemistry*, 2000, **39**, 14371.
21. G.M. MacDonald and B.A. Barry, *Biochemistry*, 1992, **31**, 9848.
22. T. Noguchi, T. Ono and Y. Inoue, *Biochemistry*, 1992, **31**, 5953.

23. J. Behr, P. Hellwig, W. Mäntele and H. Michel, *Biochemistry*, 1998, **37**, 7400.

24. P. Hellwig, B. Rost, U. Kaiser, C. Ostermeier, H. Michel and W. Mäntele, *FEBS Lett.*, 1996, **385**, 53.

25. P. Hellwig, D. Scheide, S. Bungert, W. Mäntele and T. Friedrich, *Biochemistry*, 2000, **39**, 10884.

26. P. Hellwig, U. Pfitzner, J. Behr, B. Rost, R.P. Pesavento, W.V. Donk, R.B. Gennis, H. Michel, B. Ludwig and W. Mäntele, *Biochemistry*, 2002, **41**, 9116.

27. M. Iwaki, A. Puustinen, M. Wikström and P.R. Rich, *Biochemistry*, 2003, **42**, 8809.

28. M. Iwaki, L. Giotta, A.O. Akinsiku, H. Schagger, N. Fisher, J. Breton and P.R. Rich, *Biochemistry*, 2003, **42**, 11109.

29. M. Iwaki, A. Osyczka, C.C. Moser, P.L. Dutton and P.R. Rich, *Biochemistry*, 2004, **43**, 9477.

30. J.A. Bailey, F.L. Tomson, S.L. Mecklenburg, G.M. MacDonald, A. Katsonouri, A. Puustinen, R.B. Gennis, W.H. Woodruff and R.B. Dyer, *Biochemistry*, 2002, **41**, 2675.

31. O. Einarsdottir, R.B. Dyer, D.D. Lemon, P.M. Killough, S.M. Hubig, S.J. Atherton, J.J. Lopez-Garriga, G. Palmer and W.H. Woodruff, *Biochemistry*, 1993, **32**, 12013.

32. B.H. McMahon, M. Fabian, F. Tomson, T.P. Causgrove, J.A. Bailey, F.N. Rein, R.B. Dyer, G. Palmer, R.B. Gennis and W.H. Woodruff, *Biochim. Biophys. Acta*, 2004, **1655**, 321.

33. A. Puustinen, J.A. Bailey, R.B. Dyer, S.L. Mecklenburg, M. Wikstrom and W.H. Woodruff, *Biochemistry*, 1997, **36**, 13195.

34. W.H. Woodruff, *J. Bioenerg. Biomembr.*, 1993, **25**, 177.

35. D. Heitbrink, H. Sigurdson, C. Bolwien, P. Brzezinski and J. Heberle, *Biophys. J.*, 2002, **82**, 1.

36. R.M. Nyquist, D. Heitbrink, C. Bolwien, T.A. Wells, R.B. Gennis and J. Heberle, *FEBS Lett.*, 2001, **505**, 63.

37. R.M. Nyquist, D. Heitbrink, C. Bolwien, R.B. Gennis and J. Heberle, *Proc. Natl. Acad. Sci. U.S.A.*, 2003, **100**, 8715.

38. R.M. Nyquist, K. Ataka and J. Heberle, *Chem. Biochem.*, 2004, **5**, 431.

39. T. Iwase, C. Varotsis, K. ShinzawaItoh, S. Yoshikawa and T. Kitagawa, *J. Am. Chem. Soc.*, 1999, **121**, 1415.

40. C. Koutsoupakis, E. Pinakoulaki, S. Stavrakis, V. Daskalakis and C. Varotsis, *Biochim. Biophys. Acta*, 2004, **1655**, 347.

41. E. Pinakoulaki, U. Pfitzner, B. Ludwig and C. Varotsis, *J. Biol. Chem.*, 2002, **277**, 13563.

42. M. Lübben and K. Gerwert, *FEBS Lett.*, 1996, **397**, 303.

43. M. Lübben, A. Prutsch, B. Mamat and K. Gerwert, *Biochemistry*, 1999, **38**, 2048.

44. K. Gerwert, *Curr. Opin. Struct. Biol.*, 1993, **3**, 769.

45. W. Mäntele, *Trends Biochem. Sci.*, 1993, **18**, 197.

46. R. Vogel and F. Siebert, *Curr. Opin. Chem. Biol.*, 2000, **4**, 518.

47. C. Zscherp and A. Barth, *Biochemistry*, 2001, **40**, 1875.

48. D. Moss, E. Nabedryk, J. Breton and W. Mäntele, *Eur. J. Biochem.*, 1990, **187**, 565.

49. K. Ataka and J. Heberle, *J. Am. Chem. Soc.*, 2003, **125**, 4986.

50. K. Ataka, F. Giess, W. Knoll, R. Naumann, S. Haber-Pohlmeier, B. Richter and J. Heberle, *J. Am. Chem. Soc.*, 2004, **126**, 16199.

51. K. Ataka and J. Heberle, *J. Am. Chem. Soc.*, 2004, **126**, 9445.

52. J. le Coutre, J. Tittor, D. Oesterhelt and K. Gerwert, *Proc. Natl. Acad. Sci. U.S.A.*, 1995, **92**, 4962.

53. N.J. Harrick, *Internal Reflection Spectroscopy*, John Wiley & Sons, Inc., New York 1967.

54. J. Heberle and C. Zscherp, *Appl. Spectrosc.*, 1996, **50**, 588.

55. C. Zscherp and J. Heberle, *J. Phys. Chem. B*, 1997, **101**, 10542.

56. C. Zscherp, R. Schlesinger, J. Tittor, D. Oesterhelt and J. Heberle, *Proc. Natl. Acad. Sci. U.S.A.*, 1999, **96**, 5498.

57. T. Friedrich, S. Geibel, R. Kalmbach, I. Chizhov, K. Ataka, J. Heberle, M. Engelhard and E. Bamberg, *J. Mol. Biol.*, 2002, **321**, 821.

58. C. Zscherp, R. Schlesinger and J. Heberle, *Biochem. Biophys. Res. Commun.*, 2001, **283**, 57.

59. M. Osawa, *Bull. Chem. Soc. Jpn.*, 1997, **70**, 2861.

60. R.F. Aroca, D.J. Ross and C. Domingo, *Appl. Spectrosc.*, 2004, **58**, 324A.

61. K. Ataka and M. Osawa, *Langmuir*, 1998, **14**, 951.

62. K. Ataka and M. Osawa, *J. Electroanal. Chem.*, 1999, **460**, 188.

63. K. Ataka, T. Yotsuyanagi and M. Osawa, *J. Phys. Chem.*, 1996, **100**, 10664.

64. M. Osawa, K. Ataka, K. Yoshii and T. Yotsuyanagi, *J. Electr. Spec.*, 1993, **64–5**, 371.

65. M. Osawa, in *Handbook of Vibrational Spectroscopy*, *Vol. 1*, J.M. Chalmers, P. R. Griffiths (eds.), Wiley, Chichester, 2002, 785.

66. D. Kröger, M. Liley, W. Schiweck, A. Skerra and H. Vogel, *Biosens. Bioelectron.*, 1999, **14**, 155.

67. C. Nakamura, M. Hasegawa, N. Nakamura and J. Miyake, *Biosens. Bioelectron.*, 2003, **18**, 599

68. P. Rigler, W.P. Ulrich, P. Hoffmann, M. Mayer and H. Vogel, *Chem. Physchem.*, 2003, **4**, 268.

69. E.L. Schmid, T.A. Keller, Z. Dienes and H. Vogel, *Analyt. Chem.*, 1997, **69**, 1979.

70. G.B. Sigal, C. Bamdad, A. Barberis, J. Strominger and G.M. Whitesides, *Analyt. Chem.*, 1996, **68**, 490.

71. R. Gamsjaeger, B. Wimmer, H. Kahr, A. Tinazli, S. Picuric, S. Lata, R. Tampe, Y. Maulet, H.J. Gruber, P. Hinterdorfer and C. Romanin, *Langmuir*, 2004, **20**, 5885.

72. M.G. Friedrich, F. Giess, R. Naumann, W. Knoll, K. Ataka, J. Heberle, J. Hrabakova, D. H. Murgida and P. Hildebrandt, *Chem. Commun.*, 2004, 2376.

73. F. Giess, M. Friedrich, J. Heberle, R. Naumann and W. Knoll, *Biophys. J.*, 2004, **87**, 3213

74. W. Uhmann, A. Becker, C. Taran and F. Siebert, *Appl. Spectrosc.*, 1991, **45**, 390.

75. C.J. Manning, R.A. Palmer and J.L. Chao, *Review of Scientific Instruments*, 1991, **62**, 1219.

76. O. Weidlich and F. Siebert, *Appl.Spectrosc.*, 1993, **47**, 1394.

77. R.A. Palmer, J.L. Chao, R.M. Dittmar, V.S. Gregoriuo and S.E. Plunkett, *Appl. Spectrosc.*, 1993, **47**, 1297.

78. X. Hu, H. Frei and T.G. Spiro, *Biochemistry*, 1996, **35**, 13001.
79. R. Rammelsberg, B. Heßling, H. Chorongiewski and K. Gerwert, *Appl. Spectrosc.*, 1997, **51**, 558.
80. A.K. Dioumaev, M.S. Braiman, *J. Phys. Chem. B*, 1997, **101**, 1655.
81. C. Rödig, I. Chizhov, O. Weidlich and F. Siebert, *Biophys. J.*, 1999, **76**, 2687.
82. P.Y. Chen and R.A. Palmer, *Appl. Spectrosc.*, 1997, **51**, 580.
83. U. Haupts, J. Tittor and D. Oesterhelt, *Annu. Rev. Biophys. Biomol. Struct.*, 1999, **28**, 367.
84. J.K. Lanyi, *FEBS Lett.*, 1999, **464**, 103.
85. J. Heberle, *Biochim. Biophys. Acta.*, 2000, **1458**, 135.
86. H. Luecke, B. Schobert, H.T. Richter, J.P. Cartailler and J.K. Lanyi, *J. Mol. Biol.*, 1999, **291**, 899.
87. D. Oesterhelt and B. Hess, *Eur. J. Biochem.*, 1973, **37**, 316.
88. R.H. Lozier, W. Niederberger, R.A. Bogomolni, S. Hwang and W. Stoeckenius, *Biochim. Biophys. Acta*, 1976, **440**, 545.
89. W. Sperling, P. Carl, C. Rafferty and N.A. Dencher, *Biophys. Struct. Mech.*, 1977, **3**, 79.
90. S. Grzesiek and N.A. Dencher, *FEBS Lett.*, 1986, **208**, 259.
91. T. Kitagawa and A. Maeda, *Photochem. Photobiol.*, 1989, **50**, 883.
92. T. Althaus, W. Eisfeld, R. Lohrmann and M. Stockburger, *Isr. J. Chem.*, 1995, **35**, 227.
93. J. Lugtenburg, R.A. Mathies, R.G. Griffin and J. Herzfeld, *Trends Biochem. Sci.*, 1988, **13**, 388.
94. D. Oesterhelt, L. Schuhmann and H. Gruber, *FEBS Lett.*, 1974, **44**, 257.
95. D. Oesterhelt and L. Schuhmann, *FEBS Lett.*, 1974, **44**, 262.
96. B. Aton, A.G. Doukas, R.H. Callender, B. Becher and T.G. Ebrey, *Biochemistry*, 1977, **16**, 2995.
97. A. Lewis, J. Spoonhower, R.A. Bogomolni, R.H. Lozier and W. Stoeckenius, *Proc. Natl. Acad. Sci. U.S.A.*, 1974, **71**, 4462.
98. B. Aton, A.G. Doukas, D. Narva, R.H. Callender, U. Dinur and B. Honig, *Biophys. J.*, 1980, **29**, 79.
99. M.S. Braiman, T. Mogi, T. Marti, L.J. Stern, H.G. Khorana and K.J. Rothschild, *Biochemistry*, 1988, **27**, 8516.
100. M.S. Braiman, T. Mogi, L.J. Stern, N.R. Hackett, B.H. Chao, H.G. Khorana and K.J. Rothschild, *Proteins*, 1988, **3**, 219.
101. K. Fahmy, O. Weidlich, M. Engelhard, J. Tittor, D. Oesterhelt and F. Siebert, *Photochem. Photobiol.*, 1992, **56**, 1073.
102. A.K. Dioumaev and M.S. Braiman, *J. Am. Chem. Soc.*, 1995, **117**, 10572.
103. M.S. Braiman, A.K. Dioumaev and J.R. Lewis, *Biophys. J.*, 1996, **70**, 939.
104. A.K. Dioumaev, *Biochemistry (Mosc)*, 2001, **66**, 1269.
105. K. Gerwert, B. Hess, J. Soppa and D. Oesterhelt, *Proc. Natl. Acad. Sci. U.S.A.*, 1989, **86**, 4943.
106. O. Bousché, M.S. Braiman, Y.-W. He, T. Marti, H.G. Khorana and K.J. Rothschild, *J. Biol. Chem.*, 1991, **266**, 11063.
107. M.S. Braiman, A.L. Klinger and R. Doebler, *Biophys. J.*, 1992, **62**, 56.

108. A. Maeda, J. Sasaki, Y. Shichida, T. Yoshizawa, M. Chang, B. Ni, R. Needleman and J.K. Lanyi, *Biochemistry*, 1992, **31**, 4684.

109. J. Sasaki, Y. Shichida, J.K. Lanyi and A. Maeda, *J. Biol. Chem.*, 1992, **267**, 20782

110. Y. Cao, G. Varo, A.L. Klinger, D.M. Czajkowsky, M.S. Braiman, R. Needleman and J.K. Lanyi, *Biochemistry*, 1993, **32**, 1981.

111. S. Száraz, D. Oesterhelt and P. Ormos, *Biophys. J.*, 1994, **67**, 1706.

112. A. Nilsson, P. Rath, J. Olejnik, M. Coleman and K.J. Rothschild, *J. Biol. Chem.*, 1995, **270**, 29746.

113. J. Sasaki, J.K. Lanyi, R. Needleman, T. Yoshizawa and A. Maeda, *Biochemistry*, 1994, **33**, 3178.

114. J. Riesle, D. Oesterhelt, N.A. Dencher and J. Heberle, *Biochemistry*, 1996, **35**, 6635.

115. A.K. Dioumaev, L.S. Brown, R. Needleman and J.K. Lanyi, *Biochemistry*, 1999, **38**, 10070.

116. S.Y. Venyaminov and N.N. Kalnin, *Biopolymers*, 1990, **30**, 1243.

117. K. Fahmy, O. Weidlich, M. Engelhard, H. Sigrist and F. Siebert, *Biochemistry*, 1993, **32**, 5862.

118. M.S. Braiman, T.J. Walter and D.M. Briercheck, *Biochemistry*, 1994, **33**, 1629.

119. M.S. Hutson, U. Alexiev, S.V. Shilov, K.J. Wise and M.S. Braiman, *Biochemistry*, 2000, **39**, 13189.

120. M.S. Hutson, S.V. Shilov, R. Krebs and M.S. Braiman, *Biophys. J.*, 2001, **80**, 1452.

121. T.J. Walter and M.S. Braiman, *Biochemistry*, 1994, **33**, 1724.

122. M.S. Braiman, D.M. Briercheck and K.M. Kriger, *J. Phys. Chem. B*, 1999, **103**, 4744.

123. Y. Xiao, M.S. Hutson, M. Belenky, J. Herzfeld and M.S. Braiman, *Biochemistry*, 2004, **43**, 12809.

124. M. Rüdiger, U. Haupts, K. Gerwert and D. Oesterhelt, *EMBO J.*, 1995, **14**, 1599.

125. A. Maeda, J. Sasaki, Y. Shichida and T. Yoshizawa, *Biochemistry*, 1992, **31**, 462.

126. A. Maeda, *Biochemistry (Mosc)*, 2001, **66**, 1256.

127. H. Kandori, *Biochim. Biophys. Acta*, 2004, **1658**, 72.

128. G. Zundel, *Adv. Chem. Phys.*, 2000, **111**, 1.

129. F. Garczarek, J. Wang, M.A. El-Sayed and K. Gerwert, *Biophys. J.*, 2004, **87**, 2676.

130. R. Rammelsberg, G. Huhn, M. Lübben and K. Gerwert, *Biochemistry*, 1998, **37**, 5001.

131. K. Schulten and P. Tavan, *Nature*, 1978, **272**, 85.

132. L.S. Brown, Y. Gat, M. Sheves, Y. Yamazaki, A. Maeda, R. Needleman and J.K. Lanyi, *Biochemistry*, 1994, **33**, 12001.

133. S. Druckmann, M. Ottolenghi, A. Pande, J. Pande and R.H. Callender, *Biochemistry*, 1982, **21**, 4953.

134. J. Heberle and N.A. Dencher, *Proc. Natl. Acad. Sci. U.S.A.*, 1992, **89**, 5996.

135. C. Scharnagl and S.F. Fischer, *Chem. Phys.*, 1996, **212**, 231.

136. H.J. Sass, G. Büldt, R. Gessenich, D. Hehn, D. Neff, R. Schlesinger, J. Berendzen and P. Ormos, *Nature*, 2000, **406**, 649.

137. B. Schobert, L.S. Brown and J. K. Lanyi, *J. Mol. Biol.*, 2003, **330**, 553.

138. T. Rink, J. Riesle, D. Oesterhelt, K. Gerwert and H. J. Steinhoff, *Biophys. J.* 1997, **73**, 983.

139. S. Subramaniam, I. Lindahl, P. Bullough, A.R. Faruqi, J. Tittor, D. Oesterhelt, L. Brown, J. Lanyi and R. Henderson, *J. Mol. Biol.*, 1999, **287**, 145.

140. H.J. Sass, R. Gessenich, M.H. Koch, D. Oesterhelt, N.A. Dencher, G. Büldt and G. Rapp, *Biophys. J.* 1998, **75**, 399.

141. Y. Marantz, E. Nachliel, A. Aagaard, P. Brzezinski and M. Gutman, *Proc. Natl. Acad. Sci. U.S.A.*, 1998, **95**, 8590.

142. P. Ädelroth and P. Brzezinski, *Biochim. Biophys. Acta*, 2004, **1655**, 102.

143. K.J. Rothschild, P. Roepe, J. Lugtenburg and J.A. Pardoen, *Biochemistry*, 1984, **23**, 6103.

CHAPTER 12

Intraprotein Proton Transfer–Concepts and Realities from the Bacterial Photosynthetic Reaction Center

COLIN A. WRAIGHT

Department of Biochemistry and Center for Biophysics & Computational Biology
University of Illinois at Urbana-Champaign

1   Introduction

For almost 40 years – since the discovery by Chance and coworkers[1,2] of cytochrome photooxidation at very low temperatures in the photosynthetic bacterium, *Chromatium*–the bacterial photosynthetic reaction center (RC) has been the premier testing ground for our understanding of electron transfer in biological function, and even in aperiodic systems, generally.[3] This position of influence significantly predates the publication of the X-ray structure of the RC from *Rhodopseudomonas viridis*[4,5] but the latter event propelled the theoretical and conceptual developments in this area to new heights. The critical contribution of the X-ray structures, however, was not so much the spectacular details of the atomic structure but more simply the firm establishment of distances between cofactors and, in some cases, their relative orientations.

Because of the significant distances involved, the large majority of biological electron transfers are adequately accounted for by non-adiabatic theories as developed by Marcus, Levitch and Dogonadze, Hopfield, and Jortner, among others.[6–10] The underlying physics has been described with ever increasing sophistication, which has built steady confidence in our understanding of the fundamental events. However, these refinements have had only a small impact on the quantitation necessary to understand the biological processes and, especially, the evolution and design of extant systems. Indeed, there has been a long and occasionally contentious–but for the outsider usually amusing–discourse between proponents of the atomic level and broad-brush views of biological electron transfer.[11–21] The result is that, within reason, it is a rather simple matter to understand why biological electron transfer is a robust activity, and how different behaviors can arise in natural and mutant variants.

In principle, the wavefunction that describes ET does correspond to the detailed structure of the medium and in that sense, at least, the network of bonds between a donor, D, and acceptor, A, provides the basis of the connectivity between them the electronic coupling in the Hamiltonian matrix (see below).[17,22] Furthermore, it is possible to demonstrate the influence of the bonding pathways between synthetically engineered redox sites, including, for example, better coupling along β-strands than across α-helices.[17,20,23] However, there is little or no indication that Nature utilizes these features as a design principle in any natural functions.[15,24,25] The overwhelming evolutionary approach is to hang cofactors on the protein scaffold by any suitable liganding, at separations that ensure ET rates that are non-limiting, with the intervening protein acting purely as a homogeneous medium characterized by the β-factor in the term $\exp[-\beta(r-r_0)]$, which approximates the electronic coupling in the non-adiabatic limit. The constraint of what is non-limiting is determined by the turnover time of whatever overall process the ET is functioning within. With few exceptions, the separation between donor and acceptor cofactors in a functional sequence is less than 14 Å, which provides ET times of 100 μs, or faster, for reactions that are thermodynamically neutral or even somewhat unfavorable.[25] Since there is no specific pathway involved, the ET step is insensitive to a wide range of mutations–the biological meaning of robustness. Of course, mutations that influence the energetics of the reactants or products will affect the equilibrium constant as expected, and very likely the kinetics, but in a manner unrelated to the electronic coupling.

In contrast, proton transfer is potentially very sensitive to the structural and energetic details and dynamics of the environment, and it is a significant challenge to discern whereby it achieves the robustness that often characterizes successful, naturally evolved systems.

2   Proton Transfer vs. Electron Transfer

The obvious and critical distinction between electron and proton transfer is the almost 2000-fold difference in the mass of the particle. The tiny mass of the electron allows transfer by quantum mechanical tunneling to proceed with modest driving forces at a biologically meaningful rate over distances of up to 15 Å, and non-physiologically beyond 25 Å.[15,25] At these and even much shorter distances, the donor–acceptor interaction is very weak and the electron transfer is clearly in the non-adiabatic regime.

Tunneling probabilities (rates) are proportional to $(\text{mass})^{-1/2}$ and, within the same functional time constraint of 0.1 ms, proton tunneling is limited to no more than 1 Å. (Figure 1). The necessary close approach of the heavy atom systems will result in sufficient perturbation of the wave functions that adiabatic processes will usually dominate. Indeed, even if the tunneling rate at some average non-bonding distance is adequate to support function, it is evident that it will be greatly modulated by thermal fluctuations of the distance between neighboring atoms–a fluctuation of 0.1 Å, around 1 Å, changes the tunneling rate by 1–2 orders of magnitude.[26] Thus, the elementary (pairwise) transfer is usually controlled by the dynamics of the system, regardless of the nature of the heavy atom framework.

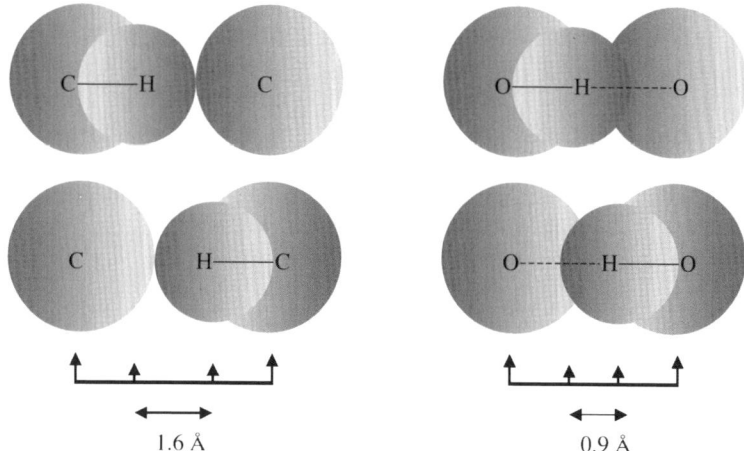

Figure 1 *Proton transfer distances in non-bonded and hydrogen-bonded pairs (schematic – charges are omitted for generality). Left: Proton (or hydrogen atom) transfer between non-bonded donor (C–H) and acceptor (C). Right: Proton transfer between hydrogen-bonded donor (O–H) and acceptor (O). Typical dimensions: heavy atom (C or O) radius ≈ 1.5 Å, neutral hydrogen atom ≈ 1.2 Å, C–H bond length ≈ 1.1 Å, O–H bond length ≈ 0.95 Å, O–O hydrogen bonding distance ≈ 2.8–3.2 Å*

In bioenergetic systems, which are designed to catalyze proton-coupled electron transfer (PCET) and net proton translocation, proton transfer is invariably between 'normal' acids and bases, as defined by Eigen, *e.g.* most oxygen- and nitrogen-containing functional groups.[a] These exhibit intrinsically fast rates of PT, partly due to their favorable electronic structure,[29,30] but also, and perhaps more importantly, because the donor and acceptor associate by hydrogen bond formation,[27] and the tunneling distance of ≈ 1 Å corresponds roughly to the separation of alternate proton positions in the hydrogen bond (Figure 1). This facilitates the formation of the acceptor bond as the donor bond is breaking and leads generally to low activation barriers and fast pairwise transfer. The question is, therefore, what and where is the rate-limiting step?

3 The Grotthuss Mechanism and Hydrogen-bonded Chains

It has long been recognized–remarkably, for 200 years–that protons have the potential for a unique mode of transport in water[31b] and, by extension, in other highly connected hydrogen bonding systems.[32–34] The Grotthuss mechanism involves a simple

---

[a] 'Normal' acids and bases are defined operationally as being those that exhibit near diffusion-controlled rates when thermodynamically favorable, with the reverse rate smaller by a factor $10^{\Delta pK}$.[27,28]

[b] It seems especially remarkable, today, that this proposal was made prior to the normally cited date for the atomic theory of matter – clearly the pressure to publish was different in those days! – and before the empirical formula of water was correctly known. Grotthuss presented his idea as a mechanism for transfer in electrolysis according to the description: OH $\cdots$ OH $\cdots$ OH $\rightarrow$ HO $\cdots$ HO $\cdots$ HO. (John Dalton's atomic theory was actually presented in public lectures at the Royal Institution in 1803, but was published only in 1808, as Vol.1 of *A New System of Chemical Philosophy*).

shift of hydrogen bonds to effectively relocate a net protonic charge from one position to another without significantly moving the mass of the proton. In water, this process contributes at least four-fifths of the measured transfer number of hydrogen ions, and the ionic mobility of H^+ is about seven times that of Na^+. Current views of the Grotthuss mechanism have the rate limiting event as the breaking of one or two critical hydrogen bonds *outside* the primary solvation sphere of the proton charge, allowing reorganization of the first shell from a hydronium, H_3O^+ (or Eigen ion, $H_9O_4^+ = H_3O^+(H_2O)_3$), to a Zundel ion, $H_5O_2^+$, as a transition intermediate (Figure 2). The proton then redistributes along its own coordinate between the two oxygen atoms and further solvation adjustments can trap it at the new oxygen, in the form of a new H_3O^+.[35] The electrostatic energies are the same in the initial and final states, of course, and are also of little consequence in the transition intermediate–the energies of the Eigen and Zundel ions are not greatly different.[36] The activation energy for the anomalous proton mobility that characterizes the Grotthuss mechanism (approx. 2.5 kcal/mol) arises from the breakage of a 'typical' water–water hydrogen bond. Recent models suggest that the rate-limiting step may involve the coordinated rearrangement of hydrogen bonds as far away as the third solvation shell.[37]

Computational studies on proton migration in water have provided a fairly good picture of the operation of the mechanism that underlies the idea of PT through a hydrogen-bonded chain or network.[37–41] Given the flatness of the potential energy surface in bulk water, it might not be surprising to encounter such a mechanism there, but even in water the result is not a long-distance concerted transfer over multiple oxygen centers. The excess mobility of the proton arises simply from the increased step size of the random walk, *i.e.* diffusion, as the proton charge is moved across the diameter of a single water molecule (≈ 2.5 Å) in about 1 ps, the Debye (rotational) relaxation time of water.[42]

Caution should be exercised in transferring this picture to the inside of a protein where donor–acceptor pairs are unlikely to be so well matched in pK_a. At the same time, however, the reduced dimensionality in structured systems increases the opportunity for hydrogen-bonded chains to provide proton wires with high conductivity and specific directionality.[34,43] Consequently, a great deal of speculation and a modest amount of computation has been expended in exploring the importance of such structures in biological processes. It was pointed out early on, by Nagle and co-workers,[44] that steady-state conduction by a hydrogen-bonded chain must involve two types of activity–a hopping mode as the proton charge moves from one end of the pre-oriented hydrogen-bond chains to the other, and a turning mode, in which the hydrogen bond network reorients to restore the original configuration (Figure 3).

Various levels of microscopic calculation and dynamic simulation have shown this in action, but the results can be misleading. While transfer of the protonic charge across hydrogen bonds undoubtedly occurs, and is responsible for the excess diffusion of protons in water, in a low dielectric environment like a protein interior, additional factors intrude, including an electrostatic (Born) energy barrier to entry, as well as site differences, which can easily dominate the structural aspects of the conduction (*i.e.* the Grotthuss mechanism) once inside. Of course, the protein structure could be designed to minimize the Born energy factor, as seen in the specialized channel proteins for K^+ and Cl^- conduction,[45,46] but such a system has not yet been described for protons. Certainly, the precise balance of these factors is still unclear and PT, like ET in earlier

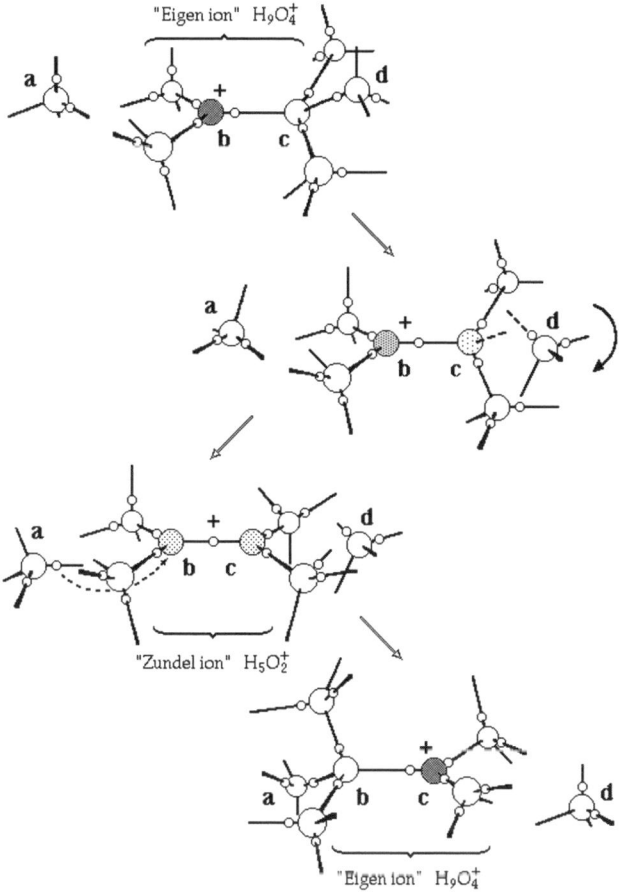

Figure 2 *The Grotthuss mechanism in water, showing Eigen and Zundel ions. From top to bottom: The hydronium ion (b) is almost planar and is solvated by three water molecules forming an Eigen ion, $H_9O_4^+$. Each solvating water is hydrogen bonded to approx. three additional neighbors – this is shown only for one solvating water (c). A hydrogen bond in the second solvation shell (c–d) is broken and the remaining ion rearranges to yield a Zundel ion, $H_5O_2^+$. The excess proton fluctuates along the "proton coordinate", between the two oxygen atoms and is trapped at either one as a new hydrogen bond (here, from a to b) reforms an Eigen ion – in this case on oxygen c. (Figure adapted and redrawn from Agmon[35])*

times, is currently enjoying its own contentious debates. This is well illustrated by the literature on gramicidin A, an ionophorous antibiotic that may come as close as we know to a proton conducting, hydrogen-bonded chain with a low Born energy factor.

Proton transport through the gramicidin A (gA) channel has been analyzed and simulated in a number of studies, using a variety of computational approaches with conflicting apparent outcomes. Pomès and Roux, for example, found the main activation barrier to H^+ movement within the gA channel to be the turning motion of the channel water molecules, while the H^+ hopping process along the ordered water

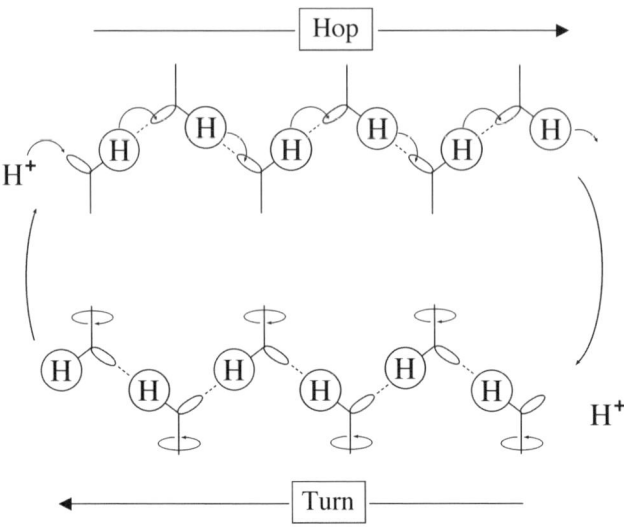

Figure 3 *The Grotthuss mechanism in a hydrogen-bonded chain, showing the distinct 'hop' and 'turn' phases*

chain was essentially activationless.[47] This surprising result implied a negligible electrostatic penalty for moving the protonic charge within and through the channel, and the authors suggested that this was due to very favorable dipole interactions with the water chains on either side of the charge site. In apparent contrast, Warshel and coworkers found a significant electrostatic penalty to the presence of the charge and believe this to be almost exclusively the source of the kinetic barrier,[48-50] *i.e.* they consider the motions necessary to ensure the hydrogen–bond connectivity and reorientation are sufficiently small as to be energetically insignificant. This, of course, is the very motion that comprises the 'turn' action of a hydrogen-bonded chain.

In fact, these two results are not comparable. Pomès and Roux calculated the motion of the charge once inside the gA channel, and did not consider the entry or exit of the proton (the cap of water molecules at each end of the water file were 'non-dissociable'). On the other hand, Warshel and coworkers calculated the proton energies relative to the bulk phase, and found a total barrier height of 5–7 kcal mol^{-1}. This is in quite good agreement with the experimental activation energy for H$^+$ conduction in gA, which is quite small. However, the estimated internal barriers were also small and, although not indicative of an activationless hopping motion, are probably not so far from those of Pomès and coworkers, given the errors and differences in both approaches.

Interestingly, Schumaker and colleagues, using a macroscopic formalism that incorporates the microscopic results of Pomès and coworkers,[51] have estimated the steady-state proton-conduction behavior of gA, including the entrance and exit events.[52] They find that the activation energy for the turn steps is lower than originally estimated, and that it is the proton's *exit* that is rate limiting. This intriguing result suggests that the channel, possibly including the induced polarization of the water file, 'holds on' to the proton as it tries to leave. NMR studies have provided structural information on the conduction of alkali metal cations through the gA channel, which may also be relevant to understanding proton conduction.[53-55] Entry to the channel is accompanied by the sequential removal of waters of hydration from the cation in three distinct steps as the ion moves from one site to the next. Ultimately, the ion is associated with only two waters–one ahead and one behind in the single file of the channel. In this same entrance (and exit) region of the channel, the ion has a significant electrostatic interaction with the four tryptophans (approx 1 $k_B T$ per indole) that comprise a well-defined collar in the membrane headgroup region of the channel.[53,56,57] Conceivably, this provides the interaction that limits proton exit from gA.

A similar discourse has developed on the topic of aquaporin, a membrane protein designed to conduct water but *not* protons. Here, the water chain that spans the channel reverses the polarity of the hydrogen-bonded water dipoles at the mid-point of the membrane span,[58,59] and this structural feature was suggested to impede proton transport by interrupting the conduction pathway.[60,61] However, subsequent calculations and molecular dynamics simulations have shown that a significant contribution to the barrier for proton conduction comes from the electrostatics and related desolvation penalty for a protonic charge penetrating into the channel interior.[62–65] The relative magnitudes of these contributions are, of course, under intense debate! In fact, the result for aquaporin is hardly surprising, but it does not contribute to the discussion about how proteins *do* conduct protons.

The conclusion from these disparate works might be that the kinetic barriers to proton transfer are small and conduction rates can be controlled by electrostatic penalties to populating internal sites, or be facilitated by a flat profile. From an experimental point of view, we need to know how to probe the mechanism by, for example, changing the proton donors and acceptors or otherwise changing the driving force for PT, and how to analyze the experimental results. These questions are not yet settled but are partly addressed in the following sections.

4   Free Energy Relationships–Marcus and Brønsted

Correlations between reaction rates and free energy changes (or equilibrium constants) are often observed in chemistry, and significance for these free energy relationships (FERs) has long been sought. In general, it is obscure except at a qualitative level, but the simplicity of ET has allowed genuinely theoretical descriptions to be developed. These are commonly referred to as Marcus theory, although many names are associated with current formulations of it. The essential ingredients of Marcus theory are that the reaction coordinate for ET is controlled by the nuclear reorganization of the environment (solvent polarization), and that the donor and acceptor are weakly interacting, so that the ET is non-adiabatic.

From the Golden Rule of perturbation theory,[c] the probability of transition between initial and final states is:

$$p(t) \propto \frac{2\pi |V_{if}|^2}{\hbar} \rho_f \qquad (1)$$

---

[c] This is commonly referred to as Fermi's Golden Rule and it is certainly true that he coined the term for his lectures at the University of Chicago, in 1947–1951,[66] indicating his great admiration for the power of this simple expression. However, the equation, itself – and the whole construct of perturbation theory that it represents – is due to Dirac, 20 years earlier.[67]

V_{if} represents the electronic coupling or matrix element that mixes the initial and final states–or, in molecular terms, the strength of interaction between the electron donor and acceptor. ρ_f is the density of states that are accessible to the transition from $i \rightarrow f$ while maintaining conservation of energy. It is commonly called the Franck–Condon factor, or thermally weighted density of states, and is the ratio of states at the transition state relative to the reactant ground state. For a thermal reaction it provides the activation energy term.

For a non-adiabatic transition, the Franck–Condon factor defines the intersection of the diabatic curves for initial and final states, and, when these are represented as harmonic oscillators, it can be easily derived and calculated. This is the familiar, classical form of Marcus theory.

$$k = \frac{2\pi}{\hbar}|V(r)|^2 \frac{1}{\sqrt{4\pi\lambda k_B T}} \exp\left(-\frac{(\lambda + \Delta G^0)^2}{4\lambda k_B T}\right) = k_0 \exp\left(-\frac{\Delta G^*}{k_B T}\right) \qquad (2)$$

where the activation free energy, $\Delta G^* = (\lambda + \Delta G^0)^2/4\lambda = \lambda/4(1 + \Delta G^0/\lambda)^2$, is given as the intersection point of two parabolas representing the diabatic curves of the reactant and product states (Figure 4). The novel concept here is the reorganization energy, λ, which is the energy gap between the reactant state at its vibrational equilibrium position and a point on the reactant curve that is vertically above the product state equilibrium. It is the equivalent, in energy units, of the difference between the equilibrium positions of reactant and product along the vibrational coordinate. The reorganization energy describes the responsiveness of the vibrational system to the electronic changes in the reaction. Together with the free energy of the reaction, ΔG^0, it quantifies the energy necessary to generate an intermediate nuclear configuration that has the same total energy before and after electron transfer.

The reorganization energy is determined both by the magnitude of the distortion and the stiffness of the bond vibrations that respond to the electronic transition. This gives rise to a crude categorization of contributions to λ–an inner component, λ_i, which reflects the molecular response of the donor and acceptor cofactors, and an outer component, λ_o, which arises from the solvent polarization response to the change in charge distribution. In many cases, the structures of biological cofactors, such as large conjugated macrocycles, minimize λ_i and the major contribution is from the solvent environment, which is primarily the protein matrix rather than the true aqueous solvent, as biological redox centers are at least partially shielded from the latter. More sophisticated versions that include quantization of the nuclear vibrations, allow the distinction between λ_i and λ_o to be made explicit.

From the Marcus equation, one can relate the activation free energy (and hence the temperature dependence) of a reaction to the standard free energy change, ΔG^0, and to the reorganization energy of the 'solvent coordinate', which includes all vibrations that are coupled to the electronic transition.

Marcus theory was originally formulated for bimolecular reactions in solution, and two additional terms were included to account for the necessity of forming an encounter complex between the reactants. These are w_r and w_p, the work involved in

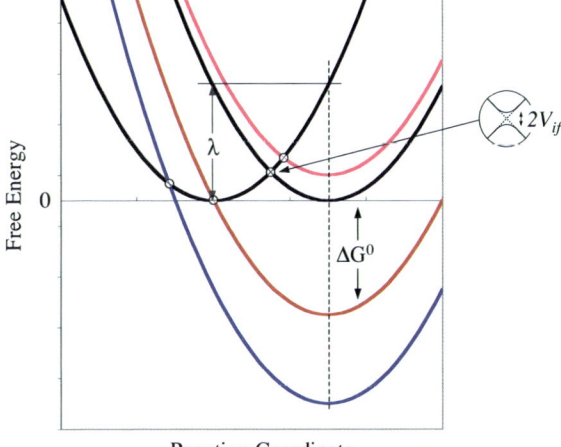

Figure 4 *Marcus theory diagram for electron transfer, showing weak electronic coupling V_{if}. A single reactant surface (black) and four possible product surfaces are represented as parabolas. The reorganization energy, λ, is given by the position on the reactant curve that lies above the equilibrium coordinate of the product curves. The product surfaces are shown for several net driving forces (ΔG^0 values). For $\Delta G^0 = 0$ (black), the intersection between reactant and product curves is at $\lambda/4$, indicating the activation free energy for the thermodynamically neutral reaction (ΔG^{0*}, the 'intrinsic' barrier). As the free energy decreases from unfavorable (pink) to favorable, the activation free energy decreases and the reaction accelerates. ΔG^* is zero and the rate is maximum when $\Delta G^0 = -\lambda$ (red). At very large (negative) driving forces ($\Delta G^0 < -\lambda$) (blue), Marcus theory predicts an 'inverted region' where the activation free energy increases again and the rate of reaction slows. At each intersection point the reactant (initial) and product (final) states interact weakly to yield two curves separated by $2V_{if}$ (shown for the black curves), where V_{if} is the coupling or interaction energy. For non-adiabatic electron transfer, the splitting is very small, yielding what is called an "avoided crossing". In solution, the formation of the reaction complex (to which Marcus theory may be applied) and dissociation of the products involve work terms, w_r and w_p, which raise the reactant and product curves– often by similar amounts*

bringing the reactants together and in separating the products, respectively. Incorporating these, the classical expression becomes:

$$\Delta G^* = w_r + \frac{\lambda}{4}\left(1 + \frac{\Delta G^0_{obs} - w_r + w_p}{\lambda}\right)^2 \tag{3}$$

where $\Delta G^0_{obs} = \Delta G^0 + w_r - w_p$, and ΔG^0 is the free energy of reaction within the encounter complex. In cases where ΔG^0 is large, the influence of w_r and w_p is small and they are commonly neglected. In biological ET, the cofactors are often preorganized and the work terms are appropriately absent.

The concept and investigation of free energy relationships (FERs) predate Marcus theory by many decades. The earliest example is the Brønsted relationship, which relates the rates of proton-transfer reactions (and hence the activation free energy) to the differences in pK_as of the donor and acceptor (the standard free energy of the reaction, ΔG^0).[68–70] A limited range of ΔG^0 yields linear FERs, but more extensive data sets invariably show curvature suggestive of a quadratic dependence. The success of Marcus theory as a mechanistically meaningful, theory-based description of a quadratic FER in ET, led to adoption of it as a basis for PT and for more complex reactions, including atom and small group transfers[71,72] and enzyme catalysis,[73,74] and it has been widely used and explored. However, while undoubtedly useful, it is fundamentally inappropriate and its success relies on correlation rather than any real mechanistic underpinnings.

A very general ('interpretation-free') analysis of the relationship between activation free energy and standard free energy of reaction is to consider an FER as a Maclaurin series expansion:

$$\Delta G^* = \Delta G^* \bigg|_{\Delta G^0 = 0} + \frac{d\Delta G^*}{d\Delta G^0}\bigg|_{\Delta G^0 = 0} \cdot \Delta G^0 + \frac{d^2 \Delta G^*}{d(\Delta G^0)^2}\bigg|_{\Delta G^0 = 0} \cdot \frac{(\Delta G^0)^2}{2} + \dots$$

$$= \Delta G_0^* + \alpha_0 \Delta G^0 + \alpha' \frac{(\Delta G^0)^2}{2}$$

(4)

ΔG_0^* is the 'intrinsic' activation free energy encountered for a reaction with no driving force ($\Delta G^0 = 0$). $\alpha = d\Delta G^*/d\Delta G^0$ is the Brønsted coefficient, and reflects the change in activation free energy as the driving force changes, and α_0 is the value of α when $\Delta G^0 = 0$. When recast in the form of the original Marcus theory, we find:

$$\Delta G^* = \frac{\lambda}{4}\left(1 + \frac{\Delta G^0}{\lambda}\right)^2 = \Delta G_0^*\left(1 + \frac{\Delta G^0}{4\Delta G_0^*}\right)^2$$

(5)

i.e. $\Delta G_0^* = \lambda/4$, $\alpha_0 = 1/2$, $\alpha' = d\alpha/dG^0 = 1/(8.\Delta G_0^*) = 1/(2\lambda)$

In the language of physical organic chemistry, the activation free energy, ΔG^*, is considered to comprise two distinct contributions–a 'kinetic' component or intrinsic activation free energy, ΔG_0^*, and a 'thermodynamic' component, ΔG^0 (Figure 5). ΔG_0^*, being equal to the free energy required to distort the reactant or product states to the transition state configuration for a reaction with $\Delta G^0 = 0$, is identical in concept to the reorganization energy, since $\lambda = 4\Delta G_0^*$ under these circumstances. Like λ, ΔG_0^* reflects the response of the environment and any molecular vibrations coupled to the PT reaction.

Although possibly no more than phenomenological, mechanistic significance has long been sought for free-energy relationships. The interpretations are mostly of a qualitative nature, the most notable being the Hammond postulate,[75] from which it follows that the slope (α) of the FER (ΔG^* *vs.* ΔG^0) reflects the position of the transition intermediate along the reaction coordinate. This in turn is interpreted as indicating whether the structure of the transition intermediate is more like the reactants ($\alpha < 0.5$) or the products ($\alpha > 0.5$). The change in slope (α) along the quadratic curve suggests that the transition state of exergonic reactions resembles reactants, while that of endergonic reactions resembles products.

By comparison with the ET formulation, it is clear that the intrinsic activation free energy ($\Delta G_0^* = \lambda/4$) is determined by the point of intersection of the diabatic curves. For non-adiabatic ET this is a reasonable approximation as the electronic coupling

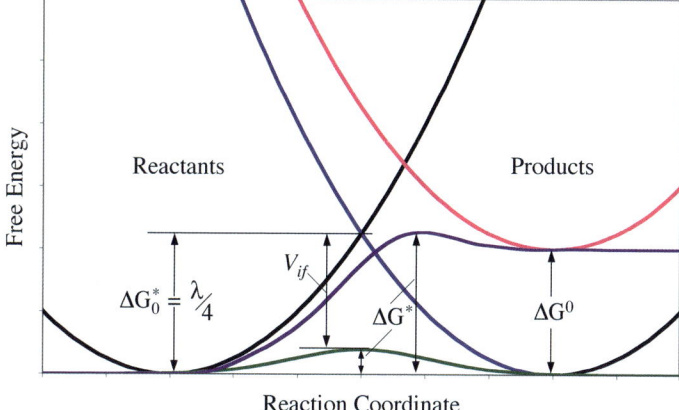

Figure 5 *Marcus formalism and adiabatic modification for facile proton transfer. Diabatic curves (parabolas) are shown for reactants (black) and for products for neutral (blue) and endergonic (pink) reactions. The intersection of the black and blue diabats gives the intrinsic barrier of the reaction, ΔG_0^* or $\lambda/4$. The actual activation free energy, ΔG^*, on the corresponding adiabatic surface (green) is much smaller and is modified by the coupling energy, V_{if}, which is substantial. For the endergonic reaction, ΔG^* on the adiabatic surface (magenta) is also much smaller than the equivalent diabatic intersection, although not by the same amount – V_{if} is not constant but varies along the reaction coordinate. Note, also, that the peak (ΔG^*) is not at the same position on the two adiabatic surfaces – the transition state is closer to the product for the endergonic reaction (Hammond's postulate)*

is small due to the substantial distance between electron donor and acceptor. For PT, however, the close approach necessary for bond breaking and formation (even if the proton tunnels), means that substantial electronic perturbations occur between proton donor and acceptor. Thus, the coupling is large and often of the same order as ΔG_0^* itself, so the true activation free energy is much less than that indicated by the Marcus equation (Figure 5). If this adiabatic character is not accounted for, the analysis can only yield a small activation energy by virtue of an improbably small reorganization energy, and it therefore grossly misrepresents the microscopic events.

Nevertheless, a beguiling correlation remains because the adiabatic transition state energy varies in parallel with the diabatic intersection. The absolute magnitudes of ΔG^0 and ΔG^* may not be related by any simple FER, but a linear relationship between $\delta\Delta G^0$ and $\delta\Delta G^*$ still holds so long as the coupling energy is insensitive to ΔG^0. Thus, any change in the diabatic intersection is reflected in ΔG^* even after correction by V_{if}.

A characteristic feature of Marcus theory for ET, and for the Marcus (quadratic) formalism, generally, is the prediction of an 'inverted region' at very large driving forces, *i.e.* when ΔG^0 becomes more negative than $-\lambda$, the rate slows down. The rate peaks at $\Delta G^0 = -\lambda$ and the width of the log (rate) *vs.* ΔG^0 parabola is proportional to $\sqrt{\lambda k_B T}$. This prediction for ET was finally confirmed after 20 years.[76] For proton transfer, it was long debated whether an inverted region was possible, and certainly attempts to observe it were commonly confounded by diffusional limitations.

However, inverted region behavior has been reported for PT between the radical cation of dimethylaniline and a series of substituted benzophenone radical anions in aprotic solvents.[77,78] A more convincing demonstration has come from studies using an electron photoinjection method at an electrode interface.[79]

5   Proton Transfer in Biology

PT is of major importance in two distinct areas of biochemistry–acid-base catalysis in enzyme activity, and proton coupled electron transfer in bioenergetics. In the former, the PT events are generally highly localized and the critical purpose is to transfer a proton between adjacent groups, for example an active-site amino acid and a substrate. Water is generally excluded from actives sites (except as a reactant), but there are many examples where a chain of water molecules maintains a specific connection to the bulk phase.[80] This may present a polarizable element that can respond to charge shifts associated with the catalytic events, including local PT. Nevertheless, given the generally pairwise nature of PT in active-site chemistry, descriptions of general acid-base catalysis are applicable and some simple relationships might be expected. In contrast, proton transfer in bioenergetics is usually over long distances, the primary purpose being to translocate protons into and across the membrane, *e.g.* of the mitochondrion, chloroplast or bacterial cell.

An intermediate example, from active-site proton transfer, is the reaction of carbonic anhydrase (CA), where a proton is taken up (or released) as part of the stoichiometric turnover:

$$CO_2 + H_2O \leftrightarrow HCO_3^- + H^+ \qquad \text{(Scheme 1)}$$

Somewhat surprisingly, the transport of H^+ in and out of the active site is the rate-limiting step.[81,82] PT occurs over a distance of 8–10Å and is associated with the regeneration of the active site Zn^{2+}–OH^- complex:

$$Zn^{2+}\text{–}OH_2 + B \leftrightarrow Zn^{2+}\text{–}OH^- + BH^+ \qquad \text{(Scheme 2)}$$

The communication between the zinc-bound water and the aqueous phase is mediated by a short chain of two to three water molecules and an amino-acid side chain, represented by B/BH^+, which is in contact with the bulk phase. In carbonic anhydrase II, the fastest of the many mammalian isozymes, B/BH^+ is a histidine (His64 in Figure 6).

In a series of detailed and elegant studies, Silverman and coworkers have used site-directed mutations and natural isozyme variants to alter the pK_a of both Zn–OH_2 and B/BH^+, roughly over the range of $\Delta pK_a = 0 \pm 3$ units.[83–86] The measured rates have generally fit well to a quadratic relationship for log k *vs.* ΔpK_a and have exhibited significant curvature. In the Marcus formalism, curvature over such a relatively narrow range of ΔG^0 (ΔpK_a) indicates that the ΔpK_a values are in the range of ΔG_0^*, or $\lambda/4$, and that λ is small. From a Marcus analysis, the implied value for ΔG_0^* is very small and the bulk of the activation energy is provided by the work term, w_r. Silverman's interpretation of this is simple and appealing–work is required to assemble or order the water chain into a functional arrangement. However, the waters are

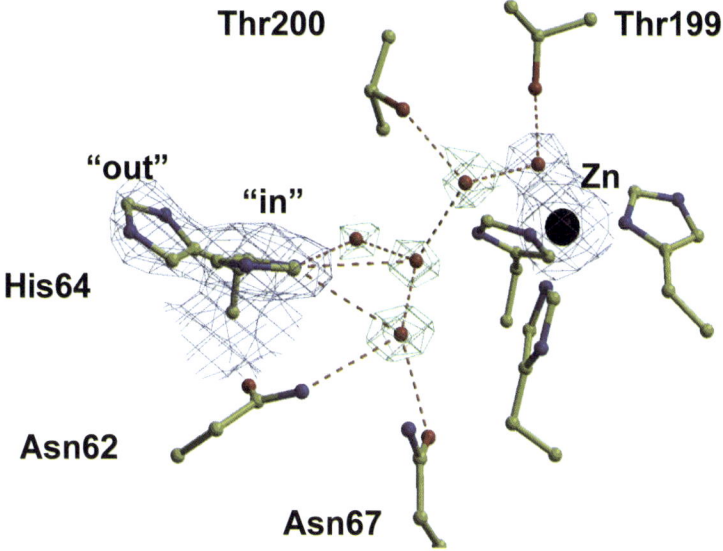

Figure 6 *The active site of carbonic anhydrase II. The reactive water/hydroxyl is bound to a zinc(II) ion (black), which is liganded by three histidines. A fourth histidine, His64 is at the entrance of the active site cleft and is observed in two distinct configurations – the 'out' position is essentially in the bulk phase. Up to four additional water molecules are seen in different crystal structures, as shown, bridging the zinc-bound water and His64 in the 'in' position.(Image courtesy of David N. Silverman: www.med.ufl.edu/pharm/facdata/silvermn/ silvermn.html)*

crystallographically resolvable and therefore cannot be highly disordered. More disturbing is the value of $\Delta G_0^* \approx 1–2$ kcal mol^{-1}, *i.e.* $\lambda \approx 5–8$ kcal mol^{-1}, whereas computational methods would suggest values at least ten times larger for a charge transfer reaction of this nature.[73,74,87-90]

From a conceptual point of view, in fact, it is curious that a process that clearly involves intermediate water molecules, with relatively extreme pK_a values, appears to conform to a Marcus analysis that considers only the pK_as of the terminal donor and acceptor, which are much more closely matched. PT in such a situation is *not* a two-state system and there is no reason why a Marcus analysis would give meaningful parameters. Computational studies of CA have yielded a very different picture, which serves well to illustrate the fundamental problems of applying a simple FER to PT.[48] These are two-fold. First, the Marcus formalism yields an intersection point for the two diabatic curves. However, the correction to this, due to the coupling between the states, is very large and the resulting transition state on the adiabatic surface is much lower energy. Second, the involvement of more than two states, i.e., with intervening water molecules between the identified donor and acceptor, invalidates the Marcus formalism altogether, as far as the overall PT is concerned (Figure 7).

The impact of the coupling or mixing between reactant and product states is so large that the free energy of the transition state between each site is comparable to the free energy differences between the sites, which are almost entirely determined

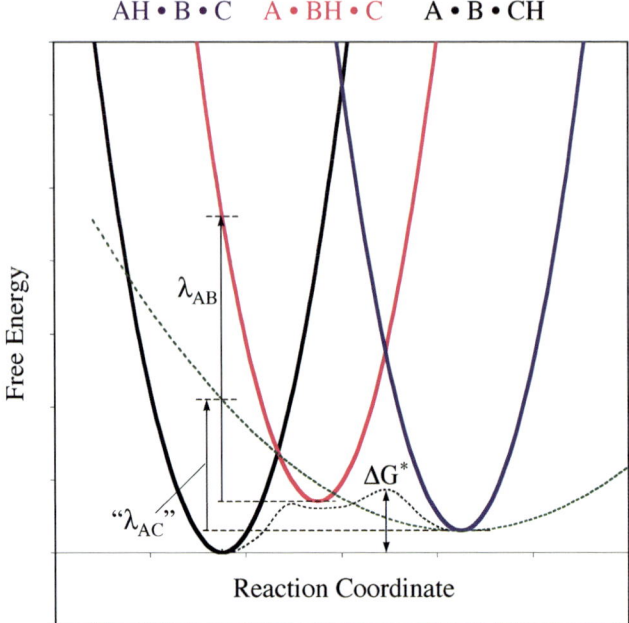

Figure 7 *Diabatic curves and adiabatic surface for sequential, multistep proton transfer. Three states are shown for the reaction sequence at the top (charges on PT donors/acceptors not shown, for generality). The true reorganization energy for the first step, λ_{AB}, is shown, and that for the second step would be similar (but not identical as the horizontal displacement is different). The adiabatic surface (dotted) shows that the actual barrier, ΔG^*, is barely related to any of the diabatic intersections. Note that the barrier height is not very different from the maximum free energy difference at the intermediate state (pink). A two-state Marcus formalism based on pK_a values for the initial donor (A) and final acceptor (C) would require a very different (and smaller) reorganisation energy, 'λ_{AC}', to yield the right magnitude for the activation free energy (dashed green parabola, shown intersecting the reactant curve at the same height as the intermediate state)*

by electrostatics. Thus, in this view, the kinetic barriers in CA are dominated by the electrostatic energies associated with protonating internal sites, rather than structural considerations such as the dynamics of hydrogen-bonded chains.[48,49]

Repeating these calculations for various mutations that resulted in the quadratic dependence of log k and the pK_a difference between the terminal donor and acceptor, *i.e.* B/BH$^+$ and Zn–OH$^-$/Zn–OH$_2$,[91] Schutz and Warshel found that the pK_a values for the intervening water molecules in the chain also change, and the effect on the total activation energy (and hence log k) rather coincidentally aligns with the ΔpK_a between terminal donor and acceptor.[48] The individual, pairwise PTs, however, are associated with large values of $\lambda \approx 80$ kcal mol^{-1}, which is considered much more realistic for this type of reaction. In contrast, the work terms are negligible, indicating that it takes very little effort to arrange the water molecules for adequate function.

This is an area of active discussion, but it is important to note that approaches that focus on pK_a values and their role in determining kinetics as well as equilibria, automatically take in to account the relevant electrostatics – that is what determines local pK_as. Furthermore, regardless of the correct balance of electrostatic *vs.* structural control of PT, long range PT is certain to be a multistep process. This is an inevitable corollary of the insignificant distances that protons can tunnel, so that long-range PT must be achieved in several, essentially adiabatic steps, not by non-adiabatic two-state tunneling as in ET.

It is conceivable that concerted PT could provide a two-state description for PT over some small number of bonds but, in general, one expects concerted processes only when the step-wise mechanism is very expensive.[92,93] To some extent, this is related to the energetics of the hopping mode of HBC activity, but the degree of coherence required is substantially greater than that implied by most computational results (but see[94]). In general, computational studies have given little support for it in bioenergetics, although the methods may yet be inadequate to the task. Nevertheless, highly polarizable hydrogen-bonded protons have been proposed as almost endemic to densely packed, hydrated, weak acid–base matrices, like proteins, giving rise to very broad – almost continuum – infrared absorbance,[95,96] and these represent a potentially two-state, long-distance PT system. It is intriguing to note that a broad infrared feature has been reported for the 'proton release complex' of bacteriorhodopsin[97] and the quinone region of reaction centers,[98] and these have been suggested to be signatures of such 'delocalized protons'.

6 'Normal' Acids and Bases

As described above, analyses based on free-energy relationships are meaningful only when the process under study is correctly identified as a single PT step. In acid–base catalysis this is often the case, but proton translocations in bioenergetics are likely to extend over several functional groups or molecules, including water. From a phenomenological view, this is similar to the situations investigated by Eigen and coworkers, working with proton donor and acceptor pairs in solution.[27,99] Here, the proton-transfer reaction is preceded by the transient association of donor and acceptor in an encounter complex:

$$AH + B \overset{k_e}{\underset{k_{-e}}{\leftrightarrow}} AH{\cdots}B \overset{k_p}{\underset{k_{-p}}{\leftrightarrow}} A{\cdots}HB \overset{k_d}{\underset{k_{-d}}{\leftrightarrow}} A + HB \qquad \text{(Scheme 3)}$$

The $\cdots$ symbol strictly implies only the solvent cage-induced association of the encounter complex, but for oxygen and nitrogen species it will almost certainly involve hydrogen bonding, too. For weak association, the concentration of complex will be very small and a steady-state analysis can be used, in which the total passage time is the sum of the component step times:

$$\tau_f = \frac{1}{k_f} = \frac{1}{k_e} + \frac{1}{k_p K_e} + \frac{1}{k_d K_e K_p} \Rightarrow k_f = \frac{k_e k_p k_d}{k_p k_d + k_{-e} k_d + k_{-e} k_{-p}} \qquad (6)$$

where $K_e = k_e/k_{-e}$ and $K_p = k_p/k_{-p}$. Often, but not always, $k_e \approx k_{-d}$ and $k_{-e} \approx k_d$. Counter examples include products with different charges from reactants, such as $AH + B$ *vs.* $A^- + BH^+$, or *vice versa*, and other sources of 'sticky products'.

The free-energy relationships discussed above apply only to the actual PT process in the encounter complex: $AH \cdots B \leftrightarrow A \cdots HB$. If this is rate limiting it will be directly observable by monitoring the reactants or products, *i.e.* for $k_p \ll k_{-e}$ and $k_{-p} \ll k_d$:

$$k_f = \frac{k_e k_p k_d}{k_p k_d + k_{-e} k_d + k_{-e} k_{-p}} \quad \Rightarrow \quad k_f = k_p K_e \tag{7}$$

K_e is generally very weak, of the order of $1\ M^{-1}$, and is also likely to not vary much for a series of reactants providing the charge types are the same.

Alternatively if PT is fast in the encounter complex, $k_p \gg k_e, k_{-e}$, we obtain the common behavior of 'normal' acids and bases:

$$k_f = \frac{k_e k_p k_d}{k_p k_d + k_{-e} k_d + k_{-e} k_{-p}} \quad \Rightarrow \quad k_f = \frac{k_e K}{K + k_e/k_{-d}} \tag{8}$$

where $K = \dfrac{k_e k_p k_d}{k_{-e} k_{-p} k_{-d}}$ but also, from a thermodynamic cycle: $K = 10^{pK_a(B)-pK_a(A)} = 10^{\Delta pK_a}$

Hence:, $k_f = \dfrac{k_e 10^{\Delta pK_a}}{k_e/k_{-d} + 10^{\Delta pK_a}} \approx \dfrac{k_e 10^{\Delta pK_a}}{1 + 10^{\Delta pK_a}}$ \hfill (9)

the latter because $k_e \approx k_{-d}$.

This predicts two distinct regimes of behavior of the rate with respect to changes in the relative strengths of the acid and base (ΔpK_a). When the donor, AH, is a stronger acid than BH^+, $\Delta pK_a = pK_a(B) - pK_a(A) > 0$ and the rate constant is equal to the association rate constant, k_e. The reaction is described as 'diffusion limited' and it is insensitive to the individual pK_a values so long as $\Delta pK_a > 0$. However, if the donor becomes relatively weak enough that $\Delta pK_a < 0$, the reaction becomes endergonic and the observed rate then slows down by a factor of ten for each unit decrease in ΔpK_a. This is nominally a Brønsted relationship with $\alpha = 1$, but the rate-limiting step is not PT and the slope relates two independent processes. The proton-transfer step is in rapid equilibrium and the equilibrium constant appears simply as a multiplier of the actual rate-limiting process, which is the formation of the encounter complex. The reorganization energy or the related concept of intrinsic activation energy, ΔG_0^*, for the PT is irrelevant. (Note that, in this regime, the reaction is diffusional but it is not 'diffusion limited' because the rapid proton-transfer equilibrium disfavors the forward reaction and not all encounters are successful.)

Although not a PT-limited process, the rate in this regime is revealing with regard to the proton-transfer reactants. For example, if a series of donors is used with an unknown acceptor, the rate dependence on the acid pK_a can provide the pK_a of the acceptor. This method has been used to advantage in studies of proton uptake by bacterial reaction centers, as described below.

7 Proton-coupled Electron Transfer in the Acceptor Quinone Function of Photosynthetic Reaction Centers

Photosynthetic reaction centers (RCs) from purple bacteria and Photosystem II take up protons as the result of quinol production: $Q + 2e^- + 2H^+ \rightarrow QH_2$.[100-104] Although RCs do not carry out transmembrane proton pumping, which is characteristic of cytochrome oxidase[105,106] and bacteriorhodopsin,[107] the uptake associated with quinone reduction constitutes the first half of a proton-translocating redox loop[108,109] that is completed with the oxidation of quinol by the cytochrome bc_1 (b_6f) complex. Furthermore, H^+ transfer to the buried quinone site is over similar long distances to those that are encountered in pumping mechanisms, and the lack of gating in the RC provides a useful simplification in the study of the essential features of proton-conduction pathway(s) in proteins. Indirect coupling of electron transfers to proton uptake is also seen in the response to the light-induced perturbation of the charge distribution of the protein, and the rich spectroscopy of the RC provides a unique system for studying the dielectric responses of proteins.

The single turnover events of the RC from *Rhodobacter (Rba.) sphaeroides* are summarized in Figure 8A, with an electron transferred from the excited state of the primary donor (P*) to the quinones, Q_A and Q_B, on the other side of the membrane. All cofactors are tightly bound except Q_B, which is in weak binding equilibrium when fully oxidized or reduced. However, when reduced to the one-electron, semiquinone state, Q_B^-, it is tightly bound. In the presence of a secondary donor (cytochrome c_2 *in vivo*), P^+ is rereduced and the RC can undergo another photoactivated turnover, which provides a second electron, also *via* Q_A^-, and Q_B is fully reduced to the quinol form (hydroquinone, QH_2) with the uptake of protons from the solution. The quinol unbinds and leaves the RC and is replaced by an oxidized quinone from the membrane pool. This returns the acceptor quinones to their original state and allows RC turnover to proceed under multiple-flash activations. Under such conditions, binary oscillations can be seen in the formation and disappearance of semiquinone and in the uptake of protons from the medium (b<1 on the first flash, 2–b on the second; for review, see[100,102]). This sequence of events is summarized by the acceptor quinone cycle shown in Figure 9, where it is clear that protons are involved in both electron transfers, although covalent attachment to Q_B only occurs on the second turnover with full reduction to quinol.

The two quinones constitute a functional 'acceptor quinone complex', organized around the central iron atom and its ligand field of four histidines and a glutamate (Figure 8B). Q_A and Q_B are both bound with the C1 carbonyl hydrogen bonded to a backbone NH, and the C4 carbonyl hydrogen bonded to the $N_\delta H$ of one of the four histidines liganded to the iron atom. In the semiquinone anion form, Q_B^- is also hydrogen bonded by the $O_\gamma H$ of Ser^{L223}. Both quinone binding pockets are predominantly non-polar, with the notable exception of Glu^{L212} and Asp^{L213} in the Q_B site. These residues are involved in the terminal delivery of protons to the quinone headgroup, during the second turnover (see below).

7.1 First Electron Transfer to Q_B and Coupled Proton Uptake (Bohr Protons)

The first electron is shared between the two quinones, according to their relative redox midpoint potentials (E_m). The negative charges of both anionic semiquinones induce pK_a shifts in ionizable residues of the protein that are close enough to sense the electric field, resulting in net proton uptake.[d] A pK_a-shifted residue may or may

---

[d] Commonly referred to as Bohr protons, after the phenomenon in haemoglobin in which H^+ uptake is coupled to oxygen binding[110,111].

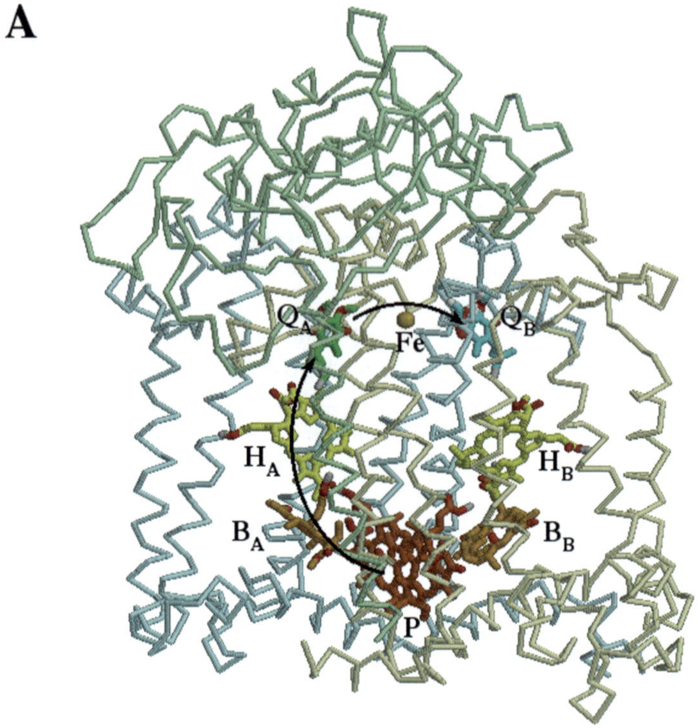

A

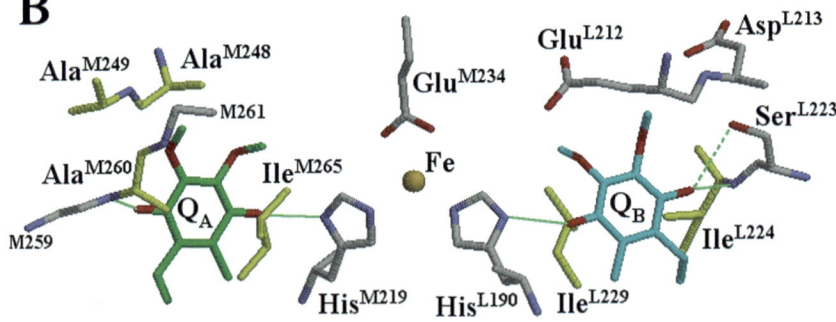

B

Figure 8 *The reaction center (RC) complex from* Rhodobacter sphaeroides *and its quinone binding sites.* **A.** *The RC comprises three subunits, a heterodimer of similar but non-identical L and M subunits, and subunit H that caps LM on the cytosolic side of the membrane. The LM dimer binds all the cofactors, while subunit H stabilizes the structure and is involved in H^+-ion uptake and transfer associated with electron transfer to the quinones. The L and M subunits and all associated cofactors are arranged around a quasi-two-fold rotational symmetry axis, normal to the plane of the membrane, and passing through the primary donor (P, a dimer of Bchl) and a ferrous (Fe^{2+}) iron midway between the two quinones. Electron transfer proceeds from the excited singlet state of the primary donor (P^*), via the monomer Bchl (B_A) to the Bph (H_A) bound to the L subunit. From H_A^-, the electron is transferred to the primary quinone, Q_A, which is bound in a fold of the M subunit, and from Q_A^- it*

not undergo change in its ionization state, depending on how close the pK_a is to the prevailing pH. This results in substoichiometric proton binding, *e.g.* $H^+/Q_B^- < 1$, that is distributed over several residues and is pH-dependent.[112,113] Net proton uptake and internal redistribution of H^+ between ionizable residues contribute substantially to the partial shielding and stabilization of the semiquinones. The difference in net H^+ uptake associated with Q_A^- and Q_B^- formation is the major determinant of the relative redox midpoint potentials (E_ms) of semiquinone states, and is what sets the first electron transfer equilibrium in favor of $Q_A Q_B^-$, *i.e.* the 'forward' direction, at physiological pH. At high pH the ET equilibrium decreases and favors $Q_A^- Q_B$ at pH > 11.

The detailed origin of this behavior resides in the structure of the Q_B domain, which features an unusually high density of ionizable residues with a striking excess of acidic groups (Figure 10). The electrostatic interactions in this domain are therefore very complex and have been reviewed recently,[100] and will be revisited here only to the extent necessary to discuss essential features of proton transfer.

Although site-directed mutagenesis has suggested specific roles for some residues in governing the properties of the semiquinone states, Q_A^- or Q_B^-, computational studies have been an important source of insight and caution. In response to the appearance of a negative charge on Q_B^-, the most prominant acidic residues in the Q_B domain – Asp^{L210}, Glu^{L212}, Asp^{L213}, Asp^{M17} and Glu^{H173} – experience significant pK_a changes. Depending on the pH, H^+ ions taken up are mostly distributed among these residues, but also with many small contributions from more weakly coupled residues. The responses to Q_A^- and to Q_B^- largely arise from the same cast of characters in the Q_B domain.[114,115] This reflects the dearth of ionizable residues around Q_A, which translates into a low effective dielectric that allows the electric field from Q_A^- to spread further.

Glu^{L212} appears to have an unusually high $pK_a \approx 8.5$ in the ground state, and this increases by a further 1 or 1.5 pH units upon appearance of Q_A^- or Q_B^-, respectively.[112–114] It is therefore expected to be fully protonated at neutral pH, but it becomes ionized at pH $\gtrsim$ 8.5 and is then the major site of light induced proton

---

crosses the symmetry axis to reach the secondary quinone, Q_B, bound to a similar fold in the L subunit. The sequence of events is:

$$3\ ps \qquad 1\ ps \qquad 0.2\ ns \qquad 10\text{–}100\ \mu s$$

$$P^* \text{------}> B_A \text{------}> H_A \text{------}> Q_A \text{------}> Q_B$$

B. *The acceptor quinone complex of the Rba. sphaeroides reaction center. Q_A (green) and Q_B (cyan) are bound around an iron-histidine ligand complex (two histidines, L230 and M266, are omitted). The view is from within the membrane plane, similar to part A. The two quinone binding sites are similar and are related by the pseudo-two-fold rotational axis of the reaction center. Q_B is shown in its 'proximal' position. Not all contact residues are shown, but both sites are predominantly non- or weakly polar, except for Glu^{L212} and Asp^{L213} in the Q_B site. Hydrophobic residues are shown in yellow (Ala^{M248}, Ala^{M249}, Ala^{M260}, Ile^{M265}, Ile^{L224}, Ile^{L229}). (Main chain atoms only are shown for M259 (asparagine) and M261 (threonine) – the side chains do not contact the quinone headgroup.) Each quinone is hydrogen bonded through its C4 carbonyl to a histidine ($N_\delta H$) and through the C1 carbonyl to a backbone amide (NH). In the semiquinone anion form, Q_B^- is also hydrogen bonded by Ser^{L223} ($O_\gamma H$) (dotted green line). Coordinates are from 1dv3.pdb*

uptake. The nominal rate constant of the first ET, $k_{AB}^{(1)}$, is pH independent below pH 8.5, but slows down progressively at higher pH, and approaches a 'well behaved' slope of -1 for log $k_{AB}^{(1)}$ *vs.* pH.[116–118] This has generally been interpreted as due to the need to (re)protonate Glu^{L212}, with a raised $pK_a \geqslant 10$ in the presence of Q_B^-. Because of the interactions between residues, the full identity of the ionized species is uncertain, but mutating Glu^{L212} to Gln (mutant L212EQ)[e] abolishes the pH dependence[117,118] and decreases H^+/Q_A^- and H^+/Q_B^- to almost zero at pH > 8,[119,120] and it is convenient and simple to refer to this residue as the responsible party.

Most calculations yield the above result for Glu^{L212}, but FTIR experiments strongly suggest that Glu^{L212} is only partially protonated at lower pH, and picks up protons in the Q_B^- state over a wide pH range.[115,121–126] Calculations also show that at neutral pH, Asp^{L213} and/or Asp^{L210} are deprotonated in the ground state, but Asp^{L213}, at least, becomes fully protonated in the Q_B^- state.[127] This is thought to be coupled to a mechanistically important structural change in which $Ser^{L223}(O_\gamma H)$ is H-bonded to $Asp^{L213}(-CO_2^-)$ in the ground state (neutral Q_B), but rotates to hydrogen bond with the C_1-O^- of Q_B^- in the Q_B^- state.[100,115,127–129] In the process, Asp^{L213} is protonated.

Note that the pH dependence of $k_{AB}^{(1)}$ in the wild type does not imply the process is PT limited, and the pH independence in the L212EQ mutant does not mean that it is not. The mutant result, however, does make it clear that H^+ *uptake* is not rate-limiting for ET to Q_B^-. Related to this point, H^+ uptake in response to the immediate formation of $P^+Q_A^-$ (with no ET to Q_B) was found to be rate-limited by some, presumed conformational, process.[130]

It is noteworthy that, although the kinetics of electron and coupled proton transfer at the Q_B site have been subject to much study, the rate determining process for the first electron transfer is still uncertain, and is generally thought to be neither electron nor proton transfer *per se*.[131] The kinetics are not simple exponential and can be considered either polyphasic[131–134] or genuinely dispersive.[135] The major rate components are insensitive to the size of the ET driving force ($\Delta E_m = E_m(Q_A) - E_m(Q_B)$), varied by using artificial quinones as Q_A but native Q-10 as Q_B.[131,136] This insensitivity indicates that ET is not rate limiting. However, when ΔE_m is large, Gunner and coworkers have observed a fast phase that *is* responsive to the driving force.[134,137] This is consistent with ET from Q_A^- directly to Q_B in an 'unprepared' state, which subsequently relaxes, for example due to structural changes and/or proton uptake.[25,100]

On-going studies by Paddock *et al.* have detected ENDOR signals in the Q_B^- spectrum that correspond to proton coupling to $Ser^{L223}(O_\gamma H)$, and have correlated these with the lifetime of the $P^+Q_B^-$ state, as an indicator of the stability of Q_B^-. Their results certainly implicate this residue in providing much of the stabilization that 'gates' the ET from Q_A^-. When $P^+Q_B^-$ is prepared at room temperature and frozen in the light, it is stable for at least a year at 100 K.[138,139] When prepared at low temperature, however, it decays in a few seconds – at least 10^7 times faster.[140f] In a mutant

---

[e] Mutations are designated by the subunit and residue number, followed by the wild type residue letter, and then the mutant residue letter. Thus, the single mutant $Glu^{L212} \rightarrow Gln$ is given as L212EQ

[f] Normal ET from Q_A^- to Q_B is blocked at low temperature in samples frozen in the dark. However, $P^+Q_B^-$ can be prepared at low temperature in mutant RCs in which the B-side chain of cofactors (see Fig 8) has been made active and the normal A-side has been inactivated by a number of mutations, and in which Q_A is lacking.[139] Yields of approx 15% $P^+Q_B^-$ per flash can be obtained.

with Ser^{L223} replaced by alanine, the charge separated state frozen in the light is much less stable. Tentatively, this is interpreted as indicating that the rotation of Ser^{L223}(O$_\gamma$H) to engage the Q$_B$ carbonyl oxygen is required to stabilize the electron on Q$_B^-$ and is the rate-limiting step in the 1st ET. However, this does not rule out proton transfer (H$^+$ redistribution) as a factor in this event, since the strongly stabilized state can only be generated in the wild type at room temperature.

7.2 Second Electron Transfer and Coupled Proton Transfer to Q$_B$ (Chemical Protons)

Full reduction of Q$_B$ on the second turnover is coupled with the delivery of two protons to the quinone head group, to form QH$_2$. As Figure 9 shows, the acceptor quinone cycle branches at this point, with two main mechanistic possibilities (a concerted process is a third possibility). In the upper branch, the second electron is transferred to yield the dianion, Q$_B^{2-}$, followed by uptake of the first proton to form Q$_B$H$^-$. In the lower branch, the first proton is taken up to give the protonated semiquinone, Q$_B$H, followed by transfer of the second electron to form Q$_B$H$^-$. In both scenarios, the second proton is taken up following formation of Q$_B$H$^-$.

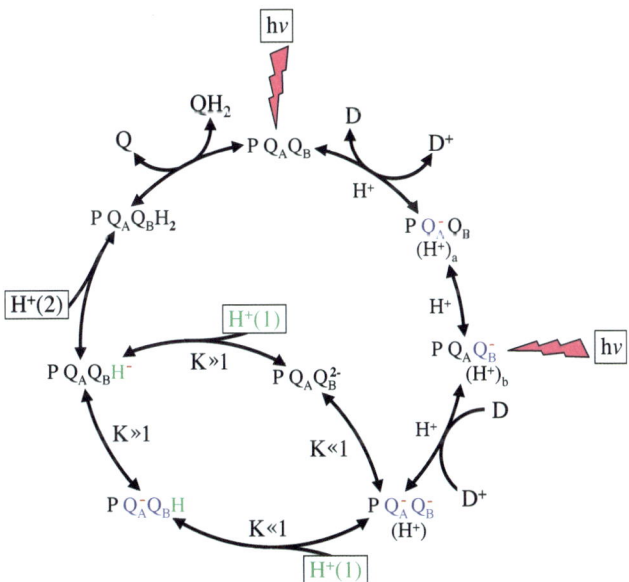

Figure 9 *The acceptor quinone cycle. Following photoactivation, the first electron is shared between the two quinones. The negative charges of the anionic semiquinones induce proton uptake to the protein, contributing to the partial shielding and stabilization of the semiquinones. Following a second photoactivation, full reduction of Q$_B$ is coupled with the delivery of two protons to the quinone head group, to form QH$_2$, which unbinds and is replaced by an oxidized quinone. Two possible routes are shown for the proton-coupled second electron transfer – the lower path (PT/ET) is the active one (see text)*

Both paths begin with an unfavorable step followed by a highly favorable one. For the upper path (ET/PT), the redox potential of the Q_B^-/Q_B^{2-} couple is not well established, but I have estimated it to be about –200 mV (vs. NHE), although Zhu and Gunner estimated a somewhat lower value of about -360 mV.[141] In either case, the electron transfer is significantly uphill from Q_A^-, with $E_m \approx -45$ mV.[142] It is also likely to be severely constrained by a large reorganization energy reflecting the substantial charge redistribution.[100] However, the pK_a of Q_B^{2-} (the second pK_a of the quinol) is expected to be high and protonation very favorable, thereby pulling the equilibrium through to Q_BH^-.

The lower route (PT/ET) in Figure 9 proceeds *via* the unfavorable protonation of Q_B^- – unfavorable because the pK_a of the semiquinone is low. The reference values for this species are reasonably well established in the literature with a few studies on ubisemiquinone (Q-10), although not in water. Two pK_a values have been reported – 6.45 in methanol[143] and 5.9 in 7 M isopropanol/1 M acetone.[144] From the known solvent (dielectric) effect on the pK_a of durosemiquinone, one can estimate an aqueous value of $pK_a = 4.9 \pm 0.1$ for ubisemiquinone.

In RCs, we might expect the Q_B semiquinone pK_a to be lower due to the local capacity to accommodate a single negative charge. In fact, experimentally, no indication of a pK_a has been detected – the Q_B (and Q_A) semiquinone is anionic at all accessible pH, even as low as pH 3.5 (R.R. Stein, PhD thesis, University of Illinois, 1985). If the pK_a of Q_BH/Q_B^- were indeed lowered due to electrostatic stabilization of the anion, one should expect a quantitatively similar effect on the midpoint potential. The $E_{m,7}$ of Q_B/Q_B^- can be estimated at $+30$ mV in isolated RCs[100,145] (it is about $+100$ mV in chromatophores[146]), which is 170 mV lower than my estimate of the E_m for Q-10 in aqueous solution ($E_m = -140$ mV, C.A.W. unpublished). This shift is consistent with a positive potential in the Q_B binding site, but may be somewhat too large.

In fact, for the second electron transfer, the relevant pK_a for Q_B^- is in the state Q_A^- Q_B^-, where the influence of the Q_A^- anion is likely to raise the pK_a of Q_B^-. This is a reasonable expectation in view of the influence that Q_A^- has on the pK_a values of amino acids in the Q_B domain. It is also supported by results with rhodoquinone acting as Q_B, which suggest the pK_a increases by about 0.7 units upon formation of $Q_A^-Q_B^-$.[147]

In the lower path of Figure 9, following protonation of Q_B^-, the second electron transfer is very favorable, involving the relatively high potential redox couple, Q_BH/Q_BH^-, with $E_m \approx +240$ mV. This would pull the reaction to completion in spite of the unfavorable, initial protonation.

The distinction between the two possible routes for the second electron transfer and coupled proton uptake was made by Graige *et al.*, who used artificial quinones as Q_A to alter the driving force for the ET without changing the properties (pK_a or E_m) of Q_B.[136] This 'driving force assay' showed clearly that the observed rate, $k_{AB}^{(2)}$, was responsive to the ET driving force, indicating that the reaction was ET limited. A full analysis showed the results to be compatible only with the PT/ET sequence, with rate-limiting ET preceded by fast PT equilibrium. Thus, the observed reaction ($Q_A^-Q_B^- \rightarrow Q_A^-Q_BH^-$) proceeds with a rate given by:

$$k_{AB}^{(2)} = k_{ET}^{(2)}.f(Q_BH) \tag{10}$$

The measured rate is pH dependent because the population of Q_BH is pH dependent. In the simplest case, one would expect the pH dependence to follow the Henderson–Hasselbalch equation:

$$f(Q_BH) = \frac{10^{pK_a - pH}}{1 + 10^{pK_a - pH}} \qquad (11)$$

However, the complex electrostatics of the protein interior do not allow such a simple response, but result in a pK_a that is pH dependent with an operational $pK_a \approx$ 4.5 at pH 7.5 (see[100,147]). The essential mechanism seems well founded and is summarized in Scheme 4.

(Scheme 4)

7.3 Proton Transfer to the Q_B Pocket

An important result of the mechanism of Scheme 4 for the 2nd ET is the implied rate of proton transfer to Q_B. In the wild type, the observed rate, $k_{AB}^{(2)}$, is about 3×10^3 s^{-1} at pH 7.0. If the pK_a for $Q_A^-Q_B^-/Q_AQ_BH$ is 4.5, this corresponds to $k_{ET}^{(2)} \sim 10^6$ s^{-1}, as estimated by Graige *et al.* (1999). Since this is rate limiting, the PT equilibrium must be established faster than 10^7 s^{-1}. The speed of equilibration is the sum of forward and reverse rates, $k_{eq} = k_{on}[H^+] + k_{off}$. In solution, this can be readily satisfied by a low pK_a, i.e., $k_{off} = k_{on}10^{-pK_a}$; so, for $k_{on} = 10^{10} - 10^{11}$ M^{-1} s^{-1}, $k_{eq} \sim k_{off} \sim 10^7$ s^{-1} when the $pK_a = 3-4$. In the RC, however, equilibration with the bulk pH must fully span the protein between Q_B and the surface, which does raise the question of how it is achieved and how it maintains substantial robustness in the face of possible mutational lesions.

Because the terminal points are well established on the quinone, the second electron transfer-coupled proton delivery is somewhat better defined than the protonation events on the first electron transfer, although the intervening pathway itself is subject to the same complex interactions. The general flavor of the pathway is shown in Figure 10, derived from studies on mutants and on the serendipitous finding that certain divalent transition metal ions, which bind in the vicinity of a histidine cluster on the surface of the H subunit,[148-151] dramatically inhibit proton entry.[151,152]

7.4 Outline of the PT Elements in the Q_B Domain

The effects on the 2nd ET of site-directed mutations in the immediate vicinity of Q_B have identified Asp^{L213} [153] and Ser^{L223} [154] as critical for the delivery of the first proton to C_1–O, and Glu^{L212} as essential for delivery of the second proton to C_4–O.[117,118] Glu^{H173} was also found to have a powerful influence on both proton deliveries.[155]

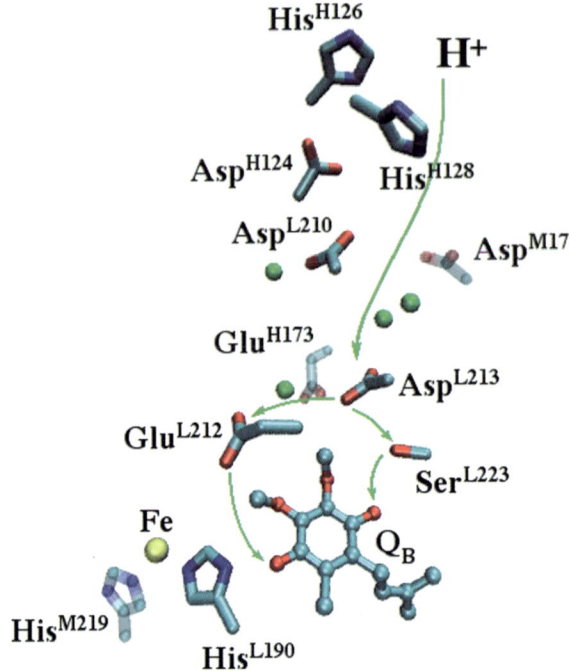

Figure 10 *Proton transfer pathway to Q_B. Side chains only are shown, and the isoprenyl tail of Q_B is truncated. Residues His^{H126}, His^{H128} and Asp^{M17} are at the surface of the protein. The path from the bulk phase to Asp^{L213} is shared by both protons delivered to Q_B. Accompanying the 1st electron to Q_B, Glu^{L212} becomes fully protonated and subsequently donates the 2nd H^+ to Q_BH^-, after the 2nd electron. The 1st H^+ is delivered in the $Q_A^-Q_B^-$ state via Ser^{L223}, prior to the 2nd electron transfer – see text. Water molecules (green) fill some but not all gaps in the putative H^+ pathway. Different positions are occupied in other structures indicating the possibility of water dynamics in the proton delivery. (Coordinates from 1aig.pdb).(Figure prepared in VMD using perspective display mode. Glu^{H173} and Asp^{M17} are shown dimmed to accentuate their positions relatively in the background.)*

No other single mutants have dramatic effects on the proton-coupled ETs to Q_B, but it is important to remember that substantial change in PT kinetics may be necessary before any readily observable effect can be seen on ET. For example, second site revertants have lent some qualification to the conclusion of absolute functional requirement for all of these residues.[119,120,156-161] In many cases, the rate of the 2nd ET in suppressor mutants is still 10–100 smaller than in the wild type. However, when the primary mutation is PT limited, even quite feeble restoration of the observed ET kinetics, if no longer PT limited, indicates several orders of magnitude restoration of the PT rates.[160]

Rather minor changes in the behavior of two single mutants, $Asp^{L210} \rightarrow$ Asn and $Asp^{M17} \rightarrow$ Asn,[162] led Paddock and coworkers to make the double mutant $Asp^{L210} \rightarrow$ Asn + $Asp^{M17} \rightarrow$ Asn (mutant L210DN/M17DN), which was dramatically inhibited.[163] The 2nd ET rate constant was inhibited by more than 200-fold ($k_{AB}^{(2)} \approx 10 \text{ s}^{-1}$) and was insensitive to the driving force for ET (the ΔE_m between Q_A and Q_B), *i.e.* the reaction is *PT limited*. Since, in the wild type, PT is faster than ET and estimated to be at least 10^7 s^{-1}, the implication is that PT in this mutant is decreased by a factor of 10^6 or more.

Inhibition of the 2nd ET in the L213DN mutant is even more dramatic, with the observed rate as slow as 0.1 s^{-1}.[153] The rate is not ET limited and so the inhibition of PT is therefore apparently of the order of 10^8 fold.[160,164] In contrast, in H173EQ mutant RCs, which exhibit observed rates similar to those seen in the L210DN/M17DN double mutant,[155] the rate *was* modified in the driving force assay,[102] indicating that it is ET limited and that PT is still fast. This mutant may be marginal with respect to its PT capabilities, but it is at least roughly described by Equation 10, implying an altered pK_a value for Q_B^-. This illustrates how misleading unqualified reaction rates can be. Thus, if one wishes to study the proton-transfer events through the lens of the coupled ET, it is important that the PT limited nature of the process be established clearly.

Finally, several divalent transition metal ions are now known to inhibit proton uptake and delivery to Q_B.[151,165,166] The binding site for these ions has been identified by crystallography[150] and also by EPR methods to be on the surface of the H subunit.[148,149] Depending on the identity of the metal, the ligands include one or both of two histidine residues (H126 and H128) and one of two aspartic acid residues (H124 or M17). The extent of inhibition is significant, although not as great as for the H173EQ mutant, but $k_{AB}^{(2)}$ is not responsive to the ET driving force so the kinetics are clearly PT limited. Identification of the metal ion binding site led to construction of a site-directed mutant lacking both histidines – $\text{His}^{H126} \rightarrow \text{Ala} + \text{His}^{H128} \rightarrow \text{Ala}$–referred to as mutant 2xHis in the orginal work[167]. This mutant was not inhibited at pH $\leqslant 7$ but it was progressively slowed at pH > 7.[152] The inhibition of the 2nd ET was only four-fold, even at pH 8.5, but the driving force assay showed it to be PT limited, as for the metal-bound wild type, which means that PT had been very substantially slowed down.

7.5 The Proton Conduction Pathway to Q_B and the Q_B Domain

The results from metal ion binding and the 2xHis mutant establish a unique entry point for H$^+$ ions that is active for proton delivery to internal protein residues for the 1st ET and to Q_B for the 2nd ET.[151,152,165] At pH $\geqslant 8$, the inhibition of proton entry by divalent metal ions inhibits the 1st ET in WT but not in L212EQ mutant RCs,[151] indicating that proton uptake on the first turnover leads to protonation (neutralization) of Glu^{L212}, which is required to favor the ET equilibrium. As Glu^{L212} was titrated, RCs in which the residue was already protonated were not dependent on light-induced H$^+$ uptake.

Similarly, Paddock *et al.*[165] found that the 2nd ET was greatly inhibited by divalent metal ions, showing that the first H$^+$ delivered to Q_B^- in the $Q_A^- Q_B^-$ state came through the same pathway. The second H$^+$ to Q_BH$^-$ is considered to be delivered internally, from Glu^{L212} protonated on the first turnover. Thus, the PT path is common to both protons, with a bifurcation within the protein that allows delivery to Glu^{L212}, on the one hand, and Q_B on the other.

With the identification of the surface histidines as the metal binding site, the major impact of mutating Glu^{H173}, Asp^{L210} and Asp^{M17}, and the terminal residues near Q_B (Glu^{L212}, Asp^{L213} Ser^{L223}), a proton conduction path is roughly mapped out (Figure 10). However, it is noteworthy that these residues do not define a continuous path, and significant gaps exist in the crystal structures. Although some gaps are bridged by water molecules in some structures, the appearance of water molecules does not seem to be a simple matter of resolution and structure refinement. More likely they are transient entities that are relatively stabilized in some states or mutants and less so in others.

7.6 Investigations of Intraprotein PT Coupled to the First and Second ETs

The nature of the PT coupled to the 1st ET has been studied in the 2xHis mutant, at pH 8.5 where PT is rate limiting. Paddock and colleagues found that the inhibition of the 1st ET that is evident in this mutant at elevated pH is reversed by a wide variety of cationic buffers, with pK_as ranging from 2 to 11 – the mutant RCs were "rescued".[168] From the slope of the rate of ET vs. buffer concentration, a second order rate constant (k_2) was obtained for the protonated buffer, and a plot of log k_2 vs. buffer pK_a gave the result expected for a 'normal' acid. The rate constant reached diffusion controlled values (approx. 10^{10} M^{-1} s^{-1}) when the buffer $pK_a \approx 2$. In a simple solution PT between an acid and a base, this would indicate the pK_a of the acceptor (base) (see Section 6 above). This is clearly too low to correspond to the expected terminal acceptor (Glu^{L212} with $pK_a \approx 8.5$), which prompted the inclusion in the model of an intermediate. This is very reasonable in view of the known structure. The existence of an intermediate changes the interpretation of the limiting pK_a value, which now reflects an acceptor pK_a of about four. This is fully consistent with the carboxylic acids – for example, Asp^{H124} or Asp^{M17}-immediately beneath the presumed cation binding site (where the two histidines (H126 and H128) would be in wild type RCs).

Paddock et al.[168] used a steady-state kinetic analysis similar to that employed widely in enzyme kinetics. From this, they obtained a steady state maximum rate (effectively k_{cat}) of 10^5 s^{-1} for imidazole, as a close analogue of the native histidine. The system is underdetermined but reasonable estimates translate to an elementary PT rate constant for transfer from imidazole to the intermediate (uphill by 5 pK units) of 10^6 s^{-1} and from the intermediate to Glu^{L212} (energetically highly favorable) of 10^{10} s^{-1}.

In the L213DN mutant, which has the most severely inhibited 2nd ET of any known mutant, Takahashi and Wraight observed partial rescue by weak neutral acids, including azide, and speculated that these agents act as proton carriers through the mutational block[169] – as has been suggested for the effect of azide on the Asp^{96}→Asn mutant of bacteriorhodopsin.[170] Subsequently, we found that H173EQ mutant RCs, also with severely inhibited 2nd ET, could be fully rescued by azide, and noted that the effect could be due to the anion binding and restoring a functional pK_a that had been depressed by the loss of charge in the Glu^{H173} → Gln mutation.[155] Given the evident strength and complexity of the interactions within the acid cluster, distinguishing between these two modes of action is difficult. Recently, we have examined the effect of weak acids on the 2nd ET using the PT-limited L210DN/M17DN double mutant.[192]

The L210DN/M17DN double mutant is readily rescued by weak neutral acids, including azide, acetate, nitrite, formate, bicarbonate, fluoride, and phosphate. The plot of log k_2 for the neutral acids vs. pK_a was linear with a slope of -1 (Figure 11), although azide was more effective and acetate was less effective than predicted by the linear fit (both with $pK_a \approx 4.75$). Results with other buffers suggested a size limitation on the activity of the weak acids, and a probable exclusion of cationic species.

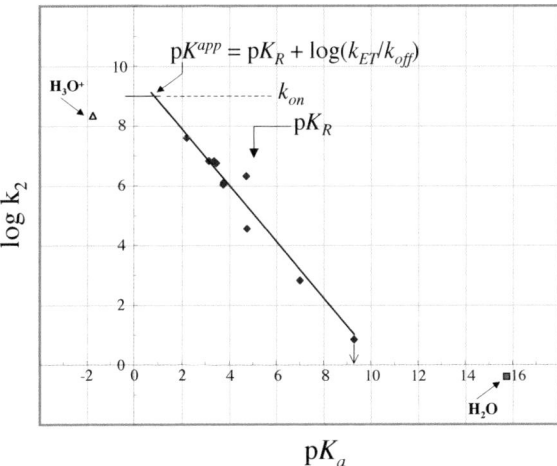

Figure 11 *Dependence on* pK_a *of the second order rate constant,* k_2*, for weak acids to rescue the second electron transfer in the L210DN/M17DN double mutant (Brønsted plot). Values for* k_2 *were obtained from the initial slope of the titration of measured rate* $(k_{AB}^{(2)})$ *as a function of total salt added,* A_T*, and converted to the free acid, [AH], using the acid* pK_a *and the prevailing pH 7.0:* $[AH] = A_T 10^{pK-pH}/(1 + 10^{pK-pH})$*. Acids shown: phosphate* $(pK_a^{(1)} = 2.15)$*, fluoride* $(pK_a = 3.16)$*, nitrite* $(pK_a = 3.37)$*, cyanate* $(pK_a = 3.46)$*, formate* $(pK_a = 3.72)$*, bicarbonate* $(pK_a = 3.58)$*, azide* $(pK_a = 4.72)$*, acetate* $(pK_a = 4.76)$*, phosphate* $(pK_a^{(2)} = 7.0)$*, ammonium* $(pK_a = 9.25)$*. The point for ammonium is an upper limit and has been omitted from the fitted line (slope* $=-1.0$*). Points are added for* H_3O^+ $(pK_a = -1.74)$ *(triangle) and* H_2O $(pK_a = 15.74)$ *(square) as potential donors, calculated for the inhibited rate (unrescued), with* $[H_3O^+] = 10^{-7}$ *M (pH 7.0) and* $[H_2O] = 55.5$ *M. Dashed line indicates the diffusion limit at* $k_{on} \approx 10^9$ M^{-1} s^{-1} *(see text)*

Notably, the maximum level of restoration of activity did not correlate simply with the acid pK_a, *e.g.* high concentrations of azide ($pK_a \approx 4.72$) restored $k_{AB}^{(2)}$ to almost wild type values, but formate ($pK_a \approx 3.72$) restored it only to about 30%. This can be accounted for by considering the protonated (neutral acid) and deprotonated (anion) forms to bind competitively:

$$A^- \cdot R^-$$

$$K_{I^-} \qquad \updownarrow$$

$$R^-$$

$$K_D \quad k_{on} \updownarrow k_{off} \quad \searrow k_2 \qquad\qquad \text{Scheme 5}$$

$$AH \cdot R^- \underset{k_{-H}}{\overset{k_H}{\rightleftharpoons}} A^- \cdot RH \xrightarrow{k_{ET}}$$

$$\xrightarrow{k_H'}$$

R^- represents a group in the proton transfer pathway to Q_B^- in the RC. K_D and K_I are the dissociation constants for the protonated (AH) and deprotonated (A^-) forms of the rescuing acid - binding of both AH and A^- is very weak. $k_{on} \approx 10^9$–10^{10} M^{-1} s^{-1} and $k_{off} \approx 10^9$–10^{11} s^{-1} are the on and off rates for acid (AH) binding and release. k_H and k_{-H} are the rate constants for forward and backward proton transfer within the RC ($k_H/k_{-H} = K_H = 10^{\Delta pK}$). k_{ET} is the rate constant for electron transfer to Q_BH ($k_{ET} \approx 10^6$ s^{-1} from estimates in wild type RCs – see above,[100,147]).

The data extrapolate to the maximum expected (diffusion controlled) rate of about 10^9–10^{10} M^{-1} s^{-1}, at an acid $pK_a = 0$–1. In the simplest case, this would be equal to the pK_a of the acceptor species (R^-), within the reaction center. However, the model predicts that the apparent pK_a will be offset from the actual pK_a of RH/R^- by log (k_{ET}/k_{off}), which is of the order of −4. Thus, the actual pK_a of the acceptor group should be in the ballpark of 4-5. This is consistent with R^- being a carboxylic acid, but it could also be Q_B^- itself, for which $pK_a \approx 4.5$ has been estimated.[100,147]

All rescuing acids shared similar features – small size, neutral acid form, very weak binding of the acid, and slightly stronger binding of the anion. This relative order of binding strengths could account for the relative inactivity of the cationic acids tested. In the absence of any added acids, the inhibited rate in this mutant could reflect the activity of H_3O^+ as donor, with $pK_a = -1.74$. At pH 7, the calculated bimolecular rate constant would be about 2×10^8 M^{-1} s^{-1}. This is significantly low on the plot in Figure 11, but it would be consistent with a disfavored electrostatic environment for cations and might provide some rationalization for the existence of a PT pathway at all. The natural Q_B-site engineering appears to be primarily directed at stabilizing the anionic semiquinone. To the extent that this is achieved by a positive electrostatic potential, this would be incompatible with the equally important function of delivering H^+ ions to Q_B for full reduction. Thus, free access by H_3O^+ is not a satisfactory solution, in spite of its strong proton–donor potential, and the pathway is based on neutral and anionic carboxylic acid/carboxylate functions.

In addition, the very low concentration of H^+ ions under physiological conditions necessitates some design functions to enhance the availability of the H^+ from solution[171–174]. The net charge of the protein is negative, which will enhance the concentration of H^+ at the surface.[130] However, the behavior of the 2xHis mutant indicates that this is insufficient and the cluster of surface histidines, H126 and H128, and possibly H68, greatly enhance the availability of protons by providing a local reservoir. The functional pK_as of residues involved in inhibitory metal binding appear to range over 5.9-7.4.[175,176] This spread of values may indicate electrostatic interactions, which serve to widen the pH range over which at least one of them is protonated.

8 Concluding Remarks

The establishment of a unique H^+ ion entry point and a defined pathway for proton delivery in RCs are worth discussing in terms of preconceptions about proteins and small molecules. It has been known for decades that proteins are not readily penetrated by small molecules, although it was a somewhat radical idea when first espoused.[177,178] It became clear from studies of myoglobin that even very weakly polar (CO) or apolar (O_2) molecules could not readily reach the heme binding site except through a specific pathway that was designed for flexibility and access.[179,180] On the other hand, there are also good indications that water does penetrate into the surface layers of proteins, contributing, among other things, to the non-homogeneous dielectric properties of proteins in a functionally important way[181–186]. From the latter, it seemed possible that the highly polar nature of the Q_B domain could permit proton delivery in a rather non-specific fashion and that no single pathway would be dominant. Water chains, identified in the crystal structures, were suggested as possible proton pathways,[187,188] but attempts to disrupt these did not result in effects that could be ascribed to proton transfer rates or equilibria.[189–191] This also seemed to lend some support to a distributed proton delivery network.

However, the identification of the surface histidine site as a unique entry point that serves both the protein residues of the Q_B domain and Q_B, itself, shows the pathway to be narrowly defined. Indeed, one can estimate that all other possible routes combined are at least 10^3 times less effective than the identified path.[163] This will not surprise anyone who is thinking in terms of proton pumps, where a unique gate must control the flow and backflow of protons across the membrane, but it is not a necessary design property for the non-transmembrane proton traffic in cofactor oxidation and reduction. It suggests, perhaps, that the requirements for an effective proton pathway, at least of significant length, are not so easily satisfied. This might include maintaining buried structures of sufficient polarity, in a generic sense, and of establishing specific pK_a values that are accessible to physiological pHs while sufficiently active in internal proton transfer, as well as designing surface properties that can aid in proton collection and injection into the protein.

Having established a functional proton transfer pathway, however, some manner of robustness can be expected from the fact that proton transfer between 'normal' acids is intrinsically fast and can be non-rate-limiting over a wide range of values. In the case of proton-coupled electron transfer in the reaction center acceptor quinone complex, proton transfer must be substantially impaired before it is even noticeable. Of course, loss of an active proton carrier, like Asp^{L213}, by mutation to a non-ionizable residue, can be catastrophic, but the net rate is quite tolerant of significant departures in the pK_a values of the functional proton carriers, due, for example, to mutations in the electrostatic environment. In apparent contrast, alterations in the pK_a of the Q_B semiquinone, as is probably the case in the H173EQ mutant, have direct and large magnitude effects. The difference is that the population of Q_BH, which is determined by its pK_a, is the multiplier of the rate–limiting electron–transfer rate constant. A change in the pK_a of an intermediate carrier, however, has its impact on a rate process that is likely to be very fast (pairwise proton transfer).

In a sense, PT pathways may be robust for the same reason as ET processes – the intervening medium is designed with overkill. For ET, the protein is treated as a homogeneous medium and distances are kept quite short, not pushing the envelope of maximum reach. For PT, the spacing and small scale mobility of the functional groups are sufficient to make the intrinsic rates very fast and the net rate relatively insensitive to the details (pK_a values) of the design.

Acknowledgements

Work from the author's laboratory and preparation of this review were supported by the National Science Foundation (MCB 03-44449) and National Institutes of Health (GM53508).

References

1. B. Chance and M. Nishimura, On the mechanism of chlorophyll-cytochrome interaction: the temperature insensitivity of light-induced cytochrome oxidation in *Chromatium, Proc. Natl. Acad. Sci. U.S.A.,* 1960, **46**, 19.
2. D. DeVault and B. Chance, Studies of photosynthesis using a pulsed laser. I. Temperature dependence of cytochrome oxidation rate in chromatium. Evidence from tunneling, *Biophys. J.,* 1966, **6**, 825.
3. D. DeVault, Quantum-mechanical tunneling in biological systems, *Q. Rev. Biophys.,* 1980, **13**, 387.
4. J. Deisenhofer, O. Epp, R. Miki, R. Huber and H. Michel, Structure of the protein subunits in the photosynthetic reaction center of *Rhodopseudomonas viridis* at 3 Å resolution, *Nature,* 1985, **318**, 618.
5. H. Michel, O. Epp and J. Deisenhofer, Pigment-protein interactions in the photosynthetic reaction center from *Rhodopseudomonas viridis, EMBO J.,* 1986 **5**, 2445.
6. R.A. Marcus, Chemical and electrochemical electron-transfer theory, *Ann. Rev. Phys. Chem.,* 1964, **15**, 155.
7. J.J. Hopfield, Electron transfer between biological molecules by thermally activated tunneling. *Proc. Natl. Acad. Sci. U.S.A.,* 1974, **71**, 3640.
8. J. Jortner, Dynamics of electron transfer in bacterial photosynthesis, *Biochim. Biophys. Acta.,* 1980, **594**, 193.
9. J. Ulstrup and J. Jortner, The effect of intramolecular quantum modes on free energy relationships for electron transfer reactions, *J. Chem. Phys.,* 1975, **63**, 4358.
10. R.A. Marcus and N. Sutin, Electron transfers in chemistry and biology, *Biochim. Biophys. Acta,* 1985, **811**, 265.
11. D.N. Beratan, J.N. Onuchic and J.J. Hopfield, Electron tunneling through covalent and non-covalent pathways in proteins, *J. Chem. Phys.,* 1987, **86**, 4488.
12. D.N. Beratan, J.N. Onuchic, J.R. Winkler and H.B. Gray, Electron-tunneling pathways in proteins, *Science,* 1992, **258**, 1740.
13. H.B. Gray and J.R. Winkler, Electron transfer in proteins, *Annu. Rev. Biochem.,* 1996, **65**, 537.
14. C.C. Moser, J.M. Keske, K. Warncke, R.S. Farid and P.L. Dutton, Nature of biological electron transfer, *Nature,* 1992, **355**, 796.
15. C.C. Page, C.C. Moser, X. Chen and P.L. Dutton, Natural engineering principles of electron tunnelling in biological oxidation-reduction, *Nature,* 1999, **402**, 47.
16. J.N. Onuchic, D.N. Beratan, J.R. Winkler and H.B. Gray, Pathway analysis of protein electron-transfer reactions, *Ann. Rev. Biophys. Biomol. Struct.,* 1992, **21**, 349.
17. P.J.F. De Rege, S.A. Williams and M.J. Therien, Direct evaluation of electronic coupling mediated by hydrogen bonds: implications for Biological electron transfer, *Science,* 1995, **269**, 1409.
18. S.S. Skourtis and D.N. Beratan, High and low resolution theories of protein electron transfer, *JBIC,* 1997, **2**, 378.

19. R.J.P. Williams, The medium in electron transfer proteins, *JBIC,* 1997, 2, 373.
20. O. Farver and I. Pecht, The role of the medium in long-range electron transfer, *JBIC,* 1997, **2**, 387.
21. P.L. Dutton and C.C. Moser, Quantum biomechanics of long-range electron transfer in protein: hydrogen bonds and reorganization energies, *Proc. Natl. Acad. Sci., U.S.A.,* 1994, **91**, 10247.
22. S. Larsson, Electron transfer in proteins, *Biochim. Biophys. Acta,* 1998, **1365**, 294.
23. R. Langen, *et al.* Electron tunneling in proteins: Coupling through a B Strand, *Science,* 1995, **268**, 1733.
24. R.S. Farid, C.C. Moser and P.L. Dutton, Electron transfer in proteins, *Curr. Opin. Struct. Biol.,* 1993, **3**, 225.
25. C.C. Moser, et al. Length, time and energy scales of photosystems, *Advances in Protein Chemistry,* 2003, **63**, 71.
26. L.I. Krishtalik, The mechanism of the proton transfer: an outline, *Biochim. Biophys. Acta,* 2000, **1458**, 6.
27. M. Eigen, Proton transfer, acid-base catalysis, and enzymatic hydrolysis, *Angew. Chem. Internat. Edit.,* 1964, **3**, 1.
28. R.P. Bell, *The Proton in Chemistry,* Cornell University Press, Ithaca, N.Y., 1973.
29. C.F. Berlasconi, Intrinsic barriers of reactions and the principle of nonperfect synchronization, *Acc. Chem. Res.,* 1987, **20**, 301.
30. A.J. Kresge, What makes proton transfer fast?, *Acc. Chem. Res.,* 1975, **8**, 354.
31. C.J.T.v. Grotthuss, Sur la decomposition de l'eau et des corps qu'elle tient en dissolution a l'aide de l'electricite galvanique (Theory of decomposition of liquids by electric currents), *Annales de Chimie,* 1806, **58**, 54.
32. J.F. Nagle and H.J. Morowitz, Molecular mechanisms for proton transport in membranes, *Proc. Natl. Acad. Sci., U.S.A.,* 1978, **75**, 298.
33. J.F. Nagle and S. Tristam-Nagle, Hydrogen bonded chain mechanisms for proton conduction and proton pumping, *J. Membrane Biol.,* 1983, **74**, 1.
34. M.L. Brewer, U.W. Schmitt and G.A. Voth, The formation and dynamics of proton wires in channel environments, *Biophys. J.,* 2001, **80**, 1691.
35. N. Agmon, The Grötthuss mechanism, *Chem. Phys. Lett.,* 1995, **244**, 456.
36. U.W. Schmitt and G.A. Voth, The computer simulation of proton transport in water, *J. Chem. Phys.,* 1999, **111**, 9361.
37. H. Lapid, N. Agmon, M.K. Petersen and G.A. Voth, A bond-order analysis of the mechanism for hydrated proton mobility in liquid water, *J. Chem. Phys.,* 2005, **122**, 014506.
38. M. Tuckerman, K. Laasonen, M. Sprik and M. Parrinello, *Ab initio* molecular dynamics simulation of the solvation and transport of hydronium and hydroxyl ions in water, *J. Chem. Phys.,* 1995, **103**, 150.
39. T.J.F. Day, U.W. Schmitt and G.A. Voth, The mechanism of hydrated proton transport in water, *J. Am. Chem. Soc.,* 2000, **122**, 12027.
40. I. Ohmine and S. Saito, Water dynamics: fluctuation, relaxation, and chemical reactions in hydrogen bond network rearrangement, *Acc. Chem. Res.,* 1999, **32**, 741.
41. A. Staib, D. Borgis and J.T. Hynes, Proton transfer in hydrogen-bonded acid–base complexes in polar solvents, *J. Chem. Phys.,* 1995, **102**, 2487.

42. N. Agmon, Proton solvation and proton mobility, *Israel J. Chem.*, 1999, **39**, 493.
43. R. Pomès, Theoretical studies of the Grotthuss mechanism in biological proton wires, *Israel J. Chem.*, 1999, **39**, 387.
44. J.F. Nagle, M. Mille and H.J. Morowitz, Theory of hydrogen bonded chains in bioenergetics, *J. Chem. Phys.*, 1980, **72**, 3959.
45. R. Dutzler, E.B. Campbell, M. Cadene, B.T. Chait and R. MacKinnon, X-ray structure of a ClC chloride channel at 3.0 Å reveals the molecular basis of anion selectivity, *Nature*, 2002, **415**, 287.
46. D.A. Doyle, *et al.* The structure of the potassium channel: molecular basis of K+ conduction and selectivity, *Science*, 1998, **280**, 69.
47. R. Pomès and B. Roux, Molecular mechanism of H^+ conduction in the single-file water chain of the gramicidin channel, *Biophys. J.*, 2002, **82**, 2304.
48. C.N. Schutz and A. Warshel, Analyzing free energy relationships for proton translocations in enzymes: carbonic anhydrase revisited, *J. Phys. Chem. B*, 2004, **108**, 2066.
49. S. Braun-Sand, M. Strajbl and A. Warshel, Studies of proton translocation in biological systems: simulating proton transport in carbonic anhydrase by EVB-based models, *Biophys. J.*, 2004, **87**, 2221.
50. S. Braun-Sand, A. Burykin, Z.T. Chu and A. Warshel, Realistic simulations of proton transport along the gramicidin channel: demonstrating the importance of solvation effects, *J. Phys. Chem. B*, 2005, **109**, 583.
51. M.F. Schumaker, R. Pomès and B. Roux, Framework model for single proton conduction through gramicidin, *Biophys. J.*, 2001, **80**, 12.
52. J.A. Gowen, *et al.* The role of Trp side chains in tuning single proton conduction through gramicidin channels, *Biophys. J.*, 2002, **83**, 880.
53. M.D. Becker, D.V. Greathouse, R.E. Koeppe II and O.S. Andersen, Amino acid sequence modulation of gramicidin channel function: effects of tryptophan-to-phenylalanine substitutions on the single channel conductance and duration, *Biochemistry*, 1991, **30**, 8830.
54. F.Tian, K.-C. Lee, W. Hu and T.A. Cross, Monovalent cation transport: lack of structural deformation upon cation binding, *Biochemistry*, 1996, **35**, 11959.
55. F. Tian and T.A. Cross, Cation transport: an example of structural based selectivity, *J. Mol. Biol.*, 1999, **285**, 1993.
56. D. Anderson, R.B. Shirts, T.A. Cross and D.D. Busath, Noncontact dipole effects on channel permeation. V. Computed potentials for fluorinated gramicidin, *Biophys. J.*, 2001, **81**, 1255.
57. W. Hu and T.A. Cross, Tryptophan hydrogen bonding and electrical dipole moments: functional roles in the gramicidin channel and implications for membrane proteins, *Biochemistry*, 1995, **34**, 14147.
58. D. Fu *et al.* Structure of a glycerol-conducting channel and the basis for its selectivity, *Science*, 2000, **290**, 481.
59. H. Sui, B.G. Han, J.K. Lee, P. Walian and B.K. Jap, Structural basis of water-specific transport through the AQP1 water channel, *Nature*, 2001, **414**, 872.
60. B. de Groot and H. Grubmüller, Water permeation across biological membranes: mechanism and dynamics of aquaporin-1 and GlpF, *Science*, 2001, **294**, 2353.

61. E. Tajkhorshid *et al.* Control of the selectivity of the aquaporin water channel family by global orientational tuning, *Science,* 2002, **296**, 525.

62. A. Burykin and A. Warshel, What really prevents proton transport through aquaporin? Charge self-energy versus proton wire proposals, *Biophys. J.,* 2003, **85**, 3696.

63. B.L. De Groot, T. Frigato, V. Helms and H. Grubmüller, The mechanism of proton exclusion in the aquaporin-1 water channel, *J. Mol. Biol.,* 2003, **333**, 279.

64. N. Chakrabarti, B. Roux and R. Pomès, Structural determinants of proton blockage in aquaporins, *J. Mol. Biol.,* 2004, **343**, 493.

65. B. Ilan, E. Tajkhorshid, K. Schulten and G.A. Voth, The mechanism of proton exclusion in aquaporin channels, *Proteins,* 2004, **55**, 223.

66. E. Fermi, *Notes on Quantum Mechanics,* University of Chicago Press, Chicago, IL, 1961.

67. P.A.M. Dirac, The quantum theory of the emission and absorption of radiation, *Proc. Roy. Soc. A,* 1927, **114**, 243.

68. J.N. Brønsted and K.J. Pedersen, Die katalytische Zersetzung des Nitramids und ihre physikalisch-chemische Bedeutung, *Z. Physik. Chem. (Leipzig),* 1924, **108**, 185.

69. J.N. Brønsted and E.A. Guggenheim, Contribution to the theory of acid and basic catalysis. Mutarotation of glucose, *J. Am. Chem. Soc.,* 1927, **49**, 2554.

70. A.J. Kresge, The Brønsted relation–recent developments, *Chem. Soc. Rev.,* 1973, **2**, 475.

71. R.A. Marcus, Theoretical relations among rate constants, barriers, and brøn sted slopes of chemical reactions, *J. Phys. Chem.,* 1968, **72**, 891.

72. A.O. Cohen and R.A. Marcus, On the slope of Free Energy plots in chemical kinetics, *J. Phys. Chem.,* 1968, **72**, 4249.

73. A. Yadav, R.M. Jackson, J.J. Holbrook and A. Warshel, Role of solvent reorganization energies in the catalytic activity of enzymes, *J. Am. Chem. Soc.,* 1991, **113**, 4800.

74. A. Warshel, *Computer Modeling of Chemical Reactions in Enzymes and Solutions,* Wiley-Interscience, New York, 1991.

75. G.S. Hammond, A correlation of reaction rates, *J. Am. Chem. Soc.,* 1955, **77**, 334.

76. J.R. Miller, Tunneling reactions of trapped electrons with added electron acceptors in alcohol glasses at 77 K, *J. Phys. Chem.,* 1978, **82**, 767.

77. K.S. Peters, A. Cashin and P. Timbers, Picosecond dynamics of nonadiabatic proton transfer: a kinetic study of proton transfer within the contact radical ion pair of substituted benzophenones/N,N-dimethylaniline, *J. Am. Chem. Soc.,* 2000, **122**, 107.

78. K.S. Peters and G. Kim, Solvent effects for nonadiabatic proton transfer in the benzophenone/N,N-dimethylaniline contact radical ion pair, *J. Phys. Chem. A,* 2001, **105**, 4177.

79. C.P. Andrieux, J. Gamby, P. Hapiot and J.-M. Saveant, Evidence for inverted region behavior in proton transfer to carbanions, *J. Am. Chem. Soc.,* 2003, **125**, 10119.

80. E. Meyer, Internal water molecules and H-bonding in biological macromolecules: a review of structural features with functional implications, *Protein Sci.,* 1992, **1**, 1543.

81. D.N. Silverman and S. Lindskog, The catalytic mechanism of carbonic anhydrase: implications of a rate-limiting protolysis of water, *Acc. Chem. Res.,* 1988, **21**, 30.

82. H. Steiner, B.-H. Jonsson and A. Lindskog, The catalytic mechanism of carbonic anhydrase, *Eur. J. Biochem.,* 1975, **59**, 253.

83. D.N. Silverman *et al.* Rate-equilibria relationships in intramolecular proton transfer in human carbonic anhydrase III, *Biochemistry,* 1993, **32**, 10757.

84. S. Lindskog and D.N. Silverman, The catalytic mechanism of mammalian carbonic anhydrases, *EXS,* 2000, **90**, 175.

85. C. Tu, R.S. Rowlett, B.C. Tripp, J.G. Ferry and D.N. Silverman, Chemical rescue of proton transfer in catalysis by carbonic anhydrases in the beta- and gamma-class, *Biochemistry,* 2002, **41**, 15429.

86. C. Tu, M. Qian, J.N. Earnhardt, P.J. Laipis and D.N. Silverman, Properties of intramolecular proton transfer in carbonic anhydrase III, *Biophys. J.,* 1998, **74**, 3183.

87. K.A. Sharp, Calculation of electron transfer reorganization energies using the finite difference Poisson–Boltzmann model, *Biophys. J.,* 1998, **73**, 1241.

88. J. Hwang and A. Warshel, Microscopic examination of free-energy relations for electron transfer in polar solvents, *J. Am. Chem. Soc.,* 1987, **109**, 715.

89. W.W. Parson, Z.T. Chu and A. Warshel, Reorganization energy of the initial electron-transfer step in photosynthetic bacterial reaction centers, *Biophys. J.,* 1998, **74**, 182.

90. G. King and A. Warshel, Investigation of the free energy functions for electron transfer reactions, *J. Chem. Phys.,* 1990, **93**, 8682.

91. D.N. Silverman, Marcus rate theory applied to enzymatic proton transfer, *Biochim. Biophys. Acta,* 2000, **1458**, 88.

92. W.J. Albery, in *Proton Transfer Reactions,* E.F. Caldin and V. Gold, (eds.), John Wiley & Sons, New York, 1975, 355.

93. J.M. Mayer, Proton-coupled electron transfer: a reaction chemist's view, *Annu. Rev. Phys. Chem.,* 2004, **55**, 363.

94. Q. Cui and M. Karplus, Is a 'proton wire' concerted or stepwise? A model study of proton transfer in carbonic anhydrase, *J. Phys. Chem. B,* 2003, **107**, 1071.

95. G. Zundel, Proton polarizability of hydrogen bonds: Infrared methods, relevance to electrochemical and biological systems, *Methods Enzymol.,* 1986 **127**, 439.

96. G. Zundel, Proton polarizability and proton transfer processes in hydrogen bonds and cation polarizabilities of other cation bonds. Their importance to understand processes in electrochemistry and biology, *Trends Phys. Chem.,* 1992, **3**, 129.

97. R. Rammelsberg, G. Huhn, M. Lübben and K. Gerwert, Bacteriorhodopsin's intramolecular proton-release pathway consists of a hydrogen-bonded network, *Biochemistry,* 1998, **37**, 5001.

98. J. Breton and E. Nabedryk, Proton uptake upon quinone reduction in bacterial reaction centers: IR signature and possible participation of a highly polarizable hydrogen bond network, *Photosynth. Res.,* 1998, **55**, 310.

99. M. Eigen and G.G. Hammes, Elementary steps in enzyme reactions (as studied by relaxation spectrometry), *Adv. Enzymol.,* 1963, **25**, 1.

100. C.A. Wraight, Proton and electron transfer in the acceptor quinone complex of bacterial photosynthetic reaction centers, *Frontiers in Bioscience,* 2004, **9**, 309.

101. V.P. Shinkarev and C.A. Wraight, in *The Photosynthetic Reaction Center,* J. Deisenhofer and J.R. Norris, (eds.), Academic Press, San Diego, 1993, 193.

102. M.Y. Okamura, M.L. Paddock, M.S. Graige and G. Feher, Proton and electron transfer in bacterial reaction centers, *Biochim. Biophys. Acta,* 2000, **1458**, 148.

103. C.A. Wraight, Electron acceptors of bacterial photosynthetic reaction centers II. H^+ binding coupled to secondary electron transfer in the quinone acceptor complex, *Biochim. Biophys. Acta,* 1979, **548**, 309.

104. M.L. Paddock, G. Feher and M.Y. Okamura, Proton transfer pathways and mechanism in bacterial reaction centers, *FEBS Lett.,* 2003, **555**, 45.

105. R.B. Gennis, Multiple proton-conducting pathways in cytochrome oxidase and a proposed role for the active-site tyrosine. *Biochim. Biophys. Acta,* 1998, **1365**, 241.

106. M. Wikström, Proton translocation by the respiratory haem-copper oxidases, *Biochim. Biophys. Acta,* 1998, **1365**, 185.

107. J.K. Lanyi, Crystallographic studies of the conformational changes that drive directional transmembrane ion movement in bacteriorhodopsin, *Biochim. Biophys. Acta,* 2000, **1459**, 339.

108. P. Mitchell, *Chemiosmotic Coupling in Oxidative and Photosynthetic Phosphorylation,* Glynn Research Ltd., Bodmin, UK, 1966.

109. P. Mitchell, Keilin's respiratory chain concept and its chemiosmotic consequences, *Science,* 1979, **206**, 1148.

110. C. Bohr, K.A. Hasselbalch and A. Krogh, Ueber einen in biologischer Beziehung wichtigen Einfluss, den die Kohlensaurespannung des Blutes auf dessen Sauerstoff-bindung übt, *Scand. Arch. Physiol.,* 1904, **16**, 402.

111. B. Chance, A.R. Crofts, M. Nishimura and B. Price, Fast membrane H^+ binding in the light-activated state of *Chromatium* chromatophores, *Eur. J. Biochem.,* 1970, **13**, 364.

112. P. Maróti and C.A. Wraight, Flash-induced H^+ binding by bacterial reaction centers: influences of the redox states of the acceptor quinones and primary donor, *Biochim. Biophys. Acta,* 1988, **934**, 329.

113. P.H. McPherson, M.Y. Okamura and G. Feher, Light-induced proton uptake by photosynthetic reaction centers from Rhodobacter sphaeroides R-26. I. Protonation of the one-electron states $D^+Q_A^-$, DQ_A^-, $D^+Q_AQ_B^-$ and $DQ_AQ_B^-$, *Biochim. Biophys. Acta,* 1988, **934**, 348.

114. J. Miksovska, M. Schiffer, D.K. Hanson and P. Sebban, Proton uptake by bacterial reaction centers: the protein complex responds in a similar manner to the reduction of either quinone, *Proc. Natl. Acad. Sci. U.S.A.,* 1999, **96**, 14348.

115. E. Alexov and M.R. Gunner, Calculated protein and proton motions coupled to electron transfer: electron transfer from Q_A^- to Q_B in bacterial photosynthetic reaction centers, Biochemistry, 1999, **38**, 8254.

116. D. Kleinfeld, M.Y. Okamura and G. Feher, Electron transfer in reaction centers of *Rhodopseudomonas sphaeroides:* I. Determination of the charge recombination pathway of $D^+Q_AQ_B^-$ and free energy and kinetic relations between $Q_A^-Q_B$ and $Q_AQ_B^-$, *Biochim. Biophys. Acta,* 1984, **766**, 126.

117. M.L. Paddock, S.H. Rongey, G Feher and M.Y. Okamura, Pathway of proton transfer in bacterial reaction centers: replacement of glutamic acid 212 in the L subunit by glutamine inhibits quinone (secondary acceptor) turnover. *Proc. Natl. Acad. Sci. U.S.A.,* 1989, **86**, 6602.

118. E. Takahashi and C.A. Wraight, Proton and electron transfer in the acceptor quinone complex of *Rhodobacter sphaeroides* reaction centers: characterization of site-directed mutants of the two ionizable residues, Glu^{L212} and Asp^{L213}, in the $Q_B$binding site, *Biochemistry,* 1992, **31**, 855.

119. J. Miksovska *et al.* In bacterial reaction centers rapid delivery of the second proton to Q_B can be achieved in the absence of L212Glu, *Biochemistry,* 1997, **36**, 12216.

120. P. Maróti, D.K. Hanson, M. Schiffer and P. Sebban, Long-range electrostatic interaction in the bacterial photosynthetic reaction centre, *Nature Struc. Biol.,* 1995, **2**, 1057.

121. C.R.D. Lancaster, M. R. Gunner and H. Michel, in *Photosynthesis: from Light to Biosphere,* vol. 1, P. Mathis, (ed.), Kluwer, Dordrecht, 1995, 903.

122. P. Beroza, D.R. Fredkin, M.Y. Okamura and R. Feher, Electrostatic calculations of amino acid titration electron transfer, $Q_A^-Q_B \rightarrow Q_AQ_B^-$, in the reaction center, *Biophys. J.,* 1995, **68**, 2233.

123. C.R.D. Lancaster, H. Michel, B. Honig and M.R. Gunner, Calculated coupling of electron and proton transfer in the photosynthetic reaction center of *Rhodopseudomonas viridis, Biophys. J.,* 1996, **70**, 2469.

124. B. Rabenstein, G.M. Ullmann and E.-W, Knapp, Calculation of protonation patterns in proteins with structural relaxation and molecular ensembles–application to the photosynthetic reaction center, *Eur. Biophys. J.,* 1998, **27**, 626.

125. B. Rabenstein, G.M. Ullmann and E.-W. Knapp, Electron transfer between the quinones in the photosynthetic reaction center and its coupling to conformational changes, *Biochemistry,* 2000, **39**, 10487.

126. H. Ishikita, G. Morra and E.-W. Knapp, Redox potential of quinones in photosynthetic reaction centers from *Rhodobacter sphaeroides:* dependence on protonation of glu-L212 and Asp-L213, *Biochemistry,* 2003, **42**, 3882.

127. E. Alexov *et al.* Modeling effects of mutations on the free energy of the first electron transfer from Q_A^- to Q_B in photosynthetic reaction centers, *Biochemistry,* 2000, **39**, 5940.

128. C.R.D. Lancaster, and H. Michel, The coupling of light-induced electron transfer and proton uptake as derived from crystal structures of reaction centres from *Rhodopseudomonas viridis* modified at the binding site of the secondary quinone, Q_B, *Structure,* 1997, **5**, 1339.

129. U. Zachariae, and C.R.D. Lancaster, Proton uptake associated with the reduction of the primary quinone Q_A influences the binding site of the secondary quinone Q_B in *Rhodopseudomonas viridis* photosynthetic reaction centres, *Biochim. Biophys. Acta*, 2001, **1505**, 280.

130. P. Maróti, and C.A. Wraight, Kinetics of H^+ ion binding by the $P^+Q_A^-$ state of bacterial photosynthetic reaction centers: rate limitation within the protein, *Biophys. J.*, 1997, **73**, 367.

131. M.S. Graige, G. Feher, and M.Y. Okamura, Conformational gating of the electron transfer reaction $Q_A^-Q_B \rightarrow Q_AQ_B^-$ in bacterial reaction centers of Rhodobacter sphaeroides determined by a driving force assay, *Proc. Natl. Acad. Sci. U.S.A.*, 1998, **95**, 11679.

132. R. Hienerwadel et al. Protonation of Glu L212 following Q_B^- formation in the photosynthetic reaction center of *Rhodobacter sphaeroides:* evidence from time-resolved infrared spectroscopy, *Biochemistry*, 1995, **34**, 2832.

133. E. Takahashi, P. Maróti and C.A. Wraight, in *Electron and Proton Transfer in Chemistry and Biology*, E. Diemann, W. Junge, A. Müller and H. Ratazczak (eds.), Elsevier, Amsterdam, 1992, 219.

134. J. Li, D. Gilroy, D.M. Tiede, and M.R. Gunner, Kinetic phases in the electron transfer from $P^+Q_A^-Q_B$ to $P^+Q_AQ_B^-$ and the associated processes in *Rhodobacter sphaeroides* R-26 reaction centers, *Biochemistry*, 1998, **37**, 2818.

135. D.M. Tiede, J. Vazquez, J. Cordova and P.A. Marone, Time-resolved electrochromism associated with the formation of quinone anions in the *Rhodobacter sphaeroides* R26 reaction center, *Biochemistry*, 1996, **35**, 10763.

136. M.S. Graige, M.L. Paddock, J.M. Bruce, G. Feher and M.Y. Okamura, Mechanism of proton-coupled electron transfer for quinone (QB) reduction in reaction centers of *Rb. Sphaeroides, J. Am. Chem. Soc.*, 1996, **118**, 9005.

137. J. Li, E. Takahashi and M.R. Gunner, -DGABo and pH dependence of the electron transfer from $P^+Q_A^-Q_B$ to $P^+Q_AQ_B^-$ in *Rhodobacter sphaeroides* reaction centers, *Biochemistry*, 2000, **39**, 7445.

138. D. Kleinfeld, M.Y. Okamura and G. Feher, Electron-transfer kinetics in photosynthetic reaction centers cooled to cryogenic temperatures in the charge separated state: evidence for light-induced structural changes, *Biochemistry*, 1984, **23**, 5780.

139. M. Paddock, R.M. Isaacson, C. Chang, G. Feher and M. Okamura, Conformations of QB$^-$. trapped by B side electron transfer in reaction centers from *Rhodobacter sphaeroides, Biophys. J.*, 2004, **86**, 11a.

140. M.L. Paddock *et al.* Hydrogen bond reorientation upon Q_B reduction revealed by ENDOR spectroscopy in reaction centers from *Rhodobacter sphaeroides, Biophys. J.*, 2005, **88**, 202a.

141. Z. Zhu and M.R. Gunner, The energetics of quinone dependent electron and proton transfers in *Rhodobacter sphaeroides* photosynthetic reaction centers, *Biochemistry*, 2005, **44**, 82.

142. P.L. Dutton, J.S. Leigh and C.A. Wraight, Direct measurement of the midpoint potential of the primary electron acceptor in *Rhodopseudomonas sphaeroides in situ* and in the isolated state: some relationships with pH and o-phenathroline, *FEBS Lett.*, 1973, **36**, 169.

143. E.J. Land and A.J. Swallow, One-electron reactions in biochemical systems as studied by pulse radiolysis, *J. Biol. Chem.,* 1970, **245**, 1890.

144. K.B. Patel and R.L. Willson, Semiquinone free radicals and oxygen: Pulse radiolysis study of one electron transfer equilibrium, *J. Chem. Soc.* (Faraday I), 1973, **69**, 814.

145. C.A. Wraight, The role of quinones in bacterial photosynthesis, *Photochem. Photobiol.,* 1979, **30**, 767.

146. A.W. Rutherford and M.C.W. Evans, Direct measurement of the redox potential of the primary and secondary quinone electron acceptors in *Rhodopseudomonas sphaeroides* (wild-type) by EPR spectrometry, *FEBS Lett.* 1980, **110**, 257.

147. M.S. Graige, M.L. Paddock, G. Feher, and M.Y. Okamura, Observation of the protonated semiquinone intermediate in isolated reaction centers from *Rhodobacter sphaeroides:* Implications for the mechanism of electron and proton transfer in proteins. *Biochemistry,* 1999, **38**, 11465.

148. L.M. Utschig, O. Poluektov, D.M. Tiede and M.C. Thurnauer, EPR investigation of Cu^{2+}-substituted photosynthetic bacterial reaction centers: evidence for histidine ligation at the surface metal site, *Biochemistry,* 2000, **39**, 2961.

149. L.M. Utschig, O. Poluektov, S.L. Schlesselman, M.C. Thurnauer and D.M. Tiede, Cu^{2+} site in photosynthetic bacterial reaction centers from *Rhodobacter sphaeroides, Rhodobacter capsulatus,* and *Rhodopseudomonas viridis, Biochemistry,* 2001, **40**, 6132.

150. H.L. Axelrod, E.C. Abresch, M.L. Paddock, G. Feher and M.Y. Okamura, Determination of the binding sites of the proton transfer inhibitors Cd^{2+} and Zn^{2+} in bacterial reaction centers, *Proc. Natl. Acad. Sci. U.S.A.,* 2000, **97**, 1542.

151. P. Ädelroth, M.L. Paddock, L.B. Sagle, G. Feher and M.Y. Okamura, Identification of the proton pathway in bacterial reaction centers: both protons associated with reduction of Q_B to Q_BH_2 share a common entry point, *Proc. Natl. Acad. Sci. U.S.A.,* 2000, **97**, 13086.

152. P. Ädelroth *et al.* Identification of the proton pathway in bacterial reaction centers: decrease of proton transfer rate by mutation of surface histidines at H126 and H128 and chemical rescue by imidazole identifies the initial proton donors, *Biochemistry,* 2001, **40**, 14538.

153. E. Takahashi and C.A. Wraight, A Crucial role for Asp^{L213} in the proton transfer pathway to the secondary quinone of reaction centers from *Rhodobacter sphaeroides. Biochim. Biophys. Acta,* 1990, **1020**, 107.

154. M.L. Paddock, P.H. McPherson, G. Feher and M.Y. Okamura, Pathway of proton transfer in bacterial reaction centers: replacement of serine-L223 by alanine inhibits electron and proton transfers associated with reduction of quinone to dihydroquinone, *Proc. Natl. Acad. Sci. U.S.A.,* 1990, **87**, 6803.

155. E. Takahashi and C.A. Wraight, Potentiation of proton transfer function by electrostatic interactions in photosynthetic reaction centers from *Rhodobacter sphaeroides:* first results from site directed mutation of the H-subunit, *Proc. Natl. Acad. Sci. U.S.A.,* 1996, **93**, 2640.

156. M.Y. Okamura, M.L. Paddock, P.H. McPherson, S.H. Rongey and G. Feher, in *Research in Photosynthesis*, N. Murata, (ed.), Kluwer Academic Publishers, Dordrecht, 1992, 349.

157. S.H. Rongey, M.L. Paddock, G. Feher and M.Y. Okamura, Pathway of proton transfer in bacterial reaction centers: second-site mutation Asn-M44 → Asp restores electron and proton transfer in reaction centers from the photosynthetically deficient Asp-L213 -> Asn mutant of *Rhodobacter sphaeroides, Proc. Natl. Acad. Sci. U.S.A.,* 1993, **90**, 1325.

158. D.K. Hanson, S.L. Nance and M. Schiffer, Second site mutation at M43 (Asn–›Asp) compensates for the loss of two acidic residues in the $Q_B$site of the reaction center, *Photosynth. Res.,* 1992, **32**, 147.

159. D.K. Hanson, D.M. Tiede, S.L. Nance, C.-H. Chang and M. Schiffer, Site-specific and compensatory mutations imply inexpected pathways for proton delivery to the Q_B binding site of the photosynthetic reaction center, *Proc. Natl. Acad. Sci. U.S.A.,* 1993, **90**, 8929.

160. M.L. Paddock *et al.* Characterization of second site mutations show that fast proton transfer to Q_B^- is restored in bacterial reaction centers of *Rhodobacter sphaeroides* containing the Asp-L213 → Asn lesion, *Photosynth. Res.,* 1998, **55**, 281.

161. J. Miksovska, M. Valerio-Lepiniec, M. Schiffer, D.K. Hanson and P. Sebban, In bacterial reaction centers, a key residue suppresses mutational blockage of two different proton transfer steps, *Biochemistry,* 1998, **37**, 2077.

162. M.L. Paddock, G. Feher, and M.Y. Okamura, Identification of the proton pathway in bacterial reaction centers: replacement of Asp-M17 and Asp-L210 with Asn reduces the proton transfer rate in the presence of Cd^{2+}, *Proc. Natl. Acad. Sci. U.S.A.,* 2000, **97**, 1548.

163. M.L. Paddock *et al.* Identification of the proton pathway in bacterial reaction centers: cooperation between Asp-M17 and Asp-L210 facilitates proton transfer to the secondary quinone (QB), *Biochemistry,* 2001, **40**, 6893.

164. M.S. Graige, M.L. Paddock, G. Feher and M.Y. Okamura, Determination of the rate limiting step in the reaction, $Q_A^- Q_B^- + H^+ \rightarrow Q_A(Q_B H)-$ in Asp-L213 → Asn mutant RCs from *Rb. sphaeroides,* Biophys. J., 1996, **70**, A11.

165. M.L. Paddock, M.S. Graige, G. Feher and M.Y.Okamura, Identification of the proton pathway in bacterial reaction centers: inhibition of proton transfer by binding of Zn^{2+} or Cd^{2+}, *Proc. Natl. Acad. Sci. U.S.A.,* 1999, **96**, 6183.

166. L.M. Utschig, Y. Ohigashi, M.C. Thurnauer and D.M. Tiede, A new metal binding site in photosynthetic bacterial reaction centers that modulates Q_A to Q_B electron transfer, *Biochemistry,* 1998, **37**, 8278.

167. S. Keller, J.T. Beatty, M.L. Paddock, J.Breton and W.Leibl, Effect of metal binding on electrogenic proton transfer associated with reduction of the secondary electron acceptor (Q_B) in *Rhodobacter sphaeroides* chromatophores, *Biochemistry,* 2001, **40**, 429.

168. M.L. Paddock, P. Ädelroth, G. Feher, M.Y. Okamura and J.T. Beatty, Determination of proton transfer rates by chemical rescue: application to bacterial reaction centers, *Biochemistry,* 2002, **41**, 14716.

169. E. Takahashi and C.A. Wraight, Small weak acids stimulate proton transfer events in site-directed mutants of the two ionizable residues, Glu^{L212} and Asp^{L213}, in the Q_B-binding site of *Rhodobacter sphaeroides* reaction centers, *FEBS,* 1991, **283**, 140.

170. J. Tittor, C. Söll, D. Oesterhelt, H.-J. Butt and E. Bamberg, A defective proton pump, point-mutated bacteriorhodopsin Asp96 → Asn, is fully reactivated by azide, *EMBO J.,* 1989, **8**, 3477.

171. M. Gutman and E. Nachlicl, The dynamics of proton exchange between bulk and surface groups, Biochim. *Biophys. Acta,* 1995, **1231**, 123.

172. V. Sacks *et al.* The dynamic feature of the proton collecting antenna of a protein surface, *Biochim. Biophys. Acta,* 1998, **1365**, 232.

173. P. Ädelroth and P. Brzezinski, Surface-mediated proton-transfer reactions in membrane-bound proteins, *Biochim. Biophys. Acta,* 2004, **1655**, 102.

174. Y. Georgievskii, E.S. Medvedev and A.A. Stuchebrukhov, Proton transport via the membrane surface, *Biophys. J.,* 2002, **82**, 2833.

175. L. Gerencsér and P. Maróti, Retardation of proton transfer caused by binding of the transition metal ion to bacterial reaction centers is due to pK_a shifts of key protonatable residues, *Biochemistry,* 2001, **40**, 1850.

176. M.L. Paddock *et al.* Mechanism of proton transfer inhibition by Cd^{2+} binding to bacterial reaction centers: determination of the pK_a of functionally important histidine residues, *Biochemistry,* 2003, **42**, 9626.

177. J.R. Lakowicz and G. Weber, Quenching of protein fluorescence by oxygen. Detection of structural fluctuations in proteins in the nanosecond time scale, *Biochemistry,* 1973, **12**, 4171.

178. G. Weber, Energetics of ligand binding to proteins, *Adv. Prot. Chem.*, 1975, **29**, 1.

179. R.H. Austin, K.W. Beeson, D.L. Eisenstein, E.H. Frauenfelder and I.C. Gunsalus, Dynamics of ligand binding to myoglobin, *Biochemistry,* 1975, **24**, 5355.

180. H. Frauenfelder, S.G. Sligar and P.G. Wolynes, The energy landscapes and motions of proteins, *Science,* 1991, **254**, 1598.

181. T. Simonson, D. Perahia and G. Bricogne, Intramolecular dielectric screening in proteins, J. Mol. Biol., 1991, **218**, 859.

182. T. Simonson and D. Perahia, Microscopic dielectric properties of cytochrome *c* from molecular dynamics simulations in aqueous solution, *J. Am. Chem. Soc.,* 1995, **117**, 7987.

183. C.N. Schutz and A. Warshel, What are the dielectric 'constants' of proteins and how to validate electrostatic models, *Proteins,* 2001, **44**, 400.

184. E.B. Garcia-Moreno *et al.* Experimental measurement of the effective dielectric in the hydrophobic care of a protein, *Biophys. Chem.,* 1997, **64**, 211.

185. G. King, F.S. Lee and A.Warshel, Microscopic simulations of macroscopic dielectric constants of solvated proteins, *J. Chem. Phys.,* 1991, **95**, 4366.

186. J.J. Dwyer *et al.* High apparent dielectric constants in the interior of a protein reflect water penetration, *Biophys. J.,* 2000, **79**, 1610.

187. E.C. Abresch *et al.* Identification of proton transfer pathways in the X-ray crystal structure of the bacterial reaction center from *Rhodobacter sphaeroides,* Photosynth. *Res.,* 1998, **55**, 119.

188. G. Fritzsch, U. Ermler and H. Michel, in *Photosynthesis: From Light to Biosphere*, vol. 1, P. Mathis, (ed.), Kluwer Academic Publishers, Dordrecht, 1995, 599.

189. A. Kuglstatter, U. Ermler, H. Michel, L. Baciou and G. Fritzsch, X-ray structure analysis of photosynthetic reaction center variants from *Rhodobacter*

sphaeroides: structural changes induced by point mutations at position L209 modulate electron and proton transfer, *Biochemistry,* 2001, **40**, 4253.

190. L. Baciou, and H. Michel, Interruption of the water chain in the reaction center from *Rhodobacter sphaeroides* reduces the rates of the proton uptake and of the second electron transfer to Q_B, *Biochemistry,* 1995, **34**, 7967.

191. E. Takahashi and C.A. Wraight, in *Photosynthesis: From Light to Biosphere,* vol. 1, P. Mathis, (ed.), Kluwer Academic Publishers, Dordrecht, The Netherlands, 1995, 691.

192. C.A. Wraight and E. Takahashi, in *Photosynthesis: Fundamental Aspects to Global Perspectives,* A. van der Est and D. Bruce (eds.), Alliance Communications Group, Lawrence, Kansas, 2005, **vol. 1**, 5.

Infrared Protein Spectroscopy as a Tool to Study Protonation Reactions Within Proteins

PETER R. RICH AND MASAYO IWAKI

Glynn Laboratory of Bioenergetics,
Department of Biology,
University College London,
Gower Street,
London WC1E 6BT,
UK

Abbreviations

ATR, attenuated total reflectance; δ, in-plane bending vibration; v, stretching vibration; γ_w and γ_r, wagging and rocking vibrations; s, symmetric; as, antisymmetric.

1 Introduction

The sensitivity and stability of modern infrared spectrometers is such that it is now quite feasible to measure changes in absorption of infrared radiation in a sample that arise from changes of a single atomic bond within a protein, despite the fact that the protein itself will have a background absorption that at some frequencies can be 10^5 times greater than that of the bond vibrational change itself. The frequencies and extinction coefficients of the IR-absorbing species are changed not just by chemical changes of the bonds concerned, but also by minute changes in environment such as distances to other atoms, steric hindrances, hydrogen-bonding strengths and local dielectric strength and polarity changes. Such information from changes in IR characteristics is particularly powerful for proteins where an atomic structural model is known from NMR or X-ray crystallography, but where dynamic mechanistic details of the catalytic cycle remain to be established. Such an extension is often not possible with X-ray or NMR methods alone and is one for which IR spectroscopy provides an ideal tool.

Because of the very large numbers and types of bonds that have IR-absorbing characteristics, the breadth of application of IR methods to aspects of protein structures and mechanism is large and spans analyses of secondary structural elements and changes of substrates and cofactors, as well as specific changes of individual amino acids themselves. In addition, modern transient methods[1] allow kinetics of IR processes to be probed even to picosecond timescales.[2,3] However, in this review, we concentrate on one specific aspect of the use of IR spectroscopy to extend mechanistic understanding of enzymological mechanisms, namely its use to detect and define the protonation state changes that are an integral part of many enzymological processes.

2   Types of Information from Protein (FT)IR Spectroscopy

A typical protein of 200 amino acids, consisting of around 3000 atoms each having three degrees of freedom, will have 9000 vibrational degrees of freedom or 'normal modes of vibration'. In practice, a normal mode of vibration usually involves the coupled movements of several linked atoms so that it is not possible to assign a vibrational band to just one specific bond. However, functional regions of specific amino acids or cofactors tend to have characteristic band patterns and, in some cases, specific bands are dominated by vibrations at a single bond–for example the stretch at $1700-1800$ cm^{-1} of $C-O$ groups that are found in diverse cofactors such as chlorophylls and biliverdin or that of the SH group at 2551 cm^{-1}. It is these types of vibrations and the ways in which they change that can provide detailed structural, mechanistic and dynamic information on proteins.

Infrared radiation will be absorbed by a normal mode vibration when its frequency equals that of the normal mode vibration itself. For each vibration, this frequency is influenced by factors such as bond strength, polarity and hydrogen bonding. Even very small changes, for example hydrogen-bonding-distance changes of less than 0.1Å, can produce frequency changes that are easily detected. Importantly, the probability of absorption, and therefore the extinction coefficient, is dependent on the polarity of the vibrating bonds, with detectable increases of extinction coefficient with even small increases in bond polarity. Hence, such changes can give functional information on catalytic processes to an accuracy that cannot be achieved with X-ray structural methods, although IR spectroscopy cannot itself provide an initial overall protein structure. Hence, IR spectroscopy is particularly applicable when studying functional aspects of proteins whose static structures have been solved. Furthermore, quite dramatic changes in the IR signature will inevitably occur when a chemical change occurs in an amino acid or cofactor, for example a protonation or redox change, radical formation or other chemical modification and, provided that bands can be assigned to specific normal modes, detailed information on the chemical changes occurring can be gained and these are often interpretable even in the absence of an atomic model of the protein.[4] In addition, a further parameter that can be measured is the bandwidth, the broadness of which increases with increased heterogeneity due to conformational flexibility. Hence, bandwidths of species within confined protein binding sites tend to be narrower than their solution counterparts. Such information may therefore provide information on bound ligand flexibility and the entropic contribution to binding energy.[5]

3   Principles of (FT)IR Spectroscopy

IR absorption is caused by the coupling of the electric field vector of electromagnetic waves with the dipole moment of molecular vibrations when the frequencies of light and of the vibration are the same. It can occur only when the dipole moment of the molecule changes during the vibration and the absorption probability, and therefore the extinction coefficient, increases with increasing bond polarity. For the mid-IR range (4000–400 cm^{-1}) at room temperature, the majority of oscillators are not thermally excited and so are in the vibrational ground state such that IR absorption leads to a transition to the first excited state. The detailed physical principles of IR absorption are outside the remit of this review, and many comprehensive textbook accounts are available.[6,7]

Modern commercially available IR machines are of excellent signal/noise and stability. Again, many good textbook accounts of spectrometer and detector design and principle are available.[6,7] In short, these machines use the Michelson interferometry principle in which the IR beam is split into two and the pathlength of one beam is varied by reflection from a continuously oscillating mirror before being added back to the second beam. Constructive or destructive interference of the beams occurs to a degree that is a function of their varying pathlength difference and an interferogram of signal intensity *versus* mirror position is recorded. The position of the mirror can be determined very precisely from the regularly spaced interferogram of a monochromatic laser beam that passes through the same optical system and this information allows accurate computation of intensity *versus* frequency by Fourier transformation of the IR interferogram. Normally, many interferograms (each governed by mirror oscillation rate and typically recorded over tens of milliseconds) are averaged before Fourier transformation in order to increase signal/noise of computed spectra, and further averaging of many such spectra is usually also necessary.

Although significant information can be obtained from absolute spectra of proteins (see below), because of the very large number of vibrational bands that are present, interpretable atomic-level information at sufficient signal/noise is usually gained by producing difference IR spectra in which the protein is changed between two states and the IR difference between absolute spectra of the two states is calculated. In this way, only those parts of the protein that respond to the induced change will change vibrational properties and, providing that the induced change is sufficiently localised, the resultant IR difference spectrum should consist of a relatively small number of changes. Usually, in order to attain the required signal/noise, the two states of the protein have to be cycled many times so that many difference spectra can be averaged and it is the development of reliable and consistent methods to do this that is the key to successful data acquisition.

4 ATR-FTIR Spectroscopy

4.1 ATR Versus Transmission FTIR Spectroscopy

Infrared spectroscopy is commonly performed in conventional transmission mode in which a sample to be analysed is placed in an IR-transmitting material or within a pair of IR-transmitting windows. The IR beam is then passed through the sample and deflected onto a suitable detector. This has advantages of ease of use, general applicability for collection of absolute absorption spectra and ability to prepare samples of defined pathlengths so that extinction coefficients can be determined. A limitation of this method for protein infrared spectroscopy, however, is the limited methods that can be used to manipulate the protein *in situ* in order to be able to acquire difference IR spectra between two states. However, elegant transmission difference IR spectroscopy studies have been performed on a wide range of proteins whose reactions can be photochemically activated[8–16] or by use of a spectroelectrochemical cell designed for IR spectroscopy.[17–22]

An alternative method of spectroscopy acquisition of IR data is attenuated total reflectance (ATR)-FTIR.[23,24] In this method the infrared beam is reflected internally (usually several times) at the surface of an IR-transmitting element such as a silicon or diamond crystal. On reflection, the evanescent wave of the IR beam penetrates the immediate environment above the prism surface to a depth determined by refractive indices and IR frequency, usually of the order of a micron or so.[23] This allows measurement of changes in the infrared spectrum of thin protein films deposited on the prism surface. ATR-FTIR spectroscopy has already been applied to several membrane proteins including rhodopsin, bacteriorhodopsin, the nicotinic acetylcholine receptor and various respiratory proteins and for these hydrophobic proteins, stable, thin samples can be adhered to the prism surface by hydrophobic interactions and buffers with varying reagents can be perfused over the protein surface to induce a range of transitions (Figure 1). Other devices can be constructed over the protein sample to allow automated redox transitions.[25] In addition, it is possible to reflect a visible beam through the protein sample and monitor visible optical changes within the sample concurrently with IR data acquisition.[27] These methods greatly increase the types of transitions in protein samples whose IR properties can be determined. In practice, a combination of both transmission and ATR methods is often optimal for comprehensive analyses.

4.2 Methods of Handling Protein Samples

The ATR mode can be used for analyses of gas, liquid and solid samples. Provided that a suitable means of initiation of changes within the protein is available (see below), it is possible to apply the method to soluble proteins in conventional biochemical buffers. In practice, in order to collect data of suitable signal/noise, however, solutions of at least 0.5 mM protein are generally required, although sample volumes may be as low as 20 μl. This is easily attainable with many proteins, but can present problems with proteins whose M_r is in excess of 100,000 or for proteins or mutants that can only be produced in small amounts.

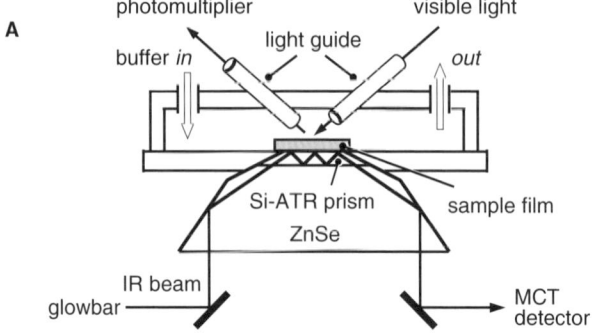

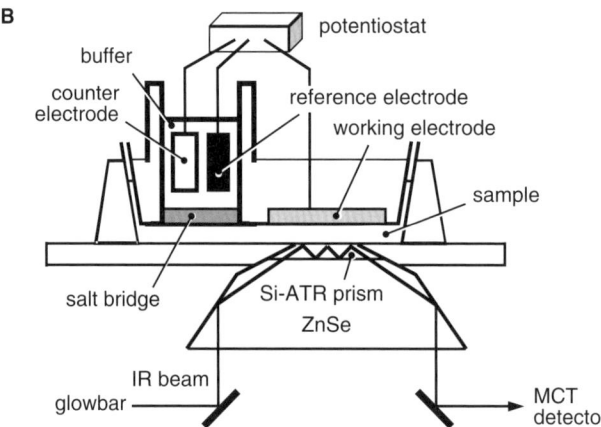

Figure 1 *Schematic of typical ATR-FTIR chambers for protein manipulations. Configuration A has a flow chamber for continuous perfusion of a buffer that can be switched to introduce or remove reactants in order to induce changes in a protein layer adhered to the ATR prism surface. Configuration B has a closed sample chamber (approx. 20 μl volume) whose top is formed from a glassy-carbon working electrode. A buffer with appropriate mediators equilibrates potential applied to the working electrode with the protein, which in this case can be a hydrophobic protein film or a soluble protein sample. In both cases, visible spectra can be recorded simultaneously with IR measurement by passage through the protein sample and reflection from the prism surface to a collecting fibre optic (shown only in configuration A)*

The method has found particular value for the study of membrane proteins because it is possible to deposit a stable, thin (several micron) layer of protein in its active hydrated state onto the hydrophobic surface of silicon or diamond ATR prism. The IR properties of the protein can then be monitored as it changes between states induced by, for example, perfusion of varying buffers over the film surface[27,28] or by electrochemical methods.[25] This greatly extends the types of transitions that are accessible to IR analyses. Typically in our laboratory, a protein film consists of 0.2–0.5 mg purified protein deposited on a 3 mm diameter silicon or diamond prism. The preparation of a stable, hydrophobic protein film is usually relatively straightforward. Proteins are purified by conventional detergent solubilisation and fractionation methods and usually are finally dissolved in aqueous buffers with salts and detergent. Removal of detergent and salt is essential in order to promote stable protein–prism interaction. In many cases, the sample can be washed in detergent-free weak buffer solution, such as 1 mM potassium phosphate, by repeated dilution and centrifugation. In other cases, especially if the original sample contains detergent with a low critical micelle concentration (CMC) or poor kinetics of exchange (for example, Tween 20/80, CMC = 0.06/0.01 mM; dodecyl maltoside, CMC = 0.2 mM; Triton X-100, CMC = 0.2 mM), prewashing with detergents of higher CMC and exchangeability (for example, cholate, CMC = 14 mM; octylglucoside, CMC = 20 mM; both used in the 0.005–0.02% w/v range) can be effective. The detergent-free protein suspension, ideally in a volume of less than 10 μl, is then dried onto the prism surface with a gentle stream of dry nitrogen, whereupon a stable interaction with the prism occurs to produce a stable protein film that can then be rehydrated with a wide range of buffers. In all cases, it is necessary to establish that the enzyme has remained in its native state throughout such procedures, and in general this is indeed the case. It is clear from the IR spectra of the 'dried' material that considerable structural water remains within the protein regardless of the length of the drying process and this is probably a key factor in retention of native conformation.

4.3   Methods of Inducing Difference Spectra

In general, averages of 10–50 difference spectra, each an average of 500–1000 interferograms, are required to obtain adequate signal/noise. Precise control of protein state interconversion is a key issue for reproducibility of spectra and we and others are still extending the types of methods that can be reliably employed. Three methods that are now routine are:

- *Buffer exchange.* This method is only applicable in the ATR mode and provides particular flexibility. A buffer is continuously perfused over the protein film surface and the protein state can be changed by switching reagents in the perfusant (see Figure 1A). When the reaction is reversible, difference spectra can be automatically recorded over many cycles by using programmable computer-controlled valves. The methods have been used to monitor changes induced by redox-state changes, pH jumps, ligand binding or interconversions between reaction intermediates. From a practical point of view, attention should be paid to degassing of buffers before use where possible to avoid bubble formation during flow and to precise matching of pH and ionic strength of buffer pairs to minimise shrinkage/swelling of the protein layer than can occur and which can produce changes in absolute IR absorbance properties that could mask the IR changes associated with the transition.
- *Electrochemically-induced transitions.* The method was firstly developed successfully for transmission FTIR spectroscopy by utilisation of an optically transparent working electrode in combination with redox mediators to equilibrate the potential applied to the working electrode with the protein.[17–22] Essentially the same principle has been developed for use with the ATR mode[25] in this case using a glassy-carbon working electrode placed several hundred microns above the protein film and again connected by a buffer containing appropriate redox mediators (see

Figure 1B). Potential is controlled automatically *via* a conventional three electrode potentiostat and this allows automated full oxidation/reduction cycles or redox potential stepping to allow redox titration or separation of individual redox centres.

- *Photochemical activation.*This was first developed successfully for transmission FTIR spectroscopy but has also been applied to great effect in ATR-FTIR studies to produce data of extremely high signal/noise.[29–31] When used in ATR mode, it is possible to combine acquisition of photolysis spectra with variation of another condition, such as pH or ionic strength, allowing sets of matched photolysis spectra to be collected on the same protein film.

5   Strategies to Assign IR Bands

As can be seen from the examples shown below, amino acids and prosthetic groups can have characteristic absolute and protonation difference spectra. As a result, tentative assignments of bands to specific types of residues can often be made by comparisons of frequencies, bandwidths and intensities of protein spectra with model compound information. In some cases, changes in certain frequency ranges can have only limited possible origins. For example, the only amino acid functional groups with bands in the $1700–1760$ cm^{-1} range are protonated forms of glutamic and aspartic acid and changes around 1100 cm^{-1} arise primarily from histidine. However, other constituents can overlay these regions: for example, the ester bond of lipids and carbonyl groups in aromatic structures can appear in the carboxylic acid region.[6,7] Hence, although reasonable tentative assignments can often be made from model compound comparisons, further tests are necessary to establish such assignments. Particularly useful is the analysis of bandshifts induced by isotope labelling. Most facile is the use of D_2O media to replace exchangeable protons with deuterons. For protein infrared spectroscopy, however, care must be taken since the rate of H/D exchange of occluded regions can be extremely slow, in some cases being resistant to exchanges even after several days of exposure.[14,23] More costly, but particularly informative, is the use of isotopically labelled materials to globally label the protein, to label one specific type of amino acid or cofactor or, in its most elegant form, to label just one specific amino acid. By combining such analyses with IR studies of specific site-directed mutations chosen by bioinformatic analyses or scrutiny of static atomic models of X-ray or NMR structures, reliable assignments of bands to specific atomic changes within the protein become feasible. Finally, considerable progress is being made in the development of simulation methods for *ab initio* calculation of normal-mode frequencies and intensities. These methods, whilst at present accurate only for relatively small molecules in simple environments, should in the future find increased applicability to the types of bandshifts of importance in protein infrared studies.

6   IR Properties of Amino Acids

6.1   Strategies to Isolate Headgroup Spectra: Amino Acids *Versus* Poly-Amino Acids

Accurate definition of the principal IR bands of amino-acid headgroups across the biologically useful range is not a trivial task. This is because the spectra of free amino acids are dominated by the v_s and v_{as} modes of the NH_3^+ and COO^- functional groups, all of which are broad with large extinction coefficients. This problem is eliminated by analyses of poly-L-amino acids (most of which are commercially available), but is replaced instead by domination of the spectra by amide I and amide II envelopes. The problem is further exacerbated for many amino acids whose neutral forms are relatively insoluble. Various attempts have been made to overcome these problems. For example, Venyaminov and Kalnin[32] subtracted spectra of simpler amino acids to remove the NH_3^+ and COO^- and Rahmelow *et al.*[33] have used an averaged amide I/II profile to subtract the majority of the amide I/II contribution to poly-amino-acid spectra. Attempts have been made to collate such information on amino acids and related model material data to provide a database of amino-acid headgroup IR bands. Most comprehensive of these are the reviews of Barth.[5,34] However, information is unavoidably incomplete as in many cases the frequencies of amino acids have to be extrapolated from related model materials, and details of extinction coefficients, bandwidths and shifts caused by specific isotope substitutions are often unavailable. Furthermore, as described below, useful IR information on proteins is usually only gained from difference spectra and the majority of literature information is given as absolute spectra, from which the appropriate difference spectra are not always easily obtained. Below we summarise such information for the five key protonatable amino acids, taken both from available literature information and from our own as yet unpublished amino acid IR database.

6.2 IR Signatures of Protonation-State Changes of Amino acid Residues

6.2.1 Lysine

Literature information on lysine protonation-induced IR changes is limited and has been summarised by Barth.[34] Figure 2 illustrates in more depth the effects of protonation change of the lysine $\varepsilon(NH_2)$ group. These data were obtained using poly-L-lysine in order to avoid interference from the strongly interfering $\beta(NH_2)$ amino group. The absolute spectra are dominated by the amide I and II envelopes and these mask the known two major broad headgroup bands, the δ_{as} and $\delta_s \varepsilon(NH_3^+)$ bands at 1630 and 1526 cm^{-1}, respectively.[32] Additional bands at lower frequencies can be observed in the absolute spectra arising in the main from δ, γ_w and $\gamma_r (CH_2)$ modes.[34] In D_2O, whereas the amide I band shifts down by only a few wavenumbers, the amide II band shifts to 1440 cm^{-1}, revealing a band at 1590 cm^{-1} that must also belong to the amino acid. Most dramatic, however, are the downshifts of the δ_{as} and $\delta_s \varepsilon(NH_3^+)$ bands to 1171 and 1062 cm^{-1}.

Most useful for studies of protonation changes, however, are the differences between spectra at low and high pH/pD that give rise to the protonated *minus* deprotonated difference spectra when the headgroup becomes protonated. In this case, broad positive δ_{as} and $\delta_s \varepsilon(NH_3^+)$ bands (1620 and 1521 cm^{-1} in H_2O and 1171 and 1062 cm^{-1} in D_2O) are clearly seen since, both in H_2O and D_2O media, equivalent bands are weak or absent in the NH_2 form.[32] In the protonated minus deprotonated difference spectrum in H_2O shown (trace E), the peaks of these broad bands are probably closer to the 1630 and 1526 cm^{-1} values quoted in ref. 32 but are slightly

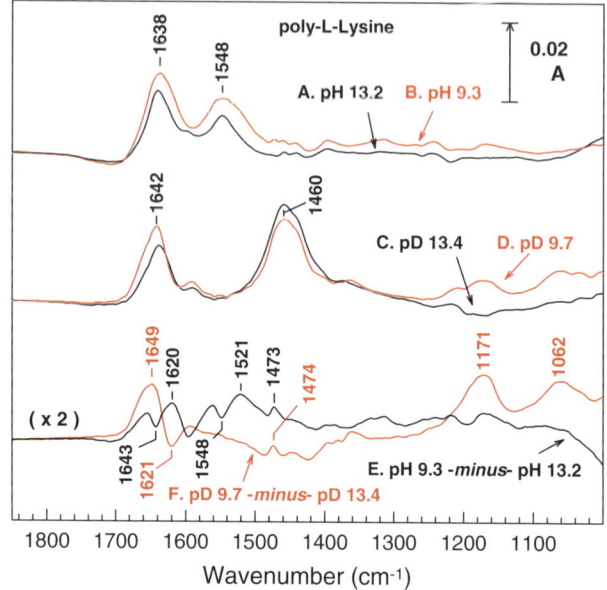

Figure 2 *Absolute and protonation ATR-FTIR difference spectra of poly-L-lysine. Poly-L-lysine was dissolved in H$_2$O or D$_2$O and the pH was adjusted with aliquots of HCl or KOH. Samples were placed on the surface of a silicon ATR microprism (3mm diameter; 3 bounce; SensIR) and spectra are the averages of 1000 interferograms at 4 cm^{-1} resolution. IR bands of H$_2$O/D$_2$O have been subtracted and difference spectra were computed by subtraction of normalized absolute spectra*

distorted by what is likely to be polypeptide amide I/II changes that appear as troughs at 1643 and 1548 cm^{-1} in the δ_{as} and δ_s ε(NH$_3^+$) peaks. These bands are shifted away from this region in the corresponding D$_2$O difference spectrum (trace F) and here only an amide I shift remains with a peak/trough at 1649/1621 cm^{-1}. Of further interest in relation to detection of lysine protonation, however, is a previously unreported H/D-insensitive band at 1473 cm^{-1} that is present only in the protonated form and therefore appears as a clear peak in both the H$_2$O and D$_2$O difference spectra. This band, together with the δ_{as} and δ_s ε(NH$_3^+$) bands at 1171 and 1062 cm^{-1} in D$_2$O media, would appear to be the best candidates for IR indicators of lysine protonation changes within proteins.

6.2.2 *Tyrosine*

The IR properties of tyrosine, tyrosinate and its radical state have been studied in particular detail, in part because this residue is particularly amenable to IR analysis and also because of its catalytic importance in many proteins. Barth[34] has collated diverse data on tyrosine and tyrosinate and the related *p*-cresol in H$_2$O and D$_2$O. A significant body of data on specific isotope effects is also available both for model compounds [12,35] and tyrosine in proteins.[12,21,36] Berthomieu and Hienerwadel[37] have recently reviewed the large body of IR data on tyrosine radicals in photosystem II and other proteins. Figure 3 illustrates the dramatic effects of protonation change of the phenolic OH group in poly-L-tyrosine on absolute and derived protonated *minus* deprotonated difference IR spectra. Consistent with published data,[21,34] the tyrosine headgroup in its protonated state (trace A) has major bands at 1614, 1598, 1515, 1455, 1241 and 1175 cm^{-1} that, apart from the 1241 cm^{-1} band that has a ν(CO) contribution, all arise from ν(CC) ring and δ(CH) modes.[34] Apart from the 1599 and 1241 cm^{-1} bands, they are relatively insensitive to H/D exchange (trace B), though do shift considerably with ring isotope substitutions.[12,35] Formation of the tyrosinate anion does have a dramatic effect on all of these bands, however, and this accounts for the principal features of the protonated *minus* deprotonated difference IR spectra shown (traces C and D). In particular, downshifts of the 1515 and 1241 cm^{-1} bands occur and the 1169 cm^{-1} band is lost altogether. All three of these changes are useful markers for monitoring tyrosine protonation changes within proteins.

6.2.3 Histidine

Histidine is another residue whose different protonation states have markedly different IR signatures and so are particularly amenable to IR analyses. Barth[34] has collated information on band positions and assignments for histidine and/or related model materials. In addition, Noguchi *et al.*[26,38] have published detailed empirical

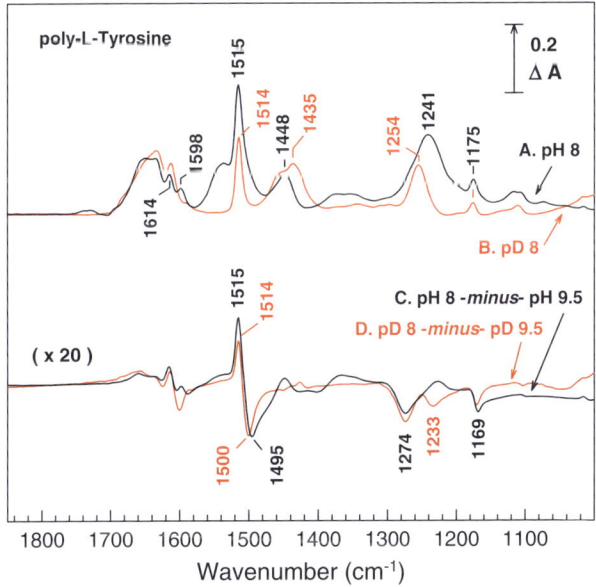

Figure 3 *Absolute and protonation ATR-FTIR difference spectra of poly-L-tyrosine. Poly-L-tyrosine was deposited as a thin film on the surface of a silicon ATR microprism (3mm diameter; 3 bounce; SensIR) and rehydrated with phosphate buffer at appropriate pH/pD. Spectra are the averages of 10000 interferograms at 4 cm^{-1} resolution. IR bands due to H$_2$O/D$_2$O have been subtracted from the absolute spectra shown and difference spectra were computed by subtraction of normalised absolute spectra*

and *ab initio* normal-mode calculations of the different protonation states of 4-methyl-imidazole and its zinc-ligated state as a model for the histidine headgroup. Iwaki *et al.*[25] have recently provided additional data on the imidazolate form of L-histidine. These data have allowed interpretation of IR signals associated with histidine structural or protonation changes in photosystems I and II, model haems, cytochromes and iron–sulfur centres.[25,39–42]

The histidine headgroup has complicated protonation chemistry. The neutral, singly protonated imidazole state has two forms in which the proton can be on the $N\pi$ (that closest to the $C\alpha$ carbon) or the $N\tau$ nitrogen atom and in its free form in aqueous media these two states are roughly equally occupied. This is reflected in the absolute IR spectra,[25,38,40] where bands arising from the two forms are present. In proteins, however, it is likely that environmental influences on individual buried histidines will be cause them to adopt just one of these forms. The usually considered protonation reaction is conversion to the doubly protonated imidazolium form (*pK* in aqueous media of 6), a reaction that is considered to be important in many enzymological processes. Figure 4 (trace A) shows the protonated (imidazolium) *minus* deprotonated (a mixture of either $N\pi$ or $N\tau$ protonated) difference spectrum for poly-L-histidine in H_2O media. Of most value for detection of this protonation change in proteins are the bands arising in part from ring C–N bonds that are found around 1100 cm^{-1}, a region that is free from major contributions from other amino acids. As pointed out by Noguchi *et al.* in their detailed studies,[26,38,40] the positions of the peaks and troughs in this region can provide information on this protonation transition and which nitrogen is protonated in the imidazole state. Also useful for detection of this protonation reaction is a sharp imidazole band at 985 cm^{-1} that is lost on protonation to the imidazolium form and so appears as a trough in trace A. Trace B shows the equivalent spectra in D_2O media. In addition to expected upshifts in the 1100 cm^{-1} region,[26,38,40] particularly useful for identification of the imidazolium/imidazole transition is the H/D sensitivity of the 985 cm^{-1} trough, which is replaced by ones at 1013 and 994 cm^{-1} in D_2O.

Although the imidazole/imidazolate transition of histidine is rarely considered in enzymology because of its high solution *pK* of 14.5, it is nevertheless a viable process within proteins, particularly when one of the histidine nitrogens is ligated to a metal centre, as occurs widely in, for example, cytochromes, chlorophylls, copper proteins, iron–sulfur proteins and superoxide dismutases. Hasegawa *et al.*[38] studied the Raman and IR properties of 4-methyl-imidazolate as a model for histidine imidazolate and showed that the imidazolate form of this compound has a particularly strong vibration at 1440 cm^{-1} arising in part from ring CN stretch, that is lost upon protonation. An equivalent H/D-insensitive band is also found in histidine imidazolate[25] at 1451 cm^{-1} and it is this that is responsible for the prominent trough at 1451 cm^{-1} in the L-histidine protonated (a mixture of $N\pi$-and $N\tau$-protonated imidazole) *minus* deprotonated (imidazolate) difference spectra shown in traces B and C (bottom). Two further notable positive bands are seen at 1323 and 1262 cm^{-1}, together with a trough at 1102 cm^{-1}. Whereas the 1451 and 1102 cm^{-1} bands are relatively H/D exchange-insensitive, the 1323 and 1262 cm^{-1} bands are both downshifted by 5–10 cm^{-1}. Equivalents of these bands have been observed in respiratory cytochrome bc_1 complex and have been used to show directly for the first time that a key imidazole/imidazolate change

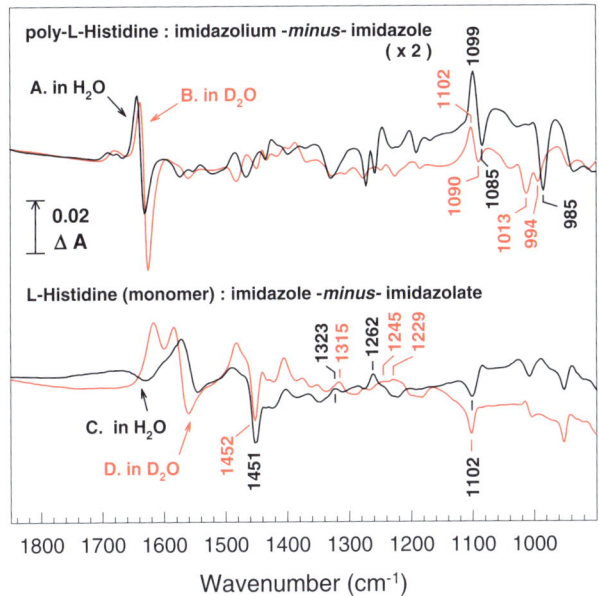

Figure 4 *Absolute and protonation ATR-FTIR difference spectra of poly-L-histidine and L-histidine. Poly-L-histidine was deposited as a thin film on the surface of a silicon ATR microprism (3mm diameter; 3 bounce; SensIR) and rehydrated with phosphate buffer. The imidazolium* minus *imidazole difference spectrum (trace A) was induced by cyclic exchange of buffers at pH 7 and 8. The spectrum shown in an average of 11 spectra, each of which was an average of 500 interferograms at 4 cm^{-1} resolution* (unpublished data of D. Marshall, London, UK).*
For the measurement in D$_2$O, poly-L-histidine was incubated in D$_2$O for 30 min prior to deposition on the prism. The imidazolium (pD 7) minus *imidazole (pD 8.5) different spectrum (trace B) was the average of eight spectra, each of which was the average of 2000 interferograms at 4 cm^{-1} resolution.. Imidazole* minus *imidazolate difference spectra in H$_2$O (trace C) and D$_2$O (trace D) were obtained from solutions of L-histidine at appropriate pH/pD and are taken from ref. 25. It should be noted that for practical reasons, traces A and B were obtained with a poly-L-histidine film and traces C and D with L-histidine in solution. Hence, relative extinction coefficients for the two transitions cannot be deduced from these data alone*

of a metal-bound histidine[43] does indeed occur in the active site of this enzyme.[25] In this case, the loss of the trough in the 1100 cm^{-1} region of the protein difference spectrum was interpreted as being characteristic of the known and relatively unusual N_π–metal-ligated histidine chemistry in this enzyme.[44]

6.2.4 Aspartic and Glutamic Acids

Perhaps the greatest successes to date in IR studies of protonation changes within proteins have been those involving protonation changes of aspartic and glutamic acids, residues that appear to be most commonly involved in intra-protein proton-transfer processes. The reason for this is that the very polar headgroup has several particularly strong absorption bands that alter dramatically with protonation state change. For the carboxylate forms (Figure 5), these are the COO⁻ v_{as} and v_s stretches at 1580/1398 cm⁻¹ (aspartate) and 1556/1399 cm⁻¹ (glutamate) with extinction coefficients [34] of around 825/450 M⁻¹cm⁻¹. Perhaps more importantly, however, is the fact that these bands are absent from the protonated carboxylic acid forms, being replaced by a $v(C=O)$ band in the 1710–1790 cm⁻¹ range with an extinction coefficient[34] above 200 M⁻¹cm⁻¹. It is these bands that give rise to the dominant features in the protonated *minus* deprotonated difference spectra shown in traces E and K and,

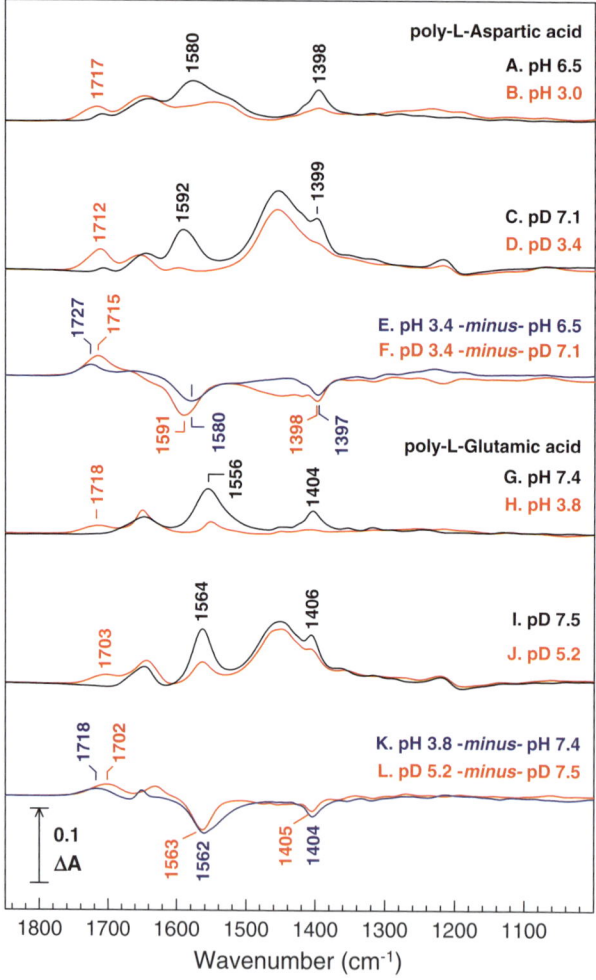

Figure 5 *Absolute and protonation ATR-FTIR difference spectra of poly-L-aspartic and poly-L-glutamic acids. Poly-L-aspartic/glutamic acid was dissolved in H_2O or D_2O and pH/pD was adjusted by addition of HCl. Spectra are the averages of 4000 interferograms at 4 cm⁻¹ resolution. IR bands due to H_2O/D_2O have been subtracted from the absolute spectra shown and ATR-FTIR difference spectra were computed by subtraction of normalised absolute spectra*

for D_2O media, in traces F and L. The major effect of H/D exchange on these spectra is the downshift of the carboxylic acid band in free solution by 12–16 cm^{-1}. The $v(C=O)$ is particularly important because it is in a region where no other protein bands occur and so can be detected at high signal/noise since there is little background absorbance and, provided that vibrational modes of other non-protein components can be ruled out, can be definitive for carboxylic group changes. Furthermore, the band position is influenced not only by H/D exchange but also by hydrogen-bonding strength and by electrostatic factors. Hence, not only can net protonation be detected (as appearance of a $v(C=O)$ peak with concomitant loss of $COO^- v_{as}$ and v_s bands) but also small changes in conformation or environment of a carboxylic acid group that retains its proton (seen as the appearance of a peak/trough in the 1710–1790 cm^{-1} region with no concomitant loss of $COO^- v_{as}$ and v_s bands). Changes of carboxylic acid bands have been used elegantly to detect multiple protonation steps in the bacteriorhodopsin photochemical cycle on microsecond timescales [8,9,31] and redox-linked protonation and conformational changes of glutamic and aspartic acid residues in a wide range of other proteins. Additionally, the split between the $COO^- v_{as}$ and v_s stretch frequencies can been used to assess whether a carboxylate functional group acts as a monodendate or bidentate ligand.[45,46]

7   Examples of Protein Infrared Spectroscopy Applications

In the following, several general illustrations are given of the types of information that can be gained by application of the above infrared methods to aspects of structure and mechanism of a range of respiratory electron transfer proteins. The interpretation of these spectra is discussed briefly and the reader is referred to the original publications for more in-depth details and to the infrared literature more widely for the vast range of applications for which protein infrared spectroscopy is a central technique.

7.1   Absolute Spectra of Proteins and Effects of Isotope Substitution

Figure 6 illustrates absolute absorption spectra of intact, purified cytochrome *c* oxidase from *Paracoccus denitrificans* in native form, after H/D exchange or global ^{15}N labelling. As with all proteins, these spectra are dominated by the broad envelopes of the many amide I and amide II bands of the polypeptide backbone (in unlabelled form centred at 1654 and 1544 cm^{-1}, respectively). The band changes of individual functional groups that are observed in typical protein difference spectra of the types discussed above are usually much less than 0.1% of the magnitude of these features. This is illustrated with an example redox difference IR spectrum of the same protein which is plotted after x20 expansion (trace D). In addition, the specific changes of amino acids and cofactors is in general accompanied by some localised polypeptide alterations around them so that these regions tend to be dominated by several overlapping amide I and II bandshifts (see trace D and Figures 7 and 8). Hence, bands of amino acids and cofactors that fall within the amide I/II envelopes can be the most

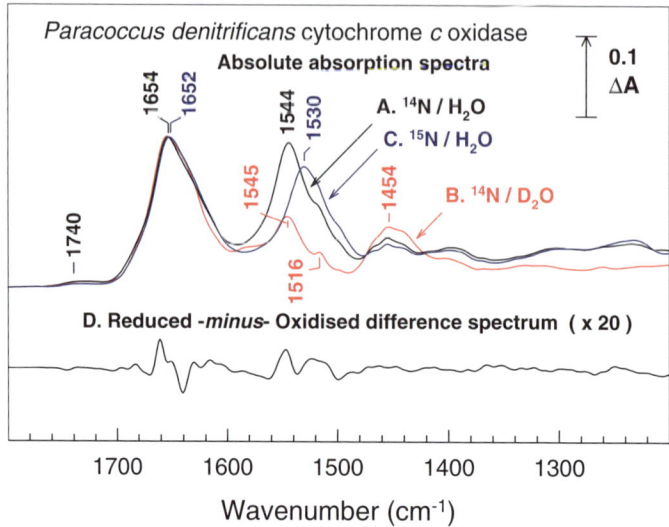

Figure 6 *Absolute spectra of* Paracoccus denitrificans *cytochrome c oxidase. Unlabelled (A, B) and universally* [15]*N-labelled (C) detergent-free purified cytochrome* c *oxidase from* Paracoccus denitrificans *(prepared and supplied by Dr. A. Puustinen, Helsinki, Finland) was deposited on a silicon ATR microprism and rehydrated with buffer at pH 9 in* H_2O *(A, C) or pD 9 in* D_2O *(B). IR bands due to buffer were subtracted from the absolute absorption spectra shown (each an average of 2000 interferograms at 4 cm^{-1} resolution). For comparison, a typical fully reduced* minus *oxidised difference spectrum of the same unlabelled material is shown (trace D), taken from ref. 52 and expanded* $\times$ *20*

difficult to distinguish both in terms of signal/noise (which is lowest in these regions because of the high background absorbances) and in terms of deconvolution from the (often dominating) overlapping amide I and II bandshifts. As a result, it is the characteristic bands of amino acids and cofactors that fall outside these regions, as summarised above, that are usually most useful. For a typical type of IR spectrometer with conventional optics, a frequency range of 4000–900 cm^{-1} is often recorded, with the majority of useful protein bands (with some notable exceptions) appearing below 1900 cm^{-1}. The amide I and II bands reflect the secondary structural composition of the polypeptide. The *P. denitrificans* cytochrome *c* oxidase is largely α-helical and this is reflected in the 1654 cm^{-1} peak position of its amide I envelope. Sophisticated deconvolution methods have been developed to quantitate these secondary structural elements from the amide I/II envelopes.[5,6,47]

Also seen in Figure 6 spectra is a band at 1740 cm^{-1}. Although protonated carboxylic acids will appear in this region, this feature actually arises predominantly from the ester bond of associated phospholipid and its intensity relative to the amide I peak may be used as a fast, simple assay of the phospholipid/protein ratio of membrane-derived protein preparations.

Since the amide I bands arise predominantly from C=O stretches, H/D exchange has little effect on the amide I envelope, shifting some of its components by

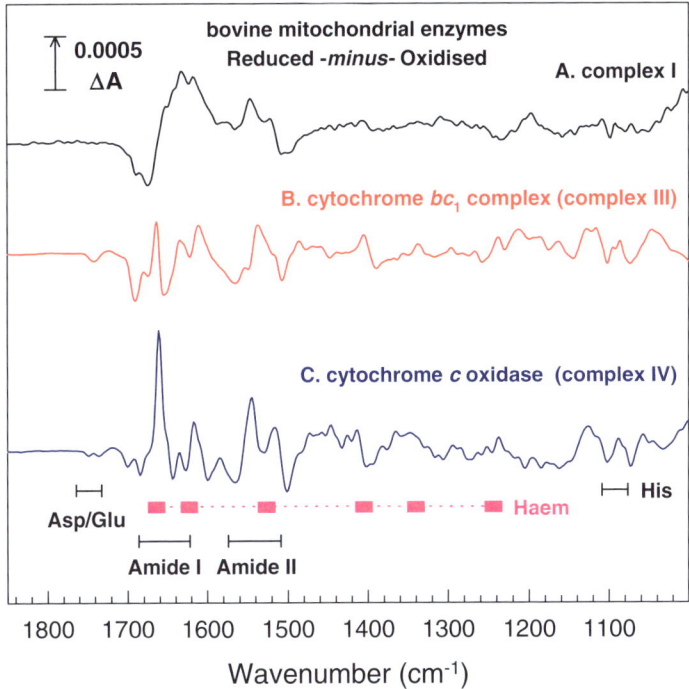

Figure 7 *Comparative redox difference spectra of bovine complex I, cytochrome* bc$_1$ *complex and cytochrome* c *oxidase. Fully reduced* minus *oxidised difference ATR-FTIR spectra of protein films of complex I (trace A, in collaboration with Drs. N. Fisher and J. Hirst) cytochrome* bc$_1$ *complex (trace B, from ref. 27) and cytochrome* c *oxidase (trace C, from ref. 52) are shown for comparison. All spectra were obtained at pH 8 by perfusion methods with oxidising and reducing buffers containing appropriate mediators*

1–2 cm^{-1}. In contrast, the amide II envelope, which is 60% N–H bending and 40% C–N stretch,[6] is shifted downwards by 14 cm^{-1} with ^{15}N labelling and a large part is shifted to around 1450 cm^{-1} on H/D exchange. These behaviours can be used to test whether bands in these regions in difference spectra are likely to arise from amide I/II changes. The shoulder on the amide II peak at 1516 cm^{-1} remains relatively unshifted and so appears more prominently after H/D exchange and arises from the protonated tyrosines within the protein.

7.2 Redox Difference Spectra

Electrochemical methods have been developed for automated generation of redox IR difference spectra for both transmission and ATR modes of FTIR spectroscopy and these are producing a wealth of information on redox-linked structural changes, including redox-linked protonation changes. Figure 7 shows a comparison of full reduced *minus* oxidised ATR-FTIR difference spectra of three of the major respiratory electron-transfer complexes isolated and purified from bovine mitochondria,

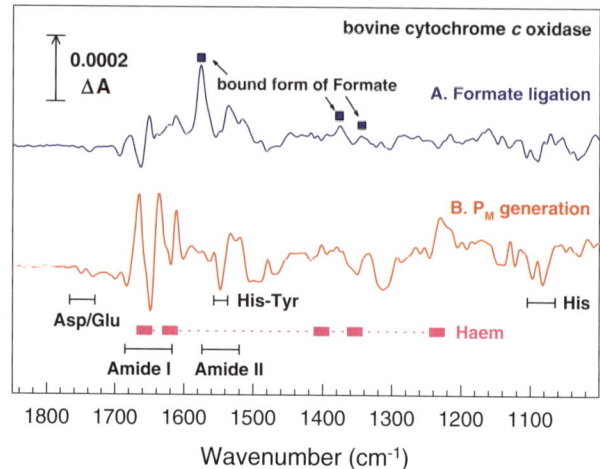

Figure 8 *Formate binding and P_M generation in bovine cytochrome c oxidase. ATR-FTIR difference spectra of formate-ligated* minus *unligated (trace A, from ref. 51) and P_M* minus *oxidized (trace B, from ref. 52) difference ATR-FTIR spectra, obtained by perfusion methods with a film of bovine cytochrome c oxidase*

namely complex I, complex III (cytochrome bc_1 complex) and complex IV (cytochrome *c* oxidase). Dominant in the spectra of complexes I and IV are large but distinctly different amide I/II changes. For complexes III and IV, many of the large bands across the 1650–1200 cm^{-1} region can be assigned to haems and the contributions of individual haems have to some extent been deconvoluted. Changes around 1100 cm^{-1} are most likely associated with histidine ligands to these haems. In complex III, a peak/trough around 1740 cm^{-1} is associated with two overlapping carboxylic acid changes and in complex IV two distinct carboxylic acid changes are evident in this region. Primary papers on such redox difference spectra[18–20,27,28,48–50] can be consulted for further details.

7.3 Ligand Binding and Catalytic Intermediates Difference Spectra

Development of perfusion and dialysis devices for use in conjunction with ATR-FTIR spectroscopy has opened the possibility of IR analyses of additional types of protein changes. Figure 8 shows an example of ligand binding (formate binding to oxidised complex IV) and of interconversion between two intermediates of a catalytic cycle (the oxidised to 'P_M' intermediate of complex IV). The formate-binding IR difference spectrum showed definitively for the first time that the bound ligand is in its deprotonated form even though it is formic acid that is taken up. The P_M *minus* O spectra have helped define the role and protonation state changes of a catalytically important histidine–tyrosine functional element and highlight a role in proton translocation for a key glutamic acid that gives rise to band changes in the 1740 cm^{-1} region. Again, the primary papers[51–54] should be consulted for details of interpretations of these spectra.

8   Outlook

Although infrared technology for studies of atomic details of protein structure and function in general is now easily available, there are presently relatively few laboratories worldwide engaged in such studies. Together with the increasing databases of protein atomic structural models, the increasing databases of model compound information and the development of accurate and reliable computation methods of IR band prediction by *ab initio* calculation, it seems likely that many more structural biology laboratories will embrace this technology in the future as a method that complements X-ray and NMR structural methods.

References

1. W. Uhmann, A. Becker, C. Taran and F. Siebert, *Applied Spectroscopy,* 1991, **45**, 390.
2. R.B. Dyer, K.A. Peterson, P.O. Stoutland and W.H. Woodruff, *Biochemistry,* 1994, **33**, 500.
3. M.L. Groot, L.J.G.W. van Wilderen, D.S. Larsen, M.A. van der Horst, I.H.M. van Stokkum, K.J. Hellingwerf and R. van Grondelle, *Biochemistry,* 2003, **42**, 10054.
4. C. Zscherp and A. Barth, *Biochemistry*, 2001, **40**, 1875.
5. A. Barth and C. Zscherp, *Q. Rev. Biophys.*, 2002, **35**, 369.
6. B.Stuart, *Infrared Spectroscopy: Fundamentals and Applications* John Wiley & Sons, Chicester, 2004.
7. H.Günzler and H.-U.Gremlich, *IR Spectroscopy: An Introduction* Wiley-VCH, Weinheim, Germany, 2002.
8. K. Gerwert, B. Hess, J. Soppa and D. Oesterhelt, *Proc. Natl. Acad. Sci. U.S.A.,* 1989, **86**, 4943.
9. M.S. Braiman, O. Bousché and K.J. Rothschild, *Proc. Natl. Acad. Sci. U.S.A.,* 1991, **88**, 2388.
10. K. Fahmy, F. Jäger, M. Beck, T.A. Zvyaga, T.P. Sakmar and F. Siebert, *Proc. Natl. Acad. Sci. U.S.A.,* 1993, **90**, 10206.
11. E. Nabedryk, J. Breton R. Hienerwadel, C. Fogel, W. Mäntele, M.L. Paddock and M. Y. Okamura, *Biochemistry,* 1995, **34**, 14722.
12. R. Hienerwadel, A. Boussac, J. Breton and C. Berthomieu, *Biochemistry,* 1997, **36**, 14712.
13. T. Noguchi and M. Sugiura, *Biochemistry,* 2002, **41**, 15706.
14. P.R. Rich and J. Breton, *Biochemistry,* 2001, **40**, 6441.
15. D.D. Lemon, M.W. Calhoun, R.B. Gennis and W.H. Woodruff, *Biochemistry,* 1993, **32**, 11953.
16. W. Mäntele, in *Biophysical Techniques in Photosynthesis* J. Amesz and A.J. Hoff, (eds.) Kluwer Academic Publishers, Dordrecht, 1996, 137–160.
17. D. Moss, E. Nabedryk, J. Breton and W. Mäntele, *Eur. J. Biochem.,* 1990, **187**, 565.
18. M. Ritter, O. Anderka, B. Ludwig, W. Mäntele and P. Hellwig, *Biochemistry,* 2003, **42**, 12391.
19. P. Hellwig, J. Behr, C. Ostermeier, O.-M.H. Richter, U. Pfitzner, A. Odenwald, B. Ludwig, H. Michel and W. Mäntele, *Biochemistry,* 1998, **37**, 7390.

20. P. Hellwig, D. Scheide, S. Bungert, W. Mäntele and T. Friedrich, *Biochemistry,* 2000, **39**, 10884.

21. P. Hellwig, U. Pfitzner, J. Behr, B. Rost, P. Pesavento, W.v. Donk, R.B. Gennis, H. Michel, B. Ludwig and W. Mäntele, *Biochemistry,* 2002, **41**, 9116.

22. P. Hellwig, B. Rost, U. Kaiser, C. Ostermeier, H. Michel and W. Mäntele, *FEBS Lett.,* 1996, **385**, 53.

23. E. Goormaghtigh, V. Raussens and J.-M. Ruysschaert, *Biochim. Biophys. Acta,* 1999, **1422**, 105.

24. S.A. Tatulian, *Biochemistry,* 2003, **42**, 11898.

25. M. Iwaki, G. Yakovlev, J. Hirst, A. Osyczka, P.L. Dutton, D. Marshall and P.R. Rich, *Biochemistry,* 2005, **44**, 4230.

26. K. Hasegawa, T.-A. Ono and T. Noguchi, *J. Phys. Chem. A,* 2002, **106**, 3377.

27. M. Iwaki, L. Giotta, A.O. Akinsiku, H. Schägger, N. Fisher, J. Breton and P.R. Rich, *Biochemistry,* 2003, **42**, 11109.

28. R.M. Nyquist, D. Heitbrink, C. Bolwien, T.A. Wells, R. Gennis and J. Heberle, *FEBS Lett.,* 2001, **505**, 63.

29. J. Heberle and C. Zscherp, *Applied Spectroscopy,* 1996, **50**, 588.

30. H. Marrero and K.J. Rothschild, *Biophys. J.,* 1987, **52**, 629.

31. C. Zscherp, R. Schlesinger, J. Tittor, D. Oesterhelt and J. Heberle, *Proc. Natl. Acad. Sci. U.S.A.,* 1999, **96**, 5498.

32. S.Y. Venyaminov and N.N. Kalnin, *Biopolymers,* 1990, **30**, 1243.

33. K. Rahmelow, W. Hübner and Th. Ackermann, *AB,* 1998, **257**, 1.

34. A. Barth, *Progress in Biophysics and Molecular Biology,* 2000, **74**, 141.

35. C. Berthomieu, C. Boullais, J.-M. Neumann and A. Boussac, *Biochim. Biophys. Acta,* 1998, **1365**, 112.

36. D. Flemming, P. Hellwig and T. Friedrich, *J. Biol. Chem.,* 2003, **278**, 3055.

37. C. Berthomieu and R. Hienerwadel, *Biochim. Biophys. Acta,* 2005, **1707**, 51.

38. K. Hasegawa, T.-A. Ono and T. Noguchi, *J. Phys. Chem. B,* 2000, **104**, 4253.

39. J. Breton, W. Xu, B.A. Diner and P.R. Chitnis, *Biochemistry,* 2002, **41**, 11200.

40. T. Noguchi, Y. Inoue and X.-S. Tang, *Biochemistry,* 1999, **38**, 10187.

41. C. Berthomieu, A. Boussac, W. Mäntele, J. Breton and E. Nabedryk, *Biochemistry,* 1992, **31**, 11460.

42. B. Schmidt, W. Hillier, J. McCracken and S. Ferguson-Miller, *Biochim. Biophys. Acta,* 2004, **1655**, 248.

43. P.R. Rich, *Biochim. Biophys. Acta,* 2004, **1658**, 165.

44. S. Iwata, M. Saynovits, T.A. Link and H. Michel, *Structure,* 1996, **4**, 567.

45. D.H. Gibson, Y. Ding, R.L. Miller, B.A. Sleadd, M.S. Mashuta and J.F. Richardson, *Polyhedron,* 1999, **18**, 1189.

46. T. Noguchi, T.-A. Ono and Y. Inoue, *Biochim. Biophys. Acta,* 1995, **1228**, 189.

47. J.L.R. Arrondo, A. Muga, J. Castresana and F.M. Goñi, *Prog. Biophys. Molec. Biol.,* 1993, **59**, 23.

48. J. Behr, P. Hellwig, W. Mäntele and H. Michel, *Biochemistry,* 1998, **37**, 7400.

49. P. Hellwig, S. Grzybek, J. Behr, H. Michel and W. Mäntele, *Biochemistry,* 1999, **38**, 1685.

50. P.R. Rich and J. Breton, *Biochemistry,* 2002, **41**, 967.

51. M. Iwaki and P.R. Rich, *J. Am. Chem. Soc.*, 2004, **126**, 2386.

52. M. Iwaki, A. Puustinen, M. Wikström and P.R. Rich, *Biochemistry*, 2003, **42**, 8809.

53. M. Iwaki, A. Puustinen, M. Wikström and P.R. Rich, *Biochemistry*, 2004, **43**, 14370.

54. R.M. Nyquist, D. Heitbrink, C. Bolwien, R.B. Gennis and J. Heberle, *Proc. Natl. Acad. Sci. U.S.A.*, 2003, **100**, 8715.

CHAPTER 14

Inhibitors of Mitochondrial F_1-ATPase

JOHN E. WALKER AND JONATHAN R. GLEDHILL

Medical Research Council Dunn Human Nutrition Unit,
Wellcome Trust/MRC Building,
Hills Road, Cambridge, CB2 2XY,
UK

1 Introduction

The mitochondrial ATP synthase (F_1F_o-ATPase, EC 3.6.1.34) is a multi-subunit assembly found in the inner membrane of the organelle. It is responsible for the synthesis of most cellular ATP. As summarised in Figure 1, the enzyme is composed of a globular catalytic domain known as F_1 (subunit composition $\alpha_3\beta_3\gamma_1\delta_1\varepsilon_1$) and a membrane-bound proton-translocating domain, F_o. The F_1 and F_o domains are linked together by central and peripheral stalks.[1-3] Respiratory complexes in the inner mitochondrial membrane use energy derived from the oxidation of nutrients to generate a transmembrane proton-motive force (PMF), which drives ATP synthesis in the F_1 domain from ADP and orthophosphate. This synthesis is coupled to the PMF by a mechanical rotary mechanism where the PMF impels the rotation of an ensemble of a ring of hydrophobic c-subunits in the F_o domain and the attached central stalk (subunits γ, δ and ε).[4-6] The c-ring rotates against another hydrophobic protein, subunit a (or ATPase-6) which is linked *via* the peripheral stalk (subunits b, d, F_6 and OSCP)[7-10] to the external surface of the F_1 domain.[11-13] The central stalk penetrates into the F_1 domain where an α-helical coiled-coil structure in the γ-subunit provides a central axis around which the three α- and three β-subunits are arranged alternately like segments of an orange. The three catalytic sites are in the β-subunits at interfaces with α-subunits[14] and their properties are modulated in a cyclical manner by the clockwise rotation of the central α-helical structure (as viewed from the membrane) to promote ATP synthesis. During ATP hydrolysis, energy released from the hydrolytic reaction drives the counterclockwise rotation of the central stalk and c-ring, and protons are ejected from the mitochondrion.

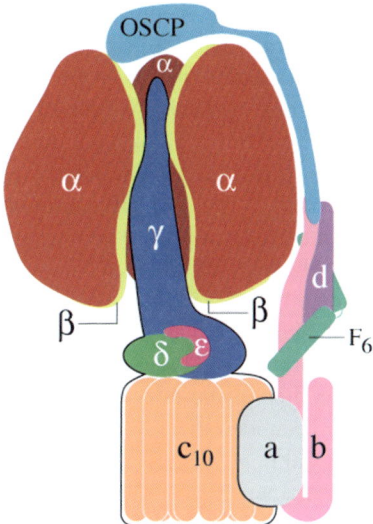

Figure 1 *Schematic of mitochondrial ATP synthase. The model is based on electron microscopy studies of single particles.[11,13] It incorporates the structure of bovine F_1-ATPase[4,14] and information from the electron density map of the F_1–c_{10} complex from* Saccharomyces cerevisiae.[6] *The β-subunit closest to the viewer is not shown for clarity. The composition, stoichiometry and arrangement of the subunits in the peripheral stalk (subunits OSCP, F_6, b and d) come from biochemical, reconstitution and electron microscopy studies.[7–10,12,131] The position of subunit a relative to the c_{10} ring was deduced from studies of the bacterial enzyme.[132] Minor subunits (e, f, g and A6L) in the F_o domain are not shown. They have no known functions in the enzyme's mechanism. The inhibitor protein, IF_1 is also not shown. The components involved in the rotor of the enzyme are outlined in black (Adapted from ref. 5)*

This rotary mechanism of ATP synthesis is consistent with the cyclic modulation of nucleotide affinity in the catalytic β-subunits required by the binding change mechanism.[1,15] In the first crystal structure of the F₁-ATPase,[14] now referred to as the 'reference state',[16] one subunit ($β_{DP}$) binds ADP (interpreted as the 'tight' state of the binding change mechanism), the second ($β_{TP}$) binds ATP (interpreted as the 'loose' state) and the third subunit ($β_E$) is in a conformation that has little or no affinity for nucleotide (corresponding to the 'empty' or 'open' state). During ATP synthesis, rotation of the c-ring brings about 120° rotational steps of the central stalk and the interconversion of the catalytic sites. Thus, each 360° rotation of the central stalk, in 120° steps, takes each catalytic site through the three nucleotide affinity states, and three ATP molecules are synthesised.

A variety of covalent and non-covalent inhibitors of mitochondrial F₁-ATPase have been identified. Covalent inhibitors have been reviewed previously,[17–19] and include 4-chloro-7-nitrobenzofurazan (NBD-Cl),[20] N,N'-dicyclohexylcarbodiimide (DCCD),[21] 8-azido-ATP,[22,23] 2-azido-ATP,[24,25] 5'-p-fluorosulphonylbenzoyladenosine[26] and 5'-p-fluorosulphonylbenzoylinosine.[27] Among the non-covalent inhibitors are non-hydrolysable substrate analogues,[1,3,14–16] azide,[1] the natural inhibitor protein

IF$_1$,[28] the efrapeptins,[29] the aurovertins,[30,31] polyphenolic phytochemicals,[32,33] non-peptidyl lipophilic cations and amphiphilic peptides.[34]

Tentoxin, a natural cyclic tetrapeptide produced by phytopathogenic fungi from *Alternaria*, is a selective inhibitor of the F$_1$-ATPase in chloropasts, but it has no effect on either mitochondrial or bacterial F-ATPases. In the chloroplast F$_1$-ATPase, tentoxin binds to a cleft close to β-Asp83 at the interface between the N-terminal domains of an α- and a β-subunit.[35] The normally tentoxin-insensitive bacterial F$_1$-ATPases can be converted into tentoxin-sensitive enzymes by specific amino-acid changes in this region.[36,37] As the tentoxin site is not present in the mitochondrial F$_1$-ATPase it will not be considered further in this review. Inhibitors of the F$_o$ domain, such as oligomycin, are not discussed either.

The diversity of mitochondrial F$_1$-ATPase inhibitors suggests the presence of many independent inhibition sites in the enzyme. As described below, four independent inhibitory sites have been identified by high-resolution structural studies of inhibited forms of mitochondrial F$_1$-ATPase (see Figure 2), and the binding sites for other inhibitors have been characterised by kinetic analysis. This review will summarise our current knowledge of the inhibitory sites in the mitochondrial F$_1$-ATPase and discuss the potential medical importance of inhibitors of ATP synthase.

2   High-resolution Structures of F$_1$-ATPase

Over the past ten years, bovine F$_1$-ATPase has been studied extensively by X-ray crystallography. The first structure (the 'reference state')[16] was determined with crystals grown in the presence of ADP and the non-hydrolysable ATP analogue 5'-adenylylimidodiphosphate (AMP-PNP).[14] In this structure, AMP-PNP was found in one β-subunit (β_{TP}), ADP but not phosphate in the second (β_{DP}) and the third (β_E) had adopted an open conformation with very low affinity for nucleotide. Since MgADP, but not phosphate, was bound to the β_{DP}-subunit, it was suggested that the structure could represent the 'ADP-inhibited state'. Subsequently, the structures of the F$_1$-ATPase inhibited by efrapeptin,[38] aurovertin B,[39] NBD-Cl,[40] DCCD,[4] ADP-AlF$_3$,[41] a large excess of AMP-PNP[42] and ADP-BeF$_3^-$ (ref. 16) have been determined. All of them were very similar to the reference state structure and they have provided overwhelming evidence that the reference structure (and related structures) represents a conformation on the catalytic pathway of ATP hydrolysis by F$_1$-ATPase.

Other crystals of F$_1$-ATPase inhibited by ADP and excess aluminium fluoride, led to a structure where the ADP-AlF$_4^-$ complex was bound to both β_{TP}- and β_{DP}-subunits; moreover, ADP and a sulphate moiety (mimicking phosphate) were found in the β_E-subunit.[43] This structure, solved at 2.0 Å resolution and with all three catalytic sites occupied simultaneously, displayed a novel 'half-closed' conformation of the β_E-subunit. This structure can be interpreted as mimicking a transition state in the catalytic pathway and probably represents a post-hydrolysis step prior to the release of products. In the structure of F$_1$-ATPase inhibited by the regulatory protein IF$_1$,[44] residues 1–37 of the inhibitor bind in the α_{DP}–β_{DP} interface, and they also contact the γ-subunit. Relative to previous structures, IF$_1$ opens the catalytic interface between α_{DP}-and β_{DP}-subunits trapping ATP in the catalytic site of subunit β_{DP}. The presence of ATP in this catalytic site implies that the inhibited state represents a

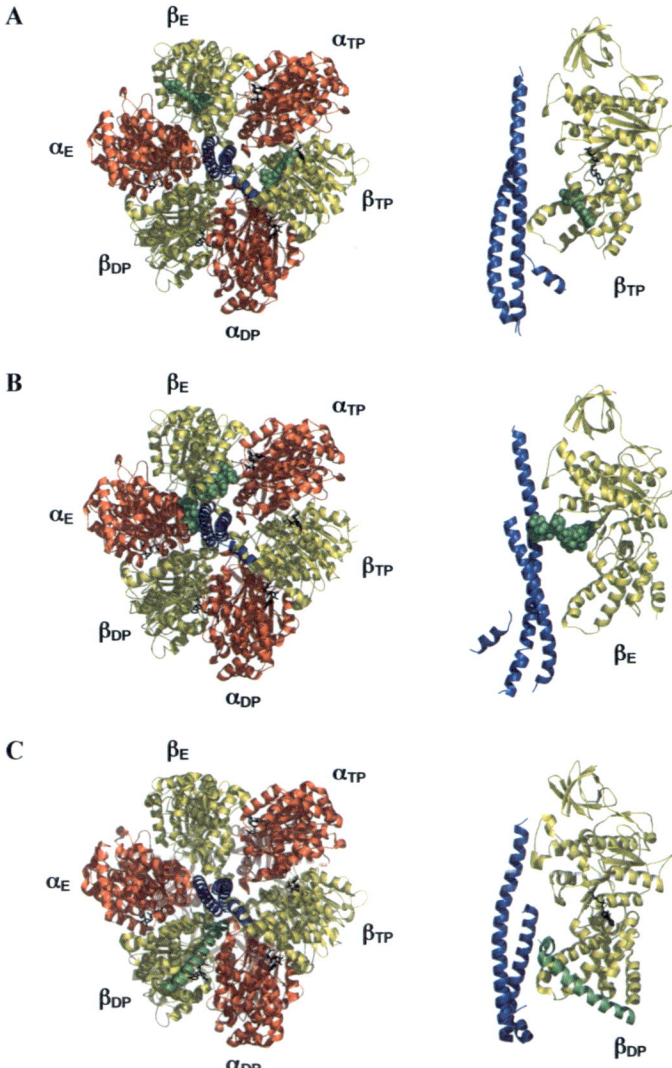

Figure 2 *Inhibitory sites in F₁-ATPase from X-ray crystallography. (A) View of the F₁-ATPase inhibited by aurovertin B[39] along the central axis (left) from the protruding region of the central stalk toward the catalytic sites. Side view (right) of the F₁-ATPase with bound aurovertin B, showing the interacting β-subunit and three α-helical segments in the γ-subunit. (B) Same views as above of the F₁-ATPase inhibited by efrapeptin.[38] (C) Same views as above of the F₁-ATPase inhibited by IF₁.[44] Each of the above inhibitors binds to a distinct site separate from the catalytic site in F₁-ATPase. The α-, β- and γ-subunits are coloured red, yellow and blue, respectively and shown in ribbon representation. Aurovertin B, efrapeptin and IF₁ are coloured green. Aurovertin B and efrapeptin are shown in space filling representation. IF₁ is shown in ribbon representation. Nucleotides are coloured black and represented in stick form. Images were produced with the program PyMOL[133]*

pre-hydrolysis state on the catalytic pathway of the enzyme. The structure of bovine F_1-ATPase inhibited with ADP-BeF$_3^-$ was the first to be solved with a completely non-hydrolysable analogue of ATP and represents one of the highest resolution structures of the $\alpha_3\beta_3$ region of the F_1 domain determined to date. In this structure, two ADP-BeF$_3^-$ complexes appear to mimic ATP bound to β_{DP}- and β_{TP}-subunits. The ADP-BeF$_3^-$ complex in the β_{DP}-subunit, rather than the complex in the β_{TP}-subunit, is poised for hydrolysis and therefore the β_{DP}-subunit, and not the β_{TP}-subunit, contains the catalytically active site.[16]

3 Characterised Sites of Inhibition

3.1 Catalytic Sites

The non-hydrolysable nucleotide triphosphate analogues AMP-PNP,[14,42] ADP-AlF$_3$,[41] ADP-AlF$_4^-$ (ref. 43) and ADP-BeF$_3^-$ (ref. 16) bind to catalytic nucleotide binding sites found at interfaces between α- and β-subunits. These catalytic sites are predominantly in β-subunits, but side chains in the adjacent α-subunits, notably α-Arg373, make essential contributions. The non-catalytic sites are predominantly in α-subunits, with some contributions from side chains in the adjacent β-subunits.[14]

AMP-PNP is almost identical to ATP except that the oxygen atom connecting the β- and γ-phosphate groups is replaced with nitrogen. The P–N bond in AMP-PNP, unlike the P–O bond in ATP, is hydrolysed very slowly, and therefore AMP-PNP serves as a relatively stable analogue of ATP able to trap ATP hydrolases in an active conformation. NDP-AlF$_3$ and NDP-AlF$_4^-$ form stable transition state analogues in a number of NTPases,[45–47] whereas NDP-BeF$_3^-$ forms a stable analogue of ATP in a pre-hydrolysis state.[48] The co-planar arrangement of the fluorine atoms in the trigonal bipyramidal ADP-AlF$_3$-OH$_2$ complex and in the octahedral ADP-AlF$_4^-$-OH$_2$ complex mimics the pentacoordinate transition state in the hydrolysis reaction involving a trigonal bipyramidal phosphoryl group.[41,43] In contrast, the tetrahedral geometry of the fluorine atoms in the ADP-BeF$_3^-$ complex makes BeF$_3^-$ an effective analogue of the tetrahedral γ-phosphate of ATP[16] (see Figure 3). The series of structures inhibited by these nucleotide triphosphate analogues has helped to define the catalytic pathway of ATP hydrolysis by F_1-ATPase. Each of these analogues has the effect of locking the enzyme in a unique conformation and the high-resolution structures determined in the presence of these inhibitors represent 'snapshots' of states that occur during a single catalytic cycle.

The covalent inhibitors of ATP hydrolysis act by modification of residues at or near the catalytic sites: 2-azido-ATP[24,25] and 5'-p-fluorosulphonylbenzoylinosine[27] react

---

Figure 3 *Chemical structures of ATP and its non-hydrolysable analogues. (A) ATP, (B) AMP-PNP, (C) the putative pentacoordinate transition state in the hydrolysis reaction involving a trigonal bipyramidal phosphoryl group, (D) the octahedral ADP-AlF$_4^-$-OH$_2$ complex mimicking the transition state, (E) the tetrahedral ADP-BeF$_3^-$ complex mimicking the ATP bound state involving a BeF$_3^-$ moiety, which acts as an effective analogue of the tetrahedral γ-phosphate in ATP. A polarised nucleophilic water molecule is also shown poised for in-line attack. Charges are not shown for clarity.*

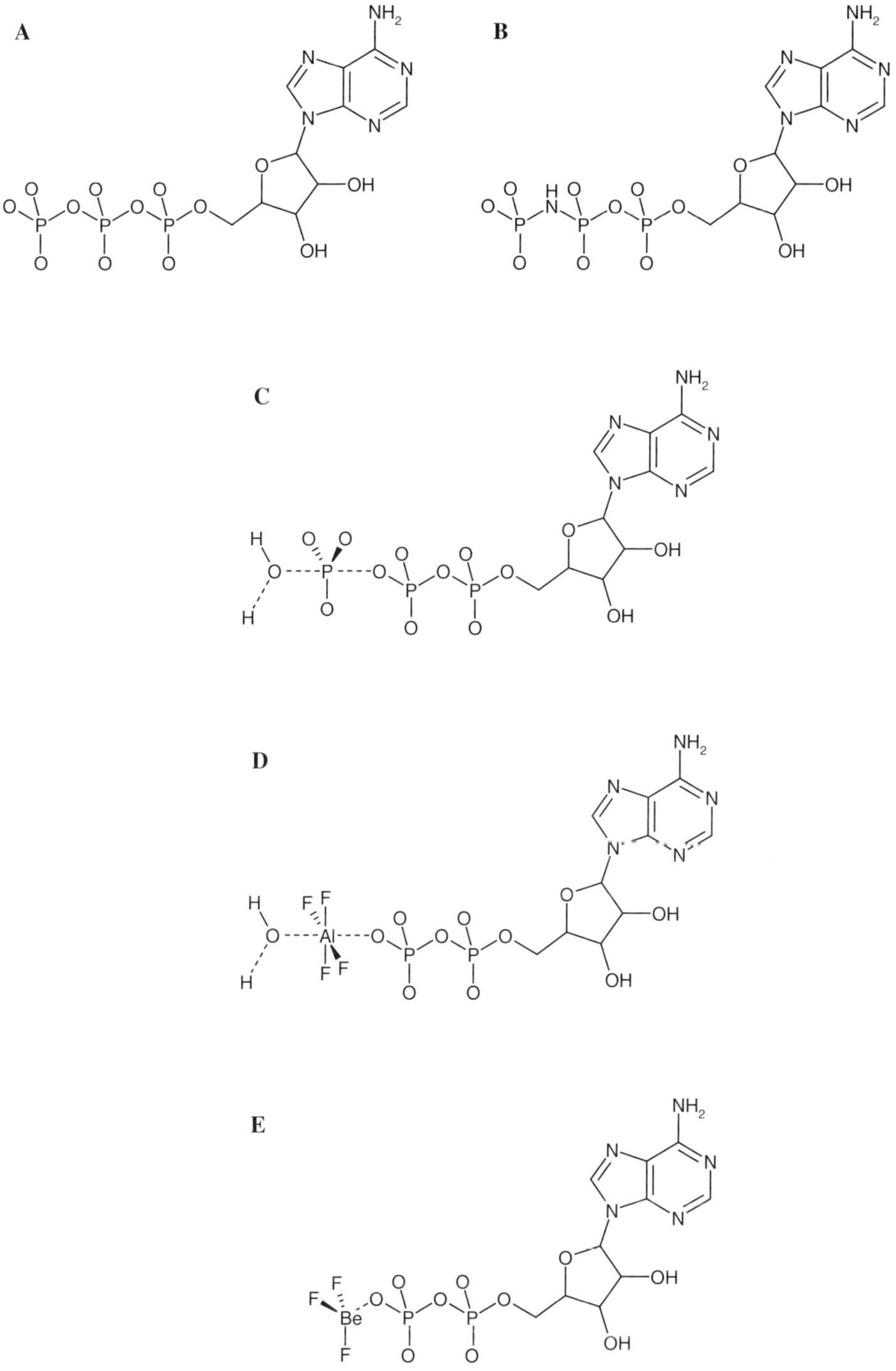

A

B

C

D

E

with β-Tyr345, which forms part of the adenine-binding pocket in the catalytic site.[14] In contrast β-Tyr311, which is one site of reaction of photoactivated 8-azido-ATP,[22,23] is 5.5 Å from the terminal phosphate at the opposite end of the binding site to the adenine pocket.[14] 5′-p-fluorosulphonylbenzoyladenosine[26] modifies β-Tyr368, which protrudes into the α-subunit nucleotide binding site.[14] NBD-Cl[40] and DCCD[4] react specifically with β-Tyr311 and β-Glu199, and the modified residues are observed in the β_E- and β_{DP}-subunits, respectively. NBD-Cl appears to inhibit the enzyme by preventing the modified β_E-subunit from adopting a nucleotide-binding conformation that is essential for catalysis to proceed. Neither β_{TP} nor β_{DP} is capable of accommodating the β-Tyr311 residue bearing an NBD group. DCCD modifies β_{DP}-Glu199 in a catalytic interface, resulting in addition of a dicyclohexyl-N-acylurea group to this residue. In the structural transition from the tight conformation of the β_{DP}-subunit to the open conformation of the β_E-subunit, there is a large movement of both the C-terminal domain and a helix in the nucleotide-binding domain relative to the region containing the covalently modified residue. The mechanism of inhibition by DCCD is probably due to steric hindrance of this conformational change by the bulky dicyclohexyl-N-acylurea moiety.

3.2 Non-catalytic Sites

3.2.1 The Aurovertin B Site

The aurovertins are a family of related antibiotics (see Figure 4) from the fungus *Calcarisporium arbuscula*[49] that block oxidative phosphorylation in mitochondria[31] and in many bacterial species.[50] They inhibit the ATP synthase by binding to β-subunits.[51] Both ATP synthesis and hydrolysis are inhibited by aurovertin[49] and the inhibition is uncompetitive with respect to nucleotides.[52]

In the crystal structure of bovine F_1-ATPase complexed with aurovertin B,[39] the antibiotic inhibitor is bound at two equivalent sites in subunits β_{TP} and β_E, in a cleft situated between the nucleotide-binding and C-terminal domains (see Figure 2A). No aurovertin B is bound to the β_{DP}-subunit because in this subunit the aurovertin B pocket is both incomplete and inaccessible to the inhibitor. The two aurovertin binding sites have different affinities (K_d values *ca* 1 and *ca* 4–6 µM).[51,53] In the structure of the F_1-ATPase-aurovertin B complex, the higher affinity site is interpreted as being in the β_{TP}-subunit and the lower affinity site in the β_E-subunit. In both subunits, the aurovertin B moiety is bound in the same pocket, but in the β_{TP}-subunit the inhibitor also interacts with α-Glu399 in the adjacent α_{TP}-subunit. Aurovertin B bound to β_E is too distant from α_E to make an equivalent interaction. Aurovertin appears to inhibit F_1-ATPase by preventing closure of the catalytic interfaces necessary for cyclic interconversion of catalytic sites. During ATP hydrolysis, aurovertin probably acts by inhibiting the conversion of the loose β_{TP} conformation to the tight β_{DP} conformation. Conversely, during synthesis, the aurovertin would inhibit the conversion of the open β_E conformation to the tight β_{DP} conformation.

A number of polyphenolic phytochemicals present in human diets inhibit the ATP synthase (see Table 1). They include several flavones (*e.g.* quercetin), isoflavones (*e.g.* genistein), stilbenes (*e.g.* resveratrol and piceatannol) and other polyphenolics such as curcumin (Figure 5).[32] Resveratrol, piceatannol and quercetin inhibit the

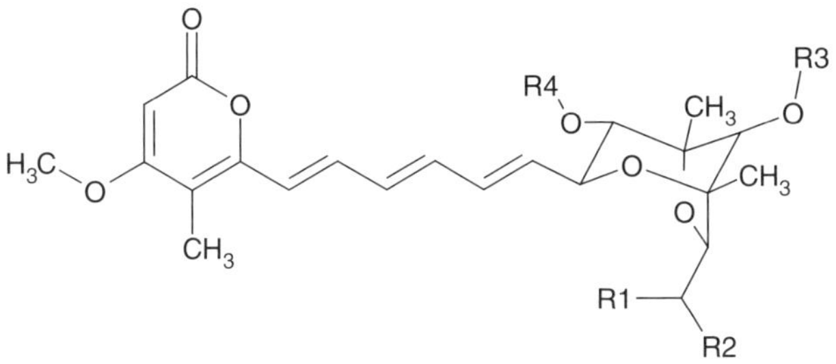

Figure 4 *The antibiotic inhibitors of F$_1$-ATPase belonging to the aurovertin family. The aurovertins consist of a (left) substituted pyrone ring linked by a rigid spacer containing conjugated double bonds to a (right) variable substituted dioxabicyclo[3,2,1]octane (or aglycone) ring. Aurovertin A: R1 = H, R2 = CH$_3$, R3 = COCH$_3$, R4 = COCH$_3$; aurovertin B: R1 = H, R2 = CH$_3$, R3 = COCH$_3$, R4 = H; aurovertin C: R1 = H, R2 = H, R3 = COCH$_3$, R4 = H; aurovertin D: R1 = OH, R2 = CH$_3$, R3 = COCH$_3$, R4 = H; aurovertin E: R1 = H, R2 = CH$_3$, R3 = H, R4 = H*

Table 1   *Inhibitory activities of the non-peptidyl lipophilic cations, amphiphilic peptides and polyphenolic phytochemicals on bovine F$_1$-ATPase*

Inhibitor	IC$_{50}$ pH 8.0 (μM)	IC$_{50}$ pH 6.7 (μM)
Lipophilic cations		
Dequalinium	8	46
Rhodamine 6G	10	27
Malachite green	14	99
Rosaniline	15	—
Acridine orange	180	—
Rhodamine 123	270	—
Rhodamine B	475	—
Coriphosphine	480	—
Quinacrine	580[†]	—
Safranin O	1140	—
Pyronin Y	1650	—
Nile blue A	> 2000	—
Amphiphilic peptides		
SynA2	0.042	0.29
SynC	0.058	0.16
Melittin	5	12
Cox IV	16	—
Polyphenolic phytochemicals		
Piceatannol	—	6.1
Resveratrol	—	6.4
Quercetin	—	66
Genistein	—	343
Curcumin	—	167

Inhibitory activities were determined using 2 mM MgATP as substrate. The values shown for pH 8.0 and pH 6.7 were determined at 30°C and 37°C, respectively.[34,62] Cox IV is the cytochrome oxidase subunit IV presequence. IC$_{50}$: inhibitor concentration that produces 50% inhibition of ATPase activity.

[†]The IC$_{50}$ value for quinacrine was determined at pH 7.0 (37°C) using 2.5 mM MgATP as substrate.[80]

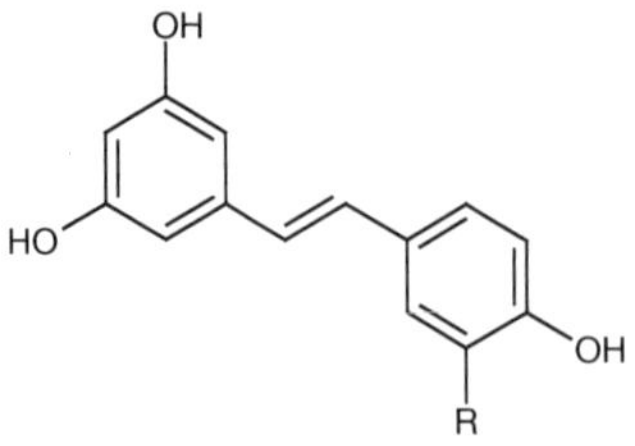

Resveratrol/Piceatannol

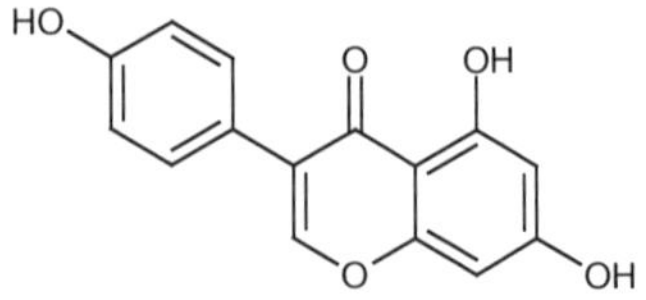

Genistein

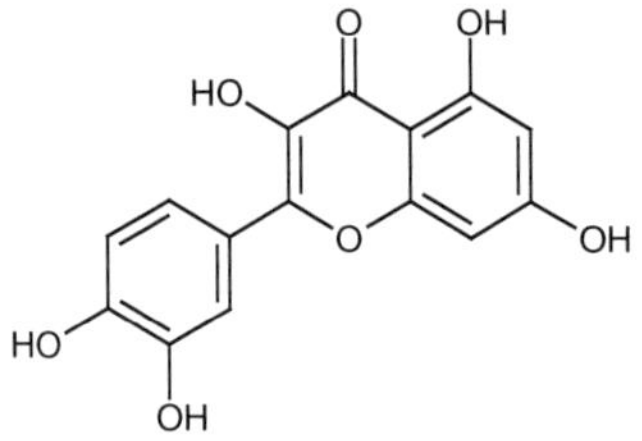

Quercetin

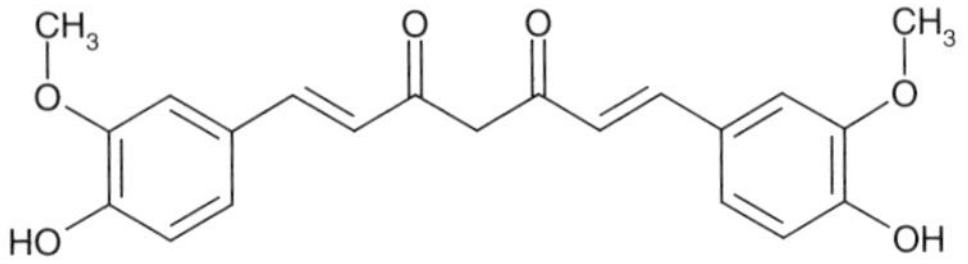

Curcumin

Figure 5 *Chemical structures of representative polyphenolic phytochemicals. Resveratrol and piceatannol consist of two phenolic rings linked by a rigid spacer containing a double bond and are examples of stilbene phytoalexins. Resveratrol: R = H; piceatannol: R = OH. Genistein and quercetin are examples of isoflavones and flavones, respectively. Curcumin consists of two phenolic rings linked by a rigid double bond spacer containing two carbonyl groups*

ATP synthase by acting on the F_1 domain, whereas genistein may preferentially interact with F_o. Curcumin appears to act on the F_1 domain, but it may affect F_o also. Resveratrol exhibits non-competitive inhibition, whereas piceatannol is a mixed inhibitor of F_1-ATPase.[32,33] Dietary phytochemicals affect numerous biological processes and they provide a range of medical benefits.[54–61]

The binding site for the polyphenolic phytochemical inhibitors resveratrol and piceatannol has been identified by multiple inhibition analysis. Both resveratrol and piceatannol exhibit mutually exclusive binding to F_1-ATPase in combination with aurovertin B.[62] These analyses suggest that resveratrol, piceatannol and other related polyphenolic phytochemicals may bind to the aurovertin B site (or sites). The chemical structures of aurovertin B and resveratrol/piceatannol have some similar features (Figure 6).

3.2.2 The Efrapeptin Site

The efrapeptins are a closely related family of modified linear peptide antibiotics from the soil hyphomycetes *Tolypocladia*. They all contain 15 amino acids, some of them non-canonical (see Figure 7).[63,64] They inhibit oxidative phosphorylation in mitochondria[29,49] and in some bacterial species,[50] and act by binding to a unique site in the F_1 domain of ATP synthase. Efrapeptin exhibits mixed inhibition with respect

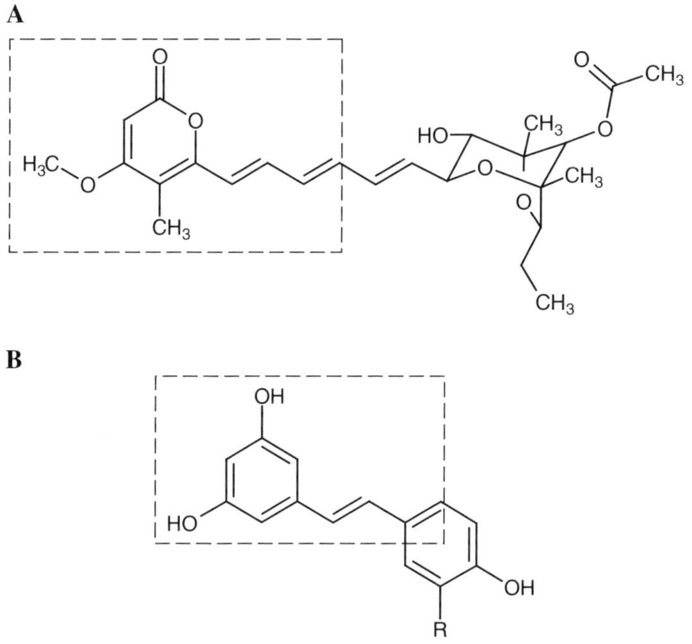

A

B

Figure 6 *Comparison of the chemical structures of aurovertin B (A) and resveratrol/piceatannol (B). Boxed areas indicate the similar features of the two molecules. Resveratrol: R = H; piceatannol: R = OH*
(Adapted from ref. 62)

	1	5	10	15

C   Ac-Pip-Aib-Pip-Aib-Aib-Leu-βAla-Gly-Aib-Aib-Pip-Aib-Gly-Leu-Aib-X

D   Ac-Pip-Aib-Pip-Aib-Aib-Leu-βAla-Gly-Aib-Aib-Pip-Aib-Gly-Leu-Iva-X

E   Ac-Pip-Aib-Pip-Iva-Aib-Leu-βAla-Gly-Aib-Aib-Pip-Aib-Gly-Leu-Iva-X

F   Ac-Pip-Aib-Pip-Aib-Aib-Leu-βAla-Gly-Aib-Aib-Pip-Aib-Ala-Leu-Iva-X

G   Ac-Pip-Aib-Pip-Iva-Aib-Leu-βAla-Gly-Aib-Aib-Pip-Aib-Ala-Leu-Iva-X

Figure 7 *The antibiotic inhibitors of F$_1$-ATPase belonging to the efrapeptin family. The primary structures of the efrapeptins C, D, E, F and G are shown. Pip, Aib, βAla and Iva are pipecolic acid, α-aminoisobutyric acid, β-alanine and isovaline, respectively. The leucine, pipecolic acid and alanine residues have the L-configuration. Substituent X is isobutyl[2,3,4,6,7,8-hexahydro-1-pyrrolo[1,2-α]pyrimidinyl]ethylamine, and Ac is acetyl*

to nucleotide during ATP hydrolysis. However, inhibition of ATP synthesis appears to be competitive with respect to ADP or phosphate.[65]

In the crystal structure of the bovine F$_1$-ATPase-efrapeptin complex,[38] the inhibitor is bound to a unique site across the central cavity of the enzyme (see Figure 2B). It interacts with the γ- and β_E-subunits, and with the two adjacent α-subunits *via* hydrophobic contacts. In addition, two intermolecular hydrogen bonds appear to exist between the enzyme and efrapeptin. The antibiotic itself has two rigid domains in its N-terminal and C-terminal regions connected by a flexible linker and stabilised by intramolecular hydrogen bonds. It appears to act by preventing the open conformation of the β_E-subunit from closing and gaining the capacity to bind nucleotides, as required by the rotary mechanism.

3.2.3   The Site of Binding of the Natural Inhibitor Protein, IF$_1$

When a cell is deprived of oxygen, for example during ischaemia, the PMF across the inner mitochondrial membrane collapses and the ATP synthase switches from ATP synthesis to hydrolysis, pumping protons in the opposite direction across the membrane. This hydrolytic activity is thought to be prevented by the natural inhibitor protein, IF$_1$.[28,66,67]

Bovine IF$_1$, an 84-residue protein, inhibits ATP hydrolysis *in vitro*, by forming a 1:1 complex with ATP synthase. Binding of the inhibitor protein requires hydrolysis of MgATP in the absence of the PMF and is optimal below pH 7.0.[66,67] Under *in vitro* conditions, the active form of bovine IF$_1$ is a dimer associated by formation of an antiparallel α-helical coiled-coil between the C-terminal regions of the monomers (see Figure 8).[68, 69] Each monomer is folded into a single cationic amphiphilic α-helix approximately 95 Å in length. This arrangement places the N-terminal minimal inhibitory sequences (residues 14–47)[70] in opposition, allowing the dimeric IF$_1$ to bind two F$_1$ domains simultaneously.[71] The active dimeric state of the inhibitor is favoured at pH values below 7.0, and at higher pH values the dimeric IF$_1$ forms inactive tetramers and higher oligomers *via* coiled-coil interactions that occlude the N-terminal inhibitory regions.[72]

In *Saccharomyces cerevisiae*, the ATP synthase is regulated by two inhibitory proteins, IF$_1$[73,74] and STF$_1$.[75] Two other proteins, STF$_2$[76] and STF$_3$,[77] have been proposed to modulate the activities of yeast IF$_1$ and STF$_1$, although alone neither STF$_2$ nor STF$_3$ has any intrinsic inhibitory activity. Both STF$_1$ and yeast IF$_1$ are 63 residues long, and their sequences are moderately homologous. Yeast IF$_1$ and STF$_1$ oligomerise in opposite directions in relation to pH suggesting that both proteins

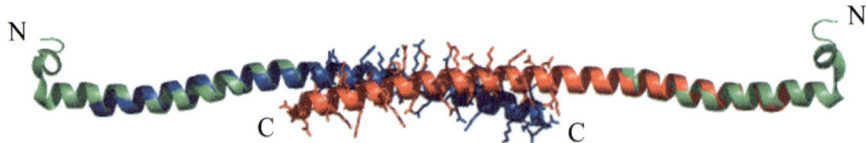

Figure 8 *Crystal structure of bovine IF$_1$ determined at 2.2 Å resolution. Ribbon diagram of the active IF$_1$ dimer (identical monomers are red and blue).[69] Dimerisation occurs via an anti-parallel α-helical coiled-coil between the C-terminal regions of monomers. This places the N-terminal inhibitory regions in opposition at distal ends of the dimer. Side chains of residues involved in coiled-coil formation are represented in stick form (residues 49–84). The superimposed N-terminal inhibitory segments that were bound to the F$_1$-ATPase complexes in the dimeric F$_1$-IF$_1$ complex are shown in green. The N- and C-termini are indicated. Image was produced with the program PyMOL[133]*

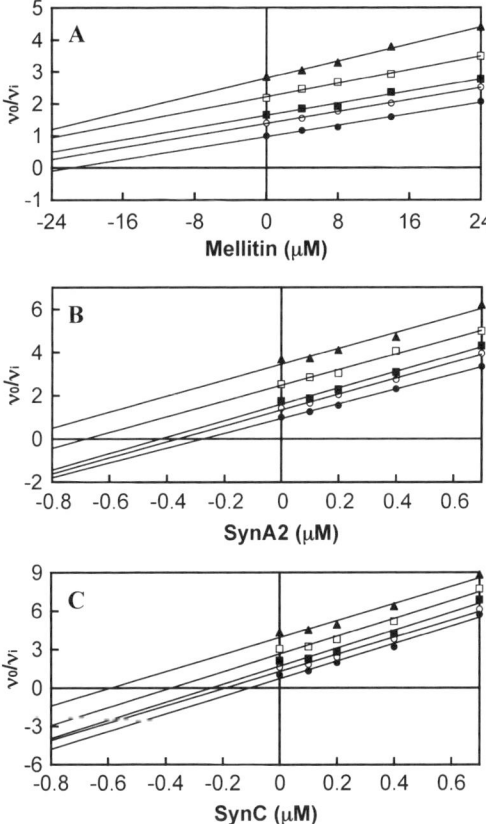

Figure 9 *Effect of the combination of IF$_1$ and the amphiphilic α-helical peptides melittin (A), SynA2 (B) and SynC (C) on bovine F$_1$-ATPase. The activity of F$_1$-ATPase was determined in the presence of combinations of each peptide and IF$_1$ and analysed by means of the Yonetani–Theorell plot.[134] For each peptide, the plots of v_0/v_i (uninhibited initial rate/inhibited initial rate) against peptide concentration at various fixed concentrations of IF$_1$ gave a series of parallel straight lines, indicating mutually exclusive binding to F$_1$-ATPase.[62] The concentrations of I$_1$ (melittin, SynA2 or SynC) are shown on the abscissa, and those of I$_2$ (IF$_1$) are: (●), 0.0 μM, (○), 0.1 μM, (■), 0.2 μM, (□), 0.5 μM, (▲), 1.0 μM. Linear regression lines are drawn through the data points*
(Adapted from ref. 62)

might regulate the yeast enzyme under different conditions.[78] The active yeast IF$_1$ is monomeric at pH values around 7.0 and does not dimerise F$_1$ domains. At higher pH values, like bovine IF$_1$, it forms inactive higher oligomers. STF$_1$ is a weaker inhibitor than yeast IF$_1$ but is capable of inhibiting the yeast enzyme completely at high concentrations (M. W. Bowler, P. J. G. Butler, D. M. Mueller and J. E. Walker, unpublished data). Inhibitor proteins from yeast and bovine mitochondria are able to inhibit the F$_1$-ATPase from both sources. Bovine IF$_1$ is an equally effective inhibitor of both bovine and yeast enzymes, but the yeast IF$_1$ is less effective against the bovine enzyme. The yeast accessory protein, STF_1, has no inhibitory effect on bovine F_1-ATPase.[78]

In the crystal structure of the dimeric bovine F_1-IF_1 complex,[44] the N-terminal regions of the dimeric inhibitor are bound to two F_1-ATPase complexes. IF_1 is bound to the catalytic interface between the C-terminal domains of the α_{DP}- and β_{DP}-subunits, and is in contact with the γ-subunit (see Figure 2C). Over residues 1–37 the inhibitor protein is in direct contact with F_1 and adopts a helix-turn-helix structure. The extent of the interaction between IF_1 and the α_{DP}- and β_{DP}-subunits compared with the rather limited interactions with the γ-subunit, suggests strongly that inhibition of hydrolysis is achieved by prevention of conformational changes in the α- and β-subunits necessary for rotary catalysis. The binding of IF_1 appears to prevent closure of the α_{DP}–β_{DP} catalytic interface, so that it resembles the α_{TP}–β_{TP} interface of the reference state structure.[14] The presence of ATP (or AMP-PNP) in the β_{DP} active site, which is occupied by ADP in the reference structure, suggests that the inhibited structure represents a pre-hydrolysis state in the catalytic cycle, where ATP is poised for hydrolysis. The IF_1 binding surface itself is extensive, with an area of *ca* 1500 Å^2, and is composed of a mixture of 60% hydrophobic and 40% hydrophilic interactions.[44]

A number of amphiphilic peptide inhibitors of F_1-ATPase have been identified (see Table 1). They include the bee venom peptide melittin and two synthetic cationic amphiphilic α-helical peptides (SynA2 and SynC) based on the mitochondrial import pre-sequence of yeast cytochrome oxidase subunit IV. Melittin and SynA2 exhibit non-competitive inhibition of F_1-ATPase, whereas SynC shows mixed inhibition.[34]

By multiple inhibition analysis, the binding site for these amphiphilic peptide inhibitors has been identified as the IF_1 binding site or a site overlapping it.[62] Each of the peptides exhibits mutually exclusive binding to F_1-ATPase in combination with IF_1 (Figure 9).

In the light of the structural similarity between the amphiphilic α-helical peptides and IF_1 (Figure 10), it is reasonable to conclude that their mutual exclusivity of binding is a consequence of competition at or near the same inhibitory site in F_1. Furthermore, binding to this region must be dependent largely on the adoption of a cationic, amphiphilic helical structure rather than sequence similarity to the inhibitor protein.[62] This suggestion is consistent with the amphiphilic nature of the F_1-IF_1 binding surface.[44]

3.2.4 The Non-peptidyl Lipophilic Cation Site

The F_1-ATPase is inhibited also by a wide variety of non-peptidyl lipophilic cations (see Table 1). Many of them are cytotoxic and they are used as antiseptics and disinfectants.[79] They include rhodamine 6G, rhodamine B, rhodamine 123, safranin O, rosaniline, malachite green, acridine orange, coriphosphine, nile blue A, pyronin Y, dequalinium, quinacrine and quinacrine mustard. In quinacrine mustard, the diethyl groups attached to the tertiary amino group of quinacrine are replaced by *bis*(chloroethyl) groups. With the exception of dequalinium and malachite green, these lipophilic cations all contain the fused tricyclic ring of the composite structure (Figure 11). Acridine orange, coriphosphine, nile blue A and pyronin lack a substituent at position R5, whereas rhodamine 6G, rhodamine B, rhodamine 123,

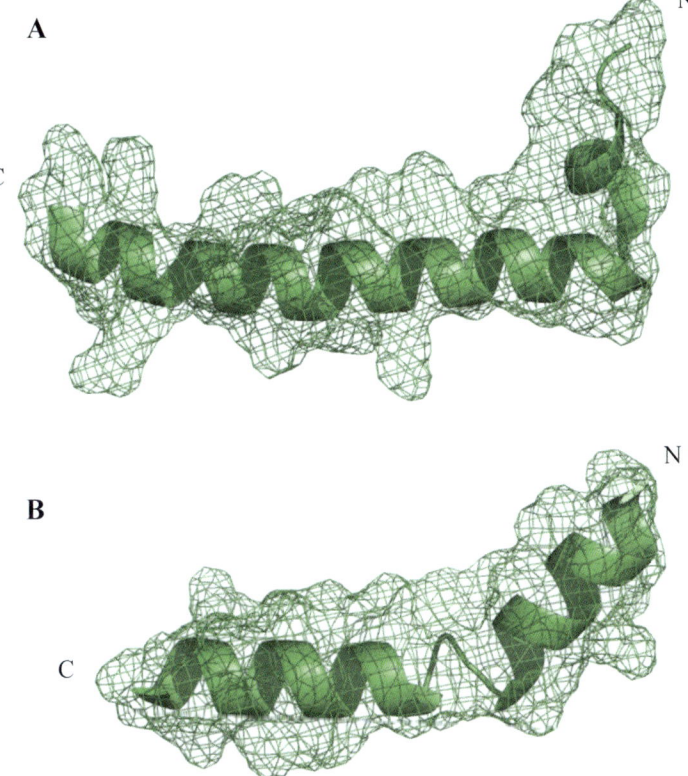

Figure 10 *High-resolution crystal structures of IF$_1$ (A) and melittin (B). Ribbon and surface mesh representations of the structures of the N terminal segment (residues 4–40) of bovine IF$_1$ taken from the F$_1$-IF$_1$ crystal structure[44] and full length bee venom melittin (residues 1–26).[135] The N- and C-termini are indicated. Both molecules adopt a cationic amphiphilic helix-turn-helix motif. Images were produced with the program PyMOL[133]*
(Adapted from ref. 62)

safranin O and rosaniline contain the fused tricyclic ring substituted with an aryl group at position R5. Malachite green shares a similar structure also substituted with an aryl group at position R5. Dequalinium consists of two aromatic quinoline rings connected by a flexible hydrophobic spacer, and the azidirinium of quinacrine mustard is substituted with a charged aliphatic group at position R5. By steady-state kinetic analyses, it has been shown that the non-peptidyl lipophilic cations, like the amphiphilic peptides, act as mixed or non-competitive inhibitors.[34]

Although quinacrine is a weak inhibitor,[80] quinacrine mustard contains an aziridinium moiety that inactivates F$_1$-ATPase by covalent reaction with one or more of the carboxylic side chains in the anionic $_{394}$DELSEED$_{400}$ segment,[81] situated in the C-terminal domain of the β-subunit.[82] Mitochondrial F$_1$-ATPase is also photoinactivated rapidly by dequalinium.[83,84] This photoinactivation is accompanied by non-specific derivatisation of either α-Phe403 or α-Phe406 in a single α-subunit. On the

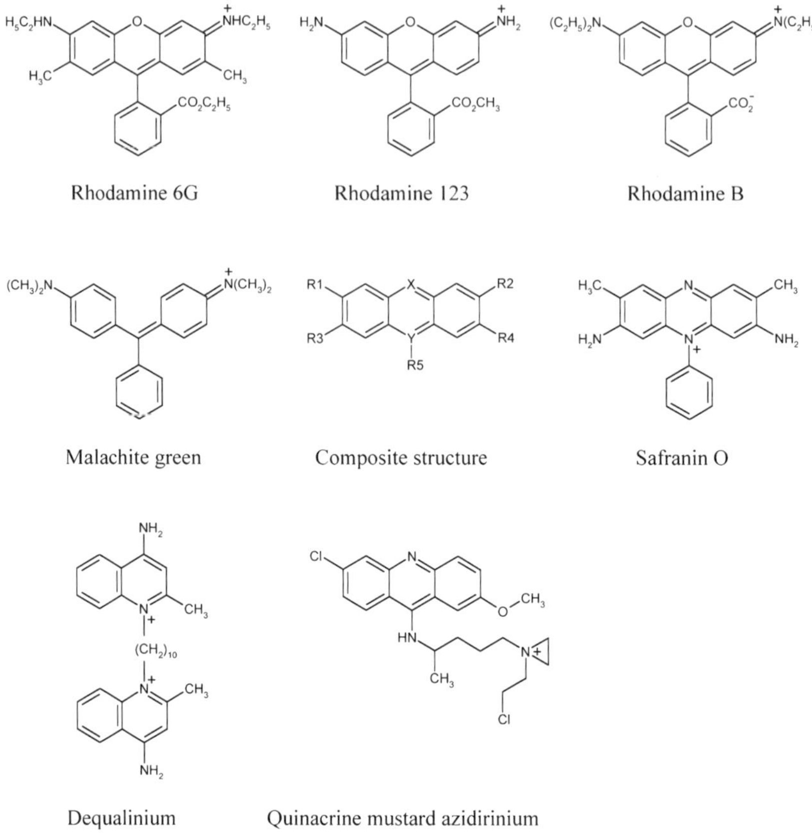

Rhodamine 6G   Rhodamine 123   Rhodamine B

Malachite green   Composite structure   Safranin O

Dequalinium   Quinacrine mustard azidirinium

Figure 11 *Chemical structures of representative non-peptidyl lipophilic cations. With the exception of dequalinium and malachite green, the non-peptidyl lipophilic inhibitors of F_1-ATPase have a common core illustrated by the composite structure. Acridine orange, coriphosphine, nile blue A and pyronin Y lack a substituent in position R5, whereas rhodamine 6G, rhodamine B, rhodamine 123, safranin O, rosaniline and malachite green are substituted in position R5 with an aryl group. Dequalinium contains two aromatic quinoline ring systems connected to each other by a flexible, hydrophobic spacer. The azidirinium of quinacrine mustard contains the fused tricyclic ring substituted with a charged aliphatic group at position R5*

basis of protection against chemical modification of F_1-ATPase with quinacrine mustard and dequalinium, it has been suggested that an interfacial region near the C-terminus of the α- and β-subunits might form part of the binding site for the amphiphilic peptides and non-peptidyl lipophilic cations.[19]

By multiple inhibition analysis, the binding site for rhodamine 6G was shown to be distinct from the binding sites for efrapeptin, aurovertin B and IF_1. Each combination of efrapeptin, aurovertin B and IF_1 with rhodamine 6G exhibits mutually non-exclusive (simultaneous) binding to F_1-ATPase. Consequently, rhodamine 6G and other structurally related non-peptidyl lipophilic cations (see Figure 11) appear to bind to a site distinct from the four inhibitory sites characterised by X-ray crystallography. However, on the basis of the mutual hindrance observed between IF_1 and rhodamine 6G on the binding of one another, it seems likely that the binding site for non-peptidyl lipophilic cations is in the C-terminal domain of the α-subunit close to, and possibly partly overlapping, the IF_1 binding site.[62]

3.2.5 Inhibition of F₁-ATPase by Azide

Azide is a much-studied non-covalent inhibitor of F_1-ATPase, but its mode of action remains both unclear and controversial. During multiple turnover of ATP hydrolysis, MgADP is prone to becoming entrapped at the catalytic site(s) in the β-subunits. This trapping produces the inactive ADP-inhibited state of the enzyme. The slow binding of ATP to the non-catalytic sites in the α-subunits has been proposed to facilitate the release of MgADP from the affected catalytic sites, thereby recovering the ATP hydrolysis activity. Therefore, during steady-state ATP hydrolysis, the F_1-ATPase is thought to be a dynamic mixture of the inhibited and uninhibited molecules.[19,85] It has been proposed that azide traps MgADP permanently at a catalytic site, possibly by acting as a phosphate analogue in a tightly bound $MgADP-N_3^-$ complex, thereby stabilising the ADP-inhibited state.[86–90] The reference structure of bovine F_1-ATPase may represent the ADP-inhibited state of the enzyme, stabilised by AMP-PNP and possibly by azide also.[14] However, azide has not yet been located in any of the crystal structures of bovine F_1-ATPase. Another proposal is that azide does not affect the hydrolytic step of catalysis *per se*, but rather that it interferes with subunit interactions and cooperativity.[91–94] High-resolution structural information about the azide binding site is required for the understanding of its mechanism of inhibition.

4 Potential Medical Significance

4.1 Development of New Antibiotic Inhibitors of F₁-ATPase

In many respects, the mitochondrial and bacterial F-ATPases are very similar to each other; the same overall structure is maintained, the sequences of key subunits, including α- and β-subunits, are highly conserved. However, there are also significant differences between F_1 domains, especially in the central stalk region.[17,18,95] In the mitochondrial enzyme, the foot of the central stalk, which interacts extensively with the c-ring in the F_o domain, contains subunits δ and ε as well as part of the γ-subunit, and the mitochondrial ε-subunit appears to stabilise the interaction of the C-terminal α-helical domain of the δ-subunit with the c-ring.[4–6] In contrast, the foot of the central stalk of the bacterial enzyme contains the same region of the γ-subunit and the ε-subunit (equivalent to the mitochondrial δ-subunit), and it has no component equivalent to the mitochondrial ε-subunit. There is evidence from cross-linking experiments,[96,97] and also from an X-ray crystallographic study of a fragment of the bacterial central stalk,[98,99] that the bacterial ε-subunit can adopt a different conformation to that observed by the equivalent δ-subunit in the bovine enzyme, where the C-terminal α-helical domain instead of being in contact with the c-ring packs alongside the α-helical coiled-coil regions of the γ-subunit and penetrates into the central cavity of the $\alpha_3\beta_3$ domain.[100,101] The mechanistic significance of this structural difference is obscure at present. However, it may be possible to exploit these differences to find inhibitors that inhibit the bacterial enzyme and not the mitochondrial enzyme, as would be required for an effective antibiotic.

So far, little research has been carried out in this area, but the amphiphilic peptides, polyphenolic phytochemicals and non-peptidyl lipophilic cations somewhat surprisingly exhibit significantly different effects on the $\alpha_3\beta_3\gamma$ subcomplex from *Bacillus* PS3 than they do on the bovine enzyme.[62] (The bacterial subcomplex lacks the bacterial ε-subunit and is very similar to the $\alpha_3\beta_3\gamma$ domain of the mitochondrial enzyme[102] to which all the inhibitors described above bind.) In addition, bovine IF$_1$, yeast IF$_1$, resveratrol and piceatannol have no inhibitory activity on the bacterial enzyme subcomplex, and melittin and rhodamine 6G stimulate the ATPase activity of the bacterial subcomplex. The stimulatory effect seen for melittin is much weaker than that observed for rhodamine 6G, which appears to stimulate ATPase activity by more than three-fold at low concentrations. As the concentrations are increased, the stimulatory effect of both melittin and rhodamine 6G decreases until, at sufficiently high concentrations, the compounds act as inhibitors. SynA2 and SynC have much weaker inhibitory activity on the bacterial enzyme than on the bovine enzyme. In contrast, dequalinium has an inhibitory activity on the bacterial enzyme that is approximately two-fold greater than on the bovine enzyme. These preliminary results indicate that further study in this area might be rewarding.

4.2 Potential Uses of the Inhibitor Protein in Medicine

Numerous studies have been conducted into the pathophysiological significance of IF$_1$ and the ATP synthase, primarily in the context of myocardial ischaemia and tumour growth.[67] However, it is conceivable that inhibitors of ATP synthase play a crucial role in the pathophysiology of any disease involving ATP metabolism. The mitochondrial ATP synthase contributes the vast majority of ATP required for contractile function in cardiac tissue, but under ischaemic conditions it has the potential to become a potent hydrolase, wasting cellular ATP.[103–105] The rate of onset of irreversible tissue damage is determined by the rate of energy depletion and is augmented by futile ATP hydrolysis.[67] The importance of this hydrolase activity in ATP depletion varies according to species and is probably dependent on IF$_1$ activity and expression. In slow heart rate species, such as man, futile hydrolysis is reduced significantly during ischaemia by IF$_1$ binding reversibly to the ATP synthase. Under such conditions, glycolysis remains the only source of cellular ATP. The high rate of glycolysis results in a reduction of pH in both the cytosol and mitochondrial matrix, promoting inhibition of ATP hydrolysis by IF$_1$ to preserve the ATP pool.[106–110]

Recently, several small molecule inhibitors of ATP synthase have been described that may have utility as cardioprotective agents.[111–113] In addition, diazoxide, a cardioprotective drug, has been proposed to promote the binding of IF$_1$ to ATP synthase and this mechanism may explain its action.[114] Ischaemic conditions are also found in tumour cells, where oxygen levels vary from the perivascular regions to the anoxic necrotic centres.[67,115] Investigations into the significance of IF$_1$ and tumour growth suggest that the ATPase activity of ATP synthase might be more tightly regulated in tumour cells.[116–119] Therefore, IF$_1$ may offer a survival mechanism in solid tumours, by helping conserve ATP under conditions of oxygen deprivation. As the amphiphilic peptide inhibitors appear to compete with the natural inhibitor protein, they may have therapeutic potential.

4.3   The Ectopic ATPase as a Target for Therapy

The mammalian ATP synthase in mitochondria is a complex of 29 polypeptides of 16 different kinds.[120] They are the F$_1$ subunits α, β, γ, δ and ε, the peripheral stalk subunits OSCP, b, d and F$_6$, the F$_o$ subunits c, a, A6L, e, f and g, and IF$_1$. The assembled mammalian complex contains three copies of both α- and β-subunits, and most probably ten copies of subunit c, as observed in the F$_1$–c$_{10}$ subcomplex of ATP synthase in the mitochondria of *Saccharomyces cerevisiae*.[6] All other subunits are present in single copies.

Recently, various subunits of the mitochondrial enzyme have been reported to be present on the external cell surface of human endothelial cells,[121–124] hepatocytes,[125] several human tumour cell lines[126] and in liver plasma membrane lipid rafts.[127] The α- and β-subunits of ATP synthase were identified on the endothelial cell surface as the major binding site for the anti-angiogenic plasminogen fragment, angiostatin.[122] Subsequently, both ATP hydrolase and ATP synthetic activities were said to be present, and angiostatin inhibited both activities.[121] The angiostatin-mediated inhibition appeared also to correlate with inhibition of endothelial cell proliferation. IF$_1$ has been proposed to attenuate the anti-angiogenic response to angiostatin on endothelial cells where it may serve a protective role for these cells in the tumour microenvironment. It has also been argued that there is a relationship between the binding sites of IF$_1$ and angiostatin on ATP synthase.[128] In a separate study, the α-subunit was proposed to bind the C-terminal domain of the cytokine p43 on endothelial cells and inhibit cell proliferation.[124] However, it remains to be determined whether the C-terminal domain of p43 inhibits the activity of ATP synthase directly, and whether it competes with angiostatin for binding to ATP synthase. Nonetheless, the interaction with the α-subunit of ATP synthase could provide a mechanism for the inhibition of endothelial cell growth and play a regulatory role in the tumour vasculature. A fragment of plasminogen (kringles 1–5) is also a potent and specific inhibitor of angiogenesis and tumour growth and it is reported to bind to the endothelial cell surface ATP synthase. Its anti-angiogenic effect in mouse tumour models has been attributed, at least in part, to its interaction with, and inhibition of the activity of, the ectopic ATP synthase.[129]

The α- and β-subunits of ATP synthase have been found also on the cell surface of hepatocytes where the β-subunit acts as a high-affinity receptor for the high density lipoprotein (HDL), apolipoprotein A-I (apoA-I). This putative protein complex hydrolyses ATP, and the binding of apoA-I stimulated its ATPase activity and also the subsequent endocytosis of *holo*-HDL. The process is inhibited by bovine IF$_1$.[125]

Several outstanding issues concerning ectopic ATPase remain to be resolved. For example, exactly what are the subunit compositions of the ectopic ATPases in the various cell types? If the enzyme is to function as an ATPase in a similar way to the mitochondrial enzyme, it will probably require most if not all of the mitochondrial subunits. So far, it has not been shown conclusively that most subunits of the mitochondrial enzyme are present on the cell surface, and, for example, there is no convincing evidence for the F_1 subunits γ, δ and ε, for the peripheral stalk subunits OSCP, b, d and F_6, for the F_o subunits c, a, A6L, e, f and g and for IF_1 on the cell surface in the ATP hydrolytic or synthetic complex. It is possible that the subunit composition of ectopic ATP synthase differs from the enzyme found in the inner mitochondrial membrane, but such a difference remains to be demonstrated.

Another important bioenergetic issue that depends on the subunit composition of ectopic ATPase is whether the energy from ATP hydrolysis is used to generate a PMF across the plasma membrane, and a related issue is how the proposed synthesis of ATP is energised. The direction of the membrane potential difference across the plasma membrane (negative inwards) makes it unlikely that the enzyme is capable of synthesising ATP by a rotary mechanism using the PMF as an energy source.

Another major issue concerns the biogenesis of the ectopic complex. Two of the subunits of mitochondrial ATP synthase are the products of the mitochondrial genome, and the rest are nuclear gene products, some, but not all, with mitochondrial import sequences that are removed during import into the organelle in order to produce the mature proteins that are assembled into the enzyme complex. So how do the various subunits make their way to the plasma membrane surface, and how could the putative enzyme complex be assembled there? Given the complexity of the assembly of the mitochondrial enzyme and the requirement for many ancilliary proteins to process mitochondrial import sequences and to assemble the complex, it is unlikely that this complex series of events, or a significant subset of them, will be duplicated at the cell surface. It seems much more likely that if an ATP synthase complex of the mitochondrial type is present on the cell surface, the subunits of the ectopic ATPase that have been detected there originate from the mitochondrion, and that if there is an ectopic ATPase resembling the mitochondrial enzyme that it has been assembled in the mitochondrion, and that subsequently the enzyme, wholly or partially, has been transported to the plasma membrane surface by an unknown pathway.

However, perhaps the most fundamental need is to demonstrate that the ATPase and ATP synthase activities are really associated with the mitochondrial ATPase subunits found on the cell surface. Until the active cell surface enzyme has been purified and studied in isolation, it remains distinctly possible that ATP hydrolysis and synthesis are being carried out by other hitherto unidentified cell surface enzymes, and that the cell surface α- and β-subunits are not related to this role. It is also clearly important to resolve what is the function of the ectopic ATPase. Is it, as has been suggested,[125,130] part of some signalling pathway(s) involving extracellular adenine nucleotides? The proposed roles for ectopic ATPase in HDL catabolism and in the promotion of angiogenesis raise possibilities for the control of cholesterol and the inhibition of tumour growth, both of them fundamental issues in cardiovascular and cancer research, respectively. However, until the basic parameters of the cell surface enzyme have been established convincingly, it is premature to try to discuss therapeutic approaches based upon knowledge of the well-characterised mitochondrial enzyme.

5 Concluding Remarks

A detailed knowledge of the atomic resolution structures of the respiratory enzyme complexes from mammalian mitochondria is likely to be essential for understanding the molecular basis of human disease states involving their dysfunction. The mitochondrial F_1-ATPase has been studied extensively in several inhibited states by X-ray crystallography. These investigations have defined the catalytic pathway of ATP hydrolysis and identified inhibitory sites on the enzyme. A detailed understanding of the inhibitors of ATP synthase and their modes and sites of action on the enzyme is of significant medical interest. Further structural studies of several of these inhibitors in complex with F_1-ATPase may enable rational development of therapeutic agents to act as novel antibiotics against bacterial ATP synthases or for the treatment of several disorders linked to the regulation of the ATP synthase, such as ischaemia-reperfusion injury, some cancers and cholesterolemia.

Abbreviations

AMP-PNP, 5′-adenylylimidodiphosphate;
ApoA-I, apolipoprotein A-I;
DCCD, N,N'-dicyclohexylcarbodiimide;
HDL, high-density lipoprotein;
IC_{50}, inhibitor concentration that produces 50% inhibition of ATPase activity;
NBD-Cl, 4-chloro-7-nitrobenzofurazan;
PMF, proton-motive force.

Acknowledgements

This work was supported by The Medical Research Council.

Note added in proof. Recently the structure of bovine F_1-ATPase inhibited by piceatannol has been solved by X-ray crystallography (J.G. Gledhill, A.G.W. Leslie and J.E. Walker, unpublished results).

References

1. P.D. Boyer, *Annu. Rev. Biochem.*, 1997, **66**, 717.
2. J.E. Walker, *Angew. Chem. Int. Edn. Engl.*, 1998, **37**, 2309.
3. A.E. Senior, S. Nadanaciva and J. Weber, *Biochim. Biophys. Acta*, 2002, **1553**, 188.
4. C. Gibbons, M.G. Montgomery, A.G. Leslie and J.E. Walker, *Nat. Struct. Biol.*, 2000, **7**, 1055.
5. D. Stock, C. Gibbons, I. Arechaga, A.G. Leslie and J.E. Walker, *Curr. Opin. Struct. Biol.*, 2000, **10**, 672.
6. D. Stock, A.G. Leslie and J.E. Walker, *Science*, 1999, **286**, 1700.
7. I.R. Collinson, M.J. van Raaij, M.J. Runswick, I.M. Fearnley, J.M. Skehel, G.L. Orriss, B. Miroux and J.E. Walker, *J. Mol. Biol.*, 1994, **242**, 408.

8. I.R. Collinson, J.M. Skehel, I.M. Fearnley, M.J. Runswick and J.E. Walker, *Biochemistry*, 1996, **35**, 12640.
9. J.E. Walker and I.R. Collinson, *FEBS Lett.*, 1994, **346**, 39.
10. I.R. Collinson, M.J. Runswick, S.K. Buchanan, I.M. Fearnley, J.M. Skehel, M.J. van Raaij, D.E. Griffiths and J.E. Walker, *Biochemistry*, 1994, **33**, 7971.
11. S. Karrasch and J.E. Walker, *J. Mol. Biol.*, 1999, **290**, 379.
12. J. Rubinstein and J.E. Walker, *J. Mol. Biol.*, 2002, **321**, 613.
13. J.L. Rubinstein, J.E. Walker and R. Henderson, *EMBO J.*, 2003, **22**, 6182.
14. J.P. Abrahams, A.G. Leslie, R. Lutter and J.E. Walker, *Nature*, 1994, **370**, 621.
15. P.D. Boyer, *Biochim. Biophys. Acta*, 1993, **1140**, 215.
16. R. Kagawa, M.G. Montgomery, K. Braig, A.G. Leslie and J.E. Walker, *EMBO J.*, 2004, **23**, 2734.
17. J.E. Walker, A.L. Cozens, M.R. Dyer, I.M. Fearnley, S.J. Powell and M.J. Runswick, *Chem. Scripta*, 1987, **27B**, 97.
18. J.E. Walker, I.M. Fearnley, R. Lutter, R.J. Todd and M.J. Runswick, *Philos. Trans. R. Soc. Lond. B. Biol. Sci.*, 1990, **326**, 367.
19. W.S. Allison, J.M. Jault, S. Zhuo and S.R. Paik, *J. Bioenerg. Biomembr.*, 1992, **24**, 469.
20. S.J. Ferguson, W.J. Lloyd, M.H. Lyons and G.K. Radda, *Eur. J. Biochem.*, 1975, **54**, 117.
21. F.S. Esch, P. Bohlen, A.S. Otsuka, M. Yoshida and W.S. Allison, *J. Biol. Chem.*, 1981, **256**, 9084.
22. M. Hollemans, M.J. Runswick, I.M. Fearnley and J.E. Walker, *J. Biol. Chem.*, 1983, **258**, 9307.
23. J. Garin, M. Vincon, J. Gagnon and P. Vignais, *Biochemistry*, 1994, **33**, 3772.
24. J. Garin, F. Boulay, J.P. Issartel, J. Lunardi and P.V. Vignais, *Biochemistry*, 1986, **25**, 4431.
25. R.L. Cross, D. Cunningham, C.G. Miller, Z.X. Xue, J.M. Zhou and P.D. Boyer, *Proc. Natl. Acad. Sci. U.S.A.*, 1987, **84**, 5715.
26. F.S. Esch and W.S. Allison, *J. Biol. Chem.*, 1978, **253**, 6100.
27. D.A. Bullough and W.S. Allison, *J. Biol. Chem.*, 1986, **261**, 14171.
28. M.E. Pullman and G.C. Monroy, *J. Biol. Chem.*, 1963, **238**, 3762.
29. H. Lardy, P. Reed and C.H. Lin, *Fed. Proc.*, 1975, **34**, 1707.
30. J.L. Connelly and H.A. Lardy, *Biochemistry*, 1964, **19**, 1969.
31. H.A. Lardy, J.L. Connelly and D. Johnson, *Biochemistry*, 1964, **19**, 1961.
32. J. Zheng and V.D. Ramirez, *Br. J. Pharmacol.*, 2000, **130**, 1115.
33. J. Zheng and V.D. Ramirez, *Biochem. Biophys. Res. Commun.*, 1999, **261**, 499.
34. D.A. Bullough, E.A. Ceccarelli, D. Roise and W.S. Allison, *Biochim. Biophys. Acta*, 1989, **975**, 377.
35. G. Groth, *Proc. Natl. Acad. Sci. U.S.A.*, 2002, **99**, 3464.
36. G. Groth, T. Hisabori, H. Lill and D. Bald, *J. Biol. Chem.*, 2002, **277**, 20117.
37. C. Schnick, N. Kortgen and G. Groth, *J. Biol. Chem.*, 2002, **277**, 51003.
38. J.P. Abrahams, S.K. Buchanan, M.J. van Raaij, I.M. Fearnley, A.G. Leslie and J.E. Walker, *Proc. Natl. Acad. Sci. U.S.A.*, 1996, **93**, 9420.

39. M.J. van Raaij, J.P. Abrahams, A.G. Leslie and J.E. Walker, *Proc. Natl. Acad. Sci. U.S.A.*, 1996, **93**, 6913.
40. G.L. Orriss, A.G. Leslie, K. Braig and J.E. Walker, *Structure*, 1998, **6**, 831.
41. K. Braig, R.I. Menz, M.G. Montgomery, A.G. Leslie and J.E. Walker, *Structure Fold. Des.*, 2000, **8**, 567.
42. R.I. Menz, A.G. Leslie and J.E. Walker, *FEBS Lett.*, 2001, **494**, 11.
43. R.I. Menz, J.E. Walker and A.G. Leslie, *Cell*, 2001, **106**, 331.
44. E. Cabezon, M.G. Montgomery, A.G. Leslie and J.E. Walker, *Nat. Struct. Biol.*, 2003, **10**, 744.
45. Y.W. Xu, S. Morera, J. Janin and J. Cherfils, *Proc. Natl. Acad. Sci. U.S.A.*, 1997, **94**, 3579.
46. J. Sondek, D.G. Lambright, J.P. Noel, H.E. Hamm and P.B. Sigler, *Nature*, 1994, **372**, 276.
47. D.E. Coleman, A.M. Berghuis, E. Lee, M.E. Linder, A.G. Gilman and S.R. Sprang, *Science*, 1994, **265**, 1405.
48. G.A. Petsko, *Proc. Natl. Acad. Sci. U.S.A.*, 2000, **97**, 538.
49. P.E. Linnett and R.B. Beechey, *Methods Enzymol.*, 1979, **55**, 472.
50. T. Saishu, Y. Kagawa and R. Shimizu, *Biochem. Biophys. Res. Commun.*, 1983, **112**, 822.
51. G.J. Verschoor, P.R. van der Sluis and E.C. Slater, *Biochim. Biophys. Acta*, 1977, **462**, 438.
52. J.P. Issartel and P.V. Vignais, *Biochemistry*, 1984, **23**, 6591.
53. J.P. Issartel, G. Klein, M. Satre and P.V. Vignais, *Biochemistry*, 1983, **22**, 3492.
54. K.D. Setchell, *Am. J. Clin. Nutr.*, 1998, **68**, 1333S.
55. D.M. Tham, C.D. Gardner and W.L. Haskell, *J. Clin. Endocrinol. Metab.*, 1998, **83**, 2223.
56. L. Fremont, *Life Sci.*, 2000, **66**, 663.
57. Y.J. Surh, *Nat. Rev. Cancer*, 2003, **3**, 768.
58. K.P. Bhat and J.M. Pezzuto, *Ann. N.Y. Acad. Sci.*, 2002, **957**, 210.
59. J. Gusman, H. Malonne and G. Atassi, *Carcinogenesis*, 2001, **22**, 1111.
60. K. Roemer and M. Mahyar-Roemer, *Drugs Today*, 2002, **38**, 571.
61. J.F. Savouret and M. Quesne, *Biomed. Pharmacother.*, 2002, **56**, 84.
62. J.R. Gledhill and J.E. Walker, *Biochem. J.*, 2005, **386**, 591.
63. S. Gupta, S.B. Krasnoff, D.W. Roberts, J.A.A. Renwick, L.S. Brinen and J. Clardy, *J. Org. Chem.*, 1992, **57**, 2306.
64. S.B. Krasnoff and S. Gupta, *J. Chem. Ecol.*, 1991, **17**, 1953.
65. R.L. Cross and W.E. Kohlbrenner, *J. Biol. Chem.*, 1978, **253**, 4865.
66. J.E. Walker, *Curr. Opin. Struct. Biol.*, 1994, **4**, 912.
67. D.W. Green and G.J. Grover, *Biochim. Biophys. Acta*, 2000, **1458**, 343.
68. D.J. Gordon-Smith, R.J. Carbajo, J.C. Yang, H. Videler, M.J. Runswick, J.E. Walker and D. Neuhaus, *J. Mol. Biol.*, 2001, **308**, 325.
69. E. Cabezon, M.J. Runswick, A.G. Leslie and J.E. Walker, *EMBO J.*, 2001, **20**, 6990.
70. M.J. van Raaij, G.L. Orriss, M.G. Montgomery, M.J. Runswick, I.M. Fearnley, J.M. Skehel and J.E. Walker, *Biochemistry*, 1996, **35**, 15618.

71. E. Cabezon, I. Arechaga, P. Jonathan, G. Butler and J.E. Walker, *J. Biol. Chem.*, 2000, **275**, 28353.

72. E. Cabezon, P.J. Butler, M.J. Runswick and J.E. Walker, *J. Biol. Chem.*, 2000, **275**, 25460.

73. H. Matsubara, T. Hase, T. Hashimoto and K. Tagawa, *J. Biochem. (Tokyo)*, 1981, **90**, 1159.

74. T. Hashimoto, Y. Negawa and K. Tagawa, *J. Biochem. (Tokyo)*, 1981, **90**, 1151.

75. A. Akashi, Y. Yoshida, H. Nakagoshi, K. Kuroki, T. Hashimoto, K. Tagawa and F. Imamoto, *J. Biochem. (Tokyo)*, 1988, **104**, 526.

76. Y. Okada, T. Hashimoto, Y. Yoshida and K. Tagawa, *J. Biochem. (Tokyo)*, 1986, **99**, 251.

77. S. Hong and P.L. Pedersen, *Arch. Biochem. Biophys.*, 2002, **405**, 38.

78. E. Cabezon, P.J. Butler, M.J. Runswick, R.J. Carbajo and J.E. Walker, *J. Biol. Chem.*, 2002, **277**, 41334.

79. M.J. Weiss, J.R. Wong, C.S. Ha, R. Bleday, R.R. Salem, G.D. Steele, Jr. and L.B. Chen, *Proc. Natl. Acad. Sci. U.S.A.*, 1987, **84**, 5444.

80. P.K. Laikind and W.S. Allison, *J. Biol. Chem.*, 1983, **258**, 11700.

81. D.A. Bullough, E.A. Ceccarelli, J.G. Verburg and W.S. Allison, *J. Biol. Chem.*, 1989, **264**, 9155.

82. M.J. Runswick and J.E. Walker, *J. Biol. Chem.*, 1983, **258**, 3081.

83. S. Zhuo and W.S. Allison, *Biochem. Biophys. Res. Commun.*, 1988, **152**, 968.

84. S. Zhuo, S.R. Paik, J.A. Register and W.S. Allison, *Biochemistry*, 1993, **32**, 2219.

85. J.M. Jault and W.S. Allison, *J. Biol. Chem.*, 1993, **268**, 1558.

86. E.A. Vasilyeva, I.B. Minkov, A.F. Fitin and A.D. Vinogradov, *Biochem. J.*, 1982, **202**, 15.

87. M.B. Murataliev, Y.M. Milgrom and P.D. Boyer, *Biochemistry*, 1991, **30**, 8305.

88. D.J. Hyndman, Y.M. Milgrom, E.A. Bramhall and R.L. Cross, *J. Biol. Chem.*, 1994, **269**, 28871.

89. J.M. Jault, C. Dou, N.B. Grodsky, T. Matsui, M. Yoshida and W.S. Allison, *J. Biol. Chem.*, 1996, **271**, 28818.

90. C. Dou, N.B. Grodsky, T. Matsui, M. Yoshida and W.S. Allison, *Biochemistry*, 1997, **36**, 3719.

91. J.G. Wise, L.R. Latchney, A.M. Ferguson and A.E. Senior, *Biochemistry*, 1984, **23**, 1426.

92. T. Noumi, M. Maeda and M. Futai, *FEBS Lett.*, 1987, **213**, 381.

93. D.A. Harris, *Biochim. Biophys. Acta*, 1989, **974**, 156.

94. J. Weber and A.E. Senior, *J. Biol. Chem.*, 1998, **273**, 33210.

95. J.E. Walker, G. Falk, N.J. Gay and V.L. Tybulewicz, *Biochem. Soc. Trans.*, 1984, **12**, 234.

96. S.P. Tsunoda, A.J. Rodgers, R. Aggeler, M.C. Wilce, M. Yoshida and R.A. Capaldi, *Proc. Natl. Acad. Sci. U.S.A.*, 2001, **98**, 6560.

97. R. Aggeler, M.A. Haughton and R.A. Capaldi, *J. Biol. Chem.*, 1995, **270**, 9185.

98. A.J. Rodgers and M.C. Wilce, *Nat. Struct. Biol.*, 2000, **7**, 1051.

99. A.C. Hausrath, G. Gruber, B.W. Matthews and R.A. Capaldi, *Proc. Natl. Acad. Sci. U.S.A.*, 1999, **96**, 13697.

100. A.C. Hausrath, R.A. Capaldi and B.W. Matthews, *J. Biol. Chem.*, 2001, **276**, 47227.
101. T. Suzuki, T. Murakami, R. Iino, J. Suzuki, S. Ono, Y. Shirakihara and M. Yoshida, *J. Biol. Chem.*, 2003, **278**, 46840.
102. Y. Shirakihara, A.G. Leslie, J.P. Abrahams, J.E. Walker, T. Ueda, Y. Sekimoto, M. Kambara, K. Saika, Y. Kagawa and M. Yoshida, *Structure*, 1997, **5**, 825.
103. W. Rouslin, J.L. Erickson and R.J. Solaro, *Am. J. Physiol.*, 1986, **250**, H503.
104. D.A. Harris and A.M. Das, *Biochem. J.*, 1991, **280**, 561.
105. R.B. Jennings and K.A. Reimer, *Annu. Rev. Med.*, 1991, **42**, 225.
106. W. Rouslin, *J. Biol. Chem.*, 1987, **262**, 3472.
107. W. Rouslin, *Am. J. Physiol.*, 1987, **252**, H622.
108. W. Rouslin and M.E. Pullman, *J. Mol. Cell. Cardiol.*, 1987, **19**, 661.
109. W. Rouslin, *J. Biol. Chem.*, 1983, **258**, 9657.
110. W. Rouslin and C.W. Broge, *J. Biol. Chem.*, 1989, **264**, 15224.
111. K.S. Atwal, S. Ahmad, C.Z. Ding, P.D. Stein, J. Lloyd, L.G. Hamann, D.W. Green, F.N. Ferrara, P. Wang, W.L. Rogers, L.M. Doweyko, A.V. Miller, S.N. Bisaha, J.B. Schmidt, L. Li, K.J. Yost, H.J. Lan and C.S. Madsen, *Bioorg. Med. Chem. Lett.*, 2004, **14**, 1027.
112. K.S. Atwal, P. Wang, W.L. Rogers, P. Sleph, H. Monshizadegan, F.N. Ferrara, S. Traeger, D.W. Green and G.J. Grover, *J. Med. Chem.*, 2004, **47**, 1081.
113. L.G. Hamann, C.Z. Ding, A.V. Miller, C.S. Madsen, P. Wang, P.D. Stein, A.T. Pudzianowski, D.W. Green, H. Monshizadegan and K.S. Atwal, *Bioorg. Med. Chem. Lett.*, 2004, **14**, 1031.
114. S. Contessi, G. Metelli, I. Mavelli and G. Lippe, *Biochem. Pharmacol.*, 2004, **67**, 1843.
115. R.M. Sutherland, *Science*, 1988, **240**, 177.
116. K. Luciakova and S. Kuzela, *FEBS Lett.*, 1984, **177**, 85.
117. E.W. Yamada and N.J. Huzel, *Biochim. Biophys. Acta*, 1992, **1139**, 143.
118. B.V. Chernyak, V.N. Dedov and V.L. Gabai, *FEBS Lett.*, 1994, **337**, 56.
119. F. Capuano, F. Guerrieri and S. Papa, *J. Bioenerg. Biomembr.*, 1997, **29**, 379.
120. J.E. Walker, R. Lutter, A. Dupuis and M.J. Runswick, *Biochemistry*, 1991, **30**, 5369.
121. T.L. Moser, D.J. Kenan, T.A. Ashley, J.A. Roy, M.D. Goodman, U.K. Misra, D.J. Cheek and S.V. Pizzo, *Proc. Natl. Acad. Sci. U.S.A.*, 2001, **98**, 6656.
122. T.L. Moser, M.S. Stack, I. Asplin, J.J. Enghild, P. Hojrup, L. Everitt, S. Hubchak, H.W. Schnaper and S.V. Pizzo, *Proc. Natl. Acad. Sci. U.S.A.*, 1999, **96**, 2811.
123. T.L. Moser, M.S. Stack, M.L. Wahl and S.V. Pizzo, *Thromb. Haemost.*, 2002, **87**, 394.
124. S.Y. Chang, S.G. Park, S. Kim and C.Y. Kang, *J. Biol. Chem.*, 2002, **277**, 8388.
125. L.O. Martinez, S. Jacquet, J.P. Esteve, C. Rolland, E. Cabezon, E. Champagne, T. Pineau, V. Georgeaud, J.E. Walker, F. Terce, X. Collet, B. Perret and R. Barbaras, *Nature*, 2003, **421**, 75.
126. B. Das, M.O. Mondragon, M. Sadeghian, V.B. Hatcher and A.J. Norin, *J. Exp. Med.*, 1994, **180**, 273.

127. T.J. Bae, M.S. Kim, J.W. Kim, B.W. Kim, H.J. Choo, J.W. Lee, K.B. Kim, C.S. Lee, J.H. Kim, S.Y. Chang, C.Y. Kang, S.W. Lee and Y. G. Ko, *Proteomics*, 2004, **4**, 3536.

128. N.R. Burwick, M.L. Wahl, J. Fang, Z. Zhong, R.A. Capaldi, D.J. Kenan and S.V. Pizzo, *J. Biol. Chem.*, 2004.

129. N. Veitonmaki, R. Cao, L.H. Wu, T.L. Moser, B. Li, S.V. Pizzo, B. Zhivotovsky and Y. Cao, *Cancer Res.*, 2004, **64**, 3679.

130. L.O. Martinez, S. Jacquet, F. Terce, X. Collet, B. Perret and R. Barbaras, *Cell. Mol. Life Sci.*, 2004, **61**, 2343.

131. J.L. Rubinstein, V.K. Dickson, M.J. Runswick and J.E. Walker, *J. Mol. Biol.*, 2005, **345**, 513.

132. W. Jiang and R.H. Fillingame, *Proc. Natl. Acad. Sci. U.S.A.*, 1998, **95**, 6607.

133. W.L. DeLano, The PyMOL Molecular Graphics System, (DeLano Scientific, San Carlos, CA, U.S.A., 2002) www.pymol.org.

134. T. Yonetani and H. Theorell, *Arch. Biochem. Biophys.*, 1964, **106**, 243.

135. T.C. Terwilliger and D. Eisenberg, *J. Biol. Chem.*, 1982, **257**, 6016.

CHAPTER 15

The Passion of the Permease

H. RONALD KABACK

Departments of Physiology and Microbiology,
Immunology & Molecular Genetics, Molecular Biology Institute,
University of California Los Angeles,
Los Angeles, California 9005-1662,USA

1   Introduction

More than 20% of the genomes sequenced to date apparently encode polytopic transmembrane proteins involved in a plethora of essential functions, particularly energy and signal transduction. Many are important with regard to human disease (e.g. depression, diabetes, drug resistance), and many drugs are targeted to membrane transport proteins (e.g. fluoxetine and omeprazole). However, the number of crystal structures of membrane proteins, especially ion-coupled transporters, is very limited. Recently, an inward-facing conformer of the *Escherichia coli* lactose permease (LacY), a paradigm for the major facilitator superfamily, which contains over 3500 members, was solved at about 3.5 Å resolution in collaboration with Jeff Abramson and So Iwata at Imperial College London. This well-studied membrane transport protein is composed of two symmetrical six-helix bundles with a large internal cavity containing bound sugar and open to the cytoplasm only. Based on the structure and a large body of biochemical and biophysical evidence, a mechanism is here proposed in which the binding site is alternatively accessible to either side of the membrane.

2   Background

As postulated by Peter Mitchell (1,2) and demonstrated conclusively in bacterial membrane vesicles (3–5), transport of many solutes against a concentration gradient is driven by an electrochemical H^+ gradient ($\Delta\bar{\mu}_{H^+}$; interior negative *and/or* alkaline). This fundamental process (secondary active transport) is found in all living organisms and plays an important role in many aspects of cell function, such as nutrient uptake, signal transduction, as well extrusion of drugs and toxic substances into the environment. Understanding the mechanism by which membrane transport proteins transduce free energy stored in an ion electrochemical gradient across a membrane into a concentration gradient of substrate is a fundamental question in biology.

The lactose permease of *Escherichia coli* (LacY) is encoded by *lacY*, the second structural gene in the *lac* operon (6) and was the first gene encoding a membrane transport protein to be cloned into a recombinant plasmid, overexpressed (7) and sequenced (8). This accomplishment opened secondary active transport to studies at the molecular level. LacY was the first protein of its class to be solubilized and purified in a completely functional state (9,10), thereby demonstrating that the product of the *lacY* gene is solely responsible for all the translocation reactions catalyzed by this transport system in *E. coli*. It has also been shown that LacY is both structurally and functionally a monomer in the membrane (see 11,12,13); however, the experimental evidence supporting this contention will not be discussed here.

LacY is a cytoplasmic integral membrane protein that belongs to the major facilitator superfamily (MFS) (14), an increasingly large group of membrane transport proteins thought to be evolutionarily related and found from archaea to the mammalian central nervous system. LacY utilizes free energy released from downhill translocation of H^+ to drive the stoichiometric accumulation of galactosides against a concentration gradient. In the absence of $\Delta\bar{\mu}_{H^+}$, LacY catalyzes the converse reaction, utilizing free energy released from downhill translocation of sugar to drive uphill translocation of H^+ with generation of $\Delta\bar{\mu}_{H^+}$, the polarity of which depends upon the direction of the substrate concentration gradient (Figure 1). In the absence of substrate, LacY does not translocate H^+. However, substrate gradients in and of themselves generate H^+ electrochemical gradients of either polarity. Therefore, it

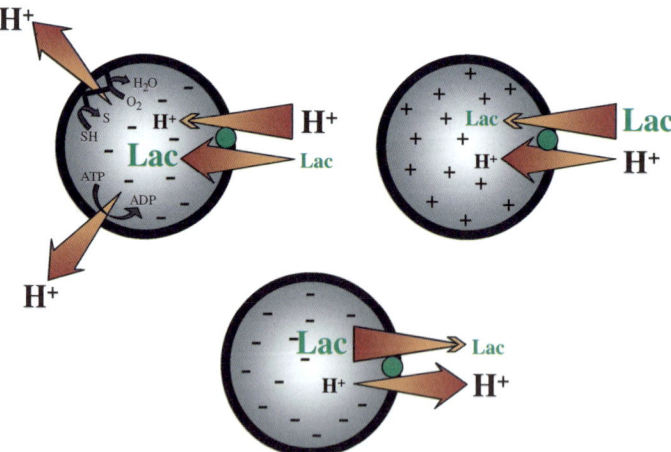

Figure 1 *Lactose/H^+ symport. In the absence of substrate, LacY does not tranlocate H^+; substrate gradients generate electrochemical H^+ gradients, the polarity of which depends upon the direction of the substrate concentration gradient*

seems intuitively likely that the primary driving force for turnover is binding and dissociation of sugar on either side of the membrane.

LacY contains 417 amino acid residues, is 80–85% helical, as shown initially and most accurately by circular dichroism (15), and has 12 helices that traverse the membrane in zig-zag fashion connected by relatively hydrophilic loops with both N and C termini on the cytoplasmic face (16–18) (Figure 2). LacY has been studied by electrospray ionization-mass spectrometry (ESI-MS) (19–24). The molecular weight reconstruction from ESI-MS of LacY with a 6-His affinity tag at the carboxyl terminus shows the purified protein to be homogeneous, and the computed mass is within 0.01% of the mass derived from the DNA sequence with a formyl group on the initiating Met. Although the formyl group is usually removed from native LacY (25), the deformylase may be saturated by overexpression of LacY.

Use of molecular biology techniques to engineer LacY for site-directed biochemical and biophysical studies has provided important information about structure and mechanism (Figure 2) (26). In addition to other site-directed mutants, functional LacY devoid of eight native Cys residues has been constructed (C-less LacY) and used for cysteine-scanning mutagenesis (27). Analysis of mutants at every position in the protein has led to the following observations (reviewed in 26,27,28):

(1) Only six side chains are irreplaceable with respect to active transport: Glu126 (helix IV) and Arg144 (helix V) which are crucial for substrate binding; Glu269 (helix VIII) which is likely involved in both substrate binding and H^+ translocation; and Arg302 (helix IX), His322 (helix X) and Glu325 (helix X) which play irreplaceable roles in H^+ translocation.

(2) Residues in addition to Glu126 and Arg144 that are important determinants for sugar binding and recognition have also been identified.

(3) Substrate-induced changes in the reactivity of side chains with various chemical modification reagents, site-directed fluorescence and spin-labeling suggest widespread conformational changes during turnover.

Based on these observations, a model for the transport mechanism has been postulated which is supported by the structure to be discussed. A possible helix packing model was also proposed from distance constraints obtained from thiol cross-linking experiments and engineered Mn(II) binding sites (29) which is consistent with many of the local interactions revealed, but does not accurately describe certain global aspects of the structure. Although the results provided a crude approximation of the structure and mechanism, high-resolution structures are required to clarify the transport reaction, which include substrate binding, coupling between substrate and H^+ translocation and large-scale conformational changes during turnover. However, all crystallization trials failed until lately, due in all probability to the conformational flexibility of wild-type LacY.

Recently, the X-ray structure of a LacY thermostable mutant, C154G, which binds ligand but does very little transport, was solved at a resolution of 3.5 Å in collaboration with So Iwata and Jeff Abramson at Imperial College London (30). The structure confirms many of the biochemical and biophysical studies and also reveals a number of unexpected, novel findings.

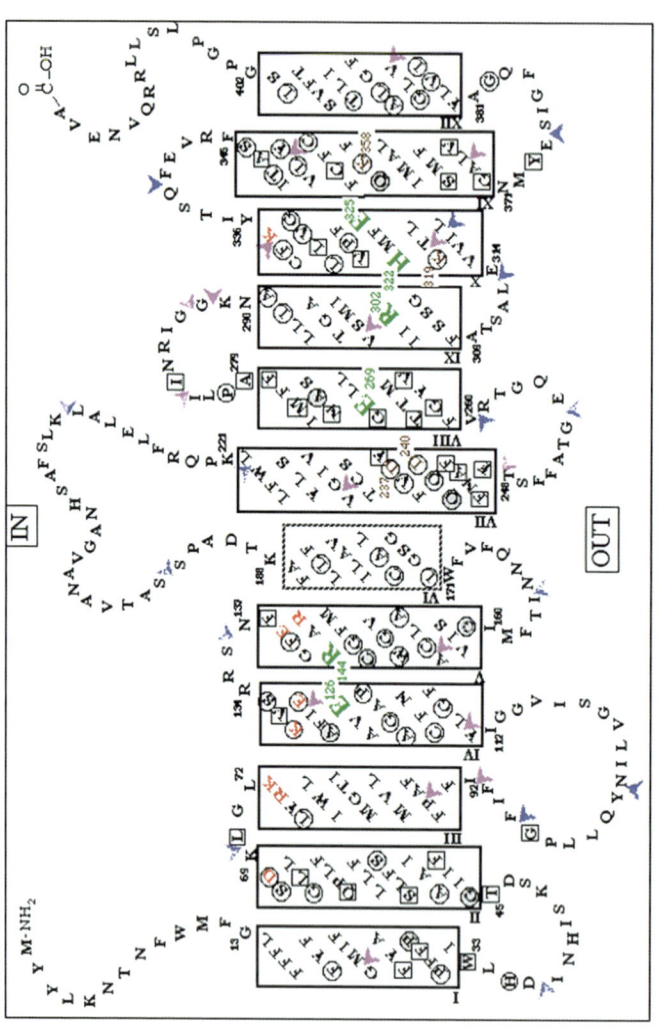

Figure 2 *Secondary structure model of LacY derived from hydropathy and deletion analyses. The single-letter amino acid code is used, residues irreplaceable for active transport are highlighted in large green type. Charge pairs Asp237/Lys358 and Asp240/Lys319 are shown in small brown type. Solid rectangles represent helical regions defined by deletion analysis. The dashed helix VI represents the region defined by a decrease in downhill transport assessed by phenotype on indicator media. Orange letters represent ionizable residues predicted to be within the cytoplasmic ends of transmembrane helices II, III, IV and V by deletion analysis. Squared residues represent positions where transport activity of single-cysteine replacement mutants are inhibited by N-ethylmaleimide (NEM) treatment. Circled residues represent positions where missense mutations have been shown to inhibit lactose accumulation. Residues in gray circles represent positions where both results have been observed. Two-tone arrowheads indicate locations where discontinuities in the primary sequence have been introduced and solid arrowheads indicate regions where amino acids have been inserted into LacY. Purple arrowheads heads indicate good transport activity and red arrowheads indicate little or no transport activity. From ref. 18*

3   Overall Structure of LacY

The asymmetric unit of the LacY crystal is composed of an artificial dimer (30), with two molecules oriented in opposite directions, indicating that the monomer is indeed the functional and structural unit (see 11,12,13). Viewed normal to the membrane (Figure 3, bottom), LacY has a distorted oval shape with dimensions of 30 Å x 60 Å. Parallel to the membrane (Figure 3, top), LacY is heart-shaped with a large interior

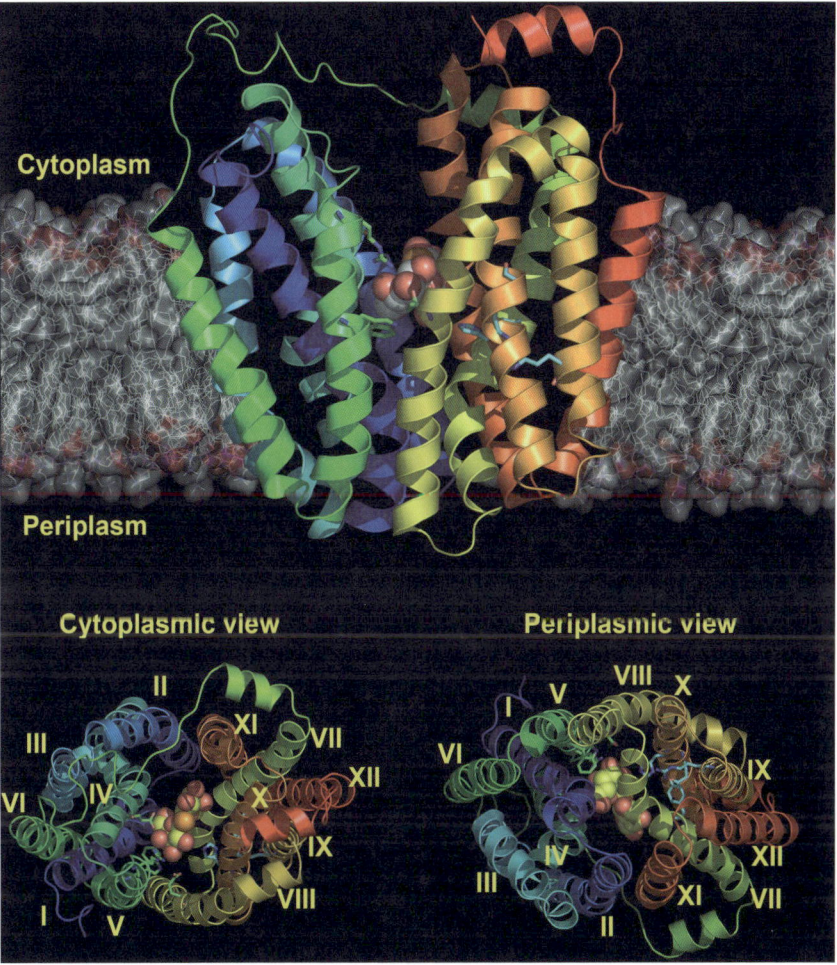

Figure 3 *Overall structure of LacY. The figures are based on the C154G mutant structure with bound β-D-galactopyranosyl-1-thio-β-D-galactospyranoside (TDG).* **Top.** *Ribbon representation of LacY viewed parallel to the membrane. The twelve transmembrane helices are colored from the N-terminus (N) in purple to the C-terminus (C) in pink with TDG represented by spheres.* **Bottom.** *Ribbon representation of LacY viewed along the membrane normal from the cytoplasmic (left) and periplasmic (right) sides. The color scheme is the same as in* **A** *and the twelve transmembrane helices are labelled with roman numerals*

hydrophilic cavity (dimensions *ca.* 25 Å x 15 Å) open on the cytoplasmic side, representing the inward-facing conformation of LacY. Within the cavity, a single bound ligand molecule (β-D-galactopyranosyl 1-thio-β-D-galactopyranoside; TDG) is observed in the middle of the membrane displaced slightly towards the cytoplasmic side. This is consistent with the idea that LacY has only one binding site that is alternatively accessible to either side of the membrane (1,31). In addition, it is noteworthy that many of the helices are highly irregular in shape.

Accessibility of single-Cys mutants to the highly impermeant thiol reagent methanethiosulfonate ethylsulfonate (MTSES) (32,33) occurs at positions lining the hydrophilic cavity (Figure 4) (see 34,35). MTSES accessibility to Cys replacements at positions leading from the hydrophilic cavity to the periplasmic side likely represent the pathway by which sugar gains access to the binding site. The pattern observed is consistent with the alternating access model, since the single-Cys mutants used in these studies catalyze significant transport activity. Thus, single-Cys residues in either the inward- or outward-facing hydrophilic cavities appear to be accessible to water during the conformational fluctuation between the two conformations. It is also highly noteworthy that the X-ray structure of LacY exhibits no re-entrance loops. Thus, the accessibility of the cytoplasmic halves of many of the helices to MTSES cannot be attributed to such a structural feature.

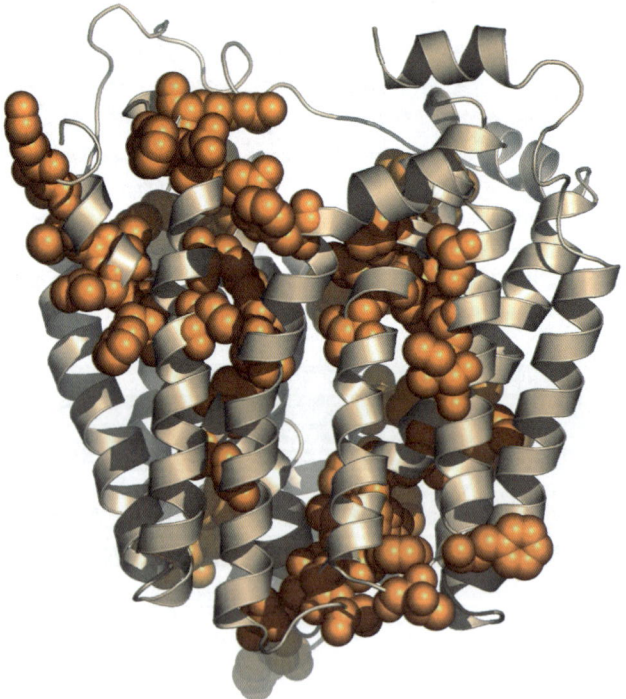

Figure 4 *Accessibility of Cys-replacement mutants to water. Individual Cys-replacement mutants were labeled with N- [^{14}C]ethylmaleimide before and after treatment with MTSES, an impermeant thiol reagent. Viewed parallel to the membrane*

The monomer is composed of 12 transmembrane helices (30), as predicted by hydropathy plots (15) and alkaline phosphatase fusion analyses (16). The N- and C-terminal six helices form two distinct bundles connected by a long cytoplasmic loop between helices VI and VII (Figures 3 & 5). The N- and C-terminal six-helix domains have a similar topology and are related by a pseudo two-fold axis of symmetry, as proposed for oxalate/formate antiporter (OxlT) and other members of the MFS (36,37). A hydrophilic cavity is formed between helices I, II, IV and V of the N-terminal domain and helices VII, VIII, X and XI of the C-terminal domain, while helices III, VI, IX and XII are largely embedded in the bilayer, as suggested from intermolecular cross-linking experiments (12). Kinks in helices I and VII, as well as IV and X, are observed, indicating that these helices may play an important role in the conformation change(s) between the outward-facing and inward-facing conformations during turnover.

It is highly noteworthy that the 3-D structure of another member of the MFS, the wild-type P_i/glycerol-3-P antiporter (GlpT), was solved at the same time as LacY (37). Although the two proteins have little sequence homology and catalyze completely different translocation reactions, their overall fold is similar. In addition, the 2-D structure of the OxlT, a third member of the MFS, also exhibits similar features, suggesting that the members of the MFS may all have a similar fold (36).

Most recently, wild-type LacY has yielded crystals after more than a decade of effort (L. Guan & HRK, unpublished data). Although the crystals are difficult to obtain because wild-type LacY tends to aggregate very readily, and they diffract to

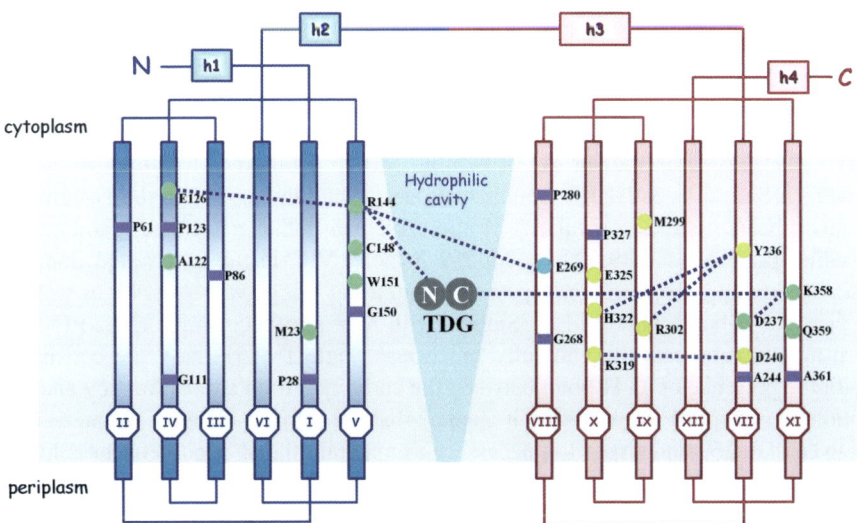

Figure 5 *Secondary structure schematic of LacY. The N- and C-terminal portions of the enzyme are colored in blue and red, respectively. Residues at the kinks in the transmembrane helices are marked with purple squares; residues marked with green and yellow circles are involved in substrate binding and proton translocation, respectively; while residue E269, colored aqua, is involved in both substrate binding and H+ translocation. The large hydrophilic cavity is designated by a blue triangle and TDG is depicted as two black circles labeled N and C*

only 4 Å, the data obtained is sufficient to model the backbone. Importantly, the overall fold of wild-type LacY exhibits an inward-facing conformation and is very similar, if not identical, to that of the C154G mutant. Since wild-type GlpT and LacY, as well as C154G LacY, have very similar overall folds and are in the inward-facing conformation, this conformation likely reflects the lowest free-energy form. However, it is not clear whether this is the preferred form in the membrane or in a crystal lattice. In any event, it is noteworthy that the hydrophilic cavity is sufficiently large that it is visualized by freeze-fracture electron microscopy of proteoliposomes reconstituted with purified LacY (38,39) or in filamentous arrays of LacY (40).

4   The Substrate-binding Site

A primary interaction is found between the irreplaceable residue Arg144 (helix V) and the O_3 and O_4 atoms of the galactopyranosyl ring *via* a bidentate H-bond (30), as suggested by biochemical findings (Figure 6) (41–44). Another irreplaceable residue Glu126 (helix IV) is in near proximity to Arg144 and may interact with the O_4, O_5 or O_6 atoms of galactopyranosyl ring *via* water molecules. Although a proposed salt bridge between Arg144 and Glu126 (45,46) is not observed, such an interaction may be present in the absence of ligand (24) or in another conformational intermediate. Interestingly, a hydrophobic interaction is observed between the bottom of the galactopyranosyl ring and the indole ring of Trp151 (helix V), as proposed (47). Fluorescence studies (48) support the conclusion from the structure that Trp151 is in a hydrophilic environment. Furthermore, phosphorescence studies in the absence and presence of ligand verify a direct interaction between the galactopyranosyl and indole rings. The C-6 atom of the galactopyranosyl ring also appears to interact hydrophobically with Met23 (helix I). A similar interaction between the C_6 atom of TDG and a His side chain is observed in the structure of an *E. coli* enterotoxin (49). However, mutant M23A exhibits the same binding affinity as wild-type LacY (I. Smirnova & HRK, unpublished data). The binding-site in the N-terminal domain bears a striking similarity to those of many other galactoside- and sugar-binding proteins (see 49,50,51). Glu269 in helix VIII in the C-terminal domain, another irreplaceable residue, appears to form a salt bridge with Arg144, as well as a possible H-bond with Trp151. Studies with *N*-bromosuccinamide (NBS) (52), a Trp-modification reagent, and fully functional single-Trp151 LacY are consistent with the presence of an H-bond between the carboxyl group at position 269 and the indole N of Trp151. Importantly, it appears that the primary charge-pair interaction between Glu269 and Arg144 is necessary to maintain the H-bond between Glu269 and Trp151, which acts to keep Trp151 in an optimal orientation. It has also been proposed that Glu269 is involved in both ligand binding and H^+ translocation (23,24,53–55). Thus, it is likely that contacts between Glu269 in the C-terminal domain and Arg144 and Trp151 in the N-terminal domain may be key to providing the important energetic link between the two helical bundles.

Fewer interactions are observed with the sugar and residues in the C-terminal domain. Helices VII (Asp237) and XI (Lys358) (Figure 6A), which are symmetrically related to the helices I and V, respectively, are also involved in TDG binding. However, it seems likely that these residues play a supporting role relative to the

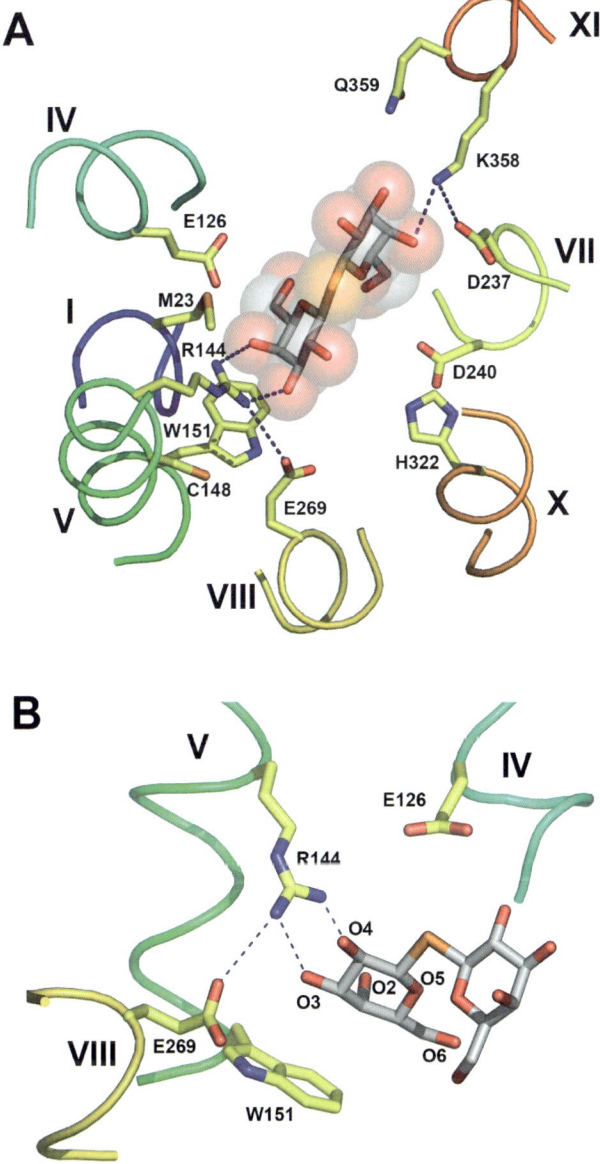

Figure 6 *Substrate binding site of LacY. Possible H-bonds and salt bridges are represented by blue broken lines. **A.** Residues involved in TDG binding viewed along the membrane normal from the cytoplasmic side. TDG is depicted as a stick model with a CPK model superimposed. **B.** Close up of the N-terminal domain of the TDG binding site*

N-terminal primary binding site by providing additional affinity for disaccharide substrates. This explains why the monosaccharide galactose has poor affinity for LacY, but behaves like any other substrate with respect to protecting Cys148 against alkylation (56). It is critical to understand that galactose is the most specific substrate for

LacY, but has very low affinity, which is increased markedly by various adducts, particularly if they are hydrophobic, at the anomeric carbon (56). In contrast to Arg144, which is absolutely required, the charge pair between Asp237 and Lys358 is interchangeable, and both can be replaced simultaneously by neutral side chains with little effect on activity (57–59). Therefore, their interactions with ligand are not absolutely required. The essential portion of the substrate-binding site with respect to specificity is in the N-terminal domain, and the residues in the C-terminal domain that interact with the adduct on the anomeric carbon of galactopyranoside increase affinity, but have little to do with specificity. Furthermore, although the C_2, C_3, and C_6 OH groups on the galactopyranosyl ring play roles in H-bonding, the C_4 OH is unequivocally the most important determinant for specificity (60). The hydrophobic interaction between the galactopyranosyl ring and Trp151 is likely to orient the ring so that important H-bonds can be realized (47). However, although the structure is consistent with and explains many biochemical observations, higher-resolution structures with different bound sugars are required to understand ligand binding in atomic detail.

5   Residues Involved in H$^+$ Translocation and Coupling

One fundamentally important problem that requires solution is the identification of H$^+$ binding site(s) and the mechanism of coupling between sugar and H$^+$ translocation. Several lines of evidence indicate that LacY is protonated prior to ligand binding (44). Most recently, it has been shown directly (J. Vazquez-Ibar, S. Schuldiner & HRK, unpublished) that addition of TDG to a concentrated solution of purified, detergent-solubilized LacY induces no change in pH, while a positive control with the antiporter EmrE (provided by Shimon Schuldiner, Hebrew University, Jerusalem) under identical conditions releases about 1 H$^+$mol^{-1} EmrE upon addition of tetraphenylphosphonium (61).

In the structure, a complex salt bridge/H-bond network is observed (Figure 7), which is composed of residues from helix VII (Tyr236 and Asp240), helix X (Lys319, His322 and Glu325) and helix IX (Arg302). Since extensive biochemical/molecular biological analysis indicates that His322, Glu325 and Arg302 are directly involved in H$^+$ translocation, it is noteworthy that Glu325 is embedded in a hydrophobic milieu formed by Met299 and Ala295 (Helix IX), Leu329 (Helix X) and Tyr236 (Helix VII), which is consistent with the notion that Glu325 is protonated in this conformation (26). Protonated Glu325 also appears to be partially stabilized by an H-bond to the Sδ atom of Met299; a similar Glu–Met interaction that stabilizes protonation of a glutamic acid residue has been reported for the D-proton pathway of bovine cytochrome c oxidase (62). Therefore, the structure represents the protonated inward-facing conformation with bound substrate. It has been suggested (63) that Arg302 could interact with Glu325 to drive H$^+$ release. However, in the current structure, the side chain of Arg302 is *ca*. 7 Å away from Glu325, which would require a large side-chain rearrangement of Arg302 to form a salt bridge. The structural data combined with biochemical/molecular biological studies (see 26) provide support for the suggestion that His322 may be the immediate H$^+$ donor to

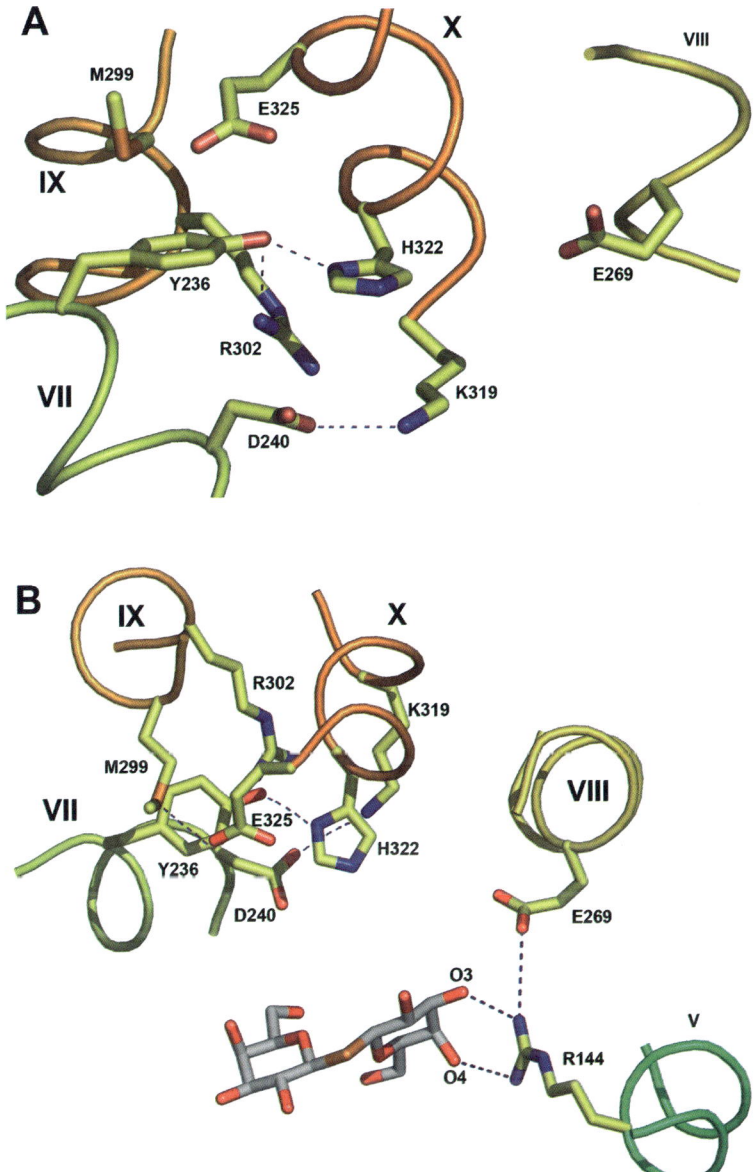

Figure 7 *Residues involved in H$^+$ translocation and coupling. H-bonds are represented by blue broken lines. **A.** View parallel to the membrane. **B.** View along the membrane normal from the cytoplasmic side*

Glu325. Since mutants with simultaneous neutral replacements for Asp240 and Lys319 maintain significant transport activity (57,58), it is unlikely that this salt bridge is directly involved in H$^+$ translocation; however, the two residues could be involved in regulation and/or stabilization of the salt bridge/H-bond network of the residues. Interestingly, when Asp240 and Lys319 are replaced with Cys residues and then cross-linked, transport is abolished (35).

The closest distance between this network and the sugar binding site is more than 6 Å, indicating that the network does not directly interact with the sugar binding site in the inward-facing conformation. Glu269 is involved in substrate binding, as discussed, and is also in proximity to His322 (closest distance 5.8 Å) (in addition, see (64–66). It is highly likely that Glu269 couples ligand binding and H^+ translocation since it is the only irreplaceable residue that appears to be involved in both sugar binding and H^+ translocation. It is also proposed that Glu269 makes direct contact with His322 in another conformation. Obviously, higher resolution structures in other conformations are required to address these important proposals.

6   Proposed Mechanism of Lactose/ H^+ Symport

The mechanism of lactose/ H^+ symport can be explained by a simple kinetic scheme (see 26,30). Briefly, influx consists of six steps: starting from the outward-facing conformation (A); protonation of LacY (B); binding of lactose (C); a conformational change that results in the inward conformation (D); release of substrate (E); release of the H^+ (F); return to the outward conformation. As discussed above, the structure corresponds to the protonated, inward-facing conformation with bound substrate (see Figure 8D).

LacY in the outward-facing conformation might be very unstable (Figure 8A) and protonated immediately as postulated (Figure 8B) (26). In this state, the H^+ is on Glu269 or shared between Glu269 and His322. Ligand is recognized initially by Trp151, Arg144 and Glu126, which may induce H^+ transfer to His322 and subsequently to Glu325 as Glu269 is recruited to complete the binding site by H-bonding to Arg144 and Trp151 (Figure 8C). This process may also trigger transition to the inward conformation (Figure 8D). Substrate is then released into the cytoplasm (Figure 8E), followed by release of the H^+ from Glu325 (Figure 8F) due probably to a decrease in *pKa* caused either by approximation to Arg302 (63) or exposure to solvent in the aqueous cavity (cytoplasmic pH is constant at 7.6). Several lines of evidence indicate that the H^+ is released from Glu325 (see 26). After releasing the H^+ inside, transition back to the outward-facing conformation is induced.

The structure of LacY exhibits a single sugar-binding site at the apex of a hydrophilic cavity open to the cytoplasm, and it has been postulated from the structure (30) that the binding site has alternating access to either side of the membrane during turnover. However, it is not clear whether $\Delta\bar{\mu}_{H^+}$ changes binding affinity, particularly with LacY, as it has been shown that $\Delta\Psi$ and ΔpH have quantitatively the same kinetic (67) and thermodynamic (68,69) effects on transport.

Although substrate protection against alkylation of Cys148 by *N*-ethylmaleimide is particularly useful for obtaining K_Ds of LacY for various substrates over a wide range of concentrations (43,44,56,60,70–72), it is difficult to obtain true K_D values on each side of the membrane for a transport protein in the presence of $\Delta\bar{\mu}_{H^+}$ because the ligands used are translocated across the membrane and may accumulate in right-side-out (RSO) vesicles thereby leading to underestimation of K_D. However, in a recent series of experiments (73), lactose or TDG protection of Cys148 against alkylation by

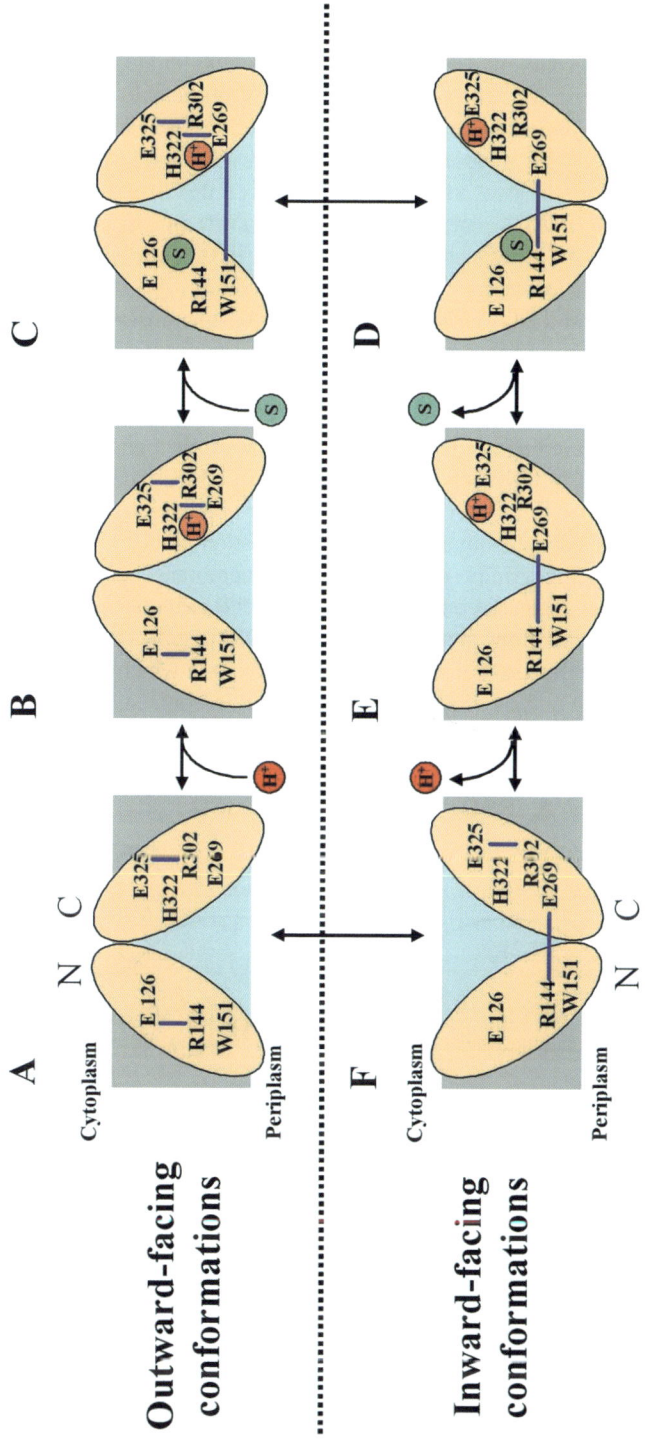

Figure 8 *A possible lactose/H⁺ symport mechanism. Key residues are labeled and H-bonds are shown with blue lines. The H⁺ and the substrate are shown in red and green circles, respectively; and the hydrophilic cavity is colored aqua*

N-ethylmaleimide were carried out on ice, which decreases substrate accumulation drastically (74). Under these experimental conditions, in the absence of $\Delta\bar{\mu}_{H^+}$, both ice-cold RSO and inside-out (ISO) vesicles likely equilibrate with the external medium. In the presence of $\Delta\bar{\mu}_{H^+}$, RSO vesicles may still be able to accumulate lactose or TDG two- to three- fold even though the reactions are carried out on ice for only 5 min. Therefore, the measured K_Ds for RSO vesicles in the presence of $\Delta\bar{\mu}_{H^+}$ may be underestimated by two- to three- fold. However, this is unlikely, as ISO vesicles in the presence of ATP generate a $\Delta\bar{\mu}_{H^+}$ of opposite polarity (interior positive and/or acid) (Figure 9) (75) which causes a decrease in the intravesicular concentration of ligand relative to the concentration in the medium (76). Remarkably, results with both lactose and TDG demonstrate that the K_D manifested by ISO vesicles exhibits less than a two-fold change in the absence or presence of $\Delta\bar{\mu}_{H^+}$. Moreover, the K_D values observed with RSO or ISO vesicles in the absence or presence of $\Delta\bar{\mu}_{H^+}$ are similar within experimental error. The results provide a strong indication that $\Delta\bar{\mu}_{H^+}$ has little or no effect on binding affinity, a conclusion that raises a number of interesting considerations regarding the mechanism by which $\Delta\bar{\mu}_{H^+}$ drives accumulation.

In the presence of $\Delta\bar{\mu}_{H^+}$ (interior negative and/or alkaline), wild-type LacY can accumulate lactose against about a 100-fold concentration gradient. Without a significant decrease in binding affinity on the inside of the membrane, how does $\Delta\bar{\mu}_{H^+}$ drive lactose accumulation against a concentration gradient? Based on the effect of D_2O on various translocation reactions, it was postulated (26,77) that the rate-limiting step for downhill transport in the *absence* of $\Delta\bar{\mu}_{H^+}$ is deprotonation which precedes return of unloaded LacY to the outer surface of the membrane; in contrast, in the *presence* of

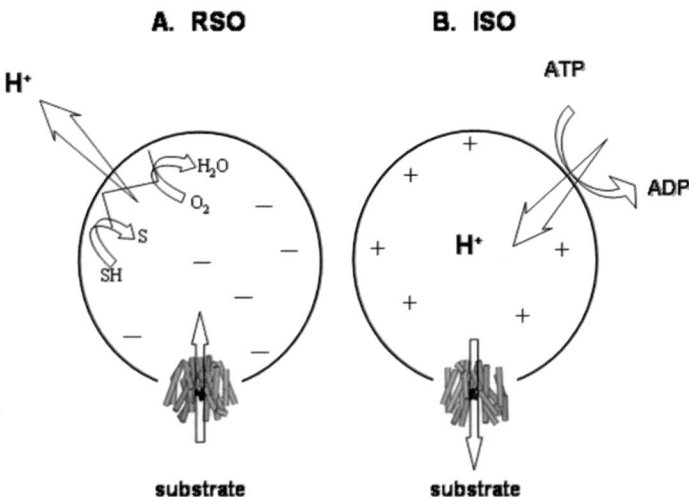

Figure 9 *Effect of $\Delta\bar{\mu}_{H^+}$ of opposite polarities on substrate translocation in RSO or ISO vesicles. A. $\Delta\bar{\mu}_{H^+}$ with RSO generated by addition of D-lactate ($\Delta\Psi$, interior negative); substrate is accumulated. B. $\Delta\bar{\mu}_{H^+}$ in ISO vesicles generated by addition of 10 mM Mg(II)ATP or 20 mM lithium D-lactate under oxygen (interior positive and/or acid), which is opposite to that of RSO vesicles; substrate effluxes from the vesicles*

$\Delta\bar{\mu}_{H^+}$, dissociation of sugar is rate-limiting. It is also noteworthy that the primary kinetic effect of $\Delta\Psi$ or ΔpH on transport is a dramatic decrease in K_m (67). Therefore, it seems reasonable to suggest that $\Delta\bar{\mu}_{H^+}$ enhances the rate of deprotonation on the inner surface of the membrane, and thereby allows unloaded LacY to return to the outward-facing conformation more rapidly. Thus, the major effect of $\Delta\bar{\mu}_{H^+}$ on active transport by LacY appears to be kinetic with little or no change in affinity for sugar.

Although biochemical and biophysical studies, as well as a single structure at a 3.5 Å resolution, can explain how the overall conformational change in LacY may be coupled to sugar binding and H^+ translocation, many fundamental questions remain: Why is protonation of LacY important for sugar binding? What is the detailed mechanism of coupling between binding and H^+ translocation? How does a lactose gradient directed inward or outward drive H^+ translocation through the same pathway? What is the time of occupancy of LacY in the outward-facing and inward-facing conformations? In addition, higher-resolution atomic detail regarding ligand–protein interactions is needed. Therefore, it is essential to obtain higher-resolution structures in different conformations, as well as dynamics, in order to fully understand the mechanism of substrate/H^+ symport by LacY.

Acknowledgements

I am indebted to Vladimir Kasho for preparing the figures used here. This work was supported in part by NIH Grant DK51131:09 to HRK.

References

1. P. Mitchell, *Biochem. Soc. Symp.* 1963, **22**, 142.
2. P. Mitchell, *Chemiosmotic coupling and energy transduction*, Glynn Research Ltd, Bodmin, England, 1968.
3. H.R. Kaback, *J. Cell Physiol.*, 1976, **89**, 575.
4. H.R. Kaback, *J. Membr Biol.*, 1983, **76**, 95.
5. H.R. Kaback, *Harvey Lectures*, 1989, **83**, 77.
6. B. Müller-Hill, *The lac Operon: A Short History of a Genetic Paradigm*, Walter de Gruyter, Berlin, New York, 1996.
7. R.M. Teather, B. Müller-Hill, U. Abrutsch, G. Aichele and P. Overath, *Molec. Gen. Genet.,* 1978 **159**, 239.
8. D.E. Büchel, B. Gronenborn and B. Müller-Hill, *Nature,* 1980, **283**, 541.
9. M.J. Newman, D.L. Foster, T.H. Wilson and H.R. Kaback, *J. Biol. Chem.,* 1981, **256**, 11804.
10. P. Viitanen, M.L. Garcia and H.R. Kaback, *Proc. Natl. Acad. Sci. U.S.A.,* 1984, **81**, 1629.
11. M. Sahin-Tóth, M.C. Lawrence and H.R. Kaback, *Proc. Natl. Acad. Sci. U.S.A.,* 1994, **91**, 5421.
12. L. Guan, F.D. Murphy and H.R. Kaback, *Proc. Natl. Acad. Sci. U.S.A.,* 2002, **99**, 3475.
13. N. Ermolova, L. Guan and H.R. Kaback, *Proc. Natl. Acad. Sci. U.S.A.,* 2003, **100**, 10187.

14. M.H. Saier, Jr. *Mol. Microbiol.,* 2000, **35**, 699.
15. D.L. Foster, M. Boublik and H.R. Kaback, *J. Biol. Chem.,* 1983, **258**, 31.
16. J. Calamia and C. Manoil, *Proc. Natl. Acad. Sci. U.S.A.,* 1990, **87**, 4937.
17. C. Wolin and H.R. Kaback, *Biochemistry,* 1999, **38**, 8590
18. C.D. Wolin and H.R. Kaback, *Biochemistry,* 2001, **40**, 1996.
19. J.P. Whitelegge, C.B. Gundersen and K.F. Faull, *Protein Science,* 1998, **7**, 1423.
20. J.P. Whitelegge, J. le Coutre, J.C. Lee, C.K. Engel, G.G. Privé, K.F. Faull and H.R. Kaback, *Proc. Natl. Acad. Sci. U.S.A.,* 1999, **96**, 10695.
21. J. le Coutre, J.P. Whitelegge, A. Gross, E. Turk, E.M. Wright, H.R. Kaback and K.F. Faull, *Biochemistry,* 2000, **39**, 4237.
22. J. le Coutre and H.R. Kaback, *Biopolymers,* 2000, **55**, 297.
23. A.B. Weinglass, J.P. Whitelegge, Y. Hu, G.E. Verner, K.F. Faull and H.R. Kaback, *EMBO J.* 2003, **22**, 1467.
24. A. Weinglass, J.P. Whitelegge, K.F. Faull and H.R. Kaback, *J. Biol. Chem.,* 2004, **279**, 41858.
25. R. Ehring, K. Beyreuther, J.K. Wright and P. Overath, *Nature,* 1980, **283**, 537.
26. H.R. Kaback, M. Sahin-Toth and A.B. Weinglass, *Nat. Rev. Mol. Cell. Biol.,* 2001, **2**, 610.
27. S. Frillingos, M. Sahin-Toth, J. Wu and H.R. Kaback, *Faseb. J.,* 1998, **12**, 1281.
28. H.R. Kaback and J. Wu, *Quart. Rev. Biophys.,* 1997, **30**, 333.
29. P.L. Sorgen, Y. Hu, L. Guan, H.R. Kaback and M.E. Girvin, *Proc. Natl. Acad. Sci. U.S.A.,* 2002, **99**, 14037.
30. J. Abramson, I. Smirnova, V. Kasho, G. Verner, H.R. Kaback and S. Iwata, *Science,* 2003, **301**, 610.
31. W.F. Widdas, *J. Physiol.,* 1952, **118**, 23.
32. A. Karlin and M.H. Akabas, *Methods Enzymol.,* 1998, **293**, 123.
33. I. Kwaw, K.C. Zen, Y. Hu and H.R. Kaback, *Biochemistry,* 2001, **40**, 10491.
34. W. Zhang, Y. Hu and H.R. Kaback, *Biochemistry,* 2003, **42**, 4904.
35. W. Zhang, L. Guan and H.R. Kaback, *J. Mol. Biol.,* 2002, **315**, 53.
36. T. Hirai, J.A. Heymann, P.C. Maloney and S. Subramaniam, *J. Bacteriol.,* 2003, **185**, 1712.
37. Y. Huang, M.J. Lemieux, J. Song, M. Auer and D.N. Wang, *Science,* 2003, **301**, 616.
38. M.J. Costello, P. Viitanen, N. Carrasco, D.L. Foster and H.R. Kaback, *J. Biol. Chem.* 1984, **259**, 15579.
39. M.J. Costello, J. Escaig, K. Matsushita, P.V. Viitanen, D.R. Menick and H.R. Kaback, *J. Biol. Chem.,* 1987, **262**, 17072.
40. J. Li and P. Tooth, *Biochemistry,* 1987, **26**, 4816.
41. S. Frillingos, A. Gonzalez and H.R. Kaback, *Biochemistry,* 1997, **36**, 14284.
42. P. Venkatesan, Y. Hu and H.R. Kaback, *Biochemistry,* 2000, **39**, 10656.
43. M. Sahin-Tóth, J. le Coutre, D. Kharabi, G. le Maire, J.C. Lee and H.R. Kaback, *Biochemistry,* 1999, **38**, 813.
44. M. Sahin-Tóth, A. Karlin and H.R. Kaback, *Proc. Natl. Acad. Sci. U.S.A.,* 2000, **97**, 10729.
45. C.D. Wolin and H.R. Kaback, *Biochemistry,* 2000, **39**, 6130.
46. M. Zhao, K.C. Zen, W.L. Hubbell and H.R. Kaback, *Biochemistry,* 1999, **38**, 7407.
47. L. Guan, Y. Hu and H.R. Kaback, *Biochemistry,* 2003, **42**, 1377.

48. J.L. Vazquez-Ibar, L. Guan, M. Svrakic and H.R. Kaback, *Proc. Natl. Acad. Sci. U.S.A.*, 2003, **100**, 12706.

49. E.A. Merritt, S. Sarfaty, I.K. Feil and W.G. Hol, *Structure*, 1997, **5**, 1485.

50. T.K. Sixma, S.E. Pronk, K.H. Kalk, B.A. van Zanten, A.M. Berghuis and W.G. Hol, *Nature*, 1992, **355**, 561.

51. F.A. Quiocho and N.K. Vyas, S.M. Hecht, (ed), *Bioorganic Chemistry: Carbohydrates*, Oxford University Press, Oxford, England, 1999, 441.

52. J.L. Vazquez-Ibar, L. Guan, A.B. Weinglass, G. Verner, R. Gordillo and H.R. Kaback, *J. Biol. Chem.*, 2004, **279**, 49214.

53. M.L. Ujwal, M. Sahin-Tóth, B. Persson and H.R. Kaback, *Mol. Membr. Biol.*, 1994, **11**, 9.

54. P.J. Franco and R.J. Brooker, *J. Biol. Chem.*, 1994, **269**, 7379.

55. A.B. Weinglass, M. Sondej and H.R. Kaback, *J. Mol. Biol.*, 2002, **315**, 561.

56. M. Sahin-Tóth, K.M. Akhoon, J. Runner and H.R. Kaback, *Biochemistry*, 2000, **39**, 5097.

57. M. Sahin-Tóth, R.L. Dunten, A. Gonzalez and H.R. Kaback, *Proc. Natl. Acad. Sci. USA*, 1992, **89**, 10547.

58. M. Sahin-Tóth and H.R. Kaback, *Biochemistry*, 1993, **32**, 10027.

59. R.L. Dunten, M. Sahin-Tóth and H.R. Kaback, *Biochemistry*, 1993, **32**, 3139.

60. M. Sahin-Tóth, M.C. Lawrence, T. Nishio and H.R. Kaback, *Biochemistry*, 2001, **43**, 13015.

61. M. Soskine, Y. Adam and S. Schuldiner, *J. Biol. Chem.*, 2004, **279**, 9951.

62. T. Tsukihara, H. Aoyama, E. Yamashita, T. Tomizaki, H. Yamaguchi, K. Shinzawa-Itoh, R. Nakashima, R. Yaono and S. Yoshikawa, *Science*, 1996, **272**, 1136.

63. M. Sahin-Tóth and H.R. Kaback, *Proc. Natl. Acad. Sci. U.S.A.*, 2001, **98**, 6068.

64. K. Jung, H. Jung, J. Wu, G.G. Privé and H.R. Kaback, *Biochemistry*, 1993, **32**, 12273.

65. K. Jung, J. Voss, M. He, W.L. Hubbell and H.R. Kaback, *Biochemistry*, 1995, **34**, 6272.

66. M. He and H.R. Kaback, *Biochemistry*, 1997, **36**, 13688.

67. D.E. Robertson, G.J. Kaczorowski, M.L. Garcia and H.R. Kaback, *Biochemistry*, 1980, **19**, 5692.

68. S. Ramos, S. Schuldiner and H.R. Kaback, *Proc. Natl. Acad. Sci. U.S.A.*, 1976, **73**, 1892.

69. S. Ramos and H.R. Kaback, *Biochemistry*, 1977, **16**, 854.

70. S. Frillingos and H.R. Kaback, *Biochemistry*, 1996, **35**, 3950.

71. P. Venkatesan and H.R. Kaback, *Proc. Natl. Acad. Sci. U.S.A.*, 1998, **95**, 9802.

72. M. Sahin-Tóth, P. Gunawan, M.C. Lawrence, T. Toyokuni and H.R. Kaback, *Biochemistry*, 2002, **41**, 13039.

73. L. Guan and H.R. Kaback, *Proc. Natl. Acad. Sci. U.S.A.*, 2004, **101**, 12148.

74. H.R. Kaback and E.M. Barnes, Jr. *J. Biol. Chem.* 1971, **246**, 5523.

75. W.W. Reenstra, L. Patel, H. Rottenberg and H.R. Kaback, *Biochemistry*, 1980, **19**, 1.

76. G.J. Kaczorowski, D.E. Robertson and H.R. Kaback, *Biochemistry*, 1979, **18**, 3697.

77. P. Viitanen, M.L. Garcia, D.L. Foster, G.J. Kaczorowski and H.R. Kaback, *Biochemistry*, 1983, **22**, 2531.

CHAPTER 16

Hydride Transfer and Proton Translocation by Nicotinamide Nucleotide Transhydrogenase

J. BAZ JACKSON, SCOTT A. WHITE AND
T. HARMA C. BRONDIJK

School of Biosciences,
University of Birmingham,
Edgbaston,
Birmingham, B15 2TT,
UK

Abbreviations

Nic^+, nicotinamide;
NicH, dihydronicotinamide;
Nic(H) nicotinamide *and* dihydronicotinamide;
NAD^+, nicotinamide adenine dinucleotide (oxidised form);
NADH, nicotinamide adenine dinucleotide (reduced form);
NAD(H), oxidised *and* reduced forms;
$NADP^+$, nicotinamide adenine dinucleotide phosphate (oxidised form), *etc.*

Residue Numbering

Unless otherwise stated, amino acid residues are given for the *Rhodospirillum rubrum* transhydrogenase according to the system described in ref (1).

1 The Function of Proton-Translocating Transhydrogenase

'Energy-linked' transhydrogenase in animal mitochondria was discovered in 1963 by Danielsson and Ernster (2). Mitchell suggested that transhydrogenase is, in fact, a proton translocator (3,4).

$$\text{NADP}^+ + \text{NADH} + \text{H}^+_{\text{out}} \Leftrightarrow \text{NADPH} + \text{NAD}^+ + \text{H}^+_{\text{in}} \tag{1}$$

Thus, the enzyme can be driven from left to right using the proton electrochemical gradient (Δp) generated by the respiratory chain, or it can operate from right to left using the redox potential difference between NADPH and NAD$^+$ to supplement Δp formation. In bacteria the equivalent enzyme was shown to contribute to the production of NADPH for use in biosynthesis and for the reduction of glutathione needed in defence against free-radical damage (5–7). Experimental work with animal mitochondria is more difficult. It was suggested that the function of transhydrogenase might be tissue specific and that the enzyme can act as a 'buffer system', either utilising Δp to drive the formation of NADPH for biosynthesis and glutathione reduction when demand is high, or generating Δp during anoxic conditions (8).

Mitochondria have an NAD-dependent isocitrate dehydrogenase (NAD-ICDH) which is thought to operate in the TCA cycle. They also have an NADP-dependent isocitrate dehydrogenase (NADP-ICDH) that is invariably overlooked in modern textbooks of biochemistry. Why do mitochondria need both a transhydrogenase and an NADP-ICDH to produce NADPH in the same cellular compartment? The question is particularly relevant to mitochondria in heart and other muscle types, where the two enzymes are highly active, but where the biosynthetic capacity of the cells is very low and there is, therefore, only a modest demand for NADPH. In answer to the question we suggested that transhydrogenase and NADP-ICDH have an additional and combined function. Together with the NAD-ICDH, they catalyse a 'micro-cycle' between isocitrate (IC) and 2-oxoglutarate (2-OG), whose purpose is to improve the fine regulation of flux through this segment of the Krebs cycle (Figure 1) (9). It was shown that the Δp generated by the respiratory chain is large enough for transhydrogenase to elevate the mitochondrial NADPH/NADP$^+$ ratio and thus drive the NADP-ICDH in 'reverse', from 2-OG to IC. Forward flux from IC to 2-OG, as widely recognised, is carried by NAD-ICDH. As explained for ATP-dependent 'substrate cycles' (*e.g.* the classical phosphofructokinase/fructose 1,6 *bis*-phosphatase cycle), this organisation increases the sensitivity of net IC $\rightarrow$ 2-OG conversion to the allosteric effectors of NAD-ICDH (10). The device also provides a controlling link between Δp and the Krebs cycle flux (increased demand for ATP is matched, through a fall in Δp, by an increase in net flux). The 'metabolic price to pay' for this improved control is consumption of Δp by transhydrogenase.

2   The Global Architecture of Transhydrogenase

Although there are differences in polypeptide composition, the global architecture of the transhydrogenase protein is essentially similar in different organisms (the enzymes from mammalian mitochondria, *Escherichia coli*, *Rhodospirillum rubrum* and *Entamoeba histolytica* have been studied in some detail) – see Figure 2. There is a dI component, which binds NAD(H), and a dIII component, which binds NADP(H); both of these protrude from the membrane (into the cytoplasm of bacteria and the matrix of mitochondria). A dII component spans the membrane. Transhydrogenase is a 'dimer' of two dI-dII-dIII 'monomers'. X-ray and NMR structures for recombinant dI and dIII (in isolation and in complex) have been solved

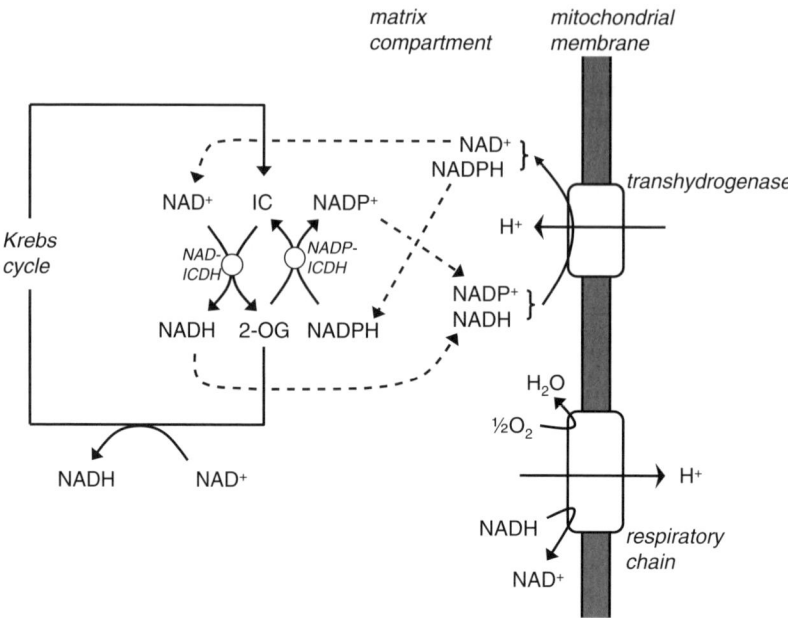

Figure 1 *Control of Krebs cycle flux by transhydrogenase. IC = iso-citrate, 2-OG = 2 oxog-lutarate, NAD-ICDH = NAD-linked iso-citrate dehydrogenase, NADP-ICDH = NADP-linked iso-citrate dehydrogenase. Other enzymes of the Krebs cycle are not shown. Dashed lines indicate nucleotide diffusion between transhydrogenase, NAD-ICDH and NADP-ICDH*

for several species with various bound ligands (1,11–18) but there is no available high-resolution structure for either isolated dII or the intact enzyme.

3   The dII Component

The membrane-spanning dII is the least understood of the three components of transhy-drogenase. Nevertheless, mutagenesis and cross-linking studies have provided useful preliminary information on the organisation of transmembrane helices in dII and on the mechanism of proton translocation through the protein (reviewed in (19–22)). Depending on the species, dII has between 12 and 14 transmembrane helices per monomer (Figure 3). In animal mitochondria there are 14 (23,24), in the *E. coli* type of transhydrogenase (25), which is expressed as two polypeptides, helix 5 is missing and, in the *R. rubrum* type, which is expressed as three polypeptides, helix 1 is also missing (24,26). The majority of conserved amino-acid residues in dII are located in helices 2–4, 9, 10, 13 and 14 and are disposed mainly towards the side of the membrane bearing the dI and dIII components. They may participate either in proton translocation or in the energy-trans-duction mechanism. Extensive cysteine labelling and cross-linking studies in the *E. coli* enzyme have led to models for the clustering of transmembrane helices (25,27–30).

All polar and most conserved amino-acid residues in the putative transmembrane helices of *E. coli* dII have been mutated and the effects of these mutations on proton translocation and 'reverse' transhydrogenation (the reduction of an NAD$^+$ analogue

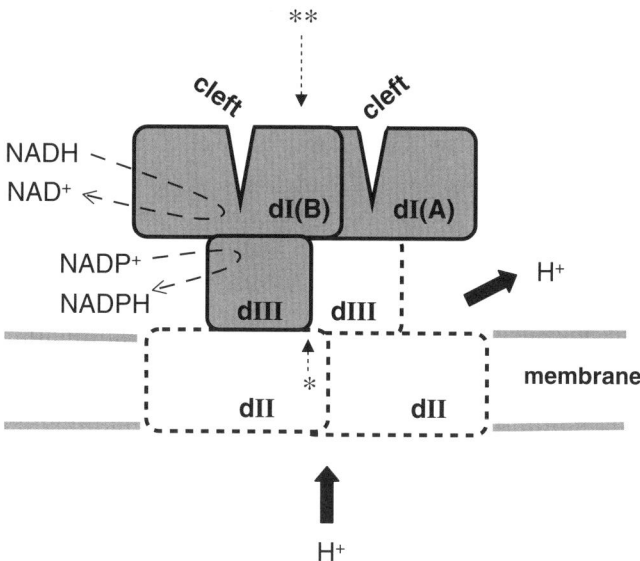

Figure 2 *Cartoon showing the global architecture of transhydrogenase. The grey-shaded regions represent the dI₂dIII₁ complex. The predicted organisation of components that are present in the intact enzyme but absent in the dI₂dIII₁ complex is shown by the dotted lines. The dashed arrows depict forward transhydrogenation and the thick solid arrows, the corresponding direction of proton translocation. The starred dotted arrows indicate the views described in Figure 4*

by NADPH, see equation 1) have been studied (31–37). Surprisingly few residues are essential. Mutations of βHis91 in helix 9, βN222 in helix 13, βG252 in helix 14 and, to a lesser extent, of βE85 in helix 9 and βS139 in helix 10 are the most notable. Much significance has been attached to the possibility that, through its acid-base properties, βHis91 might be important in proton translocation, but recent sequences show that the equivalent residue is frequently an asparagine (20% of ~160 sequences). It is looking increasingly likely that proton transfer through dII may be dependent on chains of bound water molecules. The existence of a water cavity in this component has been suggested (38).

The periplasmic loops between the putative TM helices of dII are short and quite variable in sequence. The cytoplasmic loops, some of which must interact with dIII, are more strongly conserved, especially those between helix 4 and 5, between helix 12 and 13 and in the so-called 'hinge' region between helix 14 and the dIII component. Mutation of βD213 in loop 12/13 and βR265 in the hinge region results in an enzyme that has a decreased rate of reverse transhydrogenation (33,39).

4   dI₂dIII₁ Complexes: Catalytic Properties and High-Resolution Structures

The structures and properties of the membrane-peripheral dI and dIII components of transhydrogenase have been investigated in some detail. Mixtures of recombinant dI

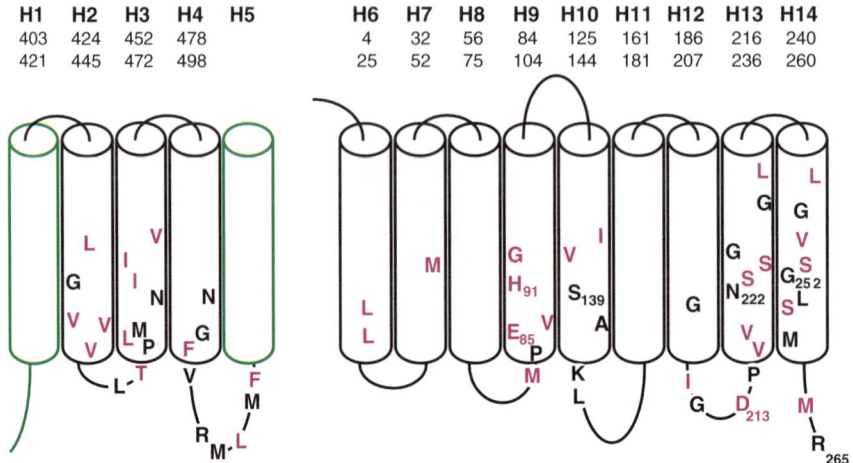

Figure 3 *Sequence conservation in the membrane-spanning dII. The putative transmembrane helices (H1–H14) are numbered on the top row according to the mammalian transhydrogenase system (23). The numbers on the rows underneath are of the N- and C-terminal residues of the E. coli helices (H1–H4 are in the α and H6–H14 are in the β subunit– see (25)). From sequence analysis, black helices appear to be transmembrane in all species; green helices do not. Residues in black are invariant in all (~160) available sequences; residues in purple are highly conserved. Numbered residues (from the E. coli sequence) are mentioned in the text. The dIII components attach to dII at the bottom of this figure*

and dIII readily form dI_2dIII_1 complexes in solution (40). A second dIII can bind (to give dI_2dIII_2) but with very low affinity (41). We discuss why, in the absence of dII, the asymmetric complex is more stable in Section 7. The dI_2dIII_1 complex catalyses fast hydride transfer between bound nucleotides (the apparent first-order rate constant for the forward reaction is $\geq 2.1 \times 10^4$ s^{-1} and, for the reverse, ≤ 590 s^{-1} (42,43). NAD(H) can rapidly associate with, and dissociate from, its binding site in the complex but NADP(H) release from its site is extremely slow, thus effectively limiting the hydride transfer reaction to a single turnover (42,44–47).

Several X-ray structures of the dI_2dIII_1 complex (from *R. rubrum*) have been solved (1,14,17) (see Figures 2 and 4). All have the same subunit organisation but there are important detailed variations in the configuration of the hydride-transfer site (see below). The two dI polypeptides, dI(A) and dI(B), each have two domains (denoted dI.1 and dI.2, both Rossmann folds) separated by a cleft and linked by two long helices. They are related by a two-fold axis of symmetry and interact with one another mainly through the helices of the dI.2 domains. The single dIII polypeptide also has the form and connectivity of the Rossmann fold. It interacts predominantly with dI(B). A binding site on dI(A) for a second dIII can be identified from the structure but computer modelling shows that conformational changes would be necessary for docking.

NADP(H) binds at the C-terminal edge of the β-sheet of dIII. Its orientation is reversed relative to that found in other Rossmann folds. Thus, the adenosine moiety of the nucleotide is located in the second of the two β-α-β structure motifs and the

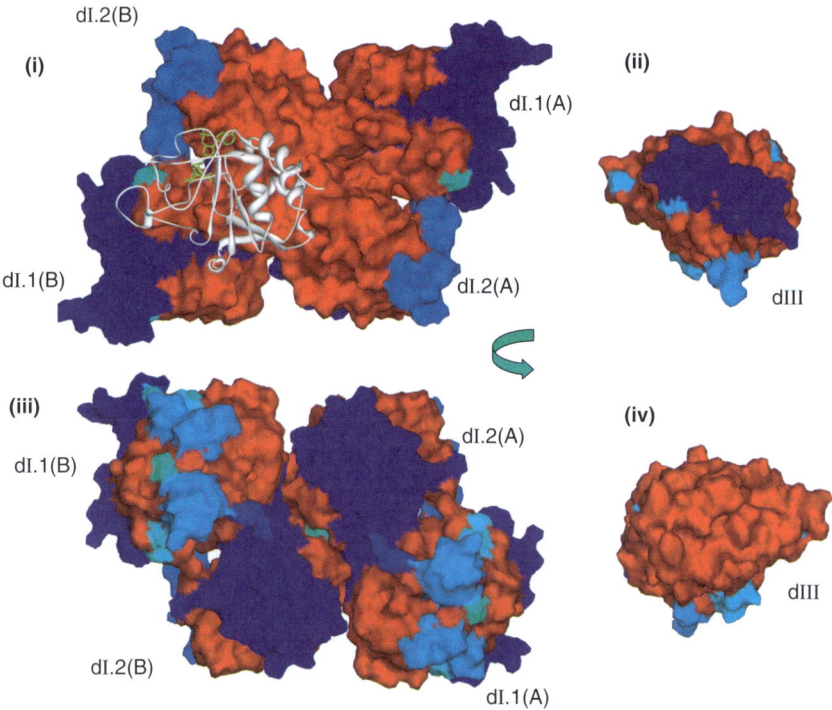

Figure 4 *Multiple-sequence alignment of transhydrogenases as an indication of subunit interactions. A representation of the dI_2dIII_1 complex with the surface coloured according to a position-specific gap penalty, calculated according to (83). Multi-sequence alignments (MSAs) were made separately for dI (95 sequences) and dIII (92 sequences) from the SwissProt database, using the program T-coffee (84). Variable (position-specific) gap penalties were calculated from the MSAs, where gaps in conserved regions are more heavily penalised than those in more variable regions, and were mapped onto the dI and dIII molecular surfaces, using the dI_2dIII_1 crystal structure as a template (1U2D). The surfaces are coloured on a red (highest gap penalty) to dark blue (lowest gap penalty) scale. Surfaces with fewer insertions and deletions, coloured red, are more likely to form interactions with other protein components. (i) A view of the dI_2dIII_1 complex, looking along the two-fold molecular symmetry axis of the dI dimer (from $*$ in Figure 2), with dI surfaces coloured as described above, and dIII shown in ribbon form with the NADP coloured in green. Most of the dI:dI contact surface and the dI:dIII contact surface is coloured red. Intriguingly, there are surface-exposed parts of dI.2 that are coloured red, but are not in contact with dIII. This might possibly indicate regions of dI that contact dIII in other conformational states of the protein. (ii) The molecular surface of dIII coloured as described above. The view of dIII is identical to that in A. The prominent dark blue surface represents the N-terminal region of dIII, where there is significant sequence variability. The lower position-specific gap penalty of this region is mostly a result of two sequences: Acetabularia acetabulum and Treponema denticola. Over-interpretation of this portion of the surface, in particular, should be avoided. (iii) a view of the same structure from the opposite side (from $**$ in Figure 2), showing that amino acid insertions and deletions are much more likely on the dI dimer surface that does not contact dIII and or dII. (iv) The view of dIII is identical to that in (iii) except that dIII the dI dimer has been removed. The surface that contacts the dI dimer (facing the viewer) is coloured entirely red*

nicotinamide mononucleotide (NMN) in the first. The Nic(H) ring of NADP(H) is held by a highly conserved loop in the polypeptide chain. NAD(H) binds at the C-terminal edge of the β-sheet of dI.2 in the orientation commonly found in Rossmann-fold structures. The NAD(H) molecule extends across the cleft between dI.2 and dI.1 such that the Nic(H) ring contacts amino-acid residues in the highly conserved 'RQD loop' of dI.1 (residues 126–136). In different subunits of the various structures, there are alterations in nucleotide and amino-acid side-chain conformation at the hydride-transfer site that we have attempted to correlate with events that take place during enzyme turnover. A recent structure of the dI$_2$dIII$_1$ complex crystallised in the presence of NADH and NADPH is strongly suggestive of the configuration of the ground state for the hydride-transfer step (17). In this configuration, the two NicH rings lie approximately parallel to, but offset from, one another with their redox-active C4 atoms approximately 3.6 Å apart (Figure 5). The orientation of the rings in this position would allow the direct transfer of the *pro-R* hydrogen of the C4$_N$ of NADH to the *re*-side of the C4$_N$ of NADP$^+$, the stereochemistry which has been observed in classical experiments with ^{3}H-labelled nucleotides (48,49). We have called this the 'proximal' position of the nucleotides. In other polypeptides, the Nic(H) ring of the NAD(H) is swung away from where the Nic(H) ring of NADP(H) would be, into a 'distal' position. The swing is achieved through rotation of bonds in the pyrophosphate and ribose-phosphate region of the NAD(H) and it would set the C4$_N$ atoms of the two nucleotides too far apart for hydride transfer. Inter-conversion of the distal and proximal forms is thought to be important in coupling the redox reaction to proton translocation (see below).

In addition to the polypeptide loops responsible for binding the Nic(H) rings of the two nucleotides, several others appear to be important in the mechanism of action of transhydrogenase. In dIII, the long loop E probably has a role in regulating the exchange of NADP(H) with nucleotide in the solvent. In all the available structures of the dI$_2$dIII$_1$ complex, loop E forms a 'lid' over the bound nucleotide with an apical, conserved Tyr residue interacting with the Nic(H) ring. As noted above, the dI$_2$dIII$_1$ complex is active in hydride transfer and, as explained below, there is a critical requirement in the intact enzyme to prevent nucleotide release from the intermediate that catalyses the hydride-transfer step. The loop E lid probably fulfils this latter function and is expected to open prior to and following hydride transfer, therefore allowing NADP$^+$ to bind to, and NADPH to dissociate from, the protein. Mutation of Asp392 underneath this region in *E. coli* transhydrogenase appears to loosen loop E and allow faster exchange of bound nucleotides (35,50,51). Adjacent to, and contacting, loop E is loop D. This feature turns sharply from the C-terminus of the fourth β strand in the sheet through a short section of helix (helix D) and then (at least in the dI$_2$dIII$_1$ complex) contacts the RQD loop of dI. A recent structure of isolated dIII shows that loop D can undergo a significant conformational change (16) which, in our view (see below), may operate a switch in dI to enable or block hydride transfer. Two other protein loops, the 'mobile' loop and the 'TAGP loop, both in dI, appear to be important in the mechanism of action of transhydrogenase. Due to its segmental mobility in the absence of bound NAD(H), the former has been identified in NMR experiments (52–56). When NAD(H) binds, the mobile loop closes down and, in some X-ray structures, it can be seen to extend from dI.2 and across the cleft

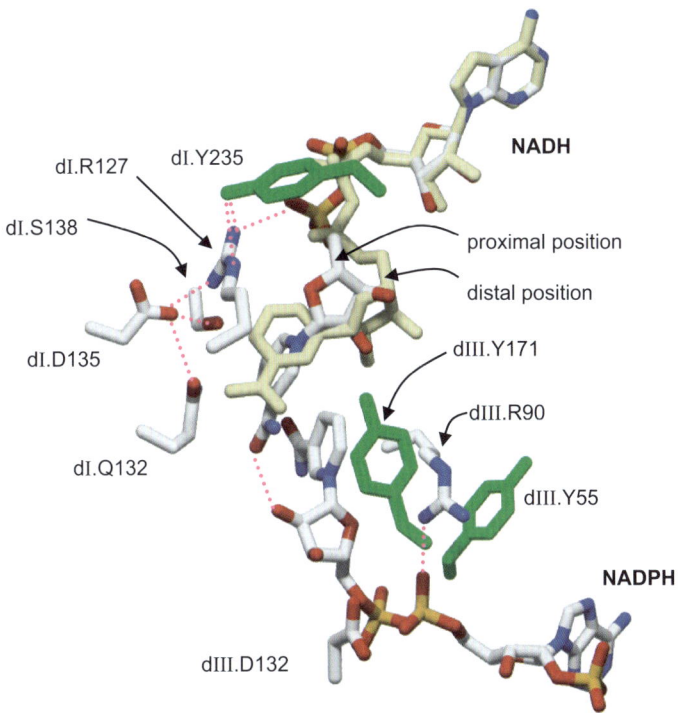

Figure 5 *The site of hydride transfer between NADH and NADP$^+$. Amino acid residues and nucleotides in conventional atom colours and the green Tyr residues are all taken from hydride-transfer site at the dI(B)-dIII interface of a structure of the dI$_2$dIII$_1$ complex crystallised in the presence of NADH and NADPH (1U2D (17)). Only selected (conserved polar and tyrosine) residues are shown. The redox active C1$_N$ atoms of the two NicH rings are 3.6 Å apart (the NADH is in its proximal position). The pink dotted lines show H-bonds. The pale yellow NADH molecule is taken from the dI(A) polypeptide of 1U2D by superimposing its adenine ring system on to that of the dI(B) NADH adenine. It is in the distal position – the distance from its C4$_N$ atom to that of the C4$_N$ of NADPH would be about 5.5 Å*

to dI.1, where its invariant Tyr235 interacts with Arg127 in the RQD loop and thus seals the hydride-transfer site from the solvent. Mutation of Tyr235 to Phe greatly lowers the rate of hydride transfer in the intact enzyme (57,58). The TAGP loop also extends from dI.2 across the cleft to dI.1 and, being more rigid, may serve to anchor the mobile loop as it closes over the bound nucleotide.

5 Considerations Relevant to the Mechanism of Proton Translocation by Transhydrogenase

The starting point in understanding the mechanism of proton translocation by transhydrogenase is that the hydride transfer reaction takes place *directly* between NADH and NADP$^+$ (46) at a site on the protein (the interface between dI and dIII) which is distant (~30 Å) from the membrane-spanning dII. Because we cannot deline-ate a proton-translocation pathway through the X-ray structure of dIII, it was con-cluded that energy coupling between the proton-translocation reactions in dII and the hydride-transfer site is achieved through long distance conformational changes. By analogy with a description that is commonly applied to the F_1F_o-ATPase, transhy-drogenase can be viewed as two (reversible) back-to-back motor/generators: in the forward direction (equation 1), proton translocation through dII generates a confor-mational change that is transmitted to the dI/dIII interface where the free energy change is conserved in the redox reaction as an elevated mass action ratio of the nucleotide products and reactants. Unlike the F_1F_o-ATPase, it is evident from the transhydrogenase structure that the motor elements involve reciprocating rather than rotary motions (Figure 6). A single proton is transferred through transhydrogenase per hydride-ion equivalent transferred between nucleotides (59).

A second feature that relates to the mechanism of coupling is the importance of changes in the binding of NADP(H). On the basis of our studies of 'cyclic transhy-drogenation' we have shown that steps occurring during proton translocation appear to regulate the binding and release of NADP$^+$ and NADPH to the enzyme rather than the hydride-transfer step itself (60–62). There is now accumulating evidence from several different types of experiments to show that mutations in *E. coli* dII (of βHis91, βAsn222, βGly252, βD213 and βR265) have an effect on the nucleotide-binding characteristics of dIII, suggesting a conformational interaction between the membrane-spanning component and the NADP(H)-binding site (31–36,39,63).

Another prime consideration in our understanding of the mechanism of transhydro-genase is the pronounced reactivity of the Nic$^+$ and NicH rings of the substrate nucleotides. Even in aqueous solution, the second-order rate constants for redox reac-tions between compounds having NicH and Nic$^+$ rings are substantial (64). The prox-imal nucleotide configuration shown in the recent crystal structure (Figure 5) is very likely similar to that responsible for hydride transfer in the intact enzyme but, if the nucleotides were to bind with this geometry in the formation of the Michaelis complex then, disastrously, the redox reaction would proceed without proton translocation. It has therefore been proposed that, in the Michaelis complex, the NADH and the NADP$^+$ bind with their NicH and Nic$^+$ rings in the distal position (17,65). Only at the appropriate step in turnover are the two rings brought together into the proximal posi-tion to effect hydride transfer. The essence of energy coupling in transhydrogenase is that the protonation and deprotonation reactions associated with proton translocation drive the enzyme alternately between a conformation in which hydride transfer is allowed (nucleotides proximal) and a conformation in which hydride transfer is blocked (nucleotides distal). A requirement of the mechanism is that either one or both nucleotides must be prevented from exchanging with nucleotide in the solvent during the hydride-transfer step. A failure to observe this would again result in a redox reac-tion uncoupled from proton translocation. The experimental evidence (see above) sug-gests that NADP(H) is the nucleotide whose exchange is restricted during hydride transfer, and that NAD(H) can bind to and dissociate from the enzyme even in confor-mations in rapid equilibrium with that in which hydride transfer takes place (60–62).

On the basis of available data we propose (below) a mechanism for the coup-ling mechanism of transhydrogenase. It is a development of our 'binding-change'

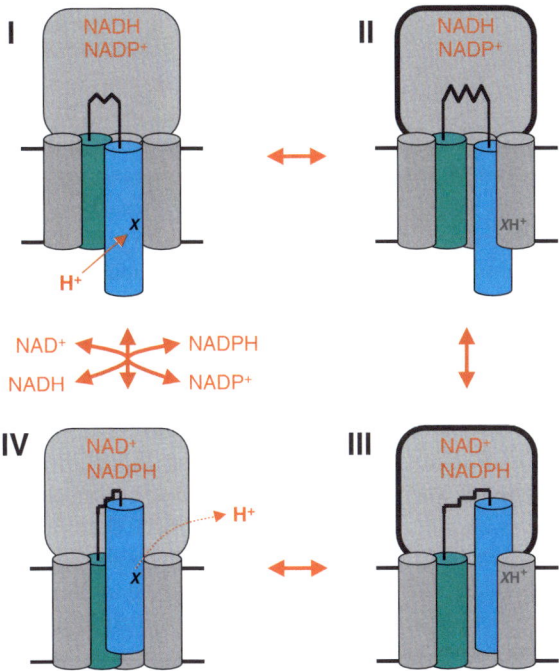

Figure 6 *The upper part of each of the four panels represents the dI/dIII part of the enzyme. A thin-lined border of dI/dIII (panels I and IV) signifies an open state, a thick-lined border (panels II and III) signifies an occluded state. The lower part of each panel represents dII embedded in the membrane – just 5 TM helices are shown. A motor helix is in blue and a stator helix in green. The zig-zag "linkage" between these two helices diagrammatically shows changes in the conformations of the redox site. The motor and stator helices are not necessarily sequential – the linkage may involve more than one segment of polypeptide chain. Changes in the linkage in the horizontal direction (as the motor helix moves parallel to the plane of the membrane, as in I ↔ II and III ↔ IV) depict inter-conversion of the open and occluded states which are stabilised/destabilised by protonation/deprotonation of X in dII. Changes in the linkage in the vertical direction (as the motor helix moves perpendicularly to the plane of the membrane) are consequent upon NADP(H) oxidation/reduction during hydride transfer in the occluded state (II ↔ III) and upon exchange of bound NADP$^+$ for NADPH (or NADPH for NADP$^+$) from the solvent in the open state (IV ↔ I). [NOTE. An equivalent model can be constructed in which the occluded state is generated by deprotonation to the cytoplasm.]*

mechanism (26,60,65–67), which itself builds on the ideas of other authors (68,69). In the first part of the description we outline the main features of the model, and its 'specificity rules' – these establish the reaction pathway and define how unwanted slip in the redox reaction and in proton conduction across the membrane are prevented (see (70)). In the second part, we briefly describe, in the context of the model, the sequence of events during Δp-driven forward transhydrogenation. In the third part we discuss in some detail the conformational events that occur at the hydride-transfer step. Recent experiments with dI_2dIII_1 complexes have provided considerable kinetic, thermodynamic and structural information about this step and enable a more rigorous analysis.

6 Hypothesis for the Mechanism of Proton Translocation by Transhydrogenase

6.1 The Specificity Rules

There are two conformations of transhydrogenase, the 'open' state and the 'occluded' state, each of which has two sub-states. In the open state (I and IV in Figure 6), hydride transfer is prevented (the Nic(H) rings are distal), and NADP(H) bound to dIII can exchange with NADP(H) in the solvent. In the occluded state (II and III), hydride transfer is enabled (Nic(H) rings are proximal), and exchange of bound NADP(H) with NADP(H) in the solvent is blocked. NAD(H) bound to dI can exchange rapidly with solvent NAD(H) in both the open and occluded states. The open state is stabilised when X, an amino-acid side chain on the blue-coloured 'motor' TM helix, is unprotonated; the occluded state is stabilised when X is protonated. Inter-conversion of the open and occluded states is shown diagrammatically as a movement in the plane of the membrane of the blue motor helix relative to a green 'stator' helix (I $\leftrightarrow$ II and III $\leftrightarrow$ IV). It involves a large change in conformation which extends to the redox site – this is represented by the alteration in the zig-zag linkage between the motor and stator helices (probable events at the redox site are described in section 3.6). In the occluded state, XH^+ is shown in grey, signifying that it is inaccessible to protons from either aqueous phase (it is located 'behind' one of the grey TM helices). In the open state, X (shown in bold) is accessible to protons from the periplasmic aqueous phase when $NADP^+$ is bound (I), and to protons from the cytoplasmic aqueous phase when NADPH is bound (IV) – hence the two 'sub-states' of the open state. Thus, there is a conformational change that arises when bound NADPH is exchanged for solvent $NADP^+$ (or *vice versa*) and this is transmitted to the membrane-spanning dII to cause the change in proton access (IV $\leftrightarrow$ I). It is shown as originating in the zig-zag linkage between the motor and stator helices and as resulting in a movement of the motor helix perpendicular to the membrane plane. In a parallel fashion in the occluded state, the reduction of $NADP^+$ (or the oxidation of NADPH) following hydride transfer causes movement of the motor helix perpendicular to the membrane plane (II $\leftrightarrow$ III) and this shifts the solvent-inaccessible XH^+ towards the cytoplasmic (or periplasmic) side – hence the two 'sub-states' of the occluded state.

The depiction of the inter-conversion of the open and occluded states as the result of a relative movement of two dII helices in the plane of the membrane is not meant to exclude other motions but it may be realistic; it is analogous to a single step in the rotary-catalysis mechanism for the relative movement of TM helices in the c and a subunits of the F_1F_o-ATPase (71) and may have an equivalent physico-chemical basis. Similarly, the suggested movements of the motor TM helix perpendicular to the membrane to drive changes in the position of X and of XH^+ are intended to be diagrammatic but represent a simple, realistic model; movements of TM helices perpendicular to the plane of the membrane have been described in X-ray structures of the Ca^{++}-ATPase

(72). The change in proton access could be achieved by relatively small movements of the helix or, alternatively, by changes in the position of the insulating barrier within dII.

6.2 The Sequence of Events During the Reduction of NADP$^+$ by NADH in Transhydrogenase Driven by Δp (see Figure 6)

(I) The protein is in the open state and it has bound NADP$^+$. NADH can bind in this state or in II (see below). Hydride transfer is not possible (the Nic(H) rings are distal to one another). X has become accessible to protons from the periplasm.

(II) Protonation of X has taken place (favoured by a positive-outside Δp). This has led to the movement of the motor helix in the plane of the membrane to give the occluded state and XH^+ has therefore become inaccessible to the solvent. The Nic(H) rings of the bound nucleotides have been shifted to the proximal position (see below) and hydride transfer becomes possible.

(III) Hydride transfer has taken place and the consequent re-polarisation of the redox site has caused a conformational change in which the motor helix has been moved perpendicularly to the membrane plane. XH^+ is therefore displaced but it remains inaccessible to the solvent.

(IV) The motor helix has moved parallel to the plane of the membrane to expose XH^+ to the cytoplasm. Deprotonation of XH^+ (favoured by Δp) has stabilised this open state. Hydride transfer is again disfavoured (the Nic(H) rings are distal). Product NADPH can dissociate. Product NAD$^+$ can dissociate here (but could also have done so from III).

6.3 Conformational Changes Associated with the Hydride-Transfer Step

Isolated dI$_2$dIII$_1$ complexes appear to be locked into a form that has the properties expected of the occluded state of intact transhydrogenase. Thus, the complexes catalyse very fast hydride transfer between bound NADP(H) and bound NAD(H), their tightly bound NADP(H) can be only very slowly released into the solvent (a feature they also share with isolated dIII protein) and their bound NAD(H) can rapidly exchange with the solvent. With some caveats, it is therefore reasonable to extrapolate from observations made on the dI$_2$dIII$_1$ complex to events that take place in the occluded state of the intact enzyme.

The crystal structures show how direct hydride transfer with the known stereochemistry can take place between the Nic$^+$ and NicH rings of nucleotides bound to the dI$_2$dIII$_1$ complex (see section 4). In the intact enzyme, the oxidation/reduction of NADP(H) that results from hydride transfer is thought to lead to a conformational change which extends across dIII into dII and causes a switch in the access to H$^+$ from one side of the membrane to the other (Figure.6). We suggest that the conformational change follows from the site re-polarisation that results from hydride transfer. There are a number of invariant charged/polar amino acids at the hydride-transfer site which may participate in this process, notably His85, Arg90 and

Asp132 in dIII and Arg127, Gln132, Asp135 and Ser138 in dI. All these amino acid residues (or their equivalents in the *E. coli* enzyme) have been the subject of mutagenesis experiments (35,50,51,73–76). There is little information on the character of the coupling conformational change but, in view of the probable rigidity of the β-sheet-core of dIII, it might involve relatively large positional changes of dI, dII and dIII. It is possible that regions with a high gap penalty in Figure.4 might indicate surfaces involved in subunit rearrangements. One possible lead is the finding that there are large chemical shift changes detected by NMR experiments in loops D and E of isolated dIII when NADP$^+$ is exchanged for NADPH (41,77). X-ray structures indicate that atomic displacements caused by the nucleotide exchange (which in these circumstances effectively simulates the redox reaction) are too small to observe in the 2.0 and 2.3 Å resolution structures (12,17). Perhaps the chemical shift changes are due primarily to electrostatic alterations in loops D and E whose consequences are blocked in isolated dIII but which, in the intact enzyme, can relax through subsequent movements of TM helices in dII (Figure 6, step II → III). The way in which amino acids in the site are re-polarised during hydride transfer is also not known. However, it may be relevant in this context that the equilibrium constant for the hydride-transfer step in the dI$_2$dIII$_1$ complex is elevated (by > 36 fold) relative to that in solution ($K_{eq} \approx 1$) implying significant interaction changes between the Nic(H) rings and the protein: the binding of NADPH and NAD$^+$ is stabilised relative to NADH and NADP$^+$ (43).

The conformations of the protein main chain and side chains, and of the nucleotide, at the NADP(H)-binding site are very similar in all available high-resolution structures of isolated dIII and dI$_2$dIII$_1$ complexes. This may reflect a precise requirement in the protein geometry at the point of origin of the subsequent large conformational change into dII. In contrast, the Nic(H) rings of NAD(H) in dI can be seen to flip between distal and proximal positions and, as indicated above, this is thought to be important in promoting/blocking hydride transfer. The distal ↔ proximal flip appears to be correlated with small (~1 Å) movements of domain dI.1 relative to dI.2, and to changes in the conformation of amino-acid side chains in the RQD loop particularly those of Arg127 and Gln132 (17). When NAD(H) takes up its proximal position, the side chain of Arg127 (which also interacts with the pyrophosphate group of the nucleotide) moves deeper into the cleft between dI.1 and dI.2 and forms H-bonds with Asp135 and Ser138. This may help to stabilise the Nic(H) ring in its proximal position. With the Nic(H) ring in the distal position, the Arg127 moves to a shallower position in the cleft and the H-bonds with Asp135 and Ser138 are lost. In conformations having NAD(H) in the proximal position, the side chain of Gln132 is folded back 'behind', and into contact with, the Nic(H) ring but, in other conformations, the Gln132 side chain extends across the dI.1-dIII interface to H-bond with the N-ribose of the NADP(H), suggesting a role in tethering the two nucleotides prior to hydride transfer (75). When Gln132 was mutated to Asn, hydride transfer was inhibited >100-fold and the modest deuterium isotope effect seen in the wild-type complex was lost. This has been taken to indicate that the Gln→Asn substitution inhibits a conformational change preceding hydride transfer, perhaps the distal ↔ proximal inter-conversion.

We argued above that, in the open state, the Nic(H) ring of NAD(H) must be kept distal. However, the X-ray structures of the dI_2dIII_1 complex suggest that, in the occluded state, the ring can adopt both the distal and the proximal positions. This may be an indication that the ring is held in the distal position during inter-conversion of the two states: only when the protein is securely in the occluded state can the ring flip into its proximal position to allow hydride transfer. This sequence of events would prevent a transient redox slip during inter-conversion (a redox reaction followed by exchange of bound NAD(H) and NADP(H) with nucleotides in the solvent). Incidentally, when NAD(H) binds to dI_2dIII_1 complexes (or to the intact enzyme in the occluded state) it must presumably first adopt the distal before the proximal position prior to hydride transfer.

The question arises as to how the Nic(H) ring of NAD(H) is maintained in the distal position in the open state. The nucleotide in the dI(A) polypeptide (the one without a partner dIII – see Figures.2 and 4) adopts the distal position in all available structures: if this reflects the situation in the intact enzyme, it could imply that cooperative interactions between the two dI components are important in establishing the relative position of the Nic(H) rings (also, see section 7). However, another possibility is suggested by a recent X-ray structure of isolated dIII in which the tip of loop D is bent (16). When this conformation is modelled into our structures of the dI_2dIII_1 complex, loop D is found to clash with the RQD loop in the B polypeptide of dI. The clash would be relieved by movement of the dI.1 domain relative to dI.2. This movement, were it to occur, would be responsible for holding dI in a conformation that favours the distal position of the Nic(H) rings. Another explanation has been put forward for the bent conformation of helix D (16) but we have indicated reasons why we think that explanation is unlikely (17).

The inter-conversion of the open and occluded states also involves a very large change in the access of bound NADP(H) to the solvent (the dissociation rate constant for these nucleotides changes by four or five orders of magnitude (78)). The raising and lowering of the loop E 'lid' in dIII would appear to be important in this change in access. The event seems not to be directly coupled to the bending of the adjacent loop D (16) but both of these processes must be contingent upon the protonation/deprotonation of the proton-translocation apparatus in dII. In dI_2dIII_1 complexes, the invariant Asp132 on loop D is buried under loop E and appears to have an elevated pK_a value (26). Changes in the ionisation state of this residue may be involved in the stabilisation of the raised and lowered positions of the lid. During NADP(H) binding in the open state, the invariant Lys164, Arg165 and the highly conserved Ser165, at the C-terminal end of loop E, form a nest of H-bond interactions with the 2′-phosphate group of the nucleotide to establish specificity (12); as the protein subsequently enters the occluded state, this region may provide a fulcrum for lowering the loop over the pyrophosphate and NMN(H) moieties. The highly conserved Tyr171 near the apex of loop E moves close to the Nic(H) ring of the bound NADP(H) in the occluded state and this may contribute to the exclusion of water from the site during hydride transfer. In the absence of NADP(H), under certain experimental conditions, dIII can weakly bind NAD(H) and its analogues (79,80). Because loop E is lowered in dI_2dIII_1 complexes, this results in the 'wrong' nucleotide being trapped long enough for hydride transfer to occur, though it is doubtful that this reaction is of physiological significance.

7 Future Directions

High-resolution structures of intact transhydrogenase are required to provide significant further understanding of the mechanism of action of the enzyme. Stemming from such structures, it should become possible to speculate about the nature of the conformational changes that link events at the hydride-transfer site with those involved in proton translocation (summarised diagrammatically in Figure 6), and thus to design better experimental tests on the coupling mechanism.

Finally, it is pertinent to consider what implications the pronounced asymmetry of the isolated dI_2dIII_1 complex might have for intact transhydrogenase. Why, when the intact enzyme is very likely a dimer of two dI-dII-dIII units, does the complex have such a low affinity for a second dIII (1,40)? Firstly, it is possible that the intact enzyme is symmetrical: each dIII component interacts similarly with its partner dI but interaction with the dII dimer limits a movement of dIII relative to dI that becomes exaggerated in the dI_2dIII_1 complex and thus sterically blocks the association of another dIII. We would assume that the tendency to drive this movement in the intact enzyme would have either mechanistic or structural importance. It is evident that, unrestricted in the dI_2dIII_1 complex, the distortion does not cause inhibition of the redox reaction. A second possibility is that the asymmetry of the dI_2dIII_1 complex reflects similar asymmetry in the intact enzyme. It has been suggested that, whereas the dIII associated with dI(B) is in the occluded state, a second dIII can associate with dI(A) only if it adopts the open state (1,26,40). This would imply that during turnover of the intact transhydrogenase, there is a compulsory alternation of sites such that the conformation of the two monomers remain 180° out of phase. A related proposal was made a number of years ago based on the results of experiments with the intact bovine enzyme, when it was concluded that complete inhibition of transhydrogenation by either N,N'-dicyclohexylcarbodiimide or 5'-[p-(fluorosulfonyl)benzoyl] adenosine was achieved with only half of the sites chemically modified (81,82). Experiments in our laboratory are currently underway to re-examine this with a molecular genetics approach.

Acknowledgements

We are grateful to the Wellcome Trust and the Biotechnology and Biological Sciences Research Council for financial support.

References

1. N.P.J. Cotton, S.A. White, S.J. Peake, S. McSweeney and J.B. Jackson, *Structure*, 2001, **9**, 165.
2. L. Danielsson and L. Ernster, *Biochem. Biophys. Res. Comm.*, 1963, **10**, 91.
3. P. Mitchell, *Biol. Rev. Cambridge Philos. Soc.*, 1966, **41**, 445.
4. J. Moyle and P. Mitchell, *Biochem. J.*, 1973, **132**, 571.

5. P.D. Bragg, P.L. Davies and C. Hou, *Biochem. Biophys. Res. Comm.*, 1972, **47**, 1248.

6. G. Ambartsoumian, R. Dari, R.T. Lin and E.B. Newman, *Microbiol.*, 1994, **140**, 1737.

7. J.W. Hickman, R.D. Barber, E.P. Skaar and T.J. Donohue, *J. Bact.*, 2002, **184**, 400.

8. J. Rydstrom and J.B. Hoek, *Biochem. J.,* 1988, **254**, 1.

9. L.A. Sazanov and J.B. Jackson, *FEBS Lett.,* 1994, **344**, 109.

10. E.A. Newsholm, R.A. Challis and B. Crabtree, *Trends Biochem. Sci.,* 1984, **9**, 277.

11. G.S. Prasad, V. Sridhar, M.Yamaguchi, Y. Hatefi and C.D. Stout, *Nature Structural Biol.*, 1999, **6**, 1126.

12. S.A. White, S.J. Peake, S. McSweeney, G. Leonard, N.N.J. Cotton and J.B. Jackson, *Structure,* 2000, **8**, 1

13. P.A. Buckley, J.B. Jackson, T. Schneider, S.A. White, D.W. Rice and P.J. Baker, *Structure,* 2000, **8**, 809.

14. A. Singh, J.D. Venning, P. Quirk, G.I. van Boxel, D.J. Rodrigues, S.A. White and J.B. Jackson, *J. Biol. Chem.,* 2003, **278**, 33208.

15. G.S. Prasad, M. Wahlberg, V. Sridhar, V. Sundaresan, M. Yamaguchi, Y. Hatefi and C.D. Stout, *Biochemistry,* 2002, **41**, 12745.

16. V. Sundaresan, M. Yamaguchi, J. Chartron and C.D. Stout, *Biochemistry,* 2003, **42**, 12143.

17. O.M. Mather, G.I. van Boxel, S.A. White and J.B. Jackson, *Biochemistry,* 2004, **43**, 10964.

18. M. Jeeves, K.J. Smith, P.G. Quirk, N.P.J. Cotton and J.B. Jackson, *Biochim. Biophys. Acta.,* 2000, **1459**, 248.

19. T. Bizouarn, J. Meuller, M. Axelsson and J. Rydstrom, *Biochim. Biophys. Acta.,* 2000, 1459, 284.

20. T. Bizouarn, O. Fjellstrom, J. Meuller, M. Axelsson, A. Bergkvist, C. Johansson, G. Karlsson and J. Rydstrom, *Biochim. Biophys. Acta.,* 2000, **1457**, 211.

21. T. Bizouarn, M. Althage, A. Pedersen, A. Tigerstrom, J. Karlsson, C. Johansson and J. Rydstrom, *Biochim. Biophys. Acta.,* 2002, **1555**, 122.

22. P.D. Bragg, *Biochim. Biophys. Acta.,* 1998, **1365**, 98.

23. M. Yamaguchi, Y. Hatefi, K. Hoch and J.A. Hoch, *J. Biol. Chem.,* 1988, **263**, 2761.

24. W.K. Studley, M. Yamaguchi, Y. Hatefi and M.H. Saier, *Microbial Compar. Genomics,* 1999, **4**, 173.

25. J. Meuller and J. Rydstrom, *J. Biol. Chem.,* 1999, **274**, 19072.

26. J.B. Jackson, S.A. White, P.G. Quirk and J.D. Venning, *Biochemistry,* 2002, **41**, 4173.

27. M. Althage, J. Karlsson, P. Gourdon, M. Levin, R.M. Bill, A. Tigerstrom and J. Rydstrom, *Biochemistry,* 2003, **42**, 10998.

28. M. Althage, T. Bizouarn, B. Kindlund, J. Mullins, J. Alander and J. Rydstrom, *Biochim. Biophy. Acta.,* 2004, **1659**, 73.

29. N.A. Glavas, C. Hou and P.D. Bragg, *Biochem. Biophys. Res. Commun.,* 1995, **214**, 230.

30. E.G. Sedgwick, J. Meuller, C. Hou, J. Rydstrom and P.D. Bragg, *Biochemistry,* 1997, **36**, 15285.
31. E. Holmberg, T. Olausson, T. Hultman, J. Rydstrom, S. Ahmad, N.A. Glavas and P.D. Bragg, *Biochemistry,* 1994, **33**, 7691.
32. N.A. Glavas, C. Hou and P.D. Bragg, *Biochemistry,* 1995, **34**, 7694.
33. M. Yamaguchi and Y Hatefi, *J. Biol. Chem.,* 1995, **270**, 1.
34. P.D. Bragg and C. Hou, *Arch. Biochem. Biophys.,* 1999, **363**, 182.
35. X. Hu, J.W. Zhang, O. Fjellstrom, T. Bizouarn and J. Rydstrom, *Biochemistry,* 1999, **38**, 1652.
36. M. Yamaguchi, C.D. Stout and Y. Hatefi, *J. Biol. Chem.,* 2002, **277**, 33670.
37. M. Yamaguchi and C.D. Stout, *J. Biol. Chem.,* 2003, **278**, 45333.
38. P.D. Bragg and C. Hou, *Arch. Biochem. Biophys.,* 2000, **380**, 141.
39. M. Althage, T. Bizouarn and J. Rydstrom, *Biochem.,* 2001, **40**, 9968.
40. J.D. Venning, D.J. Rodrigues, C.J. Weston, N.P.J. Cotton, P.G. Quirk, N. Errington, S. Finet, S.A. White and J.B. Jackson, *J. Biol. Chem.,* 2001, **276**, 30678.
41. P.G. Quirk, M. Jeeves, N.P.J. Cotton, K.J. Smith and J.B. Jackson, *FEBS Lett.,* 1999, **446**, 127.
42. J.D.Venning, S.J. Peake, P.G. Quirk and J.B. Jackson, *J. Biol. Chem.,* 2000, **275**, 19490.
43. T.J.T. Pinheiro, J.D. Venning and J.B. Jackson, *J. Biol. Chem.,* 2001, **276**, 44757.
44. J.B. Venning and J.B. Jackson, *J. Biol. Chem.,* 1999.
45. J.D. Venning, T. Bizouarn, N.P.J. Cotton, P.G. Quirk and J.B. Jackson, *Eur. J. Biochem.,* 1998, **257**, 202.
46. J.D. Venning, R.L. Grimley, T. Bizouarn, N.P.J. Cotton and J.B. Jackson, *J. Biol. Chem.,* 1997, **272**, 27535.
47. J.D. Venning and J.B. Jackson, *Biochem. J.,* 1999, **341**, 329.
48. C.P. Lee, N. Simard-Duquesne, L. Ernster and H.D. Hoberman, *Biochim. Biophys. Acta,* 1965, **105**, 397.
49. R.R. Fisher and R.J. Guillory, *J. Biol. Chem.,* 1971, **246**, 4687.
50. J. Meuller, X. Hu, C. Buntoff, T. Olausson and J. Rydstrom, *Biochim. Biophys. Acta,* 1996, **1273**, 191.
51. O. Fjellstrom, M. Axelsson, T. Bizouarn, X. Hu, C. Johansson and J. Rydstrom, *J. Biol. Chem.,* 1999, **274**, 6350.
52. C. Diggle, N.P.J. Cotton, R.L. Grimley, P.G. Quirk, C.M. Thomas and J.B. Jackson, *Eur. J. Biochem.,* 1995, **232**, 315.
53. T. Bizouarn, C. Diggle, P.G. Quirk, R.L. Grimley, N.P.J. Cotton, C.M. Thomas and J.B. Jackson, *J. Biol. Chem.,* 1996, **271**, 10103.
54. R.L. Grimley, P.G. Quirk, T. Bizouarn, C.M. Thomas and J.B. Jackson, *Biochem.,* 1997, **36**, 14762.
55. S. Gupta, P.G. Quirk, J.D. Venning, J. Slade, T. Bizouarn, R.L. Grimley, N.P.J. Cotton and J.B. Jackson, *Biochim. Biophys. Acta,* 1998, **1409**, 25.
56. P.G. Quirk, K.J. Smith, C.M. Thomas and J.B. Jackson, *Biochim. Biophys. Acta,* 1999, **1412**, 139.

57. C. Diggle, P.G. Quirk, T. Bizouarn, R.L. Grimley, N.P.J. Cotton, C.M. Thomas and J.B. Jackson, *J. Biol. Chem.*, 1996, **271**, 10109.
58. T. Bizouarn, R.L. Grimley, C. Diggle, C.M. Thomas and J.B. Jackson, *Biochim. Biophys. Acta*, 1997, **1320**, 265.
59. T. Bizouarn, L.A. Sazanov, S. Aubourg and J.B. Jackson, *Biochim. Biophys. Acta*, 1996, **1273**, 4.
60. M.N. Hutton, J.M. Day, T. Bizouarn and J.B. Jackson, *Eur. J. Biochem.*, 1994, **219**, 1041.
61. T. Bizouarn, R.L. Grimley, N.P.J. Cotton, S. Stilwell, M. Hutton and J.B. Jackson, *Biochim. Biophys. Acta*, 1995, **1229**, 49.
62. T. Bizouarn, S.N. Stilwell, J.M. Venning, N.P.J. Cotton and J.B. Jackson, *Biochim. Biophys. Acta*, 1997, **1322**, 19.
63. M. Yamaguchi and C.D. Stout, *J. Biol. Chem.*, 2003, **278**, 45333.
64. P. van Eikeren and D.L. Grier, *J. Am. Chem. Soc.*, 1977, **99**, 8057.
65. J.B. Jackson, *FEBS Lett.*, 2003, **545**, 18.
66. J.B. Jackson, P.G. Quirk, N.P.J. Cotton, J.D. Venning, S. Gupta, T. Bizouarn, S.J. Peake and C.M. Thomas, *Biochim. Biophys. Acta*, 1998, **1365**, 79.
67. J.B. Jackson, S.J. Peake and S.A. White, *FEBS Lett.*, 1999, **464**, 1.
68. J. Rydstrom, *Biochim. Biophys. Acta*, 1977, **463**, 155.
69. R.R. Fisher and S.R. Earle, Review, in *The pyridine nucleotide coenzymes*, J. Everse, B. Anderson and K.-S. You, (eds.) Academic Press, New York, 1982.
70. W.P. Jenks, *Curr. Topics Membr. Trans.*, 1983, **19**, 1.
71. W. Junge, H. Lill and E. Engelbrecht, *Trends Biochem. Sc.*, 1997, **22**, 420.
72. C. Toyoshima, H. Nomura and Y. Sugita, *FEBS Lett.*, 2003, **555**, 106.
73. P.D. Bragg and C. Hou, *Eur. J. Biochem.*, 1996, **241**, 611.
74. P.D. Bragg, N.A. Glavas and C. Hou, *Arch. Biochim. Biophys.*, 1997, **338**, 57.
75. G.I. van Boxel, P. Quirk, N.J.P. Cotton, S.A. White and J.B. Jackson, *Biochem.*, 2003, **42**, 1217.
76. T.H.C. Brondijk, G.I. van Boxel, A. Singh, O.M. Mather, H.A. White, P.G. Quirk, S.A. White and J.B. Jackson, 2005, *in preparation*.
77. A. Bergkvist, C. Johansson, T. Johansson, J. Rydstrom and B.G. Karlsson, *Biochem.*, 2000, **39**, 12595.
78. C. Diggle, T. Bizouarn, N.P.J. Cotton and J.B. Jackson, *Eur. J. Biochem.*, 1996, **241**, 162.
79. S.N. Stilwell, T. Bizouarn and J.B. Jackson, *Biochim. Biophys. Acta*, 1997, **1320**, 83.
80. A. Pedersen, J. Karlsson, M. Althage and J. Rydstrom, *Biochim. Biophy. Acta*, 2003, **1604**, 55.
81. D.C. Phelps and Y. Hatefi, *Biochemistry*, 1984, **23**, 4475.
82. D.C. Phelps and Y. Hatefi, *Biochemistry*, 1985, **24**, 3503.
83. M. Gribskov, A.D. McLachlan and D. Eisenberg, *Proc. Nat. Acad. Sci. U.S.A.*, 1987, **84**, 4355.
84. C. Notredam, D.G. Higgins and J. Heringa, *J. Mol. Biol.*, 2000, **302**, 217.

Subject Index